McGRAW-HILL YEARBOOK OF SCIENCE & TECHNOLOGY

2012

McGRAW-HILL
YEARBOOK OF
SCIENCE &
TECHNOLOGY

2012

Comprehensive coverage of recent events and research as compiled by
the staff of the McGraw-Hill Encyclopedia of Science & Technology

New York   Chicago   San Francisco   Lisbon   London   Madrid   Mexico City

Milan   New Delhi   San Juan   Seoul   Singapore   Sydney   Toronto

On the front cover
Hubble Space Telescope view of the Lagoon Nebula.
(Photo courtesy of NASA/ESA)

ISBN   978-007-177403-1
MHID   0-07-177403-3
ISSN   0076-2016

1 2 3 4 5 6 7 8 9 0   DOW/DOW   1 9 8 7 6 5 4 3 2 1

This book was printed on acid-free paper.

*It was set in Garamond Book and Neue Helvetica Black Condensed by
Aptara, Falls Church, Virginia. The art was prepared by Aptara.
The book was printed and bound by RR Donnelley.*

Contents

Consulting Editors

Dr. Milton B. Adesnik. *Department of Cell Biology, New York University School of Medicine, New York.* CELL BIOLOGY.

Dr. Vernon D. Barger. *Department of Physics, University of Wisconsin-Madison.* CLASSICAL MECHANICS; ELEMENTARY PARTICLE PHYSICS; THEORETICAL PHYSICS.

Prof. Rahim F. Benekohal. *Department of Civil and Environmental Engineering, University of Illinois, Urbana-Champaign.* TRANSPORTATION ENGINEERING.

Dr. James A. Birchler. *Division of Biological Sciences, University of Missouri, Columbia.* GENETICS.

Robert D. Briskman. *Technical Executive, Sirius XM Radio, New York.* TELECOMMUNICATIONS.

Dr. Mark W. Chase. *Molecular Systematics Section, Jodrell Laboratory, Royal Botanic Gardens, Kew, Richmond, Surrey, United Kingdom.* PLANT TAXONOMY.

Prof. Wai-Fah Chen. *Department of Civil and Environmental Engineering, University of Hawaii at Manoa, Honolulu, Hawaii.* CIVIL ENGINEERING.

Prof. Glyn A. O. Davies. *Senior Research Fellow in Aerostructures, Department of Aeronautics, Imperial College, London, United Kingdom.* AERONAUTICAL ENGINEERING AND PROPULSION.

Prof. Peter J. Davies. *Department of Plant Biology, Cornell University, Ithaca, New York.* PLANT PHYSIOLOGY.

Prof. Mohamed E. El-Hawary. *Associate Dean of Engineering, Dalhousie University, Halifax, Nova Scotia, Canada.* ELECTRICAL POWER ENGINEERING.

Dr. Gaithri A. Fernando. *Department of Psychology, California State University, Los Angeles.* CLINICAL PSYCHOLOGY.

Barry A. J. Fisher. *Retired; formerly, Director, Scientific Services Bureau, Los Angeles County Sheriff's Department, Los Angeles, California.* FORENSIC SCIENCE AND TECHNOLOGY.

Dr. Richard L. Greenspan. *The Charles Stark Draper Laboratory, Cambridge, Massachusetts.* NAVIGATION.

Prof. Joseph H. Hamilton. *Landon C. Garland Distinguished Professor of Physics, Department of Physics and Astronomy, Vanderbilt University, Nashville, Tennessee.* NUCLEAR PHYSICS.

Dr. Lisa Hammersley. *Department of Geology, California State University, Sacramento.* PETROLOGY.

Prof. Terry Harrison. *Department of Anthropology, Paleoanthropology Laboratory, New York University.* ANTHROPOLOGY AND ARCHEOLOGY.

Dr. Jason J. Head. *Department of Earth and Atmospheric Sciences, University of Nebraska-Lincoln.* VERTEBRATE PALEONTOLOGY.

Dr. Ralph E. Hoffman. *Yale Psychiatric Institute, Yale University School of Medicine, New Haven, Connecticut.* PSYCHIATRY.

Dr. Beat Jeckelmann. *Head of Electricity Section, Federal Office of Metrology (METAS), Bern-Wabern, Switzerland.* ELECTRICITY AND ELECTROMAGNETISM.

Dr. S. C. Jong. *Senior Staff Scientist and Program Director, Mycology and Protistology Program, American Type Culture Collection, Manassas, Virginia.* MYCOLOGY.

Prof. Edwin C. Kan. *Department of Electrical & Computer Engineering, Cornell University, Ithaca, New York.* PHYSICAL ELECTRONICS.

Prof. Robert E. Knowlton. *Department of Biological Sciences, George Washington University, Washington, D.C.* INVERTEBRATE ZOOLOGY.

Prof. Chao-Jun Li. *Canada Research Chair in Green Chemistry, Department of Chemistry, McGill University, Montreal, Quebec, Canada.* ORGANIC CHEMISTRY.

Prof. Donald W. Linzey. *Wytheville Community College, Virginia.* VERTEBRATE ZOOLOGY.

Dr. Dan Luss. *Cullen Professor of Engineering, Department of Chemical and Biomolecular Engineering, University of Houston, Texas.* CHEMICAL ENGINEERING.

Prof. Albert Marden. *School of Mathematics, University of Minnesota, Minneapolis.* MATHEMATICS.

Dr. Ramon A. Mata-Toledo. *Professor of Computer Science, James Madison University, Harrisonburg, Virginia.* COMPUTERS.

Prof. Krzysztof Matyjaszewski. *J. C. Warner Professor of Natural Sciences, Department of Chemistry, Carnegie Mellon University, Pittsburgh, Pennsylvania.* POLYMER SCIENCE AND ENGINEERING.

Prof. Jay M. Pasachoff. *Director, Hopkins Observatory, and Field Memorial Professor of Astronomy, Williams College, Williamstown, Massachusetts.* ASTRONOMY.

Prof. Stanley Pau. *College of Optical Sciences, University of Arizona, Tucson.* ELECTROMAGNETIC RADIATION AND OPTICS.

Dr. William C. Peters. *Professor Emeritus, Mining and Geological Engineering, University of Arizona, Tucson.* MINING ENGINEERING.

Dr. Marcia M. Pierce. *Department of Biological Sciences, Eastern Kentucky University, Richmond.* MICROBIOLOGY.

Dr. Donald Platt. *Micro Aerospace Solutions, Inc., Melbourne, Florida.* SPACE TECHNOLOGY.

Dr. Kenneth P. H. Pritzker. *Professor, Laboratory Medicine and Pathobiology, and Surgery, University of Toronto, and Pathology and Laboratory Medicine, Mount Sinai Hospital, Toronto, Ontario, Canada.* MEDICINE AND PATHOLOGY.

Prof. Justin Revenaugh. *Department of Geology and Geophysics, University of Minnesota, Minneapolis.* GEOPHYSICS.

Dr. Roger M. Rowell. *Professor Emeritus, Department of Biological Systems Engineering, University of Wisconsin, Madison.* FORESTRY.

Prof. Ali M. Sadegh. *Director, Center for Advanced Engineering Design and Development, Department of Mechanical Engineering, The City College of the City University of New York.* MECHANICAL ENGINEERING.

Prof. Joseph A. Schetz. *Department of Aerospace & Ocean Engineering, Virginia Polytechnic Institute & State University, Blacksburg.* FLUID MECHANICS.

Dr. Alfred S. Schlachter. *Advanced Light Source, Lawrence Berkeley National Laboratory, Berkeley, California.* ATOMIC & MOLECULAR PHYSICS.

Prof. Ivan K. Schuller. *Department of Physics, University of California, San Diego, La Jolla.* CONDENSED-MATTER PHYSICS.

Jonathan Slutsky. *Naval Surface Warfare Center, Carderock Division, West Bethesda, Maryland.* NAVAL ARCHITECTURE AND MARINE ENGINEERING.

Dr. Arthur A. Spector. *Department of Biochemistry, University of Iowa, Iowa City.* BIOCHEMISTRY.

Dr. Anthony P. Stanton. *Tepper School of Business, Carnegie Mellon University, Pittsburgh, Pennsylvania.* GRAPHIC ARTS AND PHOTOGRAPHY.

Dr. Michael R. Stark. *Department of Physiology, Brigham Young University, Provo, Utah.* DEVELOPMENTAL BIOLOGY.

Prof. John F. Timoney. *Maxwell H. Gluck Equine Research Center, Department of Veterinary Science, University of Kentucky, Lexington.* VETERINARY MEDICINE.

Dr. Daniel A. Vallero. *Adjunct Professor of Engineering Ethics, Pratt School of Engineering, Duke University, Durham, North Carolina.* ENVIRONMENTAL ENGINEERING.

Dr. Sally E. Walker. *Associate Professor of Geology and Marine Science, University of Georgia, Athens.* INVERTEBRATE PALEONTOLOGY.

Prof. Pao K. Wang. *Department of Atmospheric and Oceanic Sciences, University of Wisconsin, Madison.* METEOROLOGY AND CLIMATOLOGY.

Dr. Nicole Y. Weekes. *Department of Psychology, Pomona College, Claremont, California.* NEUROPSYCHOLOGY.

Prof. Mary Anne White. *Department of Chemistry, Dalhousie University, Halifax, Nova Scotia, Canada.* MATERIALS SCIENCE AND METALLURGICAL ENGINEERING.

Dr. Thomas A. Wikle. *Department of Geography, Oklahoma State University, Stillwater.* PHYSICAL GEOGRAPHY.

Article Titles and Authors

Events in our natural environment continued to capture the headlines this past year. Blizzards and record winter snowfalls in many areas of North America were followed by flooding, tornado outbreaks, and record summer heat and drought. Communities and farms in broad areas were awash in the aftermath of landfalling tropical storms. The remnants of Hurricane Katia even remained strong enough to cross the Atlantic and assault the British Isles and Northern Europe. How do we predict, prepare for, and recover from such events? Do we in fact contribute to them with the byproducts of industry and our individual use of energy, for example? Controversies over the causality, even the reality, of global climate change continue in the public and political arenas, as well as in other areas such as the best ways to ensure our future energy and food supplies. A basic understanding of the trends and developments in science and technology continues to be crucial for leaders in industry and government as well as for an informed citizenry. In this spirit, the 2012 edition of the *McGraw-Hill Yearbook of Science & Technology* continues its 50-year mission of keeping professionals and nonspecialists alike abreast of key research and development with a broad range of concise reviews invited by a distinguished panel of consulting editors and written by leaders in international science and technology.

In this edition, for example, we report on new work on the harvesting of waste energy; and on new discoveries on the responses of vertebrates to global warming in the fossil record. In other important areas of environmental science, one can learn about invasive species and their effects on native species; and about polymers from renewable resources. In the biomedical sciences, we chronicle research on dietary fructose and the regulation of body weight; HIV and bone marrow transplantation; human susceptibility to *Staphylococcus aureus*; and robotic surgery, among others. In the cell and biochemical sciences, we learn about the latest research on microRNA; ribosomal biogenesis and disease; and stem cell maintenance in embryos and adults. In the neurosciences, we report on mirror neurons and social neuroscience; oxytocin and autism; primate color vision; and paranoia mechanisms and treatment. Engineering and materials science discussions include advances in electric airplanes; HD radio; characteristics of supertall building structures; mining automation; ultrahigh-temperature ceramics; carbon nanotube materials; and cellulose nanocomposites. In the physical sciences and astrophysics, we report on the Hubble constant and dark energy; the shape of the universe; the star that changed the universe; the top quark at the Tevatron; and the first year at the Large Hadron Collider, among other developments. In computing and information technology, we chronicle developments in research on malware; Web 2.0 technologies; 3D graphic displays; and priority emergency communications. All in all, we chronicle advances in sciences and technology from astronomy to zoology.

Each contribution to the *Yearbook* is a concise yet authoritative article authored by one or more authorities in the field. The topics are selected by our consulting editors, in conjunction with our editorial staff, based on present significance and potential applications. McGraw-Hill strives to make each article as readily understandable as possible for the nonspecialist reader through careful editing and the extensive use of specially prepared graphics.

Librarians, students, teachers, the scientific community, journalists and writers, and the general reader continue to find in the *McGraw-Hill Yearbook of Science & Technology* the information they need in order to follow the rapid pace of advances in science and technology and to understand the developments in these fields that will shape the world of the twenty-first century.

Mark D. Licker
PUBLISHER

McGRAW-HILL
YEARBOOK OF
SCIENCE &
TECHNOLOGY

2012

Angiosperm Phylogeny Group (APG) classification

From its initiation in 1998, the Angiosperm Phylogeny Group (APG) has focused on the production of an ever-more stable system of classification of the flowering plants (angiosperms). Based largely on analyses of DNA sequence data, the system is compiled by a larger group of experts than any previous system and has the advantage of being testable, allowing for confidence levels in the system to be estimated for the first time.

Systems of classification. Modern plant classification emerged as a scientific discipline in the eighteenth century. Carolus Linnaeus published his classic work, *Species Plantarum*, in 1753, in which he attempted to organize plants into a system for easy identification, placing genera into groups that he acknowledged to be artificial, based on the number of pistils and stamens (the so-called sexual system) [**Fig. 1**]. Since then, many botanists have attempted to come up with workable and meaningful systems of classifying plants that, unlike Linnaeus's sexual system, reflect natural relationships, eventually taking into account such developments as Darwin's theory of evolution.

These systems, though, differed to a greater or lesser degree from each other as a result of the characters that the author (or, occasionally, authors) thought were the most important in establishing relationships. For early classifications, the characters used were mainly morphological; however, as fields such as microscopy, biochemistry, and cytogenetics developed, characteristics from these areas were considered in some cases. As a result, there were many conflicting systems running in parallel for most of the nineteenth and twentieth centuries. Although these agreed in many respects, there were significant areas of conflict between them. Importantly, because they were ultimately based on opinion (albeit with in-depth knowledge of the plants), it was not possible to disprove these systems.

One of the desired outcomes of an effective system of classification is predictability. If the pattern of relationships is clearly understood, then informed hypotheses can be made about, for example, how particular characteristics evolved or where particular chemicals might be expected to be found, and these can then be studied. However, with a wide choice of conflicting systems based on opinion, predictability was unlikely.

Use of DNA data. In the late twentieth century, a series of technological and scientific developments provided the opportunity for botanists to come up with a more stable system of classification that, importantly, could be tested and independently verified. The technology associated with DNA sequencing advanced quickly in the 1990s, and an increase in computational power allowed for the production and analysis of large matrices of DNA sequences. Phylogenetic analyses of these matrices have generated trees reflecting the genetic relationships among the taxa involved, and a range of tests (including bootstrapping and parsimony jackknifing) can be used to provide an indication of the level of confidence in relationships reflected by the trees based on DNA sequences.

First large-scale analyses. In 1993, Mark W. Chase and 41 coauthors published the first large-scale analysis of this type for the angiosperms. This was based on nearly 500 sequences for *rbcL*, the gene coding for the large subunit of ribulose-1,5-bisphosphate carboxylase/oxygenase, one of the major enzymes involved in photosynthesis. Although many relationships revealed in this analysis were similar to those proposed in various classifications, some were unexpected; in some cases, they appeared so unlikely that it was suggested that they might reflect patterns of evolution of that particular gene (which could be influenced by functional constraints imposed by the environment, for example, rather than patterns of relationships among the plants). One marked example revealed in this analysis was the group of nearly all the plants that produce mustard oils. Because of their diverse morphology, these plants had previously been suggested to form a group only by Rolf Dahlgren, a botanist with a keen interest in biochemistry. In contrast, in systems based on morphology, these plants were scattered among different groups. Arthur Cronquist, for example, placed the families involved in four orders in two subclasses. Similarly, the sacred lotus (*Nelumbo*) had been placed close to the true waterlilies (*Nymphaea* and relatives) in nearly all classifications, but in the analysis of Chase

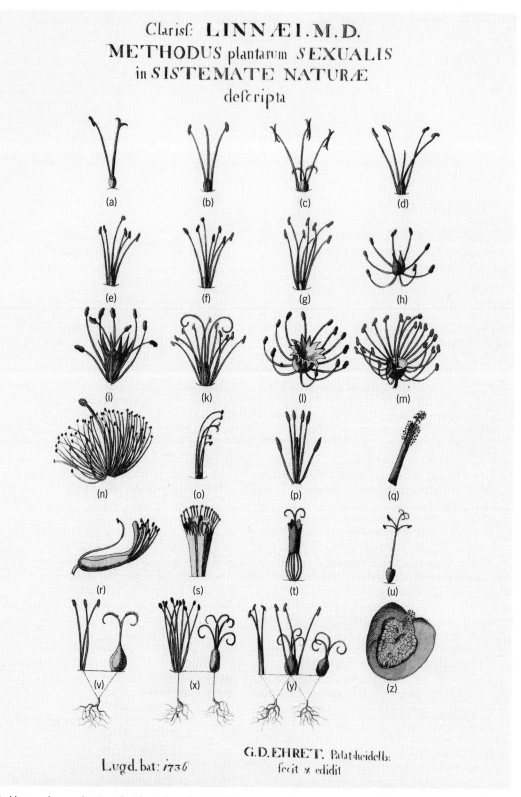

Fig. 1. Linnaeus's sexual system for plants, based principally on the number of male and female sexual organs in flowering plants, where part a has one stamen (male) and one style (female), part b has two stamens and one style, and so forth. (*Illustration by Georg Dionysius Ehret, 1736*)

and colleagues it appeared to be related to *Platanus* (plane trees or sycamores) and Proteaceae. This result was so unexpected that analyses were rerun using newly generated DNA sequences from independent DNA extractions to ensure that there had not been any confusion between samples. However, all subsequent analyses of *rbcL* (as well as many other genes) revealed the same pattern.

Over the following years, several further analyses based on different genes or regions of DNA

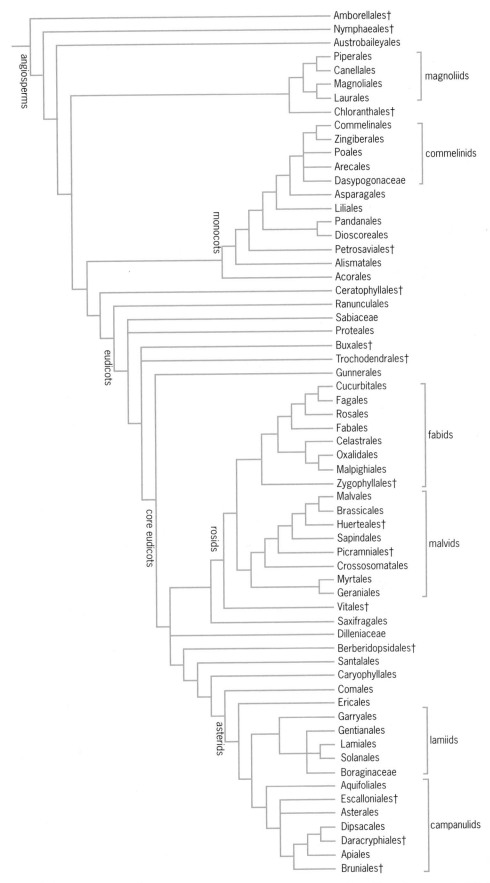

Fig. 2. The summary tree presented in APG III (from 2009) showing the interrelationships of the orders recognized. Orders that are newly recognized by the APG III are denoted by a dagger (†). (*Reprinted with permission from Angiosperm Phylogeny Group, An update of the Angiosperm Phylogeny Group classification for the orders and families of flowering plants: APG III, Bot. J. Linn. Soc., 161:105–121, 2009*)

(including 18S ribosomal DNA and *atpB*) that were involved in processes not related to *rbcL* revealed essentially similar patterns to those shown in the *rbcL* analysis. Moreover, when the results of these analyses were combined, the levels of confidence in many branches in the resulting trees increased substantially, thereby providing strong evidence that the trees reflected relationships among the plants rather than patterns of evolution of a DNA region.

Angiosperm Phylogeny Group. In 1998, a group of researchers coordinating studies on these large matrices decided that it was the appropriate time to publish a classification of the flowering plants based on these data. In contrast to previous systems, which had one or only a few authors, the resulting classification had 3 compilers (from Sweden, the United Kingdom, and the United States) and 26 contributors specializing in different plant groups. The recommended citation (APG, 1998; with APG being an abbreviation for Angiosperm Phylogeny Group) reflected this multiauthor nature by referring to the set of authors rather than the individuals. As the angiosperms were the first major group of organisms to be reclassified on the basis of DNA sequence data, the publication received much attention in the media. In the APG system, 462 families were placed in 40 orders that appeared to be monophyletic (that is, evolving from a single interbreeding population). In addition, 25 families were left unplaced by the APG in 1998 because of the lack of samples or absence of clearly supported patterns in the data.

By 2003, additional samples and data had become available, and numerous articles providing information regarding previously unplaced families had been published. As a result, the Angiosperm Phylogeny Group published an updated classification system (APG II, 2003), again with contributions from appropriate specialists (7 compilers and 20 contributors). This version refined the APG system, with many more families being placed in orders, and the system provided two alternatives in some cases, with smaller families being either recognized or combined into larger families where support existed for the larger unit. The smaller families in these cases were referred to as bracketed families. The family Asparagaceae in the broad sense, for example, included eight bracketed families, and readers could choose to recognize either one or eight families. The list of unplaced plants contained only three families (all consisting only of parasitic plants). The remainder of the list of unplaced taxa consisted of 15 genera; among these, several were also parasites. (Parasitic plants proved problematic to place because several of the DNA regions used in the analyses were involved in photosynthesis or related processes and had been lost or had become nonfunctional in these reduced plants.)

In 2009, the Angiosperm Phylogeny Group decided that it was again appropriate to publish an update (APG III, 2009; 8 compilers and 9 contributors). The summary tree for this version is shown in **Fig. 2**. New data allowed for the placement of additional families and genera. Furthermore, the system

of bracketed families used in APG II had proved unpopular with users of the APG system, so the decision was taken to recognize only the broader families in most cases. This resulted in a system with 415 families in 60 orders, with the increase in the number of orders reflecting the increase in confidence levels and the placement of previously unplaced taxa, rather than a change in the pattern of relationships. The number of unplaced taxa decreased to two families and three genera. This is the most recent update of the APG system, although minor updates are likely as a result of new data or samples becoming available. One of the families and one of the genera in the list of unplaced taxa have been placed in papers published since 2009. Specialists on other groups of plants are developing similar systems. For example, systems for other groups of vascular plants, including conifers and ferns, were published in 2011.

Benefits of APG. The stability provided by the APG and related systems has encouraged many institutions around the world to adopt the APG system for organizing their collections, both living and preserved. Major botanical gardens and herbaria have accepted it as their standard reference system. Moreover, this acceptance of one system has revolutionized communication of scientific ideas and made electronic taxonomy and related databases feasible (especially as they provide common standards for efficient operation). As a result, a number of meta-databases, combining those held by different institutions, have now been developed, and the fruits of these endeavors are already evident. For example, the Global Strategy for Plant Conservation (ratified by the Conference of the Parties to the Convention on Biological Diversity in 2002) sought to create a widely accessible working list of known plant species, which would be the initial step toward a complete list of the world's flora. The working list became a reality in 2010, and this was due largely to several major institutions being able to combine databases as a result of the increasing standardization provided by APG and related developments.

With the increase in stability provided by the Angiosperm Phylogeny Group classification and related ventures, the development of checklists of floras based on electronic databases and other similar activities has become possible, and the coming years will see many products being generated in this way.

For background information *see* CLASSIFICATION, BIOLOGICAL; DEOXYRIBONUCLEIC ACID (DNA); FLOWER; GENE; GENETIC MAPPING; GENOMICS; MAGNOLIOPHYTA; PHYLOGENY; PLANT EVOLUTION; PLANT KINGDOM; PLANT NOMENCLATURE; PLANT PHYLOGENY; PLANT TAXONOMY; SYSTEMATICS in the McGraw-Hill Encyclopedia of Science & Technology.

Michael F. Fay

Bibliography. Angiosperm Phylogeny Group, An ordinal classification of the families of flowering plants, *Ann. Missouri Bot. Gard.*, 85:531–553, 1998; Angiosperm Phylogeny Group, An update of the Angiosperm Phylogeny Group classification for the orders and families of flowering plants: APG II,

Bot. J. Linn. Soc., 141:399–436, 2003; Angiosperm Phylogeny Group, An update of the Angiosperm Phylogeny Group classification for the orders and families of flowering plants: APG III, *Bot. J. Linn. Soc.*, 161:105–121, 2009; M. W. Chase et al., Phylogenetics of seed plants: An analysis of nucleotide sequences from the plastid gene *rbcL*, *Ann. Missouri Bot. Gard.*, 80:528–580, 1993; M. J. M. Christenhusz, M. W. Chase, and M. F. Fay (eds.), Linear sequence, classification, synonymy, and bibliography of vascular plants: Lycophytes, ferns, gymnosperms and angiosperms, *Phytotaxa*, 19:1–134, 2011; E. Haston et al., The Linear Angiosperm Phylogeny Group (LAPG) III: A linear sequence of the families in APG III, *Bot. J. Linn. Soc.*, 161:128–131, 2009.

Anoxyphilic Loricifera

The oceans host life at all depths and across the widest ranges of environmental conditions (that is, temperature, salinity, oxygen content, and pressure), and the deep oceans may contain most of the world's undiscovered biodiversity. Deep-sea ecosystems host the largest hypoxic (oxygen-deficient) and anoxic (oxygen-free) regions of the biosphere and also host some of the most extreme ecosystems on Earth. As a result of the lack of molecular oxygen, some of these extreme systems were assumed to be inhabited by viruses and prokaryotes (Bacteria and Archaea), as well as protozoa, which can live under permanently anoxic conditions. In addition, unicellular eukaryotes (for example, protozoan ciliates) can be found in anoxic marine systems and recent findings have indicated that some benthic foraminiferans can be highly adapted to life without oxygen. However, until recently, there was no proof of the presence of metazoans (multicellular organisms) that can live their entire life cycle under permanently anoxic conditions.

Anoxic ecosystems lying on the deep-sea floor. Several deep hypersaline anoxic basins have been discovered in the Mediterranean Sea, the Red Sea, and the Gulf of Mexico. Nine of these basins have been identified in the eastern Mediterranean (L'Atalante, Urania, Bannock, Discovery, Tyro, La Medee, Kryos, Thetis, and Aphrodite) and these are located on the Mediterranean Ridge. The Mediterranean anoxic basins lie at depths ranging from approximately 3200 to 3600 m (10,500 to 11,800 ft) and contain brine (seawater containing a higher concentration of dissolved salt than that of the ordinary ocean), whose origin has been attributed to the dissolution of old evaporites. Brines enclosed in these basins are characterized by very high density, they create a sharp chemocline (an interface between two contrasting and predominating chemistries within a body of water), and they act as a physical barrier that hampers oxygen exchange between the anoxic sediments and the surrounding seawater. The combination of nearly saturated salt concentrations, high hydrostatic pressure, absence of light, and lack of molecular oxygen makes these basins some of the most extreme habitats on Earth. Most of the investigations carried out so far in these extreme environments have focused on the unicellular organisms inhabiting the brine and the seawater–brine interface and reveal the presence of metabolically active and highly diversified prokaryotic and unicellular eukaryotic assemblages.

Animals living in anoxic conditions. Metazoan meiofauna (multicellular organisms with sizes ranging from a few micrometers to a few millimeters) represent two-thirds of the metazoans on Earth and have a long evolutionary history and high phyletic diversity. They include 22 of the 35 animal phyla, with 6 being exclusive for this group (Gnathostomulida, Micrognathozoa, Gastrotricha, Tardigrada, Kinorhyncha, and Loricifera). The common feature of these phyla is the lack of larval dispersal in the water column and a life cycle spent entirely in sediments. These characteristics make these phyla ideal for investigating complex organisms in systems without oxygen.

The loriciferans are a phylum of recently discovered microscopic marine animals with a complex life cycle and worldwide distribution. Since the discovery of the Loricifera in the early 1980s, several hundred species have been collected in regions ranging from coastal areas [off the coast of Fort Pierce, Florida; off the coast of Roscoff, France; Tyrrhenian Sea (Mediterranean)] to deep-sea ecosystems (Bay of Biscay; Mediterranean Sea; Izu-Ogasawara Trench in the Pacific; Angola basin, Rockall, Faroe Bank, and Great Meteor Seamount in the Atlantic; New Ireland Basin, located north of Papua New Guinea; and Antarctica), where Loricifera display high species diversity. Despite their small size, Loricifera have an extraordinarily rich morphology (their body consists of a head, a neck, a thorax, and an abdomen) and complex alimentary, excretory, nervous, and muscle systems. The head of the adult can have more than 200 appendages, and the larvae always have terminal toes. The head is formed by a retractable introvert (an eversible proboscis-like structure) with a mouth cone surrounded by up to nine rings with appendages (scalids). Each scalid is independently movable and used for locomotion and as a sensory organ. The brain is the only organ containing more than 1000 cells. Loriciferans display some of the most complex life cycles among all metazoans. Their reproductive cycle differs among genera, including gametogenesis (separate sexes) and development through a series of molting larval phases, parthenogenesis (in which an egg develops without entrance of a sperm), sexual maturity of larvae, paedogenesis (reproduction by sexually mature larvae), and hermaphroditism. Loricifera are currently divided into three groups: Nanaloricidae, Pliciloricidae, and Urnaloricidae.

A long-term study conducted in the anoxic hypersaline L'Atalante basin of the Mediterranean Sea has revealed the presence of specimens belonging to the phylum of Loricifera that would be able to spend their entire life cycle without free oxygen (**Fig. 1**). Different analyses based on incubations

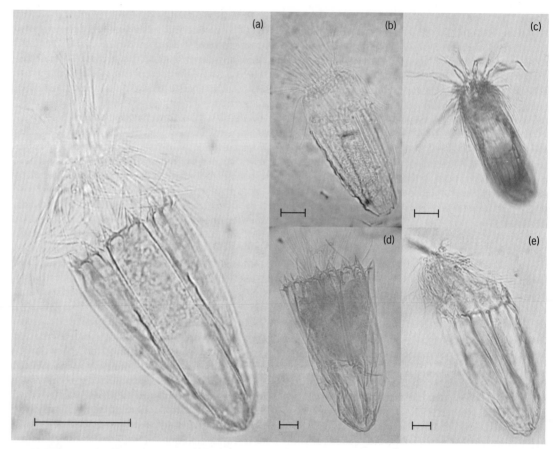

Fig. 1. Animals retrieved from the deep hypersaline anoxic L'Atalante basin. Reported are various species of Loricifera: (*a*) the genus *Spinoloricus* (stained with Rose Bengal), specifically a female with fully extended introvert, mouth cone, buccal tube, and a large oocyte showing that it is a mature female; (*b*) the genus *Rugiloricus* containing an oocyte (not stained with Rose Bengal); (*c*) the genus *Pliciloricus* (not stained with Rose Bengal); (*d*) the genus *Spinoloricus* containing an oocyte (stained with Rose Bengal); (*e*) the genus *Spinoloricus* (a molting exuvium). Scale bars: 50 μm.

with radioactive tracers and specific fluorogenic probes, analysis of the biochemical composition of the body, quantitative x-ray microanalysis, and analysis with infrared spectroscopy have revealed that these species of Loricifera are metabolically active and show specific adaptations to the extreme conditions of this deep basin. The loriciferans collected from the L'Atalante anoxic basin belong to three species, all new to science, that belong to the genera *Spinoloricus*, *Rugiloricus*, and *Pliciloricus* (Fig. 1). However, more than 80% of all known specimens belong to a new species of *Spinoloricus*. These loriciferans appear intensely colored when stained with Rose Bengal (a protein-binding stain) and consequently can be easily recognized under light microscopy. Conversely, all other animals (such as copepods or nematodes) that were dead remained quite transparent (not stained because they are devoid of proteins). Scanning electron microscopy confirmed the perfect integrity of these loriciferans, whereas all of the other meiofaunal taxa were largely degraded (**Fig. 2**). The abundance of loriciferans inhabiting this anoxic basin, ranging from 75 to 701 individuals per m², is the highest reported so far worldwide. In contrast, only two individuals of the phylum Loricifera have been found over the last 40 years in the entire deep Mediterranean Sea. Because the den-

sity of these organisms is significantly lower than the density of the brines covering the L'Atalante basin, any escape of Loricifera from these systems or their entrance from sediments surrounding the anoxic basin is virtually impossible. Additional analyses also revealed the presence of specimens of both genera *Spinoloricus* and *Rugiloricus* containing a large oocyte in their ovary, suggesting that some organisms were reproducing. Moreover, the presence of empty postlarval exuviae (exoskeletons) of loriciferans suggests the growth and molting of these organisms in this anoxic system (Fig. 1).

The viability of these loriciferans was tested under anoxic conditions (in an atmosphere of N_2), in the dark, and at an in situ temperature of about 14°C (57.2°F) with incubations with radiolabeled leucine, which was actively incorporated by these creatures. Additional experiments were conducted by incubating the Loricifera with a fluorogenic probe that reacts in the presence of active enzymes. These analyses, conducted by confocal laser microscopy, revealed a high fluorescence intensity in the Loricifera collected and maintained in anoxic conditions, thereby suggesting the presence of active metabolic functions (Fig. 2c, *d*). All of these tests provide evidence that the anoxic sediments of the L'Atalante basin are colonized by natural and living populations

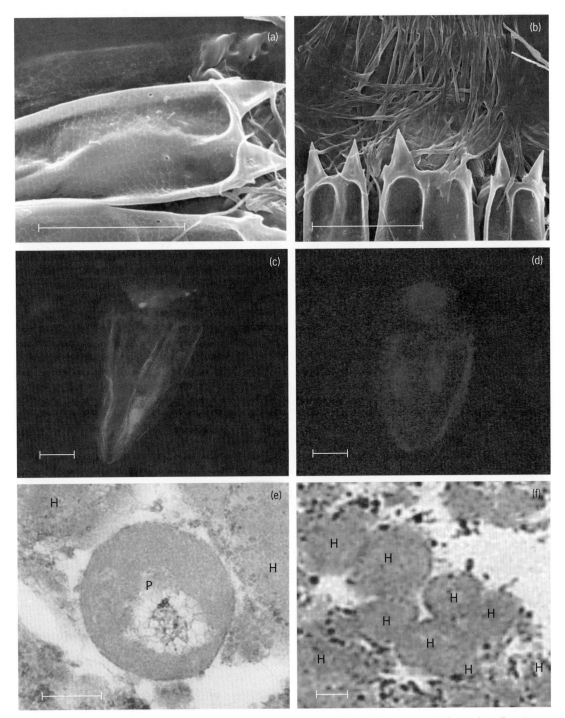

Fig. 2. Morphological, metabolic, and ultrastructural characteristics of Loricifera living in permanently anoxic sediments. (*a*) Scanning electron microscopy image of the lorica (hard protective case) of a new species of *Spinoloricus*. (*b*) Scanning electron microscopy image of the anterior edge of the lorica showing the spikes of the *Spinoloricus* species. (*c*) Living loriciferan and (*d*) killed loriciferan from anoxic sediments treated with green fluorescent chloromethyl probes (confocal laser microscopy images). (*e*) Transmission electron microscopy image of a possible prokaryote (P) in proximity to the hydrogenosome (H). (*f*) Detail of a hydrogenosome-like organelle. Scale bars: 50 μm (panels *a*, *b*, *c*, and *d*); 0.2 μm (panels *e* and *f*).

of active loriciferans. The organisms inhabiting the L'Atalante basin were also different from loriciferans inhabiting oxygenated deep-sea sediments, as suggested by the presence of different chemical and structural characteristics that were revealed using quantitative X-ray microanalysis and infrared spectroscopy. Scanning electron microscopy revealed a perfect integrity of the body of the loriciferans and the lack of prokaryotes attached to the body surface of these animals (Fig. 2*a*, *b*). Further ultrastructural analyses based on transmission electron microscopy observations on ultrafine sections have indicated the lack of mitochondria, which are key organelles for the metabolism of multicellular organisms using oxygen. These organelles are apparently replaced by hydrogenosome-like structures, possibly associated with endosymbiotic prokaryotes (Fig. 2e, f). The hydrogenosomes are organelles previously reported only in protozoa (unicellular eukaryotes) that are specifically adapted to life in the absence of oxygen. Because the hydrogenosomes have been so far reported only in obligate anaerobic eukaryotes, these results allow investigators to hypothesize that the loriciferans inhabiting the anoxic sediments of the L'Atalante basin have developed an anaerobic metabolism and specific adaptations to live without free oxygen. Although the evolutionary and adaptive mechanisms leading to the colonization of such extreme environments by these metazoans remain enigmatic, this discovery opens new perspectives for the study of metazoan life in oxygen-free habitats.

For background information see ANIMAL KINGDOM; ANOXIC ZONES; DEEP-SEA FAUNA; ECDYSOZOA; LORICIFERA; MARINE BIOLOGICAL SAMPLING; MARINE ECOLOGY; MARINE SEDIMENTS; MEDITERRANEAN SEA; OXYGEN; SEAWATER in the McGraw-Hill Encyclopedia of Science & Technology.

Roberto Danovaro; Cristina Gambi; Antonio Pusceddu; Antonio Dell'Anno; Reinhardt M. Kristensen

Bibliography. B. Boxma et al., An anaerobic mitochondrion that produces hydrogen, *Nature*, 434:74–79, 2005; R. Danovaro et al., The first metazoa living in permanently anoxic conditions, *BMC Biol.*, 8:30, 2010; J. H. P. Hackstein et al., Hydrogenosomes: Eukaryotic adaptations to anaerobic environments, *Trends Microbiol.*, 7:441–447, 1999; K. J. Hsü et al., History of the Mediterranean salinity crisis, *Nature*, 267:399–403, 1977; R. M. Kristensen, Loricifera, a new phylum with Aschelminthes characters from the meiobenthos, *Z. Zool. Syst. Evolutionforsch.*, 21:163–180, 1983.

Arctic marine transportation

The use of the Arctic Ocean for expanded marine transportation has garnered increased interest around the globe. This attention is primarily due to the historic transformation and reduction of Arctic sea-ice extent and thickness in response to a warming Earth. However, it is development of the Arctic's vast, untapped storehouse of natural resources—oil, gas, and mineral wealth—and increased marine tourism that are linking the maritime Arctic to the rest of the planet. Both are primary drivers of increasing Arctic marine traffic. Future integration of the Arctic's natural wealth with global markets and expanded polar marine tourism are expected to be dominant factors influencing Arctic marine operations. Large ships in remote areas such as the "top of the world" present a host of critical safety and environmental challenges for Arctic people and the marine environment. The physical changes in Arctic sea ice surely hold key implications for longer seasons of navigation and new access to once remote coastal regions. These changes may also allow for limited, summer transarctic voyages, beginning in the Russian maritime Arctic (**Fig. 1**).

Arctic globalization. Early in the twenty-first century the scope and breadth of Arctic marine operations is quite striking. For example, year-round navigation is maintained between Murmansk and Dudinka, a port city on the Yenisey River for the Siberian industrial mine complex at Norilsk, the world's largest nickel producer. In northwest Alaska, location of the Red Dog Mine (the world's largest zinc mine), large bulk ore carriers sail into the Chukchi Sea during a summer, ice-free navigation season to load zinc ore for carriage to Pacific markets. In the summer of 2010 a large bulk carrier, *Nordic Barents*, with iron ore sailed from Kirkenes, Norway to China across the top of Arctic Russia. Since 1991 Russian nuclear icebreakers have carried tourists to the North Pole. During recent summers icebreakers have sailed to every corner of the Arctic Ocean conducting oceanographic and geophysical research. Increasing numbers of Arctic voyages by cruise ships (including several of the world's largest liners) have been conducted off Greenland's east and west coasts, around Svalbard, and in the Canadian Arctic.

Hydrocarbon developments in the Arctic offshore, principally by Norway, Russia, and Greenland, have also stimulated increased Arctic marine traffic and operations in coastal areas. Liquefied natural gas has been shipped to markets in Spain and the U.S. east coast from the Snovit (offshore) and Hammerfest (onshore) complex in the Norwegian Arctic. Two tanker shuttle systems are operating in Russia's region of the Barents Sea. During the summers of 2010 and 2011 a Scottish company, Cairn Energy, has been using Swedish drill ships and support icebreakers to explore in seabed leases along Greenland's west coast. Shell is also poised to conduct exploratory drilling in leased areas off the northwest coast of Alaska during the summer of 2012. During the summer of 2010 a large Russian ice-class tanker completed a voyage carrying gas condensate from Murmansk to Ningo, China; this was the first tanker of more than 100,000 deadweight tons to sail the length of Russia's Northern Sea Route.

Sea-ice changes. Arctic sea-ice continues to retreat in extent and thickness early in the twenty-first century. During the past 3 decades, Arctic sea-ice coverage has declined by nearly 12% each decade. Despite this coverage change and significant sea-ice thinning, the Arctic Ocean remains fully or partially ice-covered in winter, spring, and autumn. Sea-ice simulations using global climate models also indicate the winter sea-ice cover remaining through the century and beyond. The models also show that the Arctic Ocean could be ice-free for a (short) period of time each year as early as 2030; the meaning of this is that all of the ice that once survived the seasonal melt cycle (the multiyear ice) will disappear leaving only first-year or annual sea ice. Such a future ice cover will be limited in thickness and should be more favorable for marine operations. One of the key remaining challenges for Arctic ship operations will be the high, seasonal sea-ice variability in Arctic coastal seas.

Arctic marine shipping assessment. The Arctic Council, an intergovernmental forum of the eight

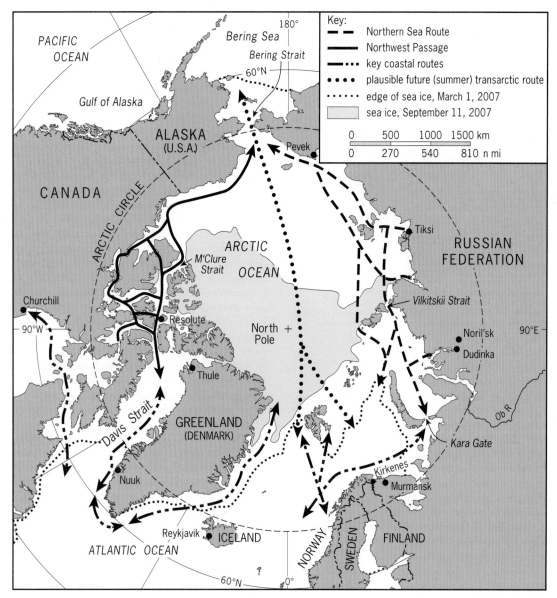

Fig. 1. Map of the Arctic Ocean, showing main Arctic marine routes and a plausible, future summer transarctic route.

Arctic states, released a large study in April 2009, the Arctic Marine Shipping Assessment (AMSA), which provides a comprehensive framework for protecting the people of the Arctic and the marine environment. This assessment can be considered a baseline of Arctic marine use (from data collected from the Arctic states), a strategic guide for a host of stakeholders, and a policy document of the Arctic Council (the study was approved by the Arctic Ministers after consensus was reached by the eight Arctic nations). Identified were more than 6000 vessels operating in the Arctic region during 2004; nearly all the voyages were destinational (ships sailing north to perform an activity and returning south) and regional (not transarctic) and the four primary types of vessels were Arctic community resupply ships, fishing vessels, bulk carriers, and marine tour ships. In a review of governance it was clear that marine navigation and marine uses in the Arctic Ocean are to be conducted within the framework provided for all oceans, the United Nations Convention on the Law of the Sea (UNCLOS). A scenarios-based approach was also used to create a set of plausible futures for Arctic marine operations. Two primary factors were identified as most influential in determining future Arctic marine traffic: governance (the degree of relative stability of rules and standards for ships in the Arctic), and the level of demand for Arctic natural resources and trade. Seventeen recommendations focused on three inter-related themes: enhancing Arctic marine safety, protecting Arctic people and the environment, and building the Arctic marine infrastructure. A lack of charts, ports, salvage, communications, environmental monitoring, aids to navigation, and more was considered a critical limitation to current and future safe navigation.

The assessment provided an overview of the issues and challenges of transarctic navigation. In keeping with the mandate of the Arctic Council, the study was careful to focus on issues related to marine safety and

Fig. 2. Varandey Terminal, an offshore platform in the eastern Barents Sea, with icebreaking shuttle tanker hooked up to receive oil. The terminal is equipped with a swivel system so the tanker can rotate or move with the ice flow. Oil is pumped from an onshore terminal to this platform and carried by tankers to another storage site in Murmansk. (*Courtesy of ConocoPhillips*)

environmental protection and not on the economic viability of the potential transarctic trade routes. Key issues considered highly important were the continuing presence of Arctic sea ice; the seasonality and reliability of such voyages (given the unpredictability of Arctic sea ice and weather); the risks of long transarctic voyages (for the ships and cargo) in a region without adequate marine infrastructure; and the question of whether ships will be escorted by icebreakers in convoy, or operated independently as many new icebreaking ships are designed to sail in ice. While crossing the Arctic Ocean may be theoretically possible today with advanced ice-capable ships, the economic and operational considerations of these routes have not yet been fully explored. Continued experimental voyages, such as those conducted during recent summers in the Russian maritime Arctic, will indicate to the international shipping community the potential and challenges of transarctic shipping.

New marine technology. Two marine transportation (shuttle) systems in the southeast corner of the Barents Sea reflect a new age of marine technology transfer to the Arctic. A three-ship icebreaking tanker operation services an offshore terminal that was developed by the American firm ConocoPhillips and Russia's Lukoil (**Fig. 2**). The icebreaking tankers, which ply between the Varandey terminal and Murmansk to the west, were built in Korea by Samsung Heavy Industries using Finnish icebreaking technology. The state-of-the art tankers are operated by Russia's largest shipping firm, Sovcomflot, and can annually deliver 12 million tons of oil. A second shuttle fleet of two tankers is being developed to deliver oil to Murmansk from an offshore oil production platform in the same region. The five tankers of these two fleets are designed to operate without icebreaker convoy; they essentially operate as icebreakers in

their own right and navigate through Arctic sea ice efficiently and safely.

A fleet of six icebreaking carriers (container ships) has also been operating year-round between the port of Dudinka in northwest Siberia and Murmansk. The icebreaking container ships primarily carry nickel plates from the mining complex Norilsk Nickel to the hub port of Murmansk for distribution to world markets. The innovation in these ships is the use of an azimuthing, steerable thruster as the main propulsion unit. Both bow and stern of the carriers are icebreaking capable, allowing the ship to break ice going ahead and astern, called the "double-acting concept" by the Finnish designers (**Fig. 3**).

Technological advances are being made in shipboard radars to better detect small icebergs and distinguish between first-year ice and more dangerous (and harder) multiyear ice. New satellite remote-sensing sensors are steadily improving so that 1–10-m resolution images can now indicate small-scale ice features (topography and ice types), information highly valuable for tactical navigation through ice fields. Special paints are being applied to icebreaking ships and stationary hull structures to provide protection from abrasion, reduce hull roughness, and also reduce the friction between the hull and snow and ice. Also, significant advances have been made in ice navigation training and ice simulators for future ice navigators.

Outlook. Two key drivers will continue to influence Arctic marine transportation in the decades ahead: the economic connections of Arctic natural resources to the globe, and, continued loss of Arctic sea ice resulting in increasing marine access and potentially longer marine navigation seasons throughout the region. Highly plausible will be future transportation of Arctic natural resources by modern, ice-capable carriers to emerging global markets. Arctic offshore hydrocarbon developments (off Greenland's west coast; in U.S. Arctic waters off northwest Alaska; in the Norwegian Arctic; and in the Barents and Kara seas of the Russian maritime Arctic) will require significant support fleets and ships to carry oil and gas out of the Arctic. Hard minerals will also be carried out by ship in the summer from northwest Alaska (zinc ore), year-round from western Siberia (nickel plates), and potentially year-round from Baffin Island to European markets (high-grade iron ore).

Major advances will be made in enhancing Arctic marine safety and environmental protection principally through mandatory rules adopted at the International Maritime Organization in London. Summer use of the Arctic Ocean for select, transarctic voyages will be conducted across Russia's Northern Sea Route. Experimental voyages in 2010 and 2011 indicate that voyages from northern Norway across the Eurasian Arctic and through Bering Strait to ports in China are technically feasible. The real challenge for future transarctic navigation will be the economic viability of such voyages using more expensive polar class ships and operating on a nonregular shipping schedule.

Fig. 3. Double acting icebreaking technology. (*a*) Icebreaking ships are designed to operate ahead in the normal mode in open water and light ice. In heavier ice, the stern of the ship becomes the bow and the ship essentially travels stern-first through the ice. (*b*) View of the MV *Norisk-Nickel*, an icebreaking container ship showing its Azipod propulsion unit. (*Aker Arctic Technology*)

For background information *see* ARCTIC OCEAN; ICEBREAKER; MERCHANT SHIP; OIL AND GAS, OFFSHORE in the McGraw-Hill Encyclopedia of Science & Technology.   Lawson W. Brigham

Bibliography. ACIA, *Arctic Climate Impact Assessment*, Cambridge University Press, 2005; *Arctic Marine Shipping Assessment 2009 Report*, Arctic Council, April 2009; L. W. Brigham, Russia opens its maritime Arctic, *Naval Instit. Proc.*, 137(5):50–54, 2011; L. W. Brigham, Think Again: The Arctic, *Foreign Policy*, September–October 2010; *Guidelines for Ships Operating in Polar Waters*, International Maritime Organization, London, Adopted 2 December 2009, January 18, 2010; V. Santos-Pedro, Toward a polar code, *Mar. Tech.*, (1):65–67, January 2011.

Arsenic-eating bacteria

Bacteria are the most diverse form of life on Earth in terms of number of species and habitats. Some species of bacteria can thrive in environments that would be deadly to most other life forms; for this reason, they are called extremophiles. One well-known extremophile is *Thermus aquaticus*, which thrives at 70°C (158°F) and generates energy via chemosynthesis. A recently discovered extremophile is *Desulforudis audaxviator*, which has been found living 2 mi (3.2 km) beneath the Earth's surface in extreme temperatures and utilizes radioactive decay from uranium to generate energy. The newest addition to the list of extremophiles is a bacterial species that lives in an arsenic-rich lake and can utilize arsenic in place of phosphorus in macromolecules (for example, DNA and proteins). This is an astounding finding because it is accepted that phosphorus is required for life and arsenic is considered poisonous. Furthermore, arsenic is chemically unstable and is theoretically unlikely to generate stable macromolecular structures. Because of these conceptual concerns, some in the scientific community have challenged the biochemical methods and data interpretation used by the group that has characterized this unique arsenicophile.

The arsenic-eating extremophile: GFAJ-1. Felisa Wolfe-Simon and colleagues cultured bacteria from lake sediments collected from Mono Lake (**Fig. 1**) in California, a region known for its naturally high levels of arsenic. The bacteria were cultured in increasing amounts (up to 5 mM) of the arsenic compound

Fig. 1. Photograph of Mono Lake. Mono Lake is located in Mono County, California, not far from Yosemite National Park. This saline lake is rich in carbonate minerals, which form the unique column structures in the lake, called tufa, a type of limestone. Although the lake is salty and alkaline, and contains arsenic, it is still teeming with life, especially at the microscopic level, including the GFAJ-1 arsenic-eating bacteria. (*Photo courtesy of Lee Collins*)

arsenate (AsO_4^{3-}), while diluting out phosphate (PO_4^{3-}) and nearly eliminating it from the growth medium. Trace impurities resulted in approximately 3 μM phosphate (1000-fold less than typical growing conditions) remaining in the growth medium. The resulting strain that grew in these conditions was named GFAJ-1 (which is an abbreviation for "Give

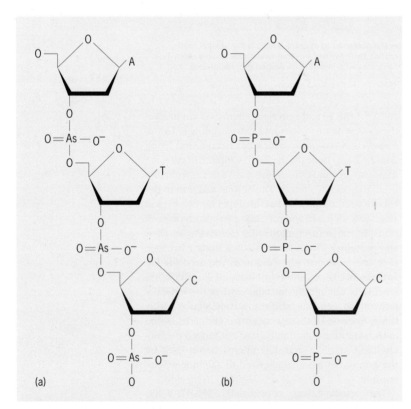

Fig. 2. Chemical structure of DNA. One of the most striking findings by Felisa Wolfe-Simon and colleagues consisted of data that strongly suggested the incorporation of arsenic (As) in place of phosphorus (P) in the DNA backbone. Because of the chemical and physical similarity of As to P, the proposed chemical structures are identical except for the substitution of As (panel *a*) for P (panel *b*). Each panel depicts three nucleotides within a strand of single-stranded DNA.

Felisa A Job"). The group was surprised that the bacteria not only survived in an environment free of phosphorus and rich in the arsenic compound, but were also reproducing. Up until now, all living systems were thought to depend on only a few essential elements (oxygen, hydrogen, sulfur, nitrogen, carbon, and phosphorus) as the basis for carrying out all physiological functions. Using the current standard (16S rRNA sequence phylogeny) for identifying bacterial species, the group determined GFAJ-1 to be a member of the Halomonadaceae family of Gammaproteobacteria, which have previously been shown to accumulate intracellular levels of arsenic. The GFAJ-1 strain is currently grown in 40 mM arsenate. The group notes that these bacteria actually grow better in the presence of phosphorus; they do not grow when phosphate or arsenate is absent.

Given the chemical similarities between arsenic and phosphorus, it was hypothesized that these bacteria could effectively substitute arsenate for phosphate (**Fig. 2**), incorporating arsenate in the structure of macromolecules such as nucleic acids, proteins, and lipids. The group demonstrated by a specialized form of mass spectrometry that the arsenic was indeed taken up by the cell. They then added a radiolabeled arsenate to the growth medium to track whether arsenate was incorporated into biological molecules. The radioactive decay of the arsenic serves as a beacon to indicate the presence of arsenic. The group observed arsenic in cellular fractions of proteins, metabolites [adenosine triphosphate (ATP), acetyl coenzyme A (acetyl CoA), and hydrogenated nicotinamide adenine dinucleotide (NADH)], lipids, and nucleic acids, suggesting that arsenic was incorporated into these biological molecules. To further confirm that arsenic was incorporated into the DNA of GFAJ-1, the group used high-resolution secondary ion mass spectrometry and intense x-ray imaging techniques. Using these methods, the scientists concluded that arsenate was associated with macromolecules, including DNA, in a manner consistent with it being a part of the macromolecular structure.

Controversy in the scientific community. Since its publication, the findings by Felisa Wolfe-Simon and her group have come under considerable criticism. Surprisingly, instead of the usual route of letters to the editor or a scientific article rebutting the claims, much of the criticism has taken place via Internet blog sites. Scientists currently have mixed feeling as to whether this is the best forum to discuss science, but this is likely the wave of the future for scientific debates. Although the authors have addressed many of the concerns about methods and interpretation of the data, a significant portion of the criticism is based on the chemistry of arsenic and the likelihood that it could substitute for phosphorus in key macromolecular structures and pathways.

Arsenic chemistry. Arsenic is the 33d element in the periodic table and belongs to the same family of elements as phosphorus. As such, arsenic and phosphorus are very similar in terms of their chemical behavior as well as their physical structure. Hence, it seems reasonable that arsenic could be used by bacteria in place of phosphorus under appropriate selective pressures. Stable changes to the DNA phosphodiester backbone have been synthesized by scientists in the laboratory and studied. For example, replacing a phosphate oxygen atom with sulfur to create phosphorothioate, or with boron to form a boranophosphate species, has been shown to be recognized by DNA polymerases and ribosomes. Evidence is available that shows that bacteria are in fact capable of integrating phosphorothioate into their genomic structure. Therefore, it is conceivable that bacteria growing in an arsenic-rich, phosphorus-poor environment would adapt by substituting arsenic for phosphorus in their nucleic acid structure, proteins, lipids, and so forth. However, arsenic is generally toxic to living systems, including bacteria.

Arsenic toxicity is attributed largely to its oxidized form, arsenate, which has the ability to utilize the same pathways that allow phosphate to enter the cell and be incorporated into substrates that are recognized by various enzymatic pathways. For example, arsenate is recognized by ATP synthase, but the resulting product is highly unstable and effectively prevents the production of ATP. The potential that arsenic-based substrates might hijack important cellular processes such as energy production may have led to the evolution of a variety of mechanisms to eliminate arsenic from the cell. These typically involve the enzymatic reduction of arsenate by arsenate reductase to an even more toxic arsenite species, which is then extruded from the bacterium. Thus, in order for a bacterium, such as GFAJ-1, to transport arsenate from its environment and incorporate it as a component of macromolecular structures, it would also have to ensure that the arsenate compound remained stable and was protected from action by a variety of arsenate reductases. Interestingly, some bacteria express arsenate reductase enzymes on their surface and utilize arsenate as electron acceptors in the respiratory chain for energy production under anaerobic conditions (that is, these bacteria can be said to "breathe arsenic").

If the cell was able to protect the arsenate from being acted on by a variety of arsenate reductase enzymes, it would still have to cope with intrinsic differences between arsenate and phosphate in terms of their chemical stability. For example, the phosphodiester bonds that make up the DNA backbone are highly resistant to attack by water, a process known as hydrolysis. Hydrolysis of these bonds is so slow that it is estimated that it would take approximately 30 million years to break half of the phosphodiester bonds in a DNA molecule by this process. This corresponds to approximately one phosphodiester linkage undergoing spontaneous hydrolysis per decade in a protobacterial genome consisting of 3 million base pairs. The stability of these bonds is not unexpected considering that all of the genetic information for an organism lies within this structure. In contrast, arsenate esters undergo rapid hydrolysis with a half-life estimated to be 60 milliseconds. Although structurally it seems plausible that arsenate could substitute for phosphate in the construction of the DNA backbone, the instability of arsenate makes this much less likely.

Conclusions. Felisa Wolfe-Simon and colleagues provided the first report of a life form, the GFAJ-1 bacterial strain of the family Halomonadaceae, that can grow by using arsenate instead of phosphate. This challenges the long-standing belief that all life forms require carbon, hydrogen, oxygen, nitrogen, sulfur, and phosphorus. Arsenic can form arsenate, which is so physicochemically similar to phosphate that it can insert in place of phosphate in many biochemical reactions. However, given the instability of arsenic-based compounds, it is difficult to understand how arsenate could truly substitute for phosphate in biochemical reactions. Nevertheless, although not yet demonstrated, it may still be possible that organisms could evolve mechanisms utilizing alternative elements, such as arsenic, for key macromolecular functions under the right selective pressures.

For background information *see* ARSENIC; ASTROBIOLOGY; BACTERIA; BACTERIAL PHYSIOLOGY AND METABOLISM; BACTERIOLOGY; BIOCHEMISTRY; BIOSPHERE; BIOSYNTHESIS; DEOXYRIBONUCLEIC ACID (DNA); MICROBIAL ECOLOGY; PHOSPHORUS in the McGraw-Hill Encyclopedia of Science & Technology.

Paul Anaya; Rebekah L. Waikel

Bibliography. D. H. Appella, Non-natural nucleic acids for synthetic biology, *Curr. Opin. Chem. Biol.*, 13:687–696, 2009; M. I. Fekry, P. A. Tipton, and K. S. Gates, Kinetic consequences of replacing the internucleotide phosphorus atoms in DNA with arsenic, *ACS Chem. Biol.*, 6:127–130, 2011; S. Silver and L. T. Phung, Novel expansion of living chemistry or just a serious mistake?, *FEMS Microbiol. Lett.*, 315:79–80, 2011; D. J. Thomas, Arsenolysis and thiol-dependent arsenate reduction, *Toxicol. Sci.*, 117:249–252, 2010; F. Wolfe-Simon et al., A bacterium that can grow by using arsenic instead of phosphorus, *Science*, 332:1163–1166, 2011.

ASCE program for sustainable infrastructure

Infrastructure includes constructed facilities and associated natural features that shelter and support human activities, such as buildings of all types, communications, energy generation and distribution, transportation of all modes, water resources, and waste treatment. An example of an associated natural feature is a wetland serving as part of a stormwater management system. Since infrastructure services are essential for human health, productivity, safety, and quality of life, infrastructure's value is equivalent to that of these human qualities. However, infrastructure also has direct economic importance. In 2008, for example, new construction put in place was $1.072 trillion, or 7.42% of the gross domestic product (GDP) of the United States. Considering that renovation, operation, and maintenance expenditures are of similar magnitude, infrastructure exceeded 10% of the GDP. Its value is a major element of wealth, representing 68% of U.S. fixed assets.

To serve humanity well, infrastructure systems often exceed a century of service life and need to be functional; affordable; environmentally benign or restorative; safe; resilient to the effects of natural, accidental, and willful hazards; and supportive of the communities they serve and affect. The American Society of Civil Engineers (ASCE) defines sustainability as, "A set of environmental, economic and social conditions in which all of society has the capacity and opportunity to maintain and improve its quality of life indefinitely without degrading the quantity, quality or the availability of natural, economic and social resources." These are the qualities pursued in ASCE's program for sustainable infrastructure.

The ASCE's program for sustainable infrastructure is rooted in its Code of Ethics. Its first fundamental canon states, "Engineers shall hold paramount the safety, health and welfare of the public and shall strive to comply with the principles of sustainable development in the performance of their professional duties." The program includes four major thrusts, as listed below.

1. Developing and implementing, in cooperation with the American Council of Engineering Companies (ACEC) and the American Public Works Association (APWA), a sustainable infrastructure project rating system called Envision™.

2. Producing and maintaining, as a nationally and internationally accredited consensus standards developing organization, the Sustainable Infrastructure Project Rating Standard.

3. Developing and operating, as a nationally and internationally accredited professional certification organization, the Council of Sustainable Infrastructure Professionals.

4. Developing and conducting a program of sustainable infrastructure education.

Brief descriptions of these program thrusts will be presented.

Sustainable infrastructure project rating system. Rating systems for the sustainability of buildings have been very successful in focusing the attention of owners, designers, builders, policy makers, and the general public on the sustainability qualities of buildings. Prominent systems for the United States and Canada are LEED of the U.S. Green Building Council and Green Globes of the Green Building Initiative. For infrastructure systems other than buildings, the Assessment and Awards Scheme for Improving Sustainability in Civil Engineering and the Public Realm (CEEQUAL) was established in the United Kingdom in 2003. For the United States, the ASCE has partnered with ACEC and APWA to establish the Institute for Sustainable Infrastructure (ISI) to develop and operate Envision, the sustainable infrastructure project rating system.

Envision provides guidance to owners and infrastructure project teams on project features that would provide higher degrees of sustainable performance and measures to rank the level of performance for each such feature. These ratings address the triple bottom line (social, economic, and environmental) using 74 objectives organized in 10 sections.

1. Project Pathway Contribution
2. Project Strategy & Management
3. Community: Long and Short Term Effects
4. Land Use and Restoration
5. Landscapes
6. Ecology and Biodiversity
7. Water Resources and Environment
8. Energy and Carbon
9. Resource Management Including Waste
10. Access and Mobility

The Project Pathway Contribution assesses how the project affects the goals and aspirations of the communities it serves and affects. Other sections address project performance. Performance levels are basic (such as meeting legal requirements), improved, superior, sustainable, and restorative. Envision's uses range from guidance of a project team to third-party evaluation and recognition.

Like CEEQUAL, Envison applies to all types of infrastructure projects, including communications, energy generation and distribution, industrial facilities, transportation of all modes, waste management, and water, stormwater, and wastewater systems. Envision was available in 2011 for use by project teams; third-party evaluations and awards will become available in 2012.

Sustainable infrastructure project rating standard. ASCE is accredited by the American National Standards Institute (ANSI) and, as such, intends to develop a sustainable infrastructure project-rating standard based on the rating criteria developed for Envision and developed following ASCE's consensus procedures, which are approved by ANSI. Once the standard is available, it can provide the rating criteria for Envision.

Products and services that comply with a consensus standard gain greater market recognition, acceptance, and use, and thereby benefit owners, producers, and users of sustainable infrastructure systems. For instance, federal law PL 104-113 requires federal agencies to use, whenever feasible, standards developed or adopted by voluntary consensus-standard bodies in lieu of developing government-unique standards or regulations.

This market recognition arises from the equitable, accessible, and responsive process followed in the development of a consensus standard. In this process, participation is open to all interested stakeholders. Balance of interests shall be sought. Consensus must be reached by representatives from materially affected and interested parties in an environment that is free from dominance by any party. Standards are required to undergo public reviews during which any member of the public may comment. Comments from the consensus body and public review period must be responded to in writing. All unresolved objections, attempts at resolution, and substantive changes to the text are provided to the consensus body for review prior to final vote. An appeals process through the standards developer to address procedural concerns is required.

The standardization process scheduled to began in 2011.

Certification of sustainable infrastructure professionals. ASCE is developing a nationally accredited, specialty certification program to recognize competence and eminence in sustainable infrastructure. The certified sustainable infrastructure professional will be a state-licensed professional engineer, or another state-licensed infrastructure professional, who has demonstrated, through examination and continuing education, competence in the principles and practices of sustainable infrastructure. The diplomate in sustainable infrastructure will be a state-licensed professional engineer, or another state-licensed infrastructure professional, who has demonstrated, through practice, examination, and continuing education, eminence in the principles and practices of sustainable infrastructure.

The ASCE's programs of specialty certification are conducted under Civil Engineering Certification, Inc. (CEC). CEC was created in 2004 by the ASCE Board of Direction to provide a mechanism for professional post-licensure certification of the various specialties within civil engineering. The CEC is accredited by the Council of Engineering & Scientific Specialty Boards (CESB).

Specialty certification in sustainable infrastructure offers a number of benefits. Certification recognizes the individual as a leader in the practice of sustainable infrastructure. Certification is an advanced qualification beyond licensure recognized by clients, employers, peers, and the public. Certification demonstrates attainment of the body of knowledge for sustainability and commitment to stay current on new technological innovations. Certification demonstrates a strong commitment to professionalism through its ethics and continuing professional development requirements. Certification provides clients with an assurance that they are engaging highly qualified participants on their projects. Certification supports the concept of qualifications based selection.

Implementation of certification for sustainable infrastructure professionals is planned for 2012.

Sustainable infrastructure education. ASCE is developing an online continuing-education program to transmit to engineers and other infrastructure professionals the bodies of knowledge for the sustainable infrastructure project rating system and for certification of sustainable infrastructure professionals.

The introductory course for the continuing education program, "Fundamentals of Sustainable Engineering, " is now available from ASCE online. It gives the engineer an understanding of sustainability, its triple bottom line and strategic contributions to projects; a certificate demonstrating understanding of the principles and concepts of sustainability; and access to the bodies of knowledge for project and professional certification.

Eleven additional courses are under development for availability in 2012. The Fundamentals of Sustainable Engineering will be prerequisite for each.

Courses are listed below and related to sections of Envision.

1. Sustainable Project Management: Project Pathway Contribution and Project Strategy and Management

2. Community Participation: Community Long and Short Term Effects

3. Land Use: Land Use and Restoration, and Landscapes

4. Ecological Issues: Ecology and Biodiversity

5. Access/Mobility: Access and Mobility

6. Water: Water Resources and Environment

7. Air: Air quality aspects of Community

8. Light: Lighting quality aspects of Community

9. Noise: Noise aspects of Community

10. Waste: Resource Management Including Waste

11. Assessments of project life-cycle impacts: Energy and carbon, plus advanced techniques of life-cycle cost and benefit assessments; life-cycle environmental assessments, and consideration of trade-offs among incommensurate qualities.

The Institute for Sustainable Infrastructure plans to develop training for its assessors in the application of Envision, also for availability in 2012. This will complement and will be coordinated with ASCE's program of Sustainable Infrastructure Education.

Role of infrastructure in climate management. Climate change is any systematic change in the long-term statistics of climate elements (such as temperature, precipitation, or winds) sustained over several decades or longer. Climate change may be caused by natural external drivers, such as changes in solar emission or in the Earth's orbit, natural internal processes of the climate system, or anthropogenic drivers, such as greenhouse-gas emissions from electric power generation and transportation.

Physical evidence and scientific community consensus suggests that the world is in a period of climate change that is leading to the alteration of weather patterns and an increase in their extremes. Human settlements, industries, and other land uses must adapt to climate change as must the infrastructure systems that serve them.

Mitigation. Mitigation of climate change involves efforts to reduce drivers of climate change, such as emissions of greenhouse gases. Some mitigation measures require additions to infrastructure, such as wind and hydroelectric generation and transmission systems and carbon capture and storage systems. Other mitigation measures are more efficient infrastructure systems that use less energy. These are recognized in the Sustainable Infrastructure Project Rating System. The Engineering Founder Societies are addressing mitigation in their Carbon Management Project and its Carbon Management Technology Conference.

Adaptation. Adaptation to climate change involves efforts to realize potential benefits and reduce losses from climate change that has and will occur. The United States has established an interagency climate change adaptation task force to address these issues: The Council on Environmental Quality. The ASCE has established a Committee on adaptation to climate change to work with its institutes, technical councils, and technical divisions on the design criteria and practices addressing climate-change effects on the extremes in temperatures; droughts; wildfires; rainfall and snowfall; flooding; wind velocities, and storm surges to which infrastructure systems must provide resilience throughout their long service lives.

Trends. This article presents a current (2011) view of the ASCE's rapidly developing program for sustainable infrastructure. Much will be learned as the project rating system, standardization, professional certification, education, and climate-change mitigation and adaptation efforts are developed and implemented; opportunities and needs for their improvements will be identified. These will be subjects for much research and can be expected to evolve and improve substantially over the years to come.

For background information *see* CIVIL ENGINEERING; ENVIRONMENTAL ENGINEERING; GLOBAL CLIMATE CHANGE; GREENHOUSE EFFECT; TRANSPORTATION ENGINEERING IN THE MCGRAW-HILL ENCYCLOPEDIA OF SCIENCE & TECHNOLOGY.

Richard N. Wright

Bibliography.   Adapting to the Impacts of Climate Change (http://www.nap.edu/catalog.php?record_id=12783); Advancing the Science of Climate Change (http://www.nap.edu/catalog.php?record_id=12782); American National Standards Institute (http://www.ansi.org/); American Society of Civil Engineers': Code of Ethics (http://www.asce.org/Leadership-and-Management/Ethics/Code-of-Ethics/); American Society of Civil Engineers': Sustainability (http://www.asce.org/sustainability/); ASCE: Fundamentals of Sustainable Engineering (http://www.asce.org/fsecourse/); Carbon Management Technology Conference (http://www.carbonmgmt.org/); CEEQUAL (http://www.ceequal.com/); Civil Engineering Certification (http://www.asce.org/PPLContent.aspx?id=2147485045); Council of Engineering & Scientific Specialty Boards (http://www.cesb.org/index.php?option=com_content&task=view&id=13&Itemid=27); Founder Societies Technologies for Carbon Management (http://www.fscarbonmanagement.org/); Green Building Initiative (http://www.thegbi.org/); Green Globes: Building Environment Assessments (http://www.greenglobes.com/); The Institute for Sustainable Infrastructure (http://www.sustainableinfrastructure.org/); Limiting the Magnitude of Future Climate Change (http://www.nap.edu/catalog.php?record_id=12785); Progress Report of the Interagency Climate Change Adaptation Task Force: Recommended Actions in Support of a National Climate Change Adaptation Strategy: 2010 (http://www.whitehouse.gov/sites/default/files/microsites/ceq/Interagency-Climate-Change-Adaptation-Progress-Report.pdf); U.S. Census Bureau: Construction & Housing: Construction Indices and Value (http://www.census.gov/compendia/statab/cats/construction_housing/construction_indices_and_value.html); U.S. Census Bureau: Construction Spending (http://www.census.gov/const/www/c30index.html); U.S. Census Bureau: The 2011 Statistical Abstract (http://www.census.gov/compendia/statab/); U.S. Green Building Council (http://www.usgbc.org/).

Australopithecus sediba

The Cradle of Humankind World Heritage Area is located northwest of Johannesburg, South Africa. This world heritage locale encompasses 47,000 hectares (116,000 acres) and contains many hominin-bearing sites. The most recently discovered site, Malapa, represents an especially rich early-hominin locality. Malapa is situated roughly 15 km (9.3 mi) north-northeast of the well-known sites of Sterkfontein, Swartkrans, and Kromdraai. Malapa contains partially articulated skeletal remains of several individuals of a newly recognized species, *Australopithecus sediba*, as well as abundant, well-preserved fauna. The discovery and geological setting of *Au. sediba* and the potential taxonomic and phylogenetic implications of this new hominin taxon are described herein.

Discovery. Malapa was first discovered by Lee R. Berger in 2008 during a geospatial survey in the dolomitic limestone region of the Cradle of Humankind World (the purpose of the survey was to identify caves and potential hominin-bearing deposits). The site, which is a deroofed cave of at least 10 × 15 m (33 × 49 ft) in area, was named Malapa, which means "homestead" in Sesotho. On further investigation of the site's fossil-bearing potential, the first hominin specimens were discovered by Matthew Berger (who was 9 years of age at the time). The original discovered block contained a juvenile hominin clavicle (collarbone), which was directly associated with a fragment of a mandible containing a canine. After preparation, the block revealed both cranial and postcranial elements, constituting one of the most complete early-hominin skeletons ever discovered: this received the catalog number MH1, which stands for Malapa Hominin 1. Later in 2008, the initial remains of a second individual were discovered in situ (in the deposit) by Lee Berger. This specimen was a well-preserved partial skeleton of an adult hominin and was labeled Malapa Hominin 2 (MH2). To date, most of the head and significant parts of the body of the juvenile MH1 have been recovered (see **illustration**). MH2 is also very complete; importantly, most of its elements are found in articulation. As parts of both MH1 and MH2 were discovered in the course of cleaning operations of the surface of the Malapa deposit, it was determined that both specimens originated from within the deposit at Malapa; in addition, at the time of their deaths, the two individuals were lying only a few centimeters apart. Other fossilized animals were also found at Malapa, including carnivores, bovids, equids, and insects.

Geology and dating of the site. Detailed geological work has given chronological context to the site and its surroundings. Malapa is situated at the north end of a series of north-to-south trending caves formed in the Lyttelton Formation of the Chuniespoort Group. This formation houses a number of other

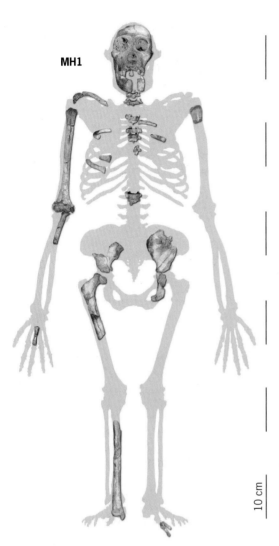

MH1

10 cm

The Malapa hominin skeleton 1 (MH1). (*Reproduced from L. R. Berger et al., Australopithecus sediba: A new species of Homo-like australopith from South Africa, Science, 328:195–204, 2010*)

fossil-bearing caves, including Gladysvale, which is located 1.5 km (0.93 mi) to the west of Malapa.

The Malapa deposit comprises calcified clastic sediments, which have five distinct sedimentary units interspersed with sheets of flowstone (deposits of calcium carbonate that accumulated against the walls or floor surfaces of a cave where water flowed on the rock). All of the sedimentary units were deposited by water action or mass flow. It was determined that the hominin skeletons and associated faunal remains were deposited within the cave during a rapid, homogeneous depositional event that occurred 1.977 million years ago (MYA). Following deposition, the sediments containing the bones and carcasses were rapidly cemented, contributing to the high quality of preservation of the Malapa fossils. Cosmogenic dating of the surrounding land surface indicates that the cave floor was approximately 30–50 m (98–164 ft) below the land surface at the time of deposition. This has led the investigating team to interpret the site as a natural death trap; in this scenario, animals likely entered through the opening of

a vertical shaft that was located several meters above the cave floor, with the animals either dying on impact from the fall or becoming trapped inside the cave and dying there.

Dating of Malapa. Dating of fossil sites in South Africa has been problematic in the past because of the nature of deposition within cave settings. However, recent advances in U-Pb dating are now rectifying this situation. U-Pb is a radiometric dating method that allows for very precisely constrained dates to be recorded for speleothems (cave formations) and flowstones associated with fossils. In addition to U-Pb, two other dating techniques were applied to the Malapa site: relative dating using fauna and paleomagnetic dating.

Relative dating was done based on the fauna associated with the hominins at Malapa, which included taxa such as *Equus* (equines) and *Tragelaphus strepsiceros* (greater kudu). These taxa first appear in the fossil record approximately 2.33 MYA in Africa, thus providing a maximum age bracket for Malapa. *Megantereon whitei* (saber-toothed cat) was also identified from the site, and this species had a last appearance datum at 1.5 MYA, thus providing a minimum age bracket. Based on the fauna, an age range of 2.33–1.5 MYA was determined for the deposit.

Samples of flowstone were obtained for radiometric dating and the U-Pb content was studied. Immediately below the MH2 skeleton is a flowstone seam that provided samples that were dated at two independent laboratories. The results from the two labs were highly concordant: 2.024 ± 0.062 MYA (Bern) and 2.026 ± 0.021 MYA (Melbourne). These dates further refine the maximum age estimate for the hominins. More recently, a capping flowstone on top of the hominin-bearing sediments was identified. Two samples derived from this flowstone returned dates of 1.975 ± 0.142 MYA and 2.075 ± 0.163 MYA, indicating that the formation of the first flowstone, the deposition of the hominin skeletons, and the formation of the second flowstone had all occurred within a very narrow time window.

Paleomagnetic dating is the analysis of normal and reversed magnetic signals recorded in the deposits. Initially, a normal polarity was identified at the base of the lower flowstone. Given the dates obtained for this flowstone, this likely represents the Huckleberry Ridge Subchron (a defined paleomagnetic event), which is dated elsewhere to 2.06 ± 0.04 MYA. Above this, a reversed polarity signal is recorded through the flowstone. Then, in the hominin-bearing sediments directly above the flowstone, a normal polarity is recorded, which initially was thought to represent the beginning of the Olduvai Subchron (1.95–1.78 MYA). Having these three paleomagnetic events recorded at Malapa allows for an initial defined age range of 2.06–1.78 MYA. However, with the discovery of the capping flowstone and errors associated with the dating, it was recognized that this depositional event could not be as young as 1.78 MYA; instead, it must represent a short 3000-year magnetic reversal that occurred between 1.977 and 1.98 MYA, placing the moment of deposition in a remarkably short 3000-year window of time.

Thus, the dates established from the three dating techniques, combined with the discovery of a capping layer of rock that could be dated, provide an exceptional level of precision for a southern African fossil deposit.

Malapa hominins. More than 200 hominin specimens have been recovered from Malapa. Most specimens have been attributed to the holotype (MH1) and paratype (MH2) skeletons. Elements of other individuals have been recovered, and these include at least one infant (MH3) and at least another adult individual (MH4).

MH1: Australopithecus sediba holotype. MH1 (also named Karabo) has a preserved cranium and most of the right side of the skeleton. The MH1 skull is missing the cranial base and much of the left side of the cranial vault. The entire face is complete. The right half of the mandible is preserved. All maxillary premolars and molars are present, with the third molars still forming in the crypt [visible in computed tomography (CT) scans as well as synchrotron scans]. The maxillary left lateral incisor is in place, whereas the isolated right central incisor and canine have been recovered. The right mandibular molars are preserved, including the third molar in the crypt, along with the left canine in a small fragment of the anterior mandible. For the MH1 postcrania, parts of the axial skeleton, shoulder girdle, upper limb, pelvic girdle, and lower limb have been recovered. Some secondary growth centers in the humerus, ulna, radius, os coxa, and femur were unfused at the time of death. Based on the state of eruption of the dental remains as well as the unfused growth centers, the developmental age of MH1 was estimated to be 12–13 years (note that MH1 could have been younger at the time of death if he grew at a faster rate than living humans, as is seen in modern apes). This is roughly comparable in ontogenetic age to the type specimen of *Homo habilis* (OH7) and the Nariokotome *H. erectus* skeleton (KNM-WT 15000). The development of MH1's brow ridges and the relatively large and rugose (roughened) muscle scars of the postcranial skeleton support the argument that MH1 was a male individual.

MH2: Australopithecus sediba paratype. MH2 is represented by a relatively complete but fragmented mandible, two isolated maxillary teeth, and significant portions of the axial skeleton, shoulder girdle, upper limb and hand, pelvic girdle, and lower limb. All of the teeth of MH2 are relatively worn. Along with the observation that the epiphyseal (fusion) lines of all observable long bones are completely fused and obliterated, this indicates that the individual was fully adult at the time of death. MH2 has an undistorted partial pelvis; however, this specimen lacks the ischium, meaning the sciatic notch (which is used in sex diagnosis) is missing. In spite of this, the pubic body of the os coxa is mediolaterally broad and square-shaped, and the muscle markings of the other postcranial remains are typically weakly to moderately rugose in comparison to MH1, which suggests that MH2 was a female.

Characteristics of Australopithecus sediba. *Australopithecus sediba* is a species not marked by any unique anatomical features; however, because of its unique assemblage of characteristics, it can be distinguished from other hominin taxa and was thus named as a new species.

Cranium and mandible. Morphologically, *Au. sediba* is probably closest in comparison to *Au. africanus* within the australopiths; at the same time, *Au. sediba* is equally comparable to specimens assigned to early members of the genus *Homo*. The two australopiths, that is, *Au. sediba* and *Au. africanus*, share many similarities in the cranium, face, palate, mandible, and teeth. *Australopithecus sediba* has a cranial capacity of only 420 cm³, but several surprisingly advanced features of the brain (based on the internal surface of the brain case) are also found. *Australopithecus sediba* possesses a weak supraorbital torus and a pronounced glabellar region, creating a defined brow ridge. Cranially, *Au. sediba* is easily set apart from *Au. africanus* because of its relatively tapered cranial vault, which is more squared. *Australopithecus sediba* also has more vertically oriented parietals. The orbital margins of *Au. africanus* are unique as they have a distinct angular indentation that is unknown in any other australopith. *Australopithecus sediba* possesses a small anterior attachment of the nasal septum, which is neither as pronounced nor as projecting as that of most fossil specimens attributed to early *Homo*, but it is more pronounced than that seen in any other australopith. The zygomatics of *Au. sediba* are not as flaring as in other australopith specimens. The less flared zygomatics of *Au. sediba* result in its squared upper facial profile, which is not known in *Au. africanus*. The mandible of *Au. sediba* has a vertical symphysis as well as a more gracile mandibular body, which is similar to that of *Au. africanus*.

Dental dimensions of *Au. sediba* indicate that the teeth are relatively small (apart from the maxillary incisors), and the dimensions, at least in the female specimen, fall outside of the range of all australopiths yet described, but in the range of the genus *Homo*. However, the upper central incisors of *Au. sediba* have features aligning *Au. sediba* with australopiths and not *Homo*. The data collected for the whole of the *Au. sediba* dentition plot at the lower end or outside the range of tooth sizes for *Au. africanus* and within the range of early *Homo*, including African *H. erectus*. The canine teeth of both individuals in particular are small and they are not australopith-like in appearance.

The overall pattern that emerges in the *Au. sediba* specimens is one in which the teeth are similar in absolute size to specimens of early *Homo*, whereas the postcanine dentition (because of the cuspal arrangement and posterior molar size increase) is more similar to *Au. africanus* and other australopiths.

Postcrania. MH1 and MH2 are comparable in size to the smaller individuals known from *Au. afarensis* and *Au. africanus* [about 30.5 kg (67 lb) and 37.4 kg (82 lb) for MH1 and MH2, respectively]. In many aspects of the postcranial skeleton, *Au. sediba* generally reflects what would typically be interpreted as a significant arboreal component to its locomotor range. These include arms that are long relative to body size, a high brachial index (an index of the proportions of the forelimb bones as determined from the length of the radius in relation to that of the humerus), large upper limb joint surfaces relative to those of the lower limb, relative pronouncement of some upper limb entheses (muscle attachment sites), and a highly mobile knee and foot. Much is still to be learned about the mode of locomotion of *Au. sediba*; in this regard, the upper limbs, hands, pelvis, and feet are being analyzed. Thus far, it is clear that many derived, *Homo*-like features are also present in some of these anatomical regions; hence, the preliminary picture appears to be one of a postcranial skeleton that is more mosaic than the skeletons of other australopiths. In particular, the pelvis of *Au. sediba* possesses a number of derived (*Homo*-like) features, suggesting that the Malapa hominins may have differed in important ways from other australopiths in their locomotor kinematics.

Overall, the craniodental and postcranial remains of *Au. sediba* show a mosaic of features that can be interpreted as transitional between australopiths and later *Homo*. The decision to place *Au. sediba* in the genus *Australopithecus* was based on such primitive features as a small brain case, a relatively small body size, a relatively long forelimb, upper limb joint dimensions that are large relative to those of the lower limb, and a relatively primitive calcaneus (heel bone).

Australopith or Homo? Australopithecus sediba is a hominin that retained a significant number of primitive characteristics in the cranium, face, arms, thorax, and feet, with perhaps the most notable among these being the low estimated adult cranial capacity of MH1. These characteristics, in conjunction with the series of derived features in the cranial, dentognathic, and pelvic remains, make these skeletons appear more derived toward *Homo* than any other australopith taxon on record. However, detailed analysis of both craniodental and postcranial remains demonstrates to us that the Malapa fossils are not yet at a *Homo* adaptive grade.

Considering the conditions that B. A. Wood and M. Collard cite as necessary for attribution of a fossil taxon to *Homo*, the Malapa fossils clearly fail in most of the criteria. The first two criteria state that both body mass and body proportions should be more similar to humans than to australopiths, which is not the case in the Malapa fossils. The third criterion stipulates that the species "should show obligate bipedalism with limited climbing ability." The fourth criterion does appear to position the Malapa hominins within *Homo* because *Au. sediba* has "teeth and jaws similar in relative size to humans." However, the small teeth from Malapa retain an australopith-like cuspal arrangement and relative sizing (at least in the posterior dentition). The remaining criterion requires that the fossils "should be more closely related to humans than to australopiths." From the fossil evidence obtained thus far, it is our opinion that *Au. sediba* does not belong in the genus *Homo* because it appears more closely related to *Au. africanus* than to *H. sapiens*.

Phylogenetic status of Australopithecus sediba. It is probable, of course, that the Malapa fossils represent a population that, in turn, samples a species that almost certainly existed for some period both earlier and later in time. Although at present there is no fossil evidence to support such a notion, the reality is that Malapa represents a single point in a biological continuum, and the species *Au. sediba* should not be considered exclusively endemic to Malapa nor to a single moment in time that occurred approximately 1.977 MYA. Given the mosaic features seen in *Au. sediba* that are shared by both *Australopithecus* and early *Homo*, and which are found in specimens in a sound temporal setting and of exceptional quality of preservation and completeness, we contend that *Au. sediba* presently stands as the best candidate for the immediate ancestor of the genus *Homo*.

Conclusions. It was initially suggested that *Au. sediba* was derived from *Au. africanus* via a cladogenetic event (cladogenesis is the branching of new species from common ancestral lineages). Support for a cladogenetic interpretation comes from the constellation of *Homo*-like characteristics in *Au. sediba*, as well as its *Australopithecus*-like traits, which push it outside the range of variability seen in the entirety of the *Au. africanus* sample. Even though *Au. sediba* is morphologically closest to *Au. africanus*, the derived appearance of aspects of the cranium and postcranium prevent the inclusion of MH1 and MH2 within *Au. africanus*. The *Au. africanus* sample is already recognized for its extremely high variability (possibly even sampling more than one species). Given that *Au. sediba* exceeds the total known morphological diversity of the *Au. africanus* sample, both cranially and postcranially, yet is temporally and geographically closest to the site of Sterkfontein, from which the largest and most diverse sample of *Au. africanus* comes, there is strong evidence for its unique specific status. As a result, although there are a few cranial features shared among *Au. africanus* and *Au. sediba*, and in fact among other australopithecines, there are still sufficient differences to warrant a specific separation between them.

Present fossil samples from across Africa allow for the generation of hypotheses of the phylogenetic position of *Au. sediba*. On present evidence, *Au. sediba* appears derived from *Au. africanus*, probably via cladogenesis. In turn, *Au. sediba* shares more derived characteristics with specimens assigned to early *Homo* than any other candidate ancestor, including *Au. afarensis*, *Au. garhi*, or *Au. africanus*. In the initial announcement of *Au. sediba*, four possible hypotheses regarding the phylogenetic position of *Au. sediba* were proposed: (1) *Au. sediba* is ancestral to *H. habilis*; (2) *Au. sediba* is ancestral to *H. rudolfensis*; (3) *Au. sediba* is ancestral to *H. erectus*; and (4) *Au. sediba* is a sister group to the ancestor of *Homo*. In a cladistic analysis, the most parsimonious cladogram placed *Au. sediba* as a stem taxon for the *Homo* clade comprising *H. habilis*, *H. rudolfensis*, and *H. erectus*. The resulting cladogram was consistent with the interpretations based on gross morphology and cranial and dental metrics. Presently, analysis of the phylogenetic status of

Au. sediba continues, along with numerous avenues of research. Although it is unlikely that the interpretations will meet with widespread acceptance, *Au. sediba* adds to the continuing expansion of our understanding of the evolutionary direction taken by the genus *Australopithecus* and adds clarity to the debate regarding the origin of the genus *Homo*.

For background information *see* ANTHROPOLOGY; AUSTRALOPITH; DATING METHODS; DENTAL ANTHROPOLOGY; EARLY MODERN HUMANS; FOSSIL; FOSSIL HUMANS; FOSSIL PRIMATES; MOLECULAR ANTHROPOLOGY; PALEOMAGNETISM; PHYSICAL ANTHROPOLOGY in the McGraw-Hill Encyclopedia of Science & Technology.   Bonita de Klerk; Darryl J. de Ruiter;
Steven E. Churchill; Lee R. Berger

Bibliography. L. R. Berger, R. S. Lacruz, and D. J. de Ruiter, Revised age estimates of *Australopithecus*-bearing deposits at Sterkfontein, South Africa, *Am. J. Phys. Anthropol.*, 119:192–197, 2002; L. R. Berger et al., *Australopithecus sediba*: A new species of *Homo*-like australopith from South Africa, *Science*, 328:195–204, 2010; K. J. Carlson et al., The endocast of MH1, *Australopithecus sediba*, *Science*, 333:1402–1407, 2011; R. J. Clarke, Latest information on Sterkfontein's *Australopithecus* skeleton and a new look at *Australopithecus*, *S. Afr. J. Sci.*, 104:443–449, 2008; P. H. G. M. Dirks et al., Geological setting and age of *Australopithecus sediba* from South Africa, *Science*, 328:205–208, 2010; H. M. McHenry and L. R. Berger, Body proportions in *Australopithecus afarensis* and *A. africanus* and the origin of the genus *Homo*, *J. Hum. Evol.*, 35:1–22, 1998; R. Pickering and J. D. Kramers, Re-appraisal of the stratigraphy and determination of new U-Pb dates for the Sterkfontein hominin site, South Africa, *J. Hum. Evol.*, 59:70–86, 2010; R. Pickering et al., *Australopithecus sediba* at 1.977 Ma and implications for the origins of the genus *Homo*, *Science*, 333:1421–1423, 2011; Y. Rak, *The Australopithecine Face*, Academic Press, New York, 1983; J. T. Stern, Climbing to the top: A personal memoir of *Australopithecus afarensis*, *Evol. Anthropol.*, 9:113–133, 2000; D. S. Strait, F. E. Grine, and M. A. Moniz, A reappraisal of early hominid phylogeny, *J. Hum. Evol.*, 32:17–82, 1997; B. A. Wood, Reconstructing human evolution: Achievements, challenges, and opportunities, *Proc. Natl. Acad. Sci. USA*, 107:8902–8909, 2010; B. A. Wood and M. Collard, The human genus, *Science*, 284:65–71, 1999; B. Zipfel and L. R. Berger, New Cenozoic fossil bearing site abbreviations for the collections of the University of the Witwatersrand, *Palaeontol. Afr.*, 44:77–80, 2010.

Bedbug infestations

Bedbugs have been reported from all types of accommodations, including sleeper coaches in trains, camper vans, ships, dormitories, and rooms in hotels and hostels of any standard. Despite the absence of evidence that bedbugs are related to poor hygiene or low socioeconomic status and despite many dozens of studies proving the absence of disease transmis-

Fig. 1. Males of the tropical bedbug *Cimex hemipterus* (*left*) and the common bedbug *Cimex lectularius* (*right*). (*Photograph courtesy of Richard Naylor*)

sion, bedbugs have retained their stigma of poverty, filth, disgust, and disease. The associated shame by human hosts now partly prevents the analysis of why bedbugs are currently spreading and increasing in numbers.

Mainly two species, which are very similar to each other in appearance and biology, are associated with humans. These two species are the common bedbug *Cimex lectularius* and the tropical bedbug *Cimex hemipterus* (**Fig. 1**). The latter is distributed across the tropics, whereas the common bedbug is found worldwide but seems rare in tropical regions.

Recent resurgence in bedbug infestations. Since approximately 2000, the number of bedbug infestations has sharply risen in several regions of the Western world, particularly North America, Europe, and Australia (**Fig. 2**). In most countries, bedbug infestations have not been systematically monitored. Therefore, their rise can only be schematically reconstructed mainly from three anecdotal sources: more complaints by hotel guests, increased inquiries about bedbugs to entomological institutes, and increased call outs for pest controllers concerning bedbugs. Taken together, however, these reports provide a consistent picture: a sharp decrease in bedbug infestations until the 1940s, a plateau phase at very low level, and a steady increase of bedbug infestations since the 1960s; then, a distinct decrease was observed in the mid-1980s followed by the current

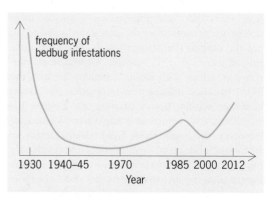

Fig. 2. Schematic reconstruction of the frequency of bedbug infestations in the Western world, combined from sources of the hospitality and pest control industries and entomology and public health institutes.

upward trend. Robust data are scarce, but annual doubling in numbers has been reported in some places, although it has not yet reached the pre-1930 levels.

Why now? Several hypotheses have been proposed for why bedbug infestations are currently increasing and spreading.

Importation from foreign travel. Bedbugs can be transported on airplanes. However, this transport is probably rare, and the current increase in bedbug infestations would require a source population of such a size that it would likely be known. The near absence of tropical bedbugs in bedbug infestations of the Western world, except Australia where this species has become established, indicates the tropics are not that source. Anecdotal reports show that bedbugs may be transported on ships and trains. Analyzing possible bedbug transport via these routes is complicated by the fact that detailed geographic analyses could interfere with the anonymity requested by the companies serving these routes.

Temperature increases accelerate bedbug development. Global warming, increased urbanization (cities are warmer than their surroundings), and more, and extended, centrally heated accommodations may lead to lengthier periods during which bedbugs can reproduce or develop faster. However, none of these temperature-related factors mirror the current trends in bedbug infestations very well.

Genetic refreshing. New bedbug infestations will often be founded by single or few females. Therefore, growing bedbug infestations consist of family members, and mating between them can have negative effects on reproduction (inbreeding depression). With an increasing spread of bedbugs, transported individuals will more often encounter places that are already colonized by other inbred families. If mating occurs between such families, increases in offspring number and survival may be observed (heterosis). In addition, single transported males, who cannot found a population, may, by genetic refreshing, now contribute to faster growth of bedbug infestations. If genetic refreshing is important, extra caution is required by persons who are frequently in contact with bedbugs (pest controllers, luggage handlers, cleaning staff, and frequent travelers).

Increased insecticide resistance. Bedbug infestations may rise because bedbugs are resistant to currently used insecticides and thus survive in places that are treated with these insecticides. Most bedbugs tested from current bedbug infestations are resistant to enormous concentrations of insecticides, particularly carbamates and pyrethroids (the concentrations levels can be up to several millionfold greater than those applied to nonresistant laboratory populations). This hypothesis is successful in explaining why bedbug infestations increasingly withstand control attempts, but it still fails to explain why bedbugs increased in the first place.

Reduced application of DDT. The ban and hence reduced application of the insecticide dichlorodiphenyltrichloroethane (DDT) has been widely sited as the main reason for the increase in bedbug infestations.

However, the evolution of DDT resistance in bedbugs within fewer than 5 few years of its introduction in the mid-1940s suggests that bringing DDT back, even in theory, would be merely a temporary solution. Note also that bedbug infestations had dramatically decreased before the introduction of DDT (Fig. 2).

Loss of folk knowledge. The loss of knowledge of the appearance, and initial control, of bedbugs during the time of their rarity may have led to late detection and spread of bedbugs from established bedbug infestations. In the United States, one in two persons living with a bedbug infestation was not aware of these pests. In the United Kingdom, 90% of people did not recognize a live bedbug.

Animals as alternative hosts. In Europe and parts of Asia, the common bedbug also parasitizes bat colonies; however, no bedbug increases in bat colonies are known that could explain current trends in bedbug infestations. In addition, up to the 1930s, as a result of certain types of houses that allowed rats to dwell between walls, it was suggested that bedbugs survived on rat blood in the absence of human hosts. This idea is paralleled by the question of whether poultry farms, which can be heavily infested, could have acted as sanctuaries for bedbug infestations during the time that they were rare among humans. If so, poultry farms might be the areas where insecticide resistance has evolved in bedbugs. The urgently necessary research on this question requires careful weighing of public health and anonymity interests.

Variation in bedbug infestations during the course of the year. In temperate regions of the Northern Hemisphere, bedbug infestations consistently peak in August and September. These peaks may be related to increased development following early summer temperature rises, or following importation from summer holiday returns, but they are unlikely related to variation in insecticide resistance or folk knowledge.

Dispersal between rooms. From an infested room, bedbugs most often spread into adjacent or opposite rooms. Adults have been found to be nine times more likely to spread into other rooms than nymphs. The rate by which bedbugs infest other rooms may be related to hunger or density of the bedbugs.

Control. As in the 1930s and 1940s, bedbug control has become a substantial industry over the past decade. A recent study of the current bedbug control measures devastatingly concluded that not much is evidence based, with most measures arising from trial and error. Several countries have issued codes of practice for pest controllers.

Basic biology. The basic biology of bedbugs contributes to their ability to thrive and cause infestations.

Feeding biology. All stages of bedbugs exclusively feed on blood. They do not live on their hosts, but aggregate in so-called harborages nearby: in cracks, behind wallpaper, behind pictures, and in electrical sockets. Little is known about the chemical cues that bedbugs use to find a host, although a certain concentration of carbon dioxide plays a role, which is perhaps related to carbon dioxide emissions from the skin and exhaled breath. Bedbugs prefer skin areas that are not covered by clothing or beddings. To withdraw blood, bedbugs inject three types of substances through the skin: a pain killer that prevents the host from feeling the bite; an anticoagulant that prevents blood clotting; and a vasodilator, which is a substance that widens the blood vessels. Blood meals take between 10 and 20 min. How bedbugs find their way back into harborages after feeding is not entirely known. However, some chemicals that bedbugs use to repel enemies have, in low concentration, an attractive effect. Other unknown attractive substances are in the fecal spots, of which there are many in the harborages. Bedbug females in an infestation feed approximately every 2–10 days.

Life cycle. Females with access to food can live and continuously lay eggs for 1 year. How long bedbugs survive in untreated bedbug infestations is unknown. At high temperatures, up to approximately 30–35°C (86–95°F), females lay more eggs. Above that temperature range, bedbug females become infertile, likely because microbes (symbionts) that provide vitamins to bedbugs and that live in the bodies of bedbugs die under these conditions. Most bedbugs die when exposed to high temperatures of 48°C (118°F) [eggs: 60°C (140°F)] or low temperatures of approximately –20°C (–4°F) for 1 h.

Freshly laid eggs are 1 mm (0.04 in.) in length and whitish in color, and they hatch within 6–15 days, depending on temperature. Bedbugs undergo five nymph stages, each lasting approximately 5 days. Without a blood meal, nymphs do not molt. At each molting, bedbug nymphs increase in size by approximately 1 mm (0.04 in.). Adults do not molt. Some insecticides, called growth regulators, prevent nymphs from turning into adults. Instead, nymphs turn into a sixth stage and sometimes a seventh stage, reaching considerable sizes.

Reproduction. Reproduction in the bedbug is very unusual. To deposit sperm, the male uses a needlelike copulatory organ to pierce through the female's ventral (belly) side rather than using the vaginal opening. The injected sperm cells move through the female body, reach the genital tract, move through the walls of the oviduct to the ovaries, and fertilize the eggs in the ovaries. At the site where the female is usually pierced by the male, the female possesses a complex organ with antiwounding, antibacterial, and sperm-activating properties.

Effect of bites on humans. People vary in their response to bedbug bites, ranging from small, localized, red, itchy papules to substantial swellings of entire body parts. Few people do not respond at all to bites. People who are bitten the first or second time in their lives show a skin response at 5–10 days after the bite; this skin response occurs faster on continued biting. Development of insensitivity to bites is not known. The medical literature on bedbug bites has a poor evidence base; however, in addition to skin responses, blisterlike bullous eruptions, mental distress, and anxiety discomfort have been noted.

For background information *see* ENTOMOLOGY, ECONOMIC; HEMIPTERA; INSECT CONTROL, BIOLOGI-CAL; INSECT PHYSIOLOGY; INSECTA; INSECTICIDE; PESTICIDE; PUBLIC HEALTH in the McGraw-Hill Encyclopedia of Science & Technology.　　　Klaus Reinhardt

Bibliography. J. Goddard and R. deShazo, Bed bugs (*Cimex lectularius*) and clinical consequences of their bites, *J. Am. Med. Assoc.*, 301:1358–1366, 2009; L. J. Pinto, R. Cooper, and S. K. Kraft, *Bed Bug Handbook: The Complete Guide to Bed Bugs and Their Control*, Pinto & Associates, Mechanicsville, MD, 2007; K. Reinhardt and M. T. Siva-Jothy, Biology of bedbugs (Cimicidae), *Annu. Rev. Entomol.*, 52:351–374, 2007; R. L. Usinger, *A Monograph of the Cimicidae*, Thomas Say Foundation/Entomological Society of America, College Park, MD, 1966.

Carbon nanotube responsive materials and applications

Carbon nanotubes (CNTs) have attracted a lot of interest in the past 20 years. Superior mechanical, electrical, and thermal properties have made CNTs the desired building blocks for hybrid materials and devices at the micro- and nanoscale. Another advantage is that different forms of CNT materials have been developed for different applications. The powder form of CNTs is low-cost and useful as filler for polymers. Substrate grown CNT forests or arrays (**Fig. 1a**) can be micrometers to centimeters in height and can be processed into various material forms. Arrays that are spinable can be pulled into a translucent belt-like CNT ribbon (Fig. 1b). When the tension on the ribbon is released, the CNTs tend to stick together to form a strand. Ribbon or strands also can be twisted using a bench-top machine into a CNT thread with a diameter of 4–10 μm or larger. Recently, the smallest synthetic threads with a diameter of 350 nm were manufactured using a scanning electron microscope (SEM). These are called nanothreads (Fig. 1c). Two or more threads can be twisted together to form a CNT yarn (Fig. 1d). Threads can also be braided to form microbraid (Fig. 1e). CNT sheet drawn from the forest is a newly developed CNT material (Fig. 1f). The various forms of CNT materials each have different properties and are useful for specific applications.

Carbon nanotube responsive properties. An interesting aspect of CNT materials is that most of their properties change with their environment. For example, the electrical resistance of CNT materials can decrease with increasing applied voltage or current. The larger current heats the material and the resistance of carbon decreases at higher temperatures (opposite to the behavior of copper). The electrical impedance of CNT thread is greater than of copper at low frequencies, and lower than of copper at high frequencies. For a single CNT, the resistance will change if the CNT is bent to different angles. For a CNT ribbon or strand, small tension will decrease the electrical resistance. The double-layer capacitance of CNT thread in a liquid depends on the ion concentration in the solution. Fluid flow along the direction of the CNT induces a current in the CNT based on the Coulomb drag mechanism. Applying

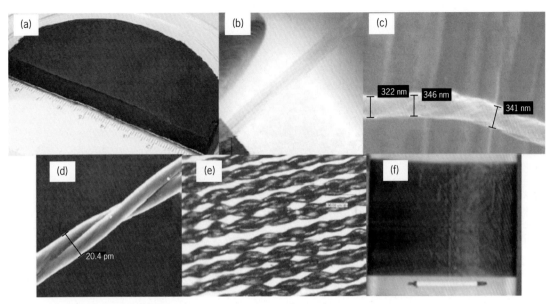

Fig. 1. Different forms of CNT material. (*a*) CNT forest or array. (*b*) CNT ribbon drawn from the forest. (*c*) CNT nanothread formed by twisting ribbon. (*d*) CNT yarn made of two threads. (*e*) CNT 4-end microbraid (*braid from Atkins & Pearce Inc.*). (*f*) CNT sheet produced from the forest.

a small voltage to CNT in an electrolyte causes the CNT to strain slightly, based on the electrochemical bond-expansion mechanism. Post-treatment methods, such as doping and annealing, will also change the electrical properties of CNT and the sensitivity to their environment. Nanotube thread can also be used as a temperature sensor or to generate a small amount of power due to the Seebeck effect. Based on such properties, nanotube materials can sense strain, chemicals, flow, and temperature, act as a wet actuator, or generate a small amount of power.

Responsive materials can be described as smart or active materials, based on different researchers' experiences. But the main point is that responsive materials change their properties (electrical, mechanical, or thermal, or their shape or function) in response to environmental stimuli. CNT materials respond to their environment in various ways and thus can be classified as responsive materials. Responsive materials might be considered to encompass a larger class of materials and effects than smart materials. Current research is focused on using CNT responsive materials to form hybrid materials and to investigate novel applications.

Applications. There are already many applications of CNTs being considered. CNTs are reported in the literature for use as *pn* junctions, transistors, computer memory, touch screens, loudspeakers, polymer reinforcement, electrical conductors, and nanomotors. Researchers are also putting CNT materials into various new and unique applications. CNT thread is being used to simultaneously reinforce polymer composites and act as a structural health monitoring sensor, for windings in a lightweight carbon electric motor, in firefighters' garments for increased heat dissipation, in biosensors as electrodes and conductors, wearable antennas, neural growth guides, and for electromagnetic shielding and devices. Some

near-term application research aimed toward commercialization is briefly discussed next.

CNT thread neural scaffold. This is being studied to determine how CNT thread can be used to promote neural tissue repair. The hypothesis is that CNT thread will promote the regeneration of damaged axonal tracts. Testing has begun with brain and peripheral nerve tissues (**Fig. 2**). The target of the application is for the CNT materials to serve as a scaffold to guide regenerating nervous tissues. The physical properties of CNT thread include characteristics that are known to promote neural tissue regeneration, including its linear geometry, high surface area-to-volume ratio, diameter similar to tissue features normally encountered by cells, surface with submicrometer roughness and its ability to conduct current. The combination of cells on the CNT thread can be considered a responsive hybrid biological-engineering material. The cells appear to align (respond) to the twist in the thread and the cells might communicate via, or be stimulated by, electrical conduction in the thread.

Chemical and biosensors. Chemical and protein sensors are being fabricated based on CNT tower electrodes. The detection of trace elements in magnesium alloys and chemical species in body fluid can be an indicator to track the degradation of biodegradable metal implants. CNT have also been used for detection of trace heavy metals by anodic stripping voltammetry. CNTs show promising potential for sensing applications in vitro and in vivo. Dissolved hydrogen gas and bone biomarkers also can be detected by potentiometry and electrochemical impedance spectroscopy (EIS) methods. The sensors might be used to understand how biodegradable metals degrade in simulated biological environments and possibly in the future to develop responsive biodegradable implants that control their own degradation.

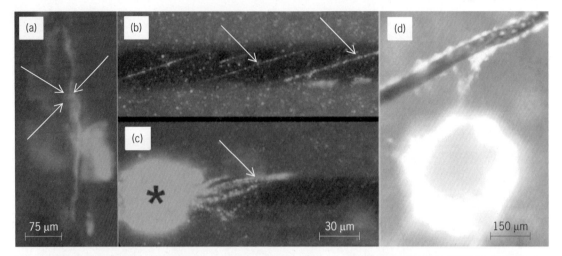

Fig. 2. CNTs as neural scaffolds. (*a*) Mouse neural stem cells align with an as-grown, vertically aligned CNT bundle (arrows show one cell, live cell stain). (*b, c*) Mouse dorsal root ganglion (asterisk in *c*) sends green neurites (arrows, immunostained for neuron-specific tubulin) along and spiraling around (*b, c*) a CNT thread. (*d*) Mouse brain neurosphere, with neural stem cells migrating onto CNT thread.

Carbon electro-mechanical systems (CEMS). Compared to devices made using metal, carbon devices have a very obvious advantage because of their light weight. The National Aeronautics and Space Administration (NASA) is seeking high-performance and light-weight electric motors and actuators. Physicians would like minimally invasive sensors and biomedical devices small enough to insert into the body. CEMS are under investigation to meet these needs. For example, a mostly carbon solenoid that will be used as the main part of electromechanical linear actuator devices is under testing. Manufacturing carbon systems is technically challenging because of the multiscales involved in transforming from nanostructured materials to micro- and macroscale materials and devices.

The operating environment of microdevices is very important because material properties and physical principles change as the devices become very small. Surface area-to-volume ratios, friction, lubrication, wear, and heat transfer change, such that macroscale devices cannot necessarily be scaled down and expected to operate at the microscale. There are two different approaches to making carbon electromechanical systems (CEMS): top down and bottom up. Top down is a method of starting from bulk material and cutting it smaller and smaller to make small devices. Top-down technology uses nanotube threads or sheets, which are easier to

handle than individual nanotubes. Bottom-up technology is starting from nanotubes or nanotube patterned arrays to build carbon devices. Manipulating individual or small bundles of CNTs is difficult and slow. Because only certain components can be fabricated by each approach, top-down and bottom-up techniques are both needed to manufacture CEMS.

In the near future, processor chips, sensors, computers, and micro-CEMS might be integrated into humans to extend our lifespan, improve our performance, and continuously monitor our health. The integration of devices into humans means the devices should be small, efficient, and precise. Small size could allow devices to be biocompatible and safely implanted in the human body. The size of human cells is about 10 μm. If we can make devices 10–100 μm in size, CEMS will be minimally invasive.

The current goal is to build a microactuator that could begin to fulfill the above requirements. A micromanipulator from Kleindiek Nanotechnik in Germany (**Fig. 3***a*) is being used to build the first device. The resolution of the micromanipulator is about 1 nm. But it is difficult to manipulate a CNT with a diameter of 10 nm. Individual CNT easily break and stick to other parts. Nanothread is a new material form used to build microdevices. Nanothread is made by twisting and drawing CNTs, whereby twisting the thread creates a small diameter and

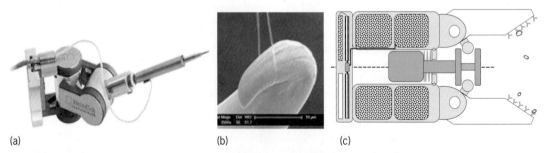

(a)   (b)   (c)

Fig. 3. Microdevice tools and design. (*a*) Main tool to make nanothread—a rotational tip. (*b*) The end of the rotational tip is a microhook that is used to easily grab a small bundle of CNTs from the array. (*c*) A proposed microdevice with a solenoid made from nanothreads.

smooth surface. A hook was made for grabbing CNTs for twisting (Fig. 3b). One possible CNT microdevice is shown in Fig. 3c. Two sets of nano-thread coils will be connected to an external power source. They will generate magnetic forces to drive the core to move from one side to the other to actuate the tweezers for use as a sensor or surgical tool in the body. Different tools could be attached to the solenoid core to perform different tasks.

Readers might be familiar with robotic animals such as the Boston Dynamics Big Dog or the Festo Robotic Seagull. These robots will be very helpful to detect and rescue survivors in earthquakes. But what if the size of a robot could be 10 or 100 times smaller than these? Tiny robots could be useful for surveillance and inspection of urban areas, to monitor our environment, and to probe and repair inside our body. Since there is a need, what limits robots from going smaller? Weight, power, and communication are the barriers. Light weight means less power is needed. CEMS might be a solution to allow development of microrobots.

Self-sensing composite materials. Polymeric composite materials are used in the space and aerospace industry because of their high in-plane mechanical performance and low density. However, interlaminar stresses and microdamage at holes, edges, and caused by impacts can lead to delamination and premature failure of composites at low in-plane stress levels. This constitutes a fundamental weakness in polymeric laminated composite materials. A further limitation of composites is the difficulty in evaluating if damage has occurred. Since the structural integrity of composite materials can be compromised by unpredictable circumstances, it will be important to continuously monitor composite structures for damage, thus providing confidence to the system operator.

An integrated and distributed sensing approach based on nanotechnology is being developed, wherein CNT thread is combined with conventional fibers in composites. The nanotube sensor thread has piezoresistive properties and can sense strain and damage. The sensor thread has recently been integrated into composite materials and used to monitor strains and detect damage including delamination. This approach for monitoring involves transverse stitching of CNT thread in laminated composites before curing the composite. The stitching penetrates multiple carbon fabric plies and ties all the layers of the laminate together and causes only minimal decrease in the fiber volume fraction of the composite. Thus the thread can reinforce the composite while acting as a damage sensor. An example of this idea is provided next.

Two CNT threads were stitched into an IM7/977-3 10-ply unidirectional composite beam sample that was subjected to three-point bending. Delamination was indicated in the resistance versus displacement curve with a sudden increase to infinity resulting from breaking of the CNT thread as shown in **Fig. 4**. The delamination was captured by the sensor in the unidirectional fabric composite. Breaking the

CNT thread means the delamination reached the CNT thread. The delamination is located somewhere along the length of the thread, but the exact position cannot be determined. Introducing more CNT threads into the laminated composite can more accurately locate damage. Based on this initial testing, self-sensing composites were determined to be very sensitive to damage, especially delamination. These materials will help to revolutionize the maintenance of structures, which can now be based on the actual condition, not just usage, of the structure.

Creating the future. New materials drive innovation and CNT materials are opening up the field of responsive materials. However, there are some limitations to CNTs that might be overcome by designing new materials. The first step in developing new and improved materials is to look at why carbon nanomaterials have such extreme properties. CNT shells are one atomic layer thick, and the strong triple sp^2 bonding of carbon combined with the hexagonal tessellated architecture of nanotubes provides high strength. The hexagon structure is the highest order polygon that tessellates and can be considered as a fundamental platform to design new atomic-layer tessellated compound and hybrid inorganic materials with 1-, 2-, or 3-dimensionality.

Looking toward applications, engineers desire materials that are almost unbreakable, with good conductivity or else an insulating capability, and the ability to survive high temperatures. To meet various design requirements, it is suggested that future research consider other elements to form tessellated structures that can provide desired properties for specific applications, such as higher-temperature applications. Boron nitride sheets and nanotubes are an immediate candidate material for high-strength and high-temperature applications. It is anticipated that sheets and nanotubes of other different

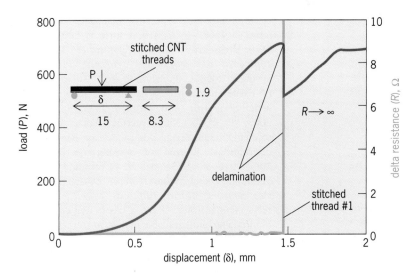

Fig. 4. Detection of delamination and debonding in a laminated composite beam. Load versus deflection and thread delta resistance (difference between actual and initial value) versus deflection curves are shown for an IM7/977-3 10-ply unidirectional composite beam subjected to three-point bending. The sample size is 15 by 8.3 by 1.9 mm. (*Reprinted from L. Jandro et al., Delamination detection with carbon nanotube thread in self-sensing composite materials, Compos. Sci. Tech., 70(7):1113–1119, copyright 2010, with permission from Elsevier*)

materials could become available in the future. A potential advantage of heteroatomic nanomaterials is they might be less likely to form defects, or less sensitive to defects, which could make scaling-up the material easier. It is up to us to create the future, a future where carbon nanomaterials along with new compound, hybrid, and inorganic responsive nanomaterials will allow designers to dream big and develop new machines that were previously impossible to build because of the limited properties of traditional materials.

[Acknowledgment. This work was sponsored by the NSF ERC for Revolutionizing Metallic Biomaterials (EEC-0812348) with program manager Dr. Leon Esterowitz, the Air Force Research Laboratory, the Office of Naval Research, and NASA.]

For background information *see* ANALYTICAL CHEMISTRY; BIOSENSOR; CARBON NANOTUBES; COMPOSITE MATERIAL; ELECTRICAL IMPEDANCE; ELECTRICAL RESISTANCE; MICROSENSOR; NANOTECHNOLOGY; POLAROGRAPHIC ANALYSIS; SEEBECK EFFECT in the McGraw-Hill Encyclopedia of Science & Technology.

Mark Schulz; Weifeng Li; Yi Song; Brad Ruff; Joe Kluener; Xuefei Guo; Julia Kuhlmann; Amos Doepke; Madhumati Ramanathan; Prashant Kumta; Gary Conroy; Kristin Simmons; J. T. Jones; Robert Koenig; Chaminda Jayasinghe; Charles Dandino; David Mast; Duke Shereen; Vesselin Shanov; Sarah Pixley; John Vennemeyer; Tracy Hopkins; Rajiv Venkatasubramanian; Anshuman Sowani

Bibliography. J. L. Abot et al., Delamination detection with carbon nanotube thread in self-sensing composite materials, *Composites Science and Technology*, 70(7):1113–1119, 2010; R. A. Freitas, *Nanomedicine*, vol. 1: *Basic Capabilities*, Landes Bioscience, 1999; X. Guo et al., Determination of trace metals by anodic stripping voltammetry using a carbon nanotube tower electrode, *Electroanalysis*, 23(5):1252–1259, 2011; C. Jayasinghe et al., Nanotube responsive materials, *MRS Bulletin*, 35(9):682–692, 2010; A. Jorio, G. Dresselhaus, M. S. Dresselhaus, *Carbon Nanotubes: Advanced Topics in the Synthesis, Structure, Properties*, Springer, 2008; C. E. Schmidt and J. B. Leach, Neural tissue engineering: Strategies for repair and regeneration, *Ann. Rev. Biomed. Eng.*, 5:293–347, 2003; M. Schulz, Nanostructured smart materials to change engineering design, *Sensor Review*, the international journal of sensing for industry, 30(4):2010; M. J. Schulz, V. N. Shanov, and Y. Yeoheung (eds.), *Nanomedicine Design of Particles, Sensors, Motors, Implants, Robots, and Devices*, Artech House, 2009; Y. Yeoheung et al., Revolutionizing biodegradable metals, *Materials Today*, 12(10):22–32, 2009.

Cell-phone use and driving

The beginning of the twentieth century roughly marks the start of a number of technologically driven revolutions in transportation and communication that have transformed our environments and everyday lives. Henry Ford's application of mass production helped to make the automobile ubiquitous, and subsequent technological innovations such as power steering and the automatic transmission have reduced the physical and mental demands associated with controlling a motor vehicle. The liberation of Alexander Graham Bell's telephone from its so-called landline allowed for the creation of wireless telephonic devices, including those that can accompany us in our automobiles. This technological combination of wireless telephony and automotive transportation has several positive features. For example, cars can incorporate wireless communications while we are driving them, allowing for improved navigation and safety; a driver or any passengers can phone for emergency assistance following an accident, which greatly improves health outcomes in cases when someone has suffered an injury; and passengers can continue to be productive while being transported in a vehicle (to the extent that communicating via telephone enhances their productivity).

Countering these positive features are the negative consequences for performance when the driver of an automotive vehicle (car, bus, or truck) uses wireless technology to carry on a conversation that is unrelated to the immediate task of driving. These consequences follow inexorably from well-known principles of human information processing. However, before discussing the mental or cognitive limitations that require that we keep our minds on the task of driving, let us first dispense with the more obvious limitations.

Structural limitations. In driver training programs and motor vehicle regulation guidebooks, there are two ubiquitous adages: "Keep your hands on the wheel" and "Keep your eyes on the road." The structural limitations implied by these adages are obvious. The steering wheel was designed to be controlled by our hands. As such, our performance in staying in lane, making planned turns, and avoiding unexpected obstacles is certainly faster and more accurate when our two hands are on the wheel compared to when we are driving with one hand, or with our elbows or forearms, or when we have relinquished all physical contact with this critical interface for controlling the direction in which our vehicle hurtles down the road. Most of the physical components of driving (such as accelerating, braking, and steering) are visually guided. The speed and accuracy with which these controlling behaviors are executed when we drive will depend critically on the timeliness and quality of the visual information on which they are based. Activities that require visual guidance, such as opening a package containing food and dialing a cell phone or texting with it, will necessarily cause detriments to driving behavior because the perception of important driving-related visual information will be delayed, if not entirely missed, when we are looking away from the road. For example, say that a visual distraction causes just a 1-s delay in when a signal requiring an emergency action is detected; then, before this action will be engaged, a vehicle

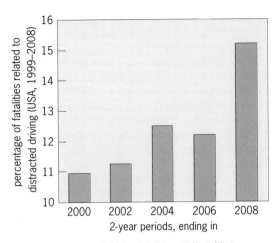

Fig. 1. Percentage of driving fatalities attributable to distracted driving in the United States. (*Replotted from F. A. Wilson and J. P. Stimpson, Trends in fatalities from distracted driving in the United States, 1999 to 2008, Am. J. Public Health, 100:2213–2219, 2010*)

going at 60 mi/h (96.5 km/h) and driven by such a distracted driver will travel an extra 88 ft (26.8 m) compared to when the driver is not distracted.

Unfortunately, a substantial percentage of serious vehicle crashes are attributed to distracted driving, and the proportion of fatalities attributed to all forms of distracted driving has been increasing steadily (to almost 16% in 2008; **Fig. 1**). Engaging in these distracted behaviors is reckless. Because they are usually visible to an outside observer (who might be a law enforcement officer), drivers can be easily warned about, or prosecuted for, engaging in these behaviors.

Cognitive limitations. Even distracting behaviors (for example, talking) that do not use the input (vision) or output (hands and feet) pathways that are crucial for driving can be dangerous. The danger is rooted in long-recognized cognitive limitations on attention. In his seminal *Principles of Psychology* (1890), William James described attention as "the taking possession by the mind, in clear and vivid form, of one out of what seem several simultaneously possible objects or trains of thought.... It implies withdrawal from some things in order to deal effectively with others."

James's phenomenological description was first implemented into a formal model (**Fig. 2**) by Donald Broadbent (1958). The fundamental component of this model was a single-channel limited-capacity mechanism that selects inputs and actions and stores information in and retrieves information from long-term memory. Whereas there are some details of Broadbent's model that are now known to be wrong, it has been argued by Neville Moray (1993) that "for the practical purpose of the designer of human machine systems in which operators must perform tasks in the data-rich world of extralaboratory tasks, Donald Broadbent's original model is not merely sufficient, but may even also be necessary. Whatever the deep structure of attention may be, its surface performance is, in the vast majority of cases, well described by a single, limited capacity chan-

nel, which is switched discretely among the various inputs."

Therefore, because we drive with our minds (as well as with our eyes, hands, and feet), activities that distract our minds from the task of driving will necessarily interfere with driving performance. Carrying on a conversation on a cell phone is an example of such an activity.

Research on cell-phone use and driving. Studies that assess the degree to which cell-phone usage while driving might be dangerous vary along a continuum, with real-world ecological validity at one end and laboratory-enabled experimental control at the other (**Fig. 3**). The results from different points along this continuum have unique strengths and weaknesses. For example, laboratory studies, through the use of experimental manipulations (including analog studies) and randomized treatments, have the potential to identify with confidence whether talking on a cell phone while driving causally disrupts driving performance (as when the participant is using a driving simulator) or causally disrupts performance on driving-like tasks. It might be argued, however, that results from such highly controlled and somewhat artificial situations may not generalize to more complex and data-rich real-world driving situations. At the other end of the continuum are studies that combine accident records with information about cell-phone usage by the driver at the time of the accident to generate relative risk statistics. Whereas such correlational analyses of real-world data certainly do not suffer from a generalizability limitation and they can identify an association between accidents and cell-phone usage, they cannot endorse a causal connection.

Field studies represent a compromise. In one type of field study, a camera is placed in the participant's car to provide continuous data on possibly distracting behaviors and any accidents or near-accidents that may occur. Aside from the knowledge that one's behavior is being monitored, there is nothing artificial about such a study. However, the participant decides if and when to engage in cell-phone activities;

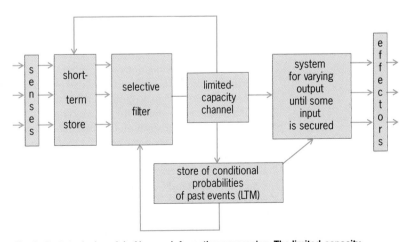

Fig. 2. Prototypical model of human information processing. The limited-capacity channel, which is required for many nonhabitual activities, including responding to unexpected events, can service only one activity at a time. It can be switched between activities, but not without costs to both. LTM = long-term memory. (*Adapted from D. E. Broadbent, Perception and Communication, Pergamon, Oxford, U.K., 1958*)

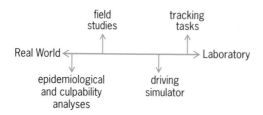

Fig. 3. The range of methods that have been applied to the question of whether carrying on a cell-phone conversation while driving is dangerous as illustrated along a continuum, with the real world at one end and the laboratory at the other.

hence, there is no experimental control. In another type of field study, a participant is asked to navigate a specific route while talking on a cell phone or engaging in some other baseline activity. Driving performance is monitored, usually by an experimenter who sits in the passenger seat. Although this method has all the benefits of experimental control and real-world complexity (suffering only from the fact that the participant knows that he or she is being observed), it is unlikely to be used for ethical and liability reasons.

When all these various methods have been applied to the question of whether driving performance is negatively affected by cell-phone usage, the results from along the entire continuum are in agreement, thus fortifying the answer that they all tend to reach: Carrying on a cell-phone conversation while driving is dangerous. Epidemiological and culpability analyses have consistently shown that talking on a cell phone while driving substantially increases the risk of having an accident (in fact, the risk can increase by up to a factor of four times). In studies that have permitted a comparison, the increased risk with a hands-free phone was about the same as that with a handheld phone. Laboratory and field studies agree with this correlational evidence and point to a causal relationship. Reaction times to detect and respond to signals while driving or performing a driving-like tracking task are significantly longer when the participant is talking on a cell phone compared to a no-phone baseline condition. Indeed, studies that have compared drunk driving with driving while using a cell phone have reported the latter to be worse than the former.

When structural limitations were minimized and when hands-free phone use and handheld phone use were compared, the two phone conversation conditions suffered equally in the reaction-time delay compared to the conversation-free baseline. In both laboratory and field studies, there is a tendency for individuals using a handheld phone to slow down a bit compared to the baseline condition; this tendency is consistently diminished when the phone used is a hands-free device. It appears, then, that some drivers may compensate, by slowing down a little, for the mental and/or structural load associated with using a handheld phone and that this tendency is reduced, if not absent, when using a hands-free phone. Because the reaction-time delay is about the same for the two types of phone, the failure to com-

pensate when using a hands-free phone may make this type of phone use while driving less safe than using a handheld phone.

Many jurisdictions around the world have made it illegal to use a handheld cell phone while driving. By excluding hands-free cell-phone use from this prohibition, these jurisdictions are conveying the false message to their citizens that hands-free cell-phone use is safe. However, scientific principles imply and research findings demonstrate that this is not the case.

For background information *see* AUTOMOBILE; AUTOMOTIVE STEERING; COGNITION; HUMAN-FACTORS ENGINEERING; INFORMATION PROCESSING (PSYCHOLOGY); MOBILE COMMUNICATIONS; PERCEPTION; PSYCHOLOGY; RISK ASSESSMENT AND MANAGEMENT; TELEPHONE in the McGraw-Hill Encyclopedia of Science & Technology. Raymond M. Klein

Bibliography. D. E. Broadbent, *Perception and Communication*, Pergamon, Oxford, U.K., 1958; W. J. Horrey and C. D. Wickens, Examining the impact of cell phone conversations on driving using meta-analytic techniques, *Hum. Factors*, 48:196–205, 2006; Y. Ishigami and R. M. Klein, Is a hands-free phone safer than a handheld phone?, *J. Saf. Res.*, 40:157–164, 2009; W. James, *The Principles of Psychology*, Harvard University Press, Cambridge, MA, 1890; A. T. McCartt, L. A. Hellinga, and K. A. Bratiman, Cell phones and driving: Review of research, *Traffic Injury Prev.*, 7:89–106, 2006; S. P. McEvoy et al., Role of mobile phones in motor vehicle crashes resulting in hospital attendance: A case-crossover study, *Br. Med. J.*, 331:428, 2005; N. Moray, Designing for attention, pp. 111–134, in A. Baddeley and L. Weiskrantz (eds.), *Attention: Selection, Awareness, and Control*, Clarendon Press, Oxford, U.K., 1993; F. A. Wilson and J. P. Stimpson, Trends in fatalities from distracted driving in the United States, 1999 to 2008, *Am. J. Public Health*, 100:2213–2219, 2010.

Cellulose nanocomposites

Cellulose is a natural polymer material made of β-D-glucopyranose units that are linked together by $(1 \rightarrow 4)$-glycosidic bonds. The length of a native cellulose molecule is at least 5000 nm, corresponding to a chain with about 10,000 glucopyranose units. Cellulose molecules are linear and are aggregated through secondary valence forces, including van der Waals forces and both intra- and intermolecular hydrogen bonds. In a plant cell wall, the linear cellulose chains, referred to as microfibrils, are approximately 3.5×10 nm in cross-sectional dimension and of indeterminate length. The microfibrils have both crystalline and amorphous regions. The linear cellulose molecules and the supermolecular microfibrils have a dominant influence on the behavior of plant cell walls as a material. The molecular organization of cellulose in the plant cell wall contributes to its unique polymer properties. For improving the interaction or interfacial adhesion of cellulose with other materials, including polymers, in the creation of nanocomposites, it is possible to chemically modify its surface. Cellulose reactivity or a lack thereof depend on its structure. To modify cellulose structure, the highly ordered hydrogen-bonded lattice must be disrupted by swelling or dissolution. The reactive sites on cellulose, which may be derivatized, are the three hydroxyl groups indicated as C-2, C-3, and C-6 in **Fig. 1**.

The general methods to produce cellulose nanofibrils (CNFs) are based on chemical, mechanical, or biological means (or any combination of these). Four different types of cellulose fibers on the nanoscale can be prepared: (1) bacterial cellulose nanofibers, (2) electrospun cellulose nanofibers, (3) microfibrillated cellulose plant cell fibers, and (4) nanorods or cellulose whiskers (**Fig. 2**). Processing techniques have a significant effect on the adhesion properties of the resulting cellulose nanofibers in composite material applications.

Nanocomposites. A growing technology that has tremendous potential for the composites industry is the development of nanoscale materials, or nanomaterials, for use in composites. Potential product applications for nanocomposites include building materials, lightweight automotive components, wind-energy and boat applications, and "green" composite materials based entirely on renewable feedstocks. Nanocomposites can include particulate, fibrous, or layered materials, with a diameter or thickness measured in nanometers (10^{-9} m). A primary advantage that cellulose nanofibers bring to polymeric composites is their extremely high ratio of surface area per unit volume. Some of the property enhancements achieved by the addition of cellulose nanofibrils to composites are mass reduction, increased stiffness, increased strength, less abrasiveness to tooling, and biodegradability.

Cellulose nanocomposites. Research on cellulose nanocomposites has captured the imagination of investigators worldwide, with significant research activity taking place in Argentina, Canada, the European Union, Japan, and the United States. Indeed, cellulose nanomaterials have been recently recognized as a signature research initiative by the National Nanotechnology Initiative in the United States.

Utilization of cellulose nanofibrils in value-added composite applications requires that their surfaces and interfaces be tailored to the specific material requirements of the application. For example, incorporation of discontinuous nanofibers into a polymer matrix for reinforcing purposes requires that the fiber surface be compatible with the polymer matrix because stress is transferred via shear along the fiber surface. Similarly, dispersion of nanofibers in a matrix or self-assembly of individual fibers requires control of interactions between the materials themselves and their environment. Therefore, the precise control of interface chemistry and functionalization to achieve desired behaviors for specific applications is crucial for composite applications of cellulose nanofibers. The surface functional groups, the surface potential, and the hydrophobicity or hydrophilicity of cellulose nanofibers can be controlled during synthesis

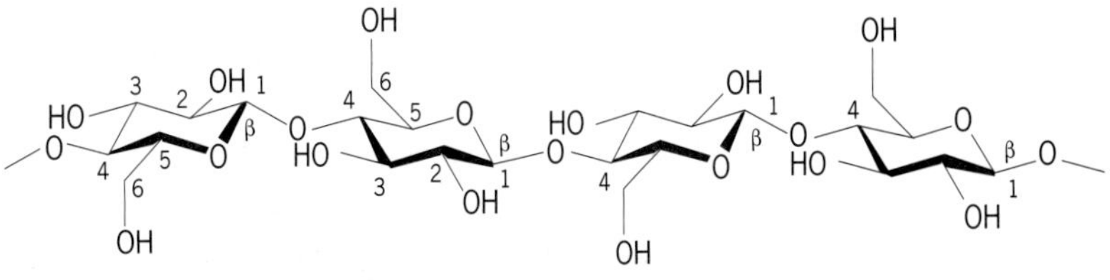

Fig. 1. Position of hydroxyl groups on the cellulose backbone.

or processing by covalent or noncovalent means.

The mechanical performance of nanocomposites is dependent on (1) the degree of dispersion of the fibers in the matrix polymer and (2) the nature and intensity of fiber–polymer adhesive interactions. An example of cellulose nanofibrils well dispersed in a hybrid sodium silicate–wood flour–CNF nanocomposite is shown in **Fig. 3**. Properly functionalized cellulose nanofibers can be applied in specific matrix polymers for optimized material properties. Cellulose nanofiber incorporation in composites utilizing thermosetting or thermoplastic matrices tends to be facilitated through the use of solvent-based systems, including aqueous dispersed polymers (that is, latexes and organic solvent-based systems). Carrier systems such as thermoplastic starch are also being explored to incorporate cellulose nanofibrils in nonpolar polymer matrices. In addition, a number of nanocomposite applications utilize melt processing techniques such as extrusion, injection molding, and fiber spinning for thermoplastic resin matrices. Successful large-scale, high-shear melt processing of cellulose nanofibrils in thermoplastic composites has yet to be realized, however, and techniques to improve cellulose nanofiber distribution in thermoplastic and thermosetting matrices remain a high priority topic of research. Extensional flow mixing (a method analogous to kneading bread dough) can be used to improve dispersion and distribution of cellulose nanofibers in highly viscous polymer matrices.

Overall, investigations of cellulose nanofiber–filled thermoplastic composites have addressed three major polymer matrix types: polyolefin thermoplastics such as polyethylene and polypropylene; engineering thermoplastics such as polyamides (nylon) and polyethylene terephthalate (PET); and bio-derived thermoplastics such as polylactic acid (PLA) and carbohydrate-derived polymers. Research on cellulose nanofiber–filled thermosetting composites has focused on epoxies, phenol-formaldehyde resin, and polyurethane systems. Cellulose nanofibers have also been applied in inorganic matrix materials, including concrete and "water glass" (sodium silicate).

Potential applications of cellulose nanocomposites. Some of the potential opportunities for applications of cellulose nanocomposites include (but are not limited to) batteries, bioplastics, coatings, smart sensors, high-efficiency filters, membranes, biomedical devices, nanopaper, and flexible displays. Further work continues to examine cellulose nanomaterial standards in order to standardize nanocellulose terminology. Challenges for "scaling up" in producing commercial quantities of cellulose nanofibers

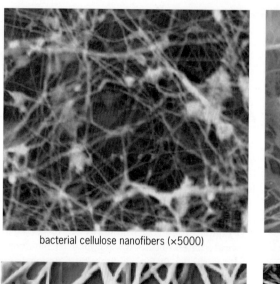
bacterial cellulose nanofibers (×5000)

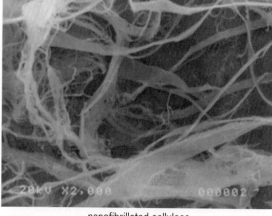
nanofibrillated cellulose

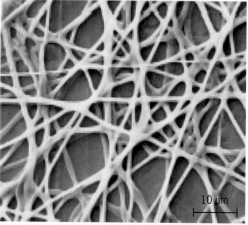
electrospun cellulose nanofibers

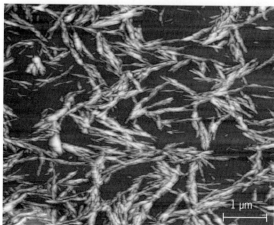
cellulose nanocrystals (whiskers)

Fig. 2. Four types of cellulose nanofibers.

include improved refining and reduced energy consumption, reduction of clogging during homogenization, controlling or elimination of nanofibril agglomeration, and reduction of production costs.

In comparison to other nanomaterial technologies, cellulose nanocomposites have a positive consumer perception in terms of sustainability, recyclability, and being a natural resource–based material. In addition, regulation issues regarding health and safety are a subject of ongoing research. Cellulose is typically considered nontoxic and in fact can be used in biomedical and drug delivery applications. The market opportunities for cellulose nanocomposites are enormous, especially with the potential to improve existing products and particularly in the area of intelligent packaging.

For background information *see* CELL WALLS (PLANT); CELLULOSE; COMPOSITE MATERIAL; NANOPARTICLES; NANOSTRUCTURE; NANOTECHNOLOGY; POLYMER; POLYMER COMPOSITES; RENEWABLE RESOURCES in the McGraw-Hill Encyclopedia of Science & Technology.   Douglas J. Gardner

Bibliography. D. J. Gardner et al., Adhesion and surface issues in cellulose and nanocellulose, *J. Adhesion Sci. Technol.*, 22:545–567, 2008; K. Oksman and M. Sain (eds.), *Cellulose Nanocomposites Processing, Characterization, and Properties* (American Chemical Society Symposium Series 938), Oxford University Press, New York, 2006; T. H. Wegner and P. E. Jones, Advancing cellulose-based nanotechnology, *Cellulose*, 13:115–118, 2006.

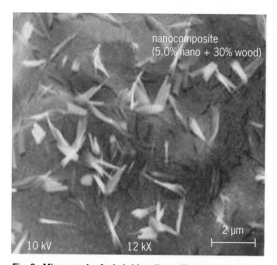
nanocomposite (5.0% nano + 30% wood)

10 kV   12 kX   2 µm

Fig. 3. Micrograph of a hybrid sodium silicate–wood flour–cellulose nanofibril composite.

Census of Marine Life

The Census of Marine Life was launched in 2000 as a 10-year program to assess and explain the diversity, distribution, and abundance of marine life throughout the world's oceans. The ultimate goal of the program was to establish a baseline of information about marine organisms against which future changes could be measured. Three questions guided this program: What lived in the oceans? What lives in the oceans? What will live in the oceans? Two novel approaches were taken to answer these questions. The first was to launch an unprecedented coordinated effort of new research and exploration of all ocean realms and taxonomic groups, from the shallow zones to the deep sea, from the coasts to the open ocean, from the tropics to the polar regions, and from microbes and invertebrates to whales (**Fig. 1**). The second was a review and synthesis of the accumulated marine biodiversity knowledge on a global scale. To accomplish these aims, an international community of more than 2700 scientists from more than 80 nations and 670 institutions came together in a collaborative effort that resulted in 540 expeditions and 9000 days at sea, and more than 3000 peer-reviewed publications reviewing 10 years of research and exploration.

What lived, what lives, and what will live in the ocean?

What lived in the ocean? Fisheries scientists, historians, and economists came together to form a project that reconstructed the history of marine animal populations since the time when human use of natural resources became important. The investigators used historical catch logs, monastery records, fish bones, and other credible materials to construct a snapshot of the past abundance of marine life. Through case studies in different regions of the world and for different types of fisheries, from shellfish to whales, it was possible to create the first reliable picture of life in the ocean prior to the establishment of fishing, which showed that the effects of humans on ocean life began much earlier and were more pronounced than previously thought. Researchers documented that overfishing and habitat destruction were, and continue to be, the top threats to marine life associated with human activities, but they also proved that recovery is possible when conservation efforts are implemented.

What lives in the ocean? To answer this question, 14 field projects were organized to explore the

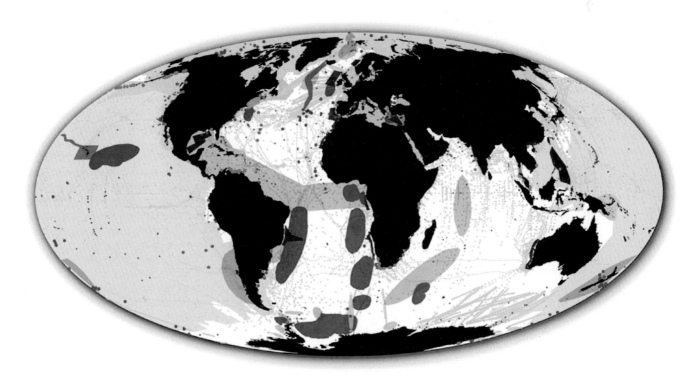

Coastal	Polar	Pelagic	Deep Sea	Global Information and Analysis
Regional ecosystem (GoMA)	Arctic Ocean (ArcOD)	Top predators (TOPP)	Vents and seeps (ChEss)	Oceans' future (FMAP)
Near shore (NaGISA)	Antarctic Ocean (CAML)	Continental shelves (POST)	Abyssal plains (CeDAMar)	Information Systems (OBIS)
Coral reefs (CReefs)		Zooplankton (CMarZ)	Seamounts (CenSeam)	Microbes (ICoMM)
			Continental margins (COMARGE)	Oceans' past (HMAP)
			Mid-Ocean Ridges (MAR-ECO)	

Fig. 1. **Census project areas sampled all major oceanic zones and realms.** (*Image courtesy of Census of Marine Life Mapping and Visualization Team*)

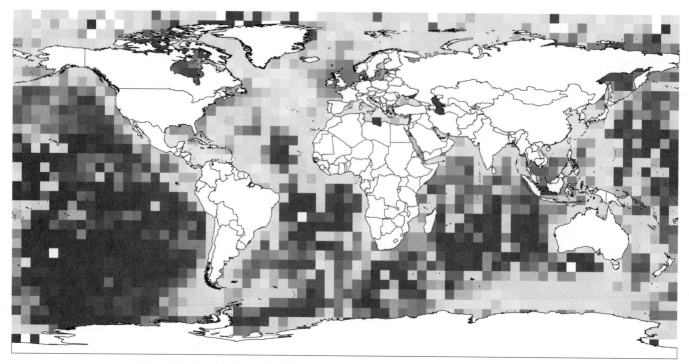

Fig. 2. Map showing the density of observations by 25 square degrees in the OBIS database. Darker squares indicate a higher density of records. Observations are most numerous in shallow waters, near the coast, and near or between developed nations. The Southern Pacific shows a huge gap in knowledge. (*Image courtesy of Edward Vanden Berghe, Ocean Biogeographic Information System; data updated in April 2011*)

major habitats and groups of species in the global ocean, as well as to assess marine life in the coastal zone (nearshore, coral reefs, and the Gulf of Maine), the open ocean (surveys of microbial and planktonic communities and populations of top predators), the deep ocean (continental margins, abyssal plains, seamounts, vents and seeps, and the Mid-Atlantic Ridge), and the polar regions [Arctic and Southern (Antarctic) oceans]. These projects implemented new or reconfigured technologies that made it possible to explore many previously unknown locations. Rapid methods of identification of species through genetic barcoding, tagging, tracking, and acoustically following animals as they moved through the oceans, along with the use of standardized sampling protocols that allowed comparisons across the global seas, were part of the technological arsenal employed to obtain knowledge about what lives in the ocean.

What will live in the ocean? To answer this question, the Future of Marine Animal Populations project integrated data from many different sources. Through numerical modeling and simulation, it was possible to make predictions about marine populations and ecosystems in the future.

The baseline of information of what lives in the ocean is contained in one of the most outstanding achievements and legacies of the Census of Marine Life: the Ocean Biogeographic Information System (OBIS) [**Fig. 2**]. OBIS is an online, open-access, interactive, and growing database that compiles data on marine species and their locations. By 2010, OBIS was the world's largest online repository of georeferenced data, containing nearly 30 million records from nearly 900 data sets. As a legacy of the Census program, OBIS is now under the umbrella of the Intergovernmental Oceanographic Commission (UNESCO) global data system, where it will continue to serve researchers, students, and policy makers, helping them to identify areas that remain unexplored and where more research might be targeted.

Diversity, distribution, and abundance of marine life. The Census of Marine Life reported that life in the ocean is more diverse, more connected, and more affected than previously thought. With regard to diversity (**Fig. 3**), the Census increased the estimate of known marine species from about 230,000 to nearly 250,000, found more than 6000 potentially new species, and formally described about 1200 of them. This known diversity represents only a fraction of the real biodiversity, not only because the vast majority of the oceans still remains unexplored (estimates range up to 95%) but because new species are still being discovered even in relatively well known areas. Census scientists estimated that there are 8.7 million species on Earth, and roughly 2 million or 91% of ocean life remain unknown. Nearly half of all the known biodiversity is represented by only three groups—crustaceans, mollusks, and fish (Fig. 3)—which are also the best-known groups, with the longest taxonomic history. Many of these species also are of commercial importance. As for microbial diversity, the numbers jump orders of magnitude. For example, more than 38,000 different kinds of bacteria were found in 1 liter of seawater, whereas 5000-19,000 were found in a gram of sand. Most of this microbial diversity is predominantly represented by rare rather than common species.

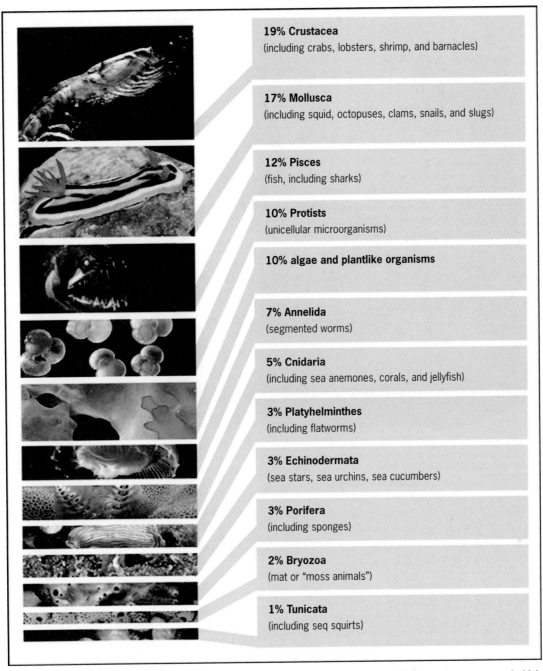

19% Crustacea
(including crabs, lobsters, shrimp, and barnacles)

17% Mollusca
(including squid, octopuses, clams, snails, and slugs)

12% Pisces
(fish, including sharks)

10% Protists
(unicellular microorganisms)

10% algae and plantlike organisms

7% Annelida
(segmented worms)

5% Cnidaria
(including sea anemones, corals, and jellyfish)

3% Platyhelminthes
(including flatworms)

3% Echinodermata
(sea stars, sea urchins, sea cucumbers)

3% Porifera
(including sponges)

2% Bryozoa
(mat or "moss animals")

1% Tunicata
(including sea squirts)

Fig. 3. A regional analysis of 25 oceanic regions shows that 92% of all known biodiversity falls into a dozen groups, of which crustaceans (for example, crabs, shrimps, and lobsters), mollusks (for example, snails, bivalves, squids, and octopus), and fish constitute almost half. Well-known vertebrates such as marine mammals (for example, seals, whales, and dolphins), reptiles (turtles and snakes), and birds account for only 2% of the species total. (*Images courtesy of Russ Hopcroft, Gary Cranitch, Julian Finn, Larry Madin, John Huisman, Katrin Iken, Bernard Picton, and Piotr Kuklinski*)

With regard to distribution (**Fig. 4**), the Census found life everywhere, even in unexpected or extreme places such as hydrothermal vents, some with temperatures that would melt lead. Anaerobic (living without oxygen) bacteria have been known for many years, but the Census discovered the first anaerobic animal (phylum Loricifera). A global analysis of the distribution of coastal species showed that species richness is higher in the tropics, and marine hot spots appear around the Philippines, Japan, China, Indonesia, Australia, India, Sri Lanka, South Africa, the Caribbean, and the southeastern United States.

On the other hand, oceanic species richness peaks in temperate latitudes. An analysis of species composition in more than 25 regions of the world showed that endemicity (species confined to a particular region) or uniqueness is high, from 22% to 48%, in South America, Australia, New Zealand, South Africa, and Antarctica, all within the Southern Hemisphere.

By tracking thousands of marine animals from 43 species as they travel through "blue highways" (migratory routes), the Census found that the world's oceans are much more connected than imagined. Animal movements may result from natural migration,

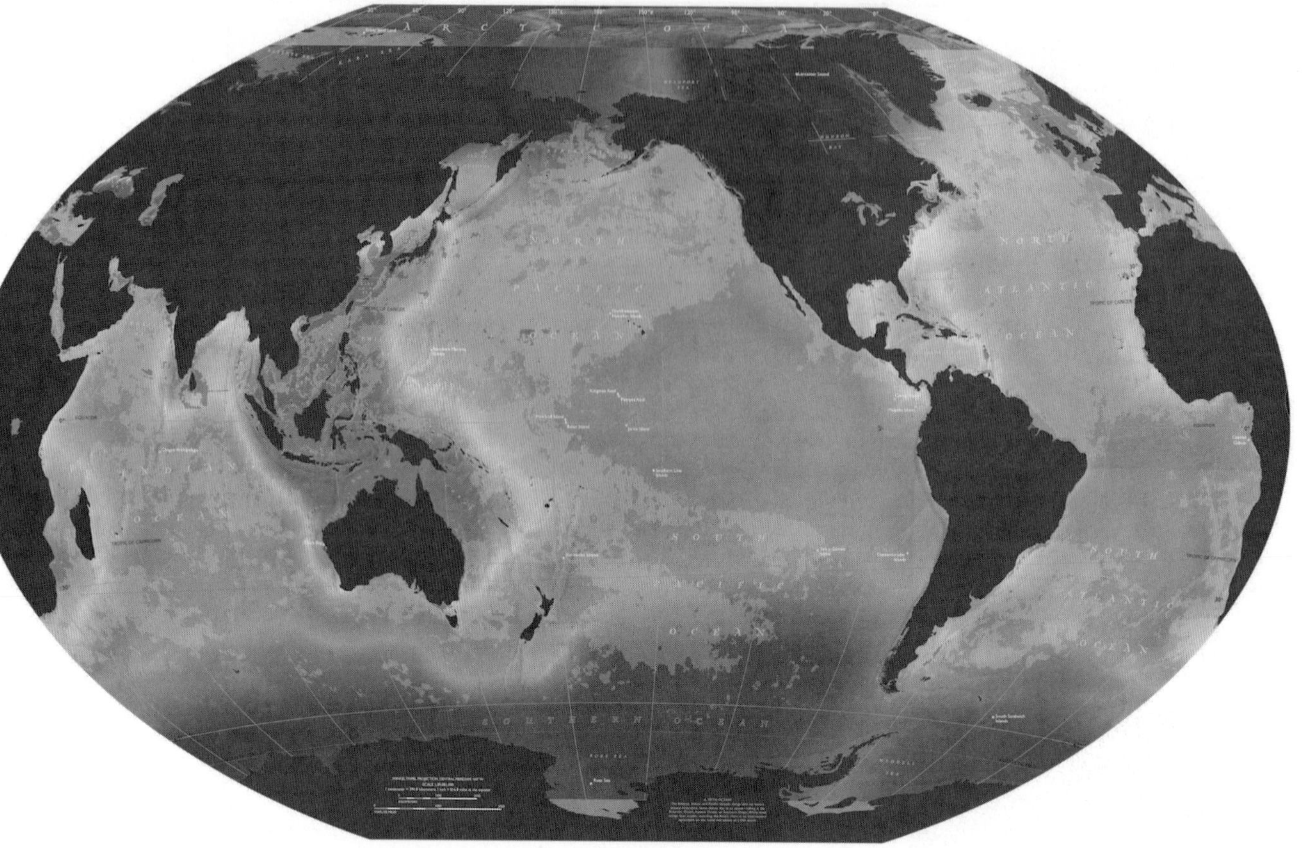

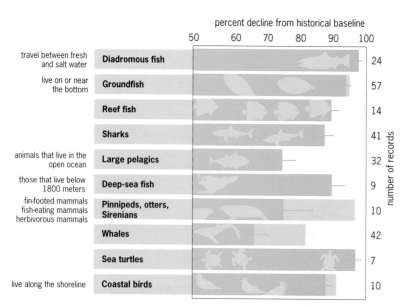

Fig. 4. The winding shores of Southeast Asia create a continuous hot spot for coastal species. Gray patches represent areas with human impact (darker gray indicates higher impact). Biodiversity and high human impacts collide in coastal areas such as the Western Pacific and North Atlantic. (*Image reproduced from National Geographic and Census of Marine Life wall map, October 2010*)

percent decline from historical baseline

travel between fresh and salt water	**Diadromous fish**	24
live on or near the bottom	**Groundfish**	57
	Reef fish	14
	Sharks	41
animals that live in the open ocean	**Large pelagics**	32
those that live below 1800 meters	**Deep-sea fish**	9
fin-footed mammals fish-eating mammals herbivorous mammals	**Pinnipeds, otters, Sirenians**	10
	Whales	42
	Sea turtles	7
live along the shoreline	**Coastal birds**	10

number of records

Fig. 5. Dark-shaded bars indicate that the estimated declines in populations of large marine animals from their historical levels average about 90%. Some recovery (shown by light-shaded bars) has been achieved by four groups, including seals, whales, birds, and such bottom-dwelling fish species as flounder and sole. These estimates for exploited species in particular, rather than all marine animals, confirm the feared declines and offer a few hopeful instances of recovery. (*Reproduced with permission from H. K. Lotze and B. Worm, Historical baselines for large marine animals, Trends Ecol. Evol., 24:254–262, 2009*)

either horizontal or vertical, or as a response to changing ocean conditions, including temperature changes. Some of the distances traveled by animals during their migrations were quite impressive, ranging from approximately 6000 km (3700 miles) [Atlantic bluefin tuna, *Thunnus thynnus*] to a round trip of 64,000 km (40,000 miles) [sooty shearwater, *Puffinus griseus*], which marked the longest-ever electronically recorded migration.

With regard to abundance, microbes constitute by far the most plentiful group. Microbes account for more than 90% of the ocean's biomass and are essential to maintaining Earth's habitability and functioning. Microbes are responsible for more than 95% of the respiration in the oceans, and they also influence climate, recycle nutrients, decompose pollutants, and constitute the base of the food web that supports all life in the seas. Some of this microbial life has been found as mats of filamentous bacteria that extend for hundreds of kilometers in the deep-sea margins of the continents. As for larger organisms, such as fish and other vertebrates, historical records indicate a decline of about 90% of top predators in comparison to past levels (**Fig. 5**). Recovery, although very slow, is possible when protective measures are taken. Despite such alarming declines, the oceans still hold plenty of marine life. For example, off the coast of New Jersey in the United States, a shoal of tens of millions of Atlantic herring, *Clupea* *harengus*, comparable in size to Manhattan Island (New York City) has been observed by using new sensor technologies. In addition, life in the seafloor depends to a large extent on the delivery of food produced by organisms near the surface.

Legacy of the Census of Marine Life. After this amazing 10-year program, Census of Marine Life scientists recognize that their work is not complete and the age of discovery is far from over. An organized international community working in collaboration has been established, and a first snapshot of life in the global oceans has been presented. These working relationships will carry on, and the baseline will serve marine research for decades to come. Some of the Census's major partner programs, for example, the Encyclopedia of Life and the World Register of Marine Species, along with OBIS, will also continue to build on the Census's legacy. The Census has contributed to public understanding of what lives in the ocean and why it is important to our very survival on Earth. Many knowledge gaps remain, but the Census has laid a foundation upon which the marine science community will build, moving forward under a common umbrella to continue to study life in a changing ocean.

For background information *see* ANIMAL KINGDOM; BIODIVERSITY; BIOGEOGRAPHY; DEEP-SEA FAUNA; FISHERIES ECOLOGY; MARINE BIOLOGICAL SAMPLING; MARINE CONSERVATION; MARINE ECOLOGY; OCEANOGRAPHY; POPULATION ECOLOGY in the McGraw-Hill Encyclopedia of Science & Technology.

Patricia Miloslavich; Darlene Trew Crist

Bibliography. V. Alexander, P. Miloslavich, and K. Yarincik, The Census of Marine Life—Evolution of worldwide marine biodiversity research, *Mar. Biodiv.*, 2011, in press; J. H. Ausubel, D. T. Crist, and P. E. Waggoner, *First Census of Marine Life: Highlights of a Decade of Discovery*, Census of Marine Life, Washington, D.C., 2010; D. T. Crist, G. Scowcroft, and J. M. Harding, Jr., *World Ocean Census*, Firefly Books, Toronto, 2009; N. Knowlton, *Citizens of the Sea: Discoveries of the Census of Marine Life*, National Geographic, Washington, D.C., 2010; A. McIntyre (ed.), *Life in the World's Oceans: Diversity, Distribution, and Abundance*, Blackwell, London, 2010; P. V. R. Snelgrove, *Discoveries of the Census of Marine Life: Making Ocean Life Count*, Cambridge University Press, London, 2010.

Characteristics of supertall building structures

Supertall building structures have unique challenges and characteristics with respect to their design and construction. Advances in analysis and design techniques, material properties, and construction methods have fostered an increase in the number of supertall structures being designed and built today. The last decade has seen a record number of these types of buildings constructed, including the tallest building in the world, the Burj Khalifa in Dubai, United Arab Emirates. This article will explore those challenges

unique to supertall building structures by focusing on the case study of the Burj Khalifa.

Successful supertall buildings. Even with the growth in supertall building construction, many proposals for supertall buildings are not realized because of a lack of understanding of what characterizes a successful supertall building. Supertall buildings are significant economic investments, consume a significant amount of resources, and take a significant amount of time to construct. Failing to identify and accommodate these issues in the design will lead to the building not being developed. As such, there are four key characteristics to supertall building structures: rational, appropriate, fast, and efficient.

Rational. Supertall structures are rational in that the engineering design takes into account and reflects the underlying physics governing the behavior of tall buildings. For example, when we study the gravity-load diagram and the wind-overturning diagram for a tall building, it becomes clear where the demand on the structure is greatest. By examining these diagrams, we can start to visualize a potential tall building form, which responds to the loading demands on the structure (**Fig. 1**).

Appropriate. Successful supertall buildings need to be appropriate responses to their environment. The primary concern in the engineering of supertall

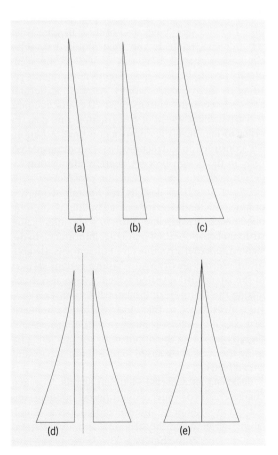

Fig. 1. Evolution of a potential tall building form. (*a*) Gravity-load diagram. (*b*) Wind-overturning diagram. (*c*) Combined gravity- and wind-load diagram for one wind direction. (*d*) Combined gravity- and wind-load diagram for multiple wind directions. (*e*) Potential tall-building form. (*Illustration courtesy of SOM, copyright © by SOM*)

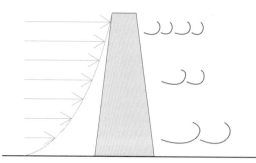

Fig. 2. Wind behavior. *(Illustration courtesy of SOM, copyright © by SOM)*

buildings is the effect of the wind on the building's structure. The shape, or massing, of a tower is often viewed in architectural terms, but is the single most important structural design parameter in supertall buildings, because of its ability to influence the effects of wind. As such, appropriate tall building forms can be achieved by choosing a massing that maximizes stiffness, while at the same time provides the key shaping components to reducing wind effects. These shaping components include measures such as changing the building's shape along its height, providing a taper to the building's profile, and providing surface treatments to reduce vibration inducing wind effects (**Fig. 2**).

Fast. Supertall buildings require a large amount of time to build simply because of their height. Time requirements relate to the time required to actually construct a given system, as well as the complexity or simplicity of the system's components relating to their constructability. As such, structural and architectural solutions must take into account constructability, speed of construction, and cost, in order to minimize the construction time. Additionally, construction and technical issues are considered early in the design process, thereby eliminating redesign iterations and construction problems.

Efficient. Structural designs are optimized for the unique structural demands of tall buildings, thereby maximizing strength and stiffness while minimizing structural quantities. Of particular importance in designing supertall structures for efficiency is the ability to use the gravity-load resisting system as part of the lateral-load resisting system, while also maximizing the footprint of the lateral system. This provides a resistance to the wind-overturning moment that uses forces already inherent within the system, thereby providing the opportunity for efficiency by using the vertical structure to resist both gravity and lateral loads.

Designing for supertall—the Burj Khalifa. The key characteristics of a successful supertall building cannot occur without an intense and cooperative collaboration between the structural and architectural requirements of the building. The design of the Burj Khalifa is an example of such collaboration. The structural engineers and architects worked closely together from the beginning of the project to determine the shape of the tower in order to provide an efficient building in terms of its structural system and its response to wind, while still maintaining the

integrity of the initial design concept, keeping the structure simple, and fostering constructability. The overall shape of the tower is an extremely efficient solution to the structural requirements of a supertall residential tower (**Fig. 3**). Starting from a slender top, the building spreads out as the gravity and wind forces accumulate. As such, the tower is a graphical representation of the structural tall building problem. As a result, even though the global forces are large, the forces in the individual members are not. These ideas led to the structure's three-wing geometry, while its 24 major setbacks produce a spiraling reduction as it ascends in height.

The Y-shaped structure can be described as a "buttressed-core" system (**Fig. 4**). Each of the building's wings buttress the others through a six-sided central core, or hexagonal hub. The central core provides the torsional resistance of the structure, acting as a strong axle and keeping the building from twisting. Corridor walls extend from the central core to near the end of each wing, terminating in thickened

Fig. 3. Burj Khalifa. *(Photograph courtesy of SOM and Nick Merrick, copyright © Hedrick Blessing)*

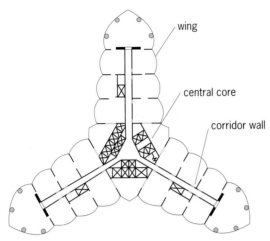

wing

central core

corridor wall

Fig. 4. Burj Khalifa typical floor plan. *(Illustration courtesy of SOM, copyright © by SOM)*

hammerhead walls. These corridor walls and hammerhead walls behave similar to the webs and flanges of a beam to resist the wind shears and moments. Perimeter columns and flat plate floor construction complete the system. At mechanical floors, outrigger walls are provided to link the perimeter columns to the interior wall system, allowing the perimeter columns to participate in the lateral load resistance of the structure. As such, one must view the Burj Khalifa as one cohesive lateral system, one in which a giant concrete beam cantilevers out of the ground so that the system works together as a single unit. Every piece of vertical concrete (and thereby all gravitational forces) is part of this giant beam, used to resist the wind. The gravitational load then helps stabilize the structure by using the weight of the building to resist the wind. The result is a very efficient structure in that no vertical concrete nor gravity load goes unused in providing resistance to the wind, the fundamental demand of a supertall structure.

Wind engineering and its effects on supertall design. Wind is the dominant demand on supertall building structures, both in terms of resisting the magnitude of its loading and resisting the effects of its induced motions. Key in resisting the impact caused by wind is the ability to manage the vortex shedding behavior of the building. Vortex shedding is the phenomena of how wind behaves as it passes an object. As wind passes a building, a vortex or eddy will be created on the side of the building. If these vortices organize over the height of the building, they can accumulate in strength so that a strong crosswind, pulsing force is created, which can increase the wind-induced loads and motions that the building experiences (Fig. 2). Disorganizing these vortices is paramount to successful supertall building design. The desired result is to "confuse" the wind by encouraging disorganized vortex shedding over the height of the building. This is accomplished through various shaping efforts as described earlier, which are then tested through a rigorous wind-tunnel testing program, accompanied by appropriate structural system refinements resulting from the testing.

With the Burj Khalifa, it was immediately apparent that for a building of this height and slenderness,

wind forces and the resulting motions in the upper levels would become dominant factors in the structural design. Over several months, the structural and architectural teams would refine the tower's shape with extensive wind-tunnel tests (**Fig. 5**). These studies were directed by Dr. Peter Irwin of Rowan Williams Davies and Irwin Inc.'s (RWDI) boundary-layer wind tunnels in Guelph, Ontario. The wind-tunnel program included rigid-model, force-balance tests, full multidegree of freedom aeroelastic model studies, measurements of localized pressures, pedestrian wind environment studies, and wind climatic studies. Through these tests, the team determined the harmonic frequency of wind gusts and eddies under various wind conditions. This information was used to set targets for the building's natural frequencies and mode slopes, thereby "tuning" it to minimize the effects of the wind. Several rounds of force-balance tests were undertaken as the geometry of the tower evolved and was refined architecturally. Initially, the tower setbacks occurred in a spiraling counterclockwise manner, then the direction of the setbacks was reversed to clockwise, and the shape of the individual setbacks were refined. After each round of wind-tunnel testing, the data were analyzed, and the building was reshaped to minimize wind effects, refine the building's harmonies, and accommodate unrelated changes in the client's program. In general, the number and spacing of the setbacks changed as did the shape of wings. This process resulted in a substantial reduction in wind forces on the tower by "confusing" the wind, in effect, encouraging disorganized vortex shedding over the height of the tower (**Fig. 6**).

Constructability issues and their effects on supertall design. Material technology and construction means and methods have a significant effect on the design of supertall building systems. Structural systems must incorporate these elements in order to provide a building that can be built in an efficient and timely manner. The key to successful construction is the ability for the contractor to efficiently deliver materials to the great heights required, and form the structure with a minimal amount of disruption. As such, structural systems must be arranged and designed to facilitate the use of construction methods and

Fig. 5. Burj Khalifa aeroelastic wind tunnel model. *(Photograph courtesy of RWDI, copyright © by RWDI)*

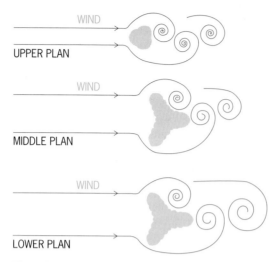

Fig. 6. Burj Khalifa wind behavior. *(Illustration courtesy of SOM, copyright © by SOM)*

formwork systems that accelerate and simplify the construction. Additionally, structural systems must use the advantages of material advancements, such as those relating to high-performance concrete, to simplify construction and minimize quantities.

Again, examining the Burj Khalifa, high-performance concrete was used for the tower, with wall and column concrete strengths ranging from C80 to C60 cube strength (80 and 60 MPa). Additionally, the C80 wall and column concrete was specified as a high-modulus concrete, in order to provide increased stiffness to the system. Using such concrete offered high stiffness, mass, and damping for controlling building motions and accelerations, which was critical in designing the world's tallest building. In fact, because of the stiffness of the system, the design team was able to design the tower to satisfy motion and acceleration criteria without the use of supplemental damping devices.

One of the most challenging concrete design issues was ensuring the pumpability of the concrete to reach the world record heights of the Tower, which necessitated that the concrete be pumped in a single-stage well more than 600 m. Four separate basic mixes were developed by the contractor to enable reduced pumping pressure as the building grew taller. The maximum allowable aggregate size decreased with building height, requiring no more than 10 mm aggregate above level 127. Prior to the start of the superstructure construction, a horizontal pumping trial ensured the pumpability of the concrete mixes (**Fig. 7**). This trial consisted of a long length of pipe with several 180° bends to simulate the pressure loss in pumping to heights more than 600 m in a single stage. The final pumping system used on site employed Putzmeister® pumps, including two of the largest in the world, capable of concrete pumping pressure up to a massive 350 bars (35 MPa) through high-pressure 150 mm pipeline.

The Burj Khalifa used the latest advancements in construction techniques and material technology. The walls were formed using an automatic, self-climbing formwork system, allowing for quick floor cycle times with a minimal amount of crane use. Dis-

cussions were held with formwork manufacturers prior to the completion of design in order to arrange the structural system so as to best use such a system. The circular nose columns were formed with circular steel forms, and the floor slabs were poured on panel formwork. Wall reinforcement was prefabricated on the ground in 8 m sections to allow for fast placement. Three primary tower cranes were located adjacent to the central core, with each continuing to various heights as required. High-speed and high-capacity construction hoists were used to

Fig. 7. Burj Khalifa concrete pumping test. *(Photograph courtesy of SOM, copyright © by SOM)*

Fig. 8. Burj Khalifa construction. *(Photograph courtesy of SOM, copyright © by SOM)*

transport workers and materials to the required heights. A specialized GPS monitoring system was developed to monitor the verticality of the structure, because of the limitations of conventional surveying techniques. The floor plate was divided into zones for organizing the construction. In each zone, the contractor used an "up-up" approach, with the walls leading the way and the slabs following. The construction sequence for the structure has the central core and slabs being cast first, in three sections; the wing walls and slabs followed behind, and the wing nose columns and slabs followed behind these. Concrete was distributed to each wing using concrete booms that were attached to the jump-form (climbing-form) system (**Fig. 8**).

Conclusion. The number of supertall buildings being constructed has advanced significantly over the last decade, as has our understanding of the design parameters and influences on the structural systems of these buildings and our ability to evaluate their performance. Keeping the structure simple, understanding the environment, and using material and construction advancements are all key to creating a successful supertall building structure.

For background information *see* ARCHITECTURAL ENGINEERING; BEAM; BUILDINGS; CANTILEVER; CONCRETE; STRUCTURAL ANALYSIS; STRUCTURAL DESIGN; STRUCTURE (ENGINEERING); WIND TUNNEL in the McGraw-Hill Encyclopedia of Science & Technology.

William F. Baker; James J. Pawlikowski

Bibliography. Aerodynamics Committee of the American Society of Civil Engineers, chaired by P. Irwin, *Outdoor Human Comfort and Its Assessment, State of the Art Report*, ASCE, Reston, Virginia, 2003; W. F. Baker, et al., Creep & shrinkage and the design of supertall buildings—A case study: The Burj Dubai Tower, *ACI SP-246: Structural Implications of Shrinkage and Creep of Concrete*, 2007; W. F. Baker, J. J. Pawlikowski, and B. S. Young, The challenges in designing the world's tallest structure: The Burj Dubai Tower, *Proceedings of the SEI/ASCE Structures Congress*, Austin, Texas, 2009.

Cholera in Haiti

A massive and deadly earthquake struck Haiti on January 12, 2010, leading to homelessness for many of the country's people. Refugee camps were established to shelter those who lost their homes in the earthquake. Ten months later, a second crisis overtook this country, when cholera broke out in the tent camps housing 1.3 million people near Port-au-Prince, the capital of the island nation. Between October 2010 and April 2011, the cholera epidemic in Haiti killed nearly 5000 individuals, with over 270,000 cases reported. Approximately 150,000 of those patients required hospitalization during the same time period, increasing the strain on this devastated nation.

Background. Cholera is caused by *Vibrio cholerae*, a species of bacteria that causes overwhelming gastrointestinal illness. *Vibrio* pathogens are gram-negative rod-shaped bacteria that have a single curve

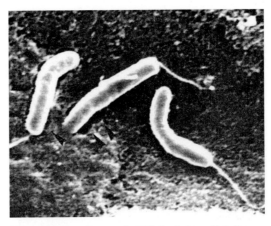

Fig. 1. Electron micrograph of *Vibrio cholerae.* Note the characteristic curved shape and single flagellum. (*Reproduced with permission from J. Willey, L. Sherwood, and C. Woolverton, eds., Prescott's Principles of Microbiology, McGraw-Hill, New York, 2008*)

in their shape (**Fig. 1**). Members of this genus are found growing naturally in marine waters and estuaries worldwide. Seven pandemics (worldwide epidemics) have been caused by *V. cholerae* since 1816. It is believed that this increase in a disease that once occurred sporadically or in localized epidemics is a result of increased travel and commerce (as well as wars) between different parts of the world. This intercontinental travel has allowed the disease to be spread by asymptomatic human carriers.

Cholera is transmitted through contaminated water and food. A very high dose of bacteria (approximately 100 million cells) must be ingested for illness to occur in normal, healthy individuals. As a result, cholera is generally seen in communities with poor sanitation, in which the drinking water is heavily contaminated with sewage. The most recent pandemic started in Asia in 1961 and spread to Africa, Europe, and other regions during the 1970s and 1980s. The strain responsible is known as *V. cholerae* O1 biotype El Tor; it spread to Peru and later to other parts of South America and Central America in 1991. Another epidemic strain, *V. cholerae* O139 Bengal, appeared in 1992 in India before spreading rapidly across Asia.

Vibrio cholerae produces cholera toxin, a secreted protein that causes massive diarrhea, resulting in severe dehydration in those who are infected. This enterotoxin can rapidly cause shock and system failure, which results in death in many patients. Cholera toxin is a complex protein that consists of two different subunits known as A and B. The B subunit binds to receptors on the intestinal epithelial cells. This is followed by internalization of the A subunit into the cell, which then interacts with proteins that control the enzyme adenylate cyclase within the host cell. This in turn leads to the catabolic conversion of adenosine triphosphate (ATP) to cyclic adenosine monophosphate (cAMP). When this occurs, the cell begins rapidly losing water and electrolytes, leading to severe diarrhea in the infected individual.

Symptoms of cholera include vomiting, a rapid heart rate, loss of skin elasticity (**Fig. 2**), dry mucous membranes, low blood pressure, thirst, muscle

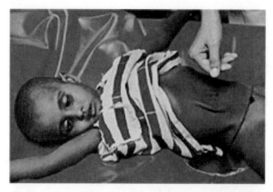

Fig. 2. A young Haitian boy with cholera being checked for dehydration. (*Courtesy of Centers for Disease Control and Prevention*)

cramps, and restlessness or irritability; however, the most diagnostic symptom is production of severe, watery diarrhea. In severe cases, patients can lose as much as 1 liter of fluid per hour during the worst stage of the disease. In fact, the diarrhea is so watery that it is referred to as "rice-water" stool because of the flecks of intestinal mucus floating in the fluid. In spite of this massive loss of fluid, the bacteria are not easily dislodged from the intestine because they bind to the mucosal cell layer using their pili, which are structures that allow for attachment to host cells, as well as other proteins used for binding. Therefore, the diarrhea loss does not flush the bacteria from the intestine, where they continue to cause fluid loss.

Haiti's cholera crisis. The first case in Haiti was diagnosed in October 2010; the number of cases grew almost exponentially during the next few months, reaching a total of more than 270,000 cases with nearly 5000 deaths by April 2011. Because of poor sanitation and unsafe drinking-water sources, this outbreak led to an epidemic within days of the appearance of the initial cases. In November 2010, the United States Centers for Disease Control and Prevention (CDC) announced that the outbreak strain had been identified as *Vibrio cholerae* serogroup O1 biotype El Tor, and it was found to be closely related to strains identified in South Asia since 2002.

Why was this disease so deadly to the Haitians? Statisticians from the CDC reported that the Haitian cholera killed its victims 11.5 times more often than the El Tor strain that spread to Peru in 1991. A number of factors are believed to have played a role in the severity of the epidemic. One was the large number of earthquake survivors living in tent cities, increasing the density of the population and leading to poor sanitation under primitive living conditions. In turn, water sources were frequently contaminated or of extremely poor quality (**Fig. 3**). Other factors included the lack of adequate medical care or nutrition. In addition, HIV infection and its resulting immune deficiencies are more prevalent in Haiti than in Peru.

However, subsequent studies have shown that these were not the only factors playing a role. When the DNA of the Haitian strain was sequenced and compared to that of other strains of cholera from around the world, it was found to be closely related to a variant South Asian strain (biotype El Tor), and

it shared two mutated genes encoding an older version of the cholera toxin. This older version of the toxin is believed to be more lethal than that found in the strains causing disease since 1961. Patients suffer from more intense diarrhea as a result of this toxin; in some cases, patients died only 2 h after first developing symptoms of the disease.

In addition to the more lethal toxin, the Haitian strain carries a third mutation that first appeared in Bangladesh in 2002. This mutation has allowed the Bangladesh strain to dominate South Asia, replacing the older strains, possibly because of its enhanced ability in causing severe diarrhea in victims. In the event that it should spread to Central or South America, the mutation in the Haitian strain may allow it to replace current American strains. The Haitian strain also carries more antibiotic resistance genes than do strains found currently in South America. This is of significance because these bacteria are well known for exchanging genetic sequences among themselves. This may lead to potentially more deadly strains of *V. cholerae* in the future.

Strategies for the future. The Haitian epidemic has devastated an already demoralized population. The United Nations first predicted that the number of cholera cases would reach 200,000; within two weeks of this initial prediction, the estimate was revised upwards to 400,000. Other studies predict that there may actually be closer to 800,000 cases by November 2011. The number of future cases, of course, will depend on what the governmental authorities do to reduce transmission. The measures include the administration of clean water to the population, the provision of cholera vaccination to at least the high-risk members of the population, and the treatment of cases with antibiotics. The most critical therapy for patients with cholera is rapid enough replacement of fluids and electrolytes to avoid system failure and shock. However, antibiotic use can reduce the length of time that individuals suffer from the disease.

Administration of vaccines to at-risk individuals can also help lower the number of cases. Typical use of vaccines involves the prevention of epidemic disease; in the past, doctors believed that vaccinating

Fig. 3. Water source in Haiti, showing lack of sanitation available to much of the population. (*Courtesy of Centers for Disease Control and Prevention*)

at-risk populations after a cholera epidemic had started was of little or no use. However, this view is being revised because cholera outbreaks have increased in length of time as well as in the severity of disease in infected individuals. A recent study of a cholera outbreak in Vietnam in 2007–2008 demonstrated that use of a killed oral vaccine was 76% effective against cholera in those vaccinated during the outbreak as it occurred. This indicates that use of a cholera vaccine in Haiti during the current epidemic should have a significant effect on the number of new cases.

For background information *see* ANTIBIOTIC; BACTERIA; CHOLERA; DIARRHEA; EARTHQUAKE; EPIDEMIC; INFECTIOUS DISEASE; MEDICAL BACTERIOLOGY; PATHOGEN; PUBLIC HEALTH; VACCINATION; VIRULENCE; WATER-BORNE DISEASE; WATER POLLUTION in the McGraw-Hill Encyclopedia of Science & Technology.   Marcia M. Pierce

Bibliography. D. D. Anh et al., Use of oral cholera vaccines in an outbreak in Vietnam: A case control study, *PLoS Negl. Trop. Dis.*, 5(1):e1006, 2011; C.-S. Chin et al., The origin of the Haitian cholera outbreak strain, *N. Engl. J. Med.*, 364:33–42, 2011; M. K. Cowan and K. P. Talaro, *Microbiology: A Systems Approach*, 2d ed., McGraw-Hill, New York, 2008; M. W. Gilmour et al., *Vibrio cholerae* in traveler from Haiti to Canada [letter], *Emerg. Infect. Dis.*, June 2011; P. R. Murray, K. S. Rosenthal, and M. A. Pfaller, *Medical Microbiology*, 6th ed., Mosby, St. Louis, 2009; E. Nester et al., *Microbiology: A Human Perspective*, 6th ed., McGraw-Hill, New York, 2008; R. Reyburn et al., The case for reactive mass oral cholera vaccinations, *PLoS Negl. Trop. Dis.*, 5(1):e952, 2011.

Circulating cancer cells

The first report of circulating tumor cells (CTCs) dates back to 1869, when Thomas Ashworth, an Australian physician, described a case of cancer in which cells similar to those in the tumors were seen in the blood after death of the patient. It is well known that most deaths from carcinomas are caused by the hematogenous dissemination (that is, dissemination via the blood) of cancer cells to distant organs and eventually the development of metastasis. When found in the bone marrow of carcinoma patients, occult epithelial cells are defined as disseminated tumor cells (DTCs); when found in the peripheral blood, they are defined as CTCs. Several investigators have provided cytogenetic evidence indicating that most detectable epithelial cells in the bone marrow or peripheral blood of patients with carcinoma are malignant. Minimal residual disease (MRD) or micrometastatic cells are defined as tumor cells that are undetectable by conventional imaging and laboratory tests used for tumor staging after curative surgery of the primary tumor. MRD is thought to be the target of adjuvant systemic treatment, that is, treatment given to patients with solid tumors (for example, women with breast cancer) to reduce their risk of relapse after primary surgery.

Detection methods. Detection of CTCs is challenging because these cells are rare, occurring at a frequency of one tumor cell per 10^6–10^7 mononuclear cells. Therefore, most CTC assays include an initial enrichment followed by a detection step. Currently, numerous assays can be employed. Based on the specific enrichment and detection technologies that are used, different CTC detection rates will result for each patient sample that is analyzed.

Enrichment of CTCs based on epithelial cell adhesion molecule (EpCAM)–positive selection is widely used, but other methods exist, including depletion of leukocytes or enrichment based on size (filtration) or density. Any enrichment results in a loss of the CTC subpopulation (or subpopulations) because of tumor cell heterogeneity; for instance, enrichment based on EpCAM results in the loss of cells with the epithelial mesenchymal transition (EMT) phenotype because of downregulation of EpCAM. After the initial enrichment step, CTC detection and characterization at the DNA, RNA, and protein levels can be performed using different technologies (see **illustration**). To date, CellSearch®, an automated immunomagnetic and immunofluorescent system, is the only method that has received approval from the United States Food and Drug Administration (FDA) for the detection of CTCs as an aid in monitoring patients with metastatic breast, colorectal, and prostate cancers. In this system, a CTC is defined as a cell staining for cytokeratins 8, 18, and 19 (markers of epithelial cells) and 4′,6-diamidino-2-phenylindole (DAPI, a marker of the cell nucleus), but not staining for CD45 (a marker of leukocytes). Another promising assay is the "CTC-chip," which is capable of efficient separation of CTCs from blood samples, mediated by the interaction of CTCs with EpCAM-coated microposts under precisely controlled laminar flow conditions. The MagSweeper® is another technology consisting of a magnetic rod that sweeps through blood to capture CTCs. Overall, because the different assays will continue to evolve rapidly, the challenge will be to prospectively evaluate the utility of each assay to address specific clinical needs.

Clinical applications. There are a number of clinical applications related to CTCs.

CTC detection in metastatic disease. In metastatic breast, colorectal, and prostate cancers, the detection of CTCs by CellSearch® prior to the starting of a new line of treatment was an independent predictor of shorter overall survival (OS). In these trials, ≥5 CTCs/7.5 mL of blood were observed in 49% of patients with metastatic breast cancer, 57% of patients with metastatic castration-resistant prostate cancer, and 18% of patients with metastatic colorectal cancer.

Several ongoing phase-III trials, particularly those involved with breast and prostate cancers, have incorporated CTC detection in their statistical design. One phase-III trial is testing the strategy of changing chemotherapy compared with continuing the same chemotherapy for metastatic breast cancer patients who have elevated CTC levels at first follow-up assessment. In metastatic prostate cancer, CTC

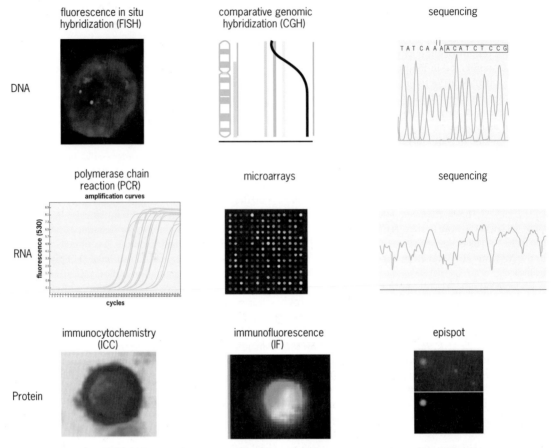

Examples of technologies for detection and characterization of circulating tumor cells (CTCs) at the DNA, RNA, and protein levels.

enumeration was included in the phase-III registration trial of abiraterone acetate (a therapeutic drug). If investigators demonstrate that early CTC response predicts the survival outcome of patients in randomized trials of new drugs, then CTCs can be a useful tool to accelerate drug development.

CTC detection in early disease. The detection of CTCs in the nonmetastatic setting is more challenging because the occurrence of these cells is a very rare event. The largest study of CTC detection in early breast cancer is the German "SUCCESS" trial. In this study, CTC detection by CellSearch® after primary surgery and before the administration of adjuvant chemotherapy was an independent predictor of shorter disease-free survival and overall survival (DFS and OS). In early breast cancer, a pooled analysis of 4703 breast cancer patients provided evidence that bone-marrow DTCs detected at the time of surgery were an independent prognostic factor for poor outcome. Partly based on these data, the American Joint Committee on Cancer has included a new M0(i+) category in the TNM (tumor, node, metastasis) staging for breast cancer. This category is defined as no clinical or radiographic evidence of distant metastases, but having deposits of molecularly or microscopically detected tumor cells [that are no larger than 0.2 mm (0.008 in.)] in blood, bone marrow, or other nonregional nodal tissue in a patient without symptoms or signs of metastases.

Liquid biopsy. Beyond enumeration, the most promising clinical application of CTCs is the real-time detection of tumor genotype (for example, detection of kinase mutations), which can predict sensitivity to a targeted agent. This is attractive for both patients and clinicians because blood can be used instead of invasive and costly biopsies that are difficult to perform. One example is the detection of a mutation that confers drug resistance in CTCs isolated using the CTC-chip from metastatic non-small cell lung cancer patients with *EGFR* gene mutations who had received tyrosine kinase inhibitors. Another example is the use of fluorescence in situ hybridization (FISH) to characterize the status of the androgen receptor and phosphatase and tensin homolog (PTEN) genes in patients with castration-resistant prostate cancer in phase-I/II clinical trials of abiraterone acetate. Despite these initial encouraging results, more data from prospective trials are needed to support the role of CTCs as a "liquid biopsy" in treatment decision algorithms.

CTCs and the biology of metastasis. CTC detection/characterization is also performed in order to better understand the biology of metastasis. Recent advances in stem cell biology have led investigators to revisit the cancer stem cell hypothesis, which argues that human cancers may be driven by a population of cells with stem cell properties. For example, investigators have provided evidence that CD44$^+$CD24$^{-/low}$ or aldehyde dehydrogenase 1 (ALDH1)–positive tumor cells have tumor-initiating properties in breast cancer. These same investigators hypothesized that the metastatic founder cells may have stem cell properties. Interestingly, a significant proportion of micrometastatic cells is resistant to chemotherapy and has a low proliferation index. Moreover, the CD44$^+$CD24$^{-/low}$ and ALDH1 markers have been identified in a significant proportion of CTCs or DTCs from patients with metastatic breast cancer.

There is clinical evidence in breast and other tumors that not all patients with detectable CTCs or DTCs will relapse. Interestingly, CTCs have been detected in one third of women without clinical evidence of disease up to 22 years after mastectomy for early breast cancer. Therefore, tumor cells can remain "dormant" over a period of time before the development of disease recurrence, and this phenomenon has been loosely defined as tumor dormancy. There is experimental evidence that several mechanisms of dormancy exist, including cellular dormancy in which DTCs enter a state of quiescence (G0–G1 arrest) and tumor mass dormancy in which DTCs divide but the lesion does not grow beyond a certain size because of limitations in blood supply or an active immune system. Several mechanisms have been suggested to control the switch between tumor dormancy and proliferation, including host-related genetic factors or exogenous factors such as stress or diet. Characterization of CTC/DTC gene pathways may lead to a better understanding of the mechanisms that regulate tumor dormancy and the escape from it.

By using single-cell comparative genomic hybridization of bone-marrow DTCs from breast cancer patients and by studying breast cancer dissemination in mouse models, investigators have proposed the parallel progression model. In this model of breast cancer metastasis, tumor cells disseminate early at ectopic sites and evolve in parallel with tumor cells in the primary site. Other investigators have also provided experimental evidence that cancer dissemination is not a unidirectional process from primary tumor to distant sites; instead, aggressive CTCs can colonize their tumors of origin, accelerating tumor growth in a process called self-seeding. Further studies are required to test these exciting models that challenge our current understanding of the metastatic process.

Perspectives. The study of CTCs may lead to novel insights into the mechanisms that regulate tumor dormancy and metastasis. Advances in detection and characterization of CTCs may provide a liquid biopsy for tailoring treatment in metastatic disease. These advances may also allow the identification of new targets for the eradication of minimal residual disease, which could increase the cure rates of patients with early-stage disease. Finally, in the future, a blood test could be developed for early cancer diagnosis.

For background information *see* BIOTECHNOLOGY; BLOOD; CANCER (MEDICINE); CELL (BIOLOGY); CHEMOTHERAPY AND OTHER ANTINEOPLASTIC DRUGS; CLINICAL PATHOLOGY; ONCOLOGY; PHARMACOLOGY; STEM CELLS in the McGraw-Hill Encyclopedia of Science & Technology.   Michail Ignatiadis

Bibliography. J. A. Guirre-Ghiso, Models, mechanisms and clinical evidence for cancer dormancy, *Nat. Rev. Cancer*, 7:834-846, 2007; M. Y. Kim et al., Tumor self-seeding by circulating cancer cells, *Cell*, 139:1315-1326, 2009; C. A. Klein, Parallel progression of primary tumours and metastases, *Nat. Rev. Cancer*, 9:302-312, 2009; K. Pantel, R. H. Brakenhoff, and B. Brandt, Detection, clinical relevance and specific biological properties of disseminating tumour cells, *Nat. Rev. Cancer*, 8:329-340, 2008; S. Riethdorf and K. Pantel, Advancing personalized cancer therapy by detection and characterization of circulating carcinoma cells, *Ann. N.Y. Acad. Sci.*, 1210:66-77, 2010.

Cognitive bias in forensic science

Forensic science has been used and accepted in courts all over the world for over a hundred years. It often provides very powerful incriminating and exculpatory evidence. However, forensic science is not always what is appears to be. Many forensic science disciplines are subjective and susceptible to a wide range of cognitive and psychological influences, such as bias.

Misunderstanding of forensic science. Forensic evidence presented in court is often misrepresented and misperceived, and thus can have excessive persuasive power. The main reasons for distortions are:

1. The portrayal of forensic science in the media and general culture (such as in Hollywood's CSI shows) is unrealistic, and overrates what forensic science can achieve. Thus, sometimes causing unrealistic expectations and value associated with forensic evidence.

2. Forensic experts who overstate the strength of the evidence. This occurs primarily because experts, across many domains, are often overconfident and overestimate their abilities. Furthermore, forensic examiners do not get appropriate feedback, because the actual ground truth in criminal cases is unknown. Adding to this problem is that in the United States, United Kingdom, and many other countries, forensic experts work within an adversarial legal system which, by its very nature, puts the forensic expert and evidence within a prosecutorial (or defense) posture. And finally, because of the nature of cognitive biases, the forensic examiners themselves are not aware of when they are influenced by them.

3. The weakness of eyewitness testimony is well known and acknowledged, thus increasing the reliance on "objective" and "scientific" evidence.

The nature of forensic evidence. Many forensic disciplines that deal with impression and trace evidence lack sufficient instruments and measurements, thereby leaving the decisions to the subjective assessment of the human examiner. Often there are two visual patterns, and the forensic examiner has to determine if the two patterns are sufficiently similar to conclude that they both came from the same source. Consider, for example, a latent fingerprint mark left at the crime scene and the fingerprint of a suspect. Even if they are from the same person, the two marks are not totally identical because of elasticity of the skin, the pressure and the surface that the fingermark was left on, and so on. Therefore, the human examiner has to decide whether they are similar and consistent enough to conclude that they originated from the same person. A similar situation is prevalent in many forensic disciplines, from firearms, tire marks, and hand writing to bite marks. Even complex DNA evidence, such mixtures or low copy number, require subjective assessments.

Without objective measurements and criteria for what constitutes sufficient similarity or consistent enough, it is up to the human forensic examiner to make such judgments. Furthermore, because forensic evidence collected at the crime scene is often far from perfect, the examiner many times has to analyze evidence in which the quality and quantity of information is limited, and with evidence that includes "noise."

The forensic expert. Forensic science has generally failed to fully acknowledge that the human examiner is the "instrument" of analysis. Because of this, forensic science has not taken sufficient steps to address the well-known weaknesses and vulnerabilities of human cognition and psychology. In contrast to great efforts in protecting and minimizing physical contamination of evidence, there is little-to-no effort in addressing potential psychological and cognitive contamination.

Often forensic examiners have been resistant to accepting the role of cognition and psychology in shaping and influencing their decision making, claiming that they are totally objective and never make identification errors. This has dramatically changed in the past few years. First, a number of cases of erroneous identifications have been made public. Most influential is the Mayfield case in Oregon, where an innocent person had been misidentified by a number of leading forensic experts. Second, the 2009 report by U.S. National Academy of Science, *Strengthening Forensic Science in the United States: A Path Forward*, has further established that many forensic disciplines are subjective and vulnerable to bias and other psychological influences.

These, along with scientific research that demonstrated bias in forensic science (see below), have resulted in a shift in attitude and culture, both in the United States [for example, the FBI and Los Angeles County Sheriff's Department (LASD) as well as in other countries (for example, the Netherlands). Nevertheless, many forensic examiners continue to deny their susceptibility to error as a result of cognitive and psychological biases.

Research findings. The past few years have seen the first empirical scientific research into bias and cognitive issues in forensic science. It is quite surprising that an important domain such as forensic science has existed for over 100 years, and has been accepted by the courts, while it was lacking basic research. This has been extensively highlighted in the U.S. National Academy of Science report. Recently the U.S. National Institute of Standards and Technology (NIST) and the U.S. National Institute of Justice (NIJ) have begun to provide extensive support for such research.

The first two scientific experiments that investigated bias in forensic science presented forensic examiners with evidence as routine casework. However, they were actually analyzing evidence that they themselves examined a few years ago. This time, the evidence was presented within different extraneous context, to scientifically observe if such contextually irrelevant information might bias their decision making. The findings showed a clear and dramatic effect: the same examiner, examining the same evidence, reached different and conflicting conclusions when the evidence was presented within different contexts. The level of such bias was mediated by the complexity of the prints, the difficulty of the comparisons, as well as the examiners and type of decision. Nevertheless, after 100 years of using forensic evidence in court, being presented as objective, totally reliable, and unbiased, scientific research has demonstrated cognitive influences and bias in forensic science.

The research was further replicated and expanded, for example, with DNA mixtures. In addition, further research showed that the lack of measures and criteria causes different examiners to perceive the same evidence differently (even without manipulating the contextual information). Moreover, the same examiner will perceive the same evidence differently when presented at different times, thus demonstrating lack of inter- and intra-consistency. Research has also shown how the use of technology in forensic science has been problematic. Technology has been introduced without sufficient care and consideration of how the technology cooperates with, and may affect, the human users. Indeed, research has shown the biasing effects technology (such as automated fingerprint identification systems) has on human forensic examiners.

The bias snowball effect. The fact that forensic evidence is not perfect or totally objective does not mean it is not a powerful and an important ally in the criminal justice system. However, it is important that the integrity and reliability of forensic evidence is maintained and improved (see below), and that it is appropriately presented and perceived in court.

The problem is not only that forensic science evidence can be biased (by what a detective tells the examiners, the context of the case, and so on), but that it can bias other lines of evidence. For example, if one piece of forensic evidence (biased or not) is known to other forensic examiners who are analyzing other forensic evidence, then their examination may be affected and biased by their knowledge of the results of the other piece of evidence (for example, a forensic examiner looking at bite marks may be influenced and biased in their examination if they know that fingerprint evidence shows the suspect is guilty).

Forensic evidence can also bias other lines of evidence. For example, eyewitnesses can be affected. In the Willingham case in Texas, eye witnesses changed their testimony after they learned that forensic evidence suggested that the fire was not accidental but the result of arson. Furthermore, suspects may confess to crimes they have not committed when they are presented with overwhelming evidence against them (such as "objective and scientific" forensic evidence), and then offered a plea bargain.

Therefore, the bias in forensic science is potentially very dangerous. Not only is forensic science evidence taken to have great weight, often overvalued (despite being subjective and vulnerable to bias), but forensic science evidence can influence and bias other elements in the legal justice system. It is critical that when all the evidence is considered that each piece of evidence is independent of each other. When they affect and influence one another, then their value and reliability is diminished. Furthermore, because one piece of evidence influences another, then greater distortive power is gathered as more evidence is affected (and affecting) other lines of evidence, causing an increasing snowball of bias.

The way forward. Forensic science evidence has an important role in the criminal justice system. However, it must take its proper place and appropriate measures to control and minimize weaknesses. Such steps are relatively easy to take, and include:

1. Educating detectives, judges, juries, and the general public as to the strength and weakness of forensic science. Portraying a realistic and scientifically based picture of forensic evidence, and specifically pointing out its subjective nature and vulnerabilities to bias.

2. Training forensic examiners to deal with and minimize bias. Bias is not an ethical issue, it results from the very making and processes carried out in the brain, and the architecture of human cognition. Bias cannot be turned off and on by mere willpower or awareness. However, proper training by cognitive experts can play an important contribution in minimizing bias in forensic science.

3. Proper procedures and protocols are essential. For example, "sequential unmasking" suggests that examiners not be exposed to irrelevant contextual information that they do not need. Another example is the correct use of technology that ensures good cooperation and distributed cognition.

For background information *see* COGNITION; CRIMINALISTICS; DECISION ANALYSIS; DECISION THEORY; FINGERPRINT; FORENSIC EVIDENCE; PSYCHOLOGY in the McGraw-Hill Encyclopedia of Science & Technology.   Itiel Dror

Bibliography. Committee on Identifying the Needs of the Forensic Science Community, *Strengthening Forensic Science in the United States: A Path Forward*, National Academies Press, Washington, D.C., 2009; I. E. Dror, The paradox of human expertise: Why experts get it wrong, in *The Paradoxical Brain*, N. Kapur (ed.), Cambridge University Press, Cambridge, U.K., pp. 177–188, 2011; I. E. Dror and R. Bucht, Psychological perspectives on problems with forensic science evidence, in *Conviction of the Innocent: Lessons from Psychological Research*, B. Cutler (ed.), American Psychological Association Press, pp. 257–276, 2011; I. E. Dror and S. Cole, The vision in 'blind' justice: Expert perception, judgment and visual cognition in forensic pattern recognition, *Psychonomic Bull. Rev.*, 17(2):161–167, 2010; I. E. Dror and G. Hampikian, Subjectivity and bias in forensic DNA mixture interpretation, *Science & Justice*, in press; I. E. Dror et al., Cognitive issues in fingerprint analysis: Inter- and intra-expert consistency and the effect of a 'target' comparison, *Forensic Sci. Int.*, 208:10–17, 2011; I. E. Dror and J. Mnookin, The use of technology in human expert domains: Challenges and risks arising from the use of automated fingerprint identification systems in forensics, *Law Probab. Risk*, 9(1):47–67, 2010.

Cognitive radio

A new emerging radio technology called cognitive radio significantly increases the utilization of the current frequency bands through spectrum-sharing approaches. Electromagnetic waves (that is, radio waves) are essential to the operation of wireless transmission systems, with frequencies in the range 3 Hz–300 GHz. Traditionally, large parts of the radio frequency bands have been licensed to authorized users by government. This static spectrum allocation mechanism causes frequency bands to be insufficient at various times and locations. Practical measurements show that most of the licensed bands are either unused or partially used in different geographical areas for much of the time. In other words, spectrum bands are used sporadically. **Figure 1** shows spectrum usage in the range from 1 Hz to 6 GHz, which was measured by the Berkeley Wireless Research Center (BWRC).

As the number of wireless connections increases, spectrum demand and spectrum congestion will become critical challenges in the forthcoming all-encompassing wireless world. In fact, future wireless networks will face spectrum scarcity as a result of the users' requirements, such as high multimedia data rate transmission over mobile networks.

For this reason, scientists have investigated flexible spectrum usage techniques or sufficient spectrum management mechanisms in order to utilize the

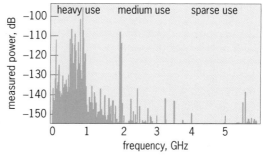

Fig. 1. Spectrum utilization, as measured by the Berkeley Wireless Research Center (BWRC).

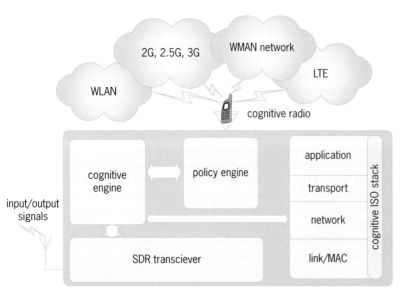

Fig. 2. Cognitive radio architecture and capability of accessing different licensed and license-exempt networks, such as wireless local-area networks (WLAN), wireless metropolitan-area networks (WMAN), long-term evolution (LTE), and second/third-generation wireless telephone technology (2G/3G). ISO = International Organization for Standardization, SDR = software-defined radio, MAC = Media Access Control.

observes the environment and intelligently adapts its radio parameters based on internal and external radio state information (**Fig. 2**).

Cognitive radio is also defined as a radio that can change its transmitter parameters based on interaction with the environment in which it operates. Thus, context awareness, in which information extracted from the radio environment and relevant user objects is used to analyze the radio situation and is able to make informed decisions about it, is the main objective of cognitive radio.

By the definition of cognitive radio, the two main characteristics of the technology are cognitive and reconfigurable capabilities. In the former, all radio functionalities in terms of context awareness and decision algorithms are managed by the cognitive engine. Moreover, the cognitive engine tries to optimize a performance goal based on inputs received from the radio's current internal and operating environment states. There is also a cognitive policy engine, which ensures that decisions provided by the cognitive engine are compatible with regulators' rules and policies. In the latter, the physical architecture of cognitive radio is implemented with reconfigurable advanced hardware technology, which is called software-defined radio (SDR).

SDR is a reconfigurable wireless communication system in which the transmission parameters (for example, operating frequency band, modulation type, and protocols) can be reconfigured dynamically through adaptation cycles. On the other hand, the transceiver parameters can be intelligently adapted based on the decision outcomes at the upper network layers.

Cognitive capability and functionality. Since cognitive radio technology has been proposed to cope with spectrum scarcity in future wireless communication, in this section we will focus on spectrum efficiency functionalities and capabilities.

Spectrum utilization is the main concept behind the cognitive radio application. To enable this, the radio should observe (sense), analyze, access, and intelligently hand off to a new radio state anywhere at any time. The cognitive cycle is shown in **Fig. 3**. The main functions of cognitive radios are spectrum sensing, spectrum management, spectrum sharing, and spectrum mobility. The cognitive capabilities of these radios are categorized as spectrum sensing, spectrum analysis, and spectrum decision.

Spectrum sensing. The aim of the spectrum-sensing module is to explore and monitor unused licensed and license-exempt frequency bands during specific times. Generally, spectrum-sensing techniques can be classified as transmitter detection (for example, energy detection, matched filter, and cyclostationary feature detection), cooperative detection, and interference-based detection. The nature of sensing can be implemented by either proactive or reactive sensing mechanisms. In the former technique, the target channels are sensed periodically. In the latter, the target channels are observed before secondary user transmission. In addition, spectrum sensing can be either centralized or distributed.

current static frequency bands more fully and protect the licensed transmissions against harmful interference. The solution to spectrum scarcity is based on a spectrum-sharing approach, which significantly improves the utilization of current underutilized spectrum bands.

The idea of cognitive radio technology (the term was first coined by J. Mitola in 1999) is believed to be capable of alleviating spectrum scarcity in the future. This technology is based on opportunistically reusing unused spectrum bands at any place and time. Dynamic spectrum access (DSA) is the main functionality of cognitive radio technology, and radio spectrum awareness and radio parameters adaptation are the main tasks of DSA. At the same time, cognitive radio technology relies on an opportunistic spectrum access mechanism in which the radio

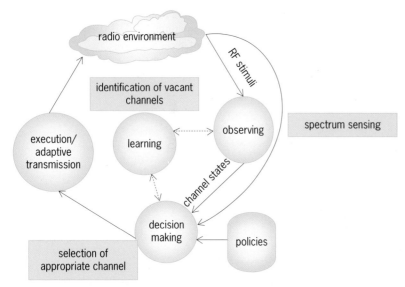

Fig. 3. Simplified cognitive system cycle and capability of spectrum utilization.

In the centralized mode, a central system, such as a base station, responds to sensing, collects channel vacancy information, and then shares these data with cognitive radio users in the coverage area. In the distributed or decentralized mode, all cognitive radio users sense the specific bands individually and respond by sharing the sensing results with one another in order to make reliable spectrum access decisions.

Spectrum analysis. The characteristics of the observed wireless channels are estimated through information collected by the sensing module. An efficient algorithm is employed to extract spectrum state information in terms of time and frequency. Also, the information represents which/where/when the licensed channels are vacant. These channel characteristics significantly improve spectrum access through these tasks.

Spectrum decision. Based on the provided spectrum analysis results and the observed radio parameters, an appropriate channel in the available unoccupied channels will be selected. In this stage, transmission characteristics such as modulation type, bandwidth, operating frequency, transmit power, and antenna characteristics are chosen based on the desired quality of service (QoS) and interference limitations. These results adapt radio transmission parameters to the new situation.

Spectrum-sharing policies. Cognitive radio is a new technology in the wireless communications context. Therefore, advanced spectrum-sharing policies are required if the new technology is to be employed. The spectrum-sharing approach can be implemented through either cooperation or coexistence. In the former case, devices with different technologies cooperate to share spectrum among them with regard to avoiding harmful interference with one another. To do this, a common protocol needs to be defined and distributed among users. In the coexistence mode, devices are allowed access to spectrum bands without providing harmful interference; hence, devices are required to sense targeted bands to avoid interference. Cognitive radio adopts a coexistence sharing technique because of the equipped sensing module in the radio.

Spectrum-sharing techniques are categorized into three main methods: sharing among licensed users, sharing among licensed and unlicensed users, and sharing among unlicensed users.

In cooperative sharing, a secondary user sends a request signal to the primary system, which confirms the availability of a licensed channel at the demand time. This cooperation guarantees the quality of service of both the primary and the secondary transmission. Also, there is a secondary spectrum market policy approach in which the secondary user is in charge of the sharing time.

Coexistence sharing includes underlay spectrum access (USA) and opportunistic spectrum access (OSA), in which cognitive radio users are allowed access to either temporarily unoccupied spectrum bands or spatially unoccupied spectrum bands. These terms will now be further explained.

In the case of temporarily unoccupied spectrum bands, licensed and unlicensed users are located in the same coverage area. Thus, cognitive users are allowed to exploit temporarily unoccupied licensed frequency bands during a specific time. In this case, a strength dynamic spectrum access needs to avoid disruption to licensed transmission.

In the case of spatially unoccupied spectrum bands, also called regional spectrum access, the cognitive radio transmitter is located far from the licensed coverage area. In this scenario, the cognitive radio transmitter is equipped with a Global Positioning System (GPS) receiver and is supported by an updated regional radio spectrum database.

Opportunistic access (the idea behind cognitive radio) requires high-accuracy, real-time spectrum sensing and an intelligent allocation algorithm, which need to be implemented on the cognitive engine. In the USA approach, the cognitive user is allowed access to licensed channels simultaneously under low transmission-power and interference-level constraints.

In all spectrum-sharing mechanisms, frequency bands are shared under an interference-level limitation, which is the most prominent issue in emerging spectrum policy. Currently the policy is implemented by opening television white spaces (that is, unused portions of the television broadcast spectrum) as an opportunity for cognitive radio usage in the forthcoming years (**Fig. 4**).

Standardization. Standardization provides a link between cognitive radio research results, implementation, and widespread deployment of the technology in both industry and the commercial world.

In 2004 the Institute of Electrical and Electronics Engineers (IEEE) initiated a set of standardization projects related to cognitive radio called IEEE P1900, which evolved in 2006 into IEEE Standards Coordinating Committee 41 (IEEE SCC41). The activities of IEEE SCC41 are aimed at facilitating the development of research ideas into standards to expedite the use of research results for public use.

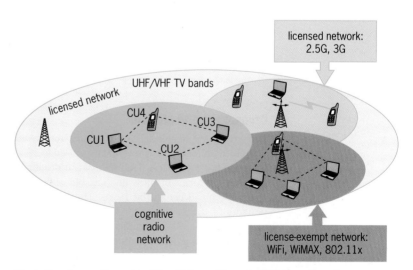

Fig. 4. Sample cognitive network in which cognitive users (CUs) are allowed access to different licensed and license-exempt frequency bands, opportunistically.

Moreover, other organizations involved in standardization activity, such as the International Telecommunication Union Radiocommunication Sector (ITU-R), the European Telecommunications Standards Institute (ETSI), the Object Management Group (OMG), the SDR Forum, and the Third-Generation Partnership Project (3GPP), are tasked with developing new cognitive radio standards. Also, the completed standards, namely, IEEE 802.15.4, IEEE 802.16, and 802.11, support some degree of cognitive radio-like functionality (dynamic frequency selection, transmit power control, coexistence, and so forth). IEEE 802.22 is believed to be the first completed cognitive radio standard in the television frequency bands.

Application. This technology is expected to revolutionize the way in which wireless communications operate and create users' experience and context. The main application areas of cognitive radio could be military wireless communication, reliable radio communication in emergency events such as earthquakes and other natural disasters, firefighting, medical emergency services including public safety, criminal control, and precision agriculture. Overall, cognitive radio technology significantly enhances the trustworthiness of radio communications by exploiting unused radio spectrum bands in the aforementioned scenarios. This cognition and flexibility can improve the spectrum efficiency of current and future wireless networks.

For background information *see* LOCAL-AREA NETWORKS; MOBILE COMMUNICATIONS; RADIO; RADIO RECEIVER; RADIO SPECTRUM ALLOCATION; RADIO TRANSMITTER; TELEVISION STANDARDS; WIRELESS FIDELITY (WI-FI) in the McGraw-Hill Encyclopedia of Science & Technology.

Mahdi Pirmoradian; Christos Politis

Bibliography. Federal Communications Commision Spectrum Policy Task Force, *Report of the Spectrum Rights and Responsibilities Working Group*, ET Docket No. 02-135, 2002; J. M. Peha, Sharing spectrum through spectrum policy reform and cognitive radio, *Proc. IEEE*, 97:708–719, 2009; M. Sherman et al., IEEE standards supporting cognitive radio and networks, dynamic spectrum access, and coexistence, *IEEE Comm. Mag.*, 46(7):72–79, July 2008; A. Shukla et al., *Cognitive Radio Technology: A Study for Ofcom—Summary Report*, QinetiQ Ltd., Farnborough, U.K., 2007.

Cryosphere and climate

The term cryosphere is derived from the Greek word *krios* meaning icy cold. It was first proposed in 1923 by the Polish scientist A. Dobrowolski to designate all terrestrial forms of snow and ice, but only came into widespread use in the 1990s. The major components of the cryosphere—snow cover, sea ice, and ice sheets—play a significant role in global climate, while glaciers, lake ice, and perennially frozen ground (permafrost) play more local roles. Seasonal snow cover accounts for up to 49% of the Northern Hemisphere land area and seasonally frozen ground up to 55%. The time scales of these same components also vary widely; seasonally frozen ground, snow cover, and floating ice last weeks to months, although some Arctic sea ice persists for several years. Glaciers have a life span of 10^2–10^4 years, while ice sheets and permafrost have time scales of 20,000 to millions of years.

Climatic roles of snow and ice. There are at least four major ways in which snow and ice cover affect the climate system. The first is albedo–temperature feedback. This operates when snow cover or sea ice shrinks, lowering the albedo (the reflectivity of a surface to solar radiation) from about 80–90% for snow cover to about 10–30% for land surfaces, and only 5% for the ocean. This reduction of the surface albedo increases the absorption of incoming solar radiation at the surface, on average, by about four times. The warming of the surface leads to a further reduction of the snow or ice cover, giving a positive feedback (amplified effect). The effect works correspondingly in reverse, with increased snow cover leading to enhanced cooling and further extension of the snow cover. This positive feedback mechanism is responsible in part for the two- to threefold amplification of the global warming signal in high northern latitudes. The second effect involves the insulation of the ground surface by snow cover or the isolation of the ocean from the atmosphere in the presence of sea ice. In each case, the transfer of heat and moisture from the underlying surface to the atmosphere (or vice versa) is cut off and this modifies the thermal regimes of the underlying ground or ocean and of the overlying atmosphere. The third effect is on the hydrological cycle caused by the temporary storage of water in snow and ice covers. This factor delays the annual snowmelt runoff peak and on long time scales (10^3–10^5 years) locks up water in ice sheets, affecting global sea level. The fourth effect involves the release or uptake of latent heat during the phase changes from vapor to liquid (2500 kJ/kg) to solid (ice) [333kJ/kg] and vice versa. Hence, evaporation requires nearly eight times more energy than melt.

Seasonal variations. Determining the climatic influence of seasonally varying cryospheric components is rather straightforward as a result of the annual cycle and interannual variability in their extent and depth. This fact gives rise to many cases, thus allowing the relationships to be established between climatic factors and snow cover or floating ice. Time series of northern hemisphere snow-cover extent derived from satellite imagery are available from 1966 and sea-ice extent in the Arctic and Antarctic from 1972. Sea-ice concentrations derived from satellite passive microwave data, which are not limited by darkness or cloud cover as visible imagery is, extend from 1979 to the present.

Anomalies in the extent of continental snow cover are inversely correlated with hemispheric air temperatures; that is, as temperatures increase, the snow cover shrinks. For example, about 50% variance in Northern Hemisphere air temperatures and March-April hemispheric snow-covered areas is common. Moreover, in spring, the anomalies of snow-covered areas correlate with negative top-of-the-atmosphere all-wavelength net radiation over northern Canada and Russia, implying a large-scale interaction involving the snow albedo.

There are large differences between surface air temperatures in the presence or absence of snow cover. Winter air temperatures are 8-10°C higher when snow cover is absent. However, ground surface temperatures are lower in winter when the ground is snow free, as the result of radiative cooling of the surface.

The relationship between sea-ice extent and temperature is complicated by the fact that the ice extent is affected by wind-driven ice motion and there is a lag of about six months between air-temperature changes and ice-extent response. Climate models indicate that the ice-area sensitivity to warming is about 400,000 to 700,000 km²/°C. In the eastern Arctic, the recent rise in spring air temperatures has been associated with a lengthening of the sea-ice melt season.

Decadal variations in regional sea-ice extent are correlated with air temperature. This has been demonstrated for Arctic sea ice off northern Eurasia for the period from 1933-2006, where two periods of retreat, separated by a partial recovery between the mid-1950s and the mid-1980s, are inversely correlated with changes in air temperature.

The extent of Arctic sea ice at its annual minimum in September for the period from 1979 to 2009 shows a linear decrease of -11.9% per decade with the three lowest years in 2007, 2008, and 2010. However the ice retreat is a complex response to ocean warming, rather than directly to air temperatures, and gives rise to enhanced bottom melt and export of multiyear ice through the Fram Strait in the East Greenland Current.

Snow cover and sea ice influence the atmosphere and climate primarily through ice–albedo feedback and the cooling effects of the snow- and ice-covered surfaces. In 1955, Hubert Lamb first demonstrated the relationship between the thickness of the lower troposphere [extending from the Earth's surface to a height of 5.5 km (the 1000–500 hPa layer), which represents the mean layer temperature] and the extent of snow cover. The snow cover sets up an extensive surface high-pressure system with a cold low-pressure system in the overlying atmosphere; this has the effect of steering cyclones southward around its margin, leading to a southward extension of the snow cover as a result of cyclonic precipitation. It has been shown that there is a downstream influence by Eurasian snow-cover extent on the planetary wave structure in the atmosphere. In the case of extensive snow cover over Eurasia, the high-pressure ridge over western North America is enhanced and vice versa. Positive snow-cover extremes in midwinter are associated with a stronger East Asian jet stream and an augmented low-pressure trough over the North Pacific Ocean.

The seasonal storage of water in mountain snowpacks plays a major role in the hydrologic cycle. Typically, snowmelt lasts about three months, from spring to early summer, with the snow melting at progressively higher elevations. This means that runoff peaks in late spring–early summer, and then gradually decreasing river flow continues through the summer. The effect of global warming is already apparent at lower elevations in many mountain ranges, where peak flow is occurring as much as a month earlier and late summer river flows are much lower, posing problems for farmers who rely on the water for irrigation. The melting and retreating of glaciers will in some areas temporarily increase runoff, but once the ice is gone the runoff will decline significantly. This is already a problem in parts of the tropical Andes of Peru, Ecuador, and Bolivia.

Long-term relationships. Glaciers respond primarily to climate forcing. They are influenced by snow accumulation in the cold season and ablation (melt) in the summer, as well as by calving, if they terminate in lakes or the ocean. The mass-balance sensitivity of glaciers to a temperature change is of the order of 1 m water equivalent per year per degree Celsius and 0.1 m per year per 10% change in snowfall. Thus, changes in temperature are generally much more important than changes in snowfall. A degree-day temperature model for 42 Arctic glaciers reveals a low sensitivity to a temperature increase in continental climates (about -0.2 m $a^{-1}K^{-1}$) and a high sensitivity (up to -2 m $a^{-1}K^{-1}$) in maritime climates. A relationship has been demonstrated between length changes on 169 glaciers and air temperature. The reconstructed warming for the first half of the twentieth century of 0.5°C is in good agreement with observations.

On time scales of 1000 years, changes in ice-sheet volume play a major role in global sea level. During the last glacial maximum, about 21,000 years ago, global sea level was about 130 m below that at present, exposing vast areas of continental shelf in the Bering Sea, the North Sea, and the seas around Indonesia, which had major consequences for human migrations into North America, the British Isles, and Australia, respectively. During the last interglacial, about 130,000 years ago, sea level was about 5-8 m above present. The ongoing global warming is causing a sea-level rise of about 2 mm/yr because of the retreat of glaciers worldwide and melting in coastal Greenland and parts of Antarctica. Projections of global sea-level rise by the year 2100 caused by ocean thermal expansion and ice melt are in the range 55-75 cm, sufficient to cause populations in low-lying coastal areas to worry.

For background information *see* ALBEDO; ANTARCTIC OCEAN; ARCTIC OCEAN; CLIMATE MODELING; CLIMATOLOGY; CYCLONE; GLACIOLOGY; GLOBAL CLIMATE CHANGE; PERMAFROST; SEA ICE; SNOW; WATER RESOURCES in the McGraw-Hill Encyclopedia of Science & Technology.   Roger G. Barry

Bibliography. R. G. Barry and T. Y. Gan, *The Global Cryosphere: Past, Present and Future*, Cambridge University Press, Cambridge, U.K., 2011; R. D. Brown and D. A. Robinson, Northern Hemisphere spring snow cover variability and change over 1922-2010 including an assessment of uncertainty,

The Cryosphere, 5:219–229, 2011; M. de Woul and R. Hock, Static mass-balance sensitivity of Arctic glaciers and ice caps using a degree-day approach, *Ann. Glaciol.*, 42:217–224, 2005; National Research Council, *Advancing the Science of Climate Change*, National Academy Press, Washington, D.C., 2010; J. Stroeve et al., Arctic sea ice decline: Faster than forecast, *Geophys. Res. Lett.*, 34:L09501, 2007; S. Vavrus, The role of terrestrial snow cover in the climate system, *Climate Dynamics*, 29:73–88, 2007.

CTRPs: novel adipokines

Obesity is a major risk factor that contributes to type 2 diabetes and cardiovascular diseases. Therefore, studies focusing on understanding the underlying mechanisms that link obesity and metabolic dysfunction may provide new pharmaceutical targets and innovative therapies for the prevention and treatment of metabolic diseases. One of the key paths for understanding obesity-related metabolic diseases is through studying the fundamental biology of adipose (fat) tissue.

Adipose tissue, adipokines, and energy balance. The traditional view of adipose tissue as merely a passive storage depot for excess calories in the form of fat (triglycerides) has undergone a dramatic transformation in the past 15 years. This shift in view is largely the result of the discovery of important hormones secreted by fat tissue (**Fig. 1**). Consequently, adipose tissue is now also considered an endocrine organ that actively secretes a variety of hormones and cytokines, collectively termed adipokines, which mediate intertissue crosstalk to maintain proper energy balance. In 1994, leptin was the first of these adipokines to be discovered. Leptin is a satiety hormone secreted by fat cells (adipocytes), circulates in blood, and acts in the hypothalamic region of the brain to suppress food intake. Circulating levels of leptin rise postprandially (following a meal) and fall during fasting. Because the circulating level of leptin is proportional to fat mass, it allows an organism to gauge the overall energy reserve stored in fat tissue and thus enables the proper allocation of energy resources. When the satiety signaling pathway controlled by leptin is disrupted, as has been documented in animals and humans harboring mutations in the leptin gene, it leads to dramatic physiological alterations, including hyperphagia (excessive food intake), morbid obesity, insulin resistance and type 2 diabetes, and delayed or complete absence of puberty. Exogenous administration of recombinant leptin strikingly reverses most of these phenotypes.

The second most widely studied adipokine is adiponectin. In contrast to the satiety signal conveyed by leptin, adiponectin has been considered a "starvation" signal that promotes the storage of triglycerides preferentially in fat tissue. Adiponectin is produced almost exclusively by adipocytes, circulates in blood, and acts in other tissues and the central nervous system to regulate whole-body energy metabolism. Specifically, adiponectin promotes fat oxidation in skeletal muscle and synergizes with insulin to suppress glucose production in the liver. These biological activities of adiponectin result in the reduction of blood glucose and free fatty acids, both of which are greatly elevated in obesity and type 2 diabetes. Interestingly, the circulating levels of adiponectin are greatly decreased in the conditions of obesity and type 2 diabetes. Increasing the circulating levels of adiponectin can reverse many of the metabolic abnormalities associated with obesity in a variety of animal models. On the other hand, despite the antidiabetic, antiatherogenic, and anti-inflammatory properties of adiponectin, animal models that lack adiponectin show surprisingly mild alterations in metabolic function. These findings suggest the existence of some other adipokine (or adipokines) that can compensate for the absence of adiponectin and led to the identification of a novel family of secreted hormones, that is, the C1q/TNF-related proteins (CTRPs).

C1q/TNF-related proteins (CTRPs) are novel adipokines. Efforts to uncover novel hormones with important metabolic function have led to the discovery of a family of secreted proteins functionally homologous to adiponectin. These proteins, designated as C1q/TNF-related proteins (CTRP1 to CTRP13), share the same structural organization as adiponectin, including a signal peptide for protein secretion, an N-terminal collagen-like domain, and a C-terminal globular domain homologous to the immune complement C1q. Both adiponectin and CTRPs belong to the C1q/TNF superfamily of proteins, having a signature C1q domain whose three-dimensional structure strikingly resembles that of tumor necrosis factor-α (TNF-α), despite a lack of sequence homology between these proteins.

In contrast to adiponectin whose expression is restricted to adipocytes, CTRPs are more widely expressed. However, adipose is the predominant tissue that expresses many of the CTRPs. The antidiabetic drug, rosiglitazone, has been shown to increase the expression of several CTRPs in mice and in cultured adipocytes, suggesting possible involvement of CTRPs in mediating some of the beneficial metabolic effects of rosiglitazone. Importantly, the circulating levels of some CTRPs are elevated in genetically engineered "knockout" mouse models

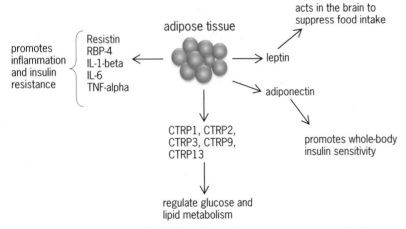

Fig. 1. Adipose tissue secretes a whole array of factors, collectively termed adipokines, which affect whole-body insulin sensitivity as well as glucose and lipid metabolism.

that lack adiponectin. This suggests that CTRPs (for example, CTRP1) may compensate for the absence of adiponectin and thus explains the remarkably mild metabolic phenotypes seen in the adiponectin knockout mice.

Common molecular features of CTRPs. All CTRPs are secreted proteins and most are found circulating in blood. Sex (male versus female), metabolic state (obese versus lean), and genetic background affect the circulating levels of CTRPs. For instance, CTRP13 exhibits a pronounced sexually dimorphic expression pattern, with females expressing considerably higher levels of CTRP13 mRNA and protein compared to male mice.

All CTRPs associate to form trimers as their basic structural unit via the globular C1q domain. Some are further assembled into higher-order structures that correspond to hexamers and high-molecular-weight oligomers via intermolecular disulfide bonds. Different oligomeric forms may have distinct functions. Furthermore, combinatorial associations between different CTRPs have been documented, highlighting a possible mechanism to generate an expanded repertoire of functionally distinct protein complexes.

Metabolic function of CTRPs. The circulating levels of CTRP1 and CTRP3 are markedly reduced in mice fed a high-fat diet, mimicking the diet-induced obesity in humans. Administration of recombinant CTRP1, CTRP3, and CTRP9 to mice has been shown to acutely lower blood glucose in normal as well as obese and diabetic mice. With the exception of CTRP3, the mechanism (or mechanisms) responsible for the glucose-lowering effect of CTRPs in vivo remains largely unknown. Although the natural target tissues acted upon by CTRPs in vivo remain to be determined, in vitro cell culture models suggest the liver and skeletal muscle as two potential sites of action where CTRPs exert their functions; this is consistent with the glucose-lowering effect observed in mice when recombinant proteins were injected.

Although data pertaining to the in vivo metabolic function for CTRP2 are currently lacking, in vitro studies demonstrated the ability of recombinant CTRP2 to increase fatty acid oxidation and glycogen deposition in cultured muscle cells, in part by activating a conserved energy-sensing pathway controlled by adenosine monophosphate–activated protein kinase (AMPK). Glycogen synthesis and fatty acid oxidation are greatly suppressed in diabetics, partly as a result of insulin resistance in the skeletal muscle. The ability of CTRP2 to promote glycogen deposition and fatty acid oxidation highlights its therapeutic potential. Because skeletal muscle tissue accounts for more than one-quarter of the human body mass, a modest enhancement in fatty acid oxidation or glucose metabolism will translate into significant improvements in overall whole-body energy metabolism over time.

Mechanism of action of CTRP3. Blood glucose levels are principally maintained by two major mechanisms: (1) glucose uptake and utilization in tissues in the fed state, and (2) de novo glucose production (gluconeogenesis) in liver in the fasted state. Dys-

regulation of either or both mechanisms results in glucose intolerance and the eventual development of type 2 diabetes. Liver is the major target tissue of CTRP3. Recombinant CTRP3 lowers blood glucose in normal and diabetic mice by suppressing hepatic gluconeogenesis. CTRP3 binds to an unidentified receptor on hepatocytes, which results in activation of the protein kinase B/Akt signaling pathway. Akt activation in turn leads to the suppression of glucose-6-phosphatase (G6Pase) and phosphoenolpyruvate carboxykinase (PEPCK) expression, which are two key enzymes involved in de novo glucose synthesis. Importantly, CTRP3 activates Akt to reduce hepatic glucose output independent of insulin, making it an attractive therapeutic approach to improve glycemic control in insulin-resistant type 2 diabetics (**Fig. 2**).

In humans, alternative splicing generates two CTRP3 isoforms; both circulate in plasma. The expression of both isoforms is coregulated; tissue that expresses one isoform invariably expresses the other. The longer isoform has 73 extra amino acids in its N-terminus, and one of these contains N-linked glycans. Thus, the two isoforms differ in size and posttranslational modification. Hetero-oligomerization between the two isoforms appears to protect the longer isoforms from proteolytic cleavage. Functionally, both isoforms are equally potent in suppressing gluconeogenesis in cultured hepatocytes.

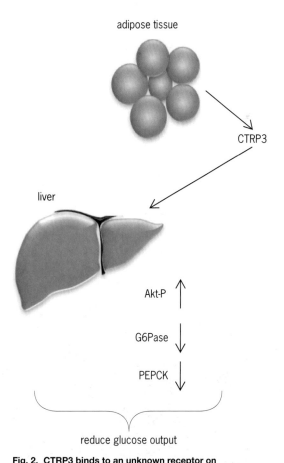

adipose tissue

CTRP3

liver

Akt-P

G6Pase

PEPCK

reduce glucose output

Fig. 2. CTRP3 binds to an unknown receptor on hepatocytes to activate protein kinase B/Akt. Activation of Akt [via phosphorylation (P)] results in the expression of enzymes (G6Pase and PEPCK) involved in de novo glucose synthesis (gluconeogenesis). The end result of this is the suppression of hepatic glucose output by CTRP3.

Future outlook. It is expected that future studies using genetically engineered gain- and loss-of-function mouse models will unravel the physiologic functions and mechanisms of actions of this fascinating family of secreted hormones. Elucidating the intertissue crosstalk mediated by CTRPs will enable a much better understanding of the endocrine circuits underlying the integrated control of whole-body energy homeostasis.

For background information *see* ADIPOSE TISSUE; AMP-ACTIVATED PROTEIN KINASE (AMPK); CYTOKINE; DIABETES; ENERGY METABOLISM; GLUCOSE; HORMONE; HUNGER; INSULIN; LEPTIN; LIPID METABOLISM; METABOLISM; OBESITY in the McGraw-Hill Encyclopedia of Science & Technology.

Jonathan M. Peterson; G. William Wong

Bibliography. A. D. Attie and P. E. Scherer, Adipocyte metabolism and obesity, *J. Lipid Res.*, 50:S395–S399, 2009; J. M. Friedman, Leptin at 14 y of age: An ongoing story, *Am. J. Clin. Nutr.*, 89:973S–979S, 2009; J. M. Peterson, Z. Wei, and G. W. Wong, C1q/TNF-related protein-3 (CTRP3), a novel adipokine that regulates hepatic glucose output, *J. Biol. Chem.*, 285:39691–39701, 2010; P. E. Scherer, Adipose tissue: From lipid storage compartment to endocrine organ, *Diabetes*, 55:1537–1545, 2006; Z. Wei, J. M. Peterson, and G. W. Wong, Metabolic regulation by C1q/TNF-related protein-13 (CTRP13): Activation of AMP-activated protein kinase and suppression of fatty acid–induced JNK signaling, *J. Biol. Chem.*, 286:15652–15665, 2011; G. W. Wong et al., A family of Acrp30/adiponectin structural and functional paralogs, *Proc. Natl. Acad. Sci. USA*, 101:10302–10307, 2004; G. W. Wong et al., Molecular, biochemical and functional characterizations of C1q/TNF family members: Adipose-tissue-selective expression patterns, regulation by PPAR-gamma agonist, cysteine-mediated oligomerizations, combinatorial associations and metabolic functions, *Biochem. J.*, 416:161–177, 2008.

Darwin's bark spider

Spiders are exceptionally diverse and abundant, being the primary predators of insects and other arthropods in many terrestrial ecosystems. Many spiders use silk traps to catch insects; in these cases, the familiar wagon wheel–shaped webs, or orb webs, are classical examples (**Fig. 1**). Spider orb webs are highly efficient and specialized traps that are thought to account for the success of web spiders. In the short term, orb webs allow spiders to catch flying insects that are not readily caught by many other kinds of predators, which may explain the ecological abundance of orb spiders. In the long term, the evolutionary origin of orb webs can explain a major radiation of spiders, resulting in the many thousands of orb spiders that are alive today. This diversity of orb spiders includes spiders that build webs of varying sizes: from webs as small as 1–2 cm (0.4–0.8 in.) in diameter, which are aimed at small flies such as fruit flies and mosquitoes, to webs that are more than 1 m (3.3 ft) in diameter, which can catch large insects and even small vertebrates. Among orb webs, however, none is larger than that of Darwin's bark spider (*Caerostris darwini*; a new species discovered in Madagascar), which can span up to 2 m (6.6 ft) across (Fig. 1). These webs are suspended along rivers and lakes, often crossing the water on bridge-lines that can span more than 20 m (66 ft). These large webs built over water allow access to prey that are not caught frequently by more typical terrestrial orbs. The prey include insects and possibly small vertebrates that use the rivers as passageways, as well as those that live part of their life in water, such as mayflies.

Exceptional spider silks. Orb spiders use extraordinary biological materials, spider silks, to spin their webs. An orb web contains two radically different types of silk threads. One is dragline silk, which is so named because most spiders use this kind of silk for safety lines that they constantly spin as they crawl about their habitats. Dragline silk is similar to steel in terms of stiffness and strength (**Fig. 2**), and forms the structural support threads, frame, and radial lines of orb webs (Fig. 1). The other type of fiber is capture spiral silk, which is used to build the adhesive spiral that sticks to prey when they hit the web. Capture spiral silk is relatively pliable and highly elastic, being almost like rubber (Fig. 2).

Among the many types of materials produced by living organisms, silks have an impressive combination of material properties. Silks are lightweight, strong, elastic, and durable fibers that readily compare to high-quality synthetic fibers in terms of desirable properties such as strength and resistance to breakage. In synthetic fibers, strength and elasticity are difficult to combine, that is, fibers can be either strong, such as steel, or elastic, such as rubber, but rarely both (Fig. 2). However, spider silks combine these typically divergent properties into fibers that can absorb more energy before breaking, a property known as toughness, compared to any other natural and most synthetic materials. Thus, the stiff dragline silks are about as strong as steel but much more elastic, and the pliable spiral silk is as stretchy as rubber but much stronger (Fig. 2). Both require about three times more kinetic energy to break per volume in comparison to high-performance synthetic polymers such as Kevlar™. However, even among spider silks, none is tougher than the silks of Darwin's bark

Fig. 1. A typical dense orb web (left), and a river-crossing web (right) of a female Darwin's bark spider, *Caerostris darwini* (inset). The riverine webs can reach 2 m (6.6 ft) in diameter.

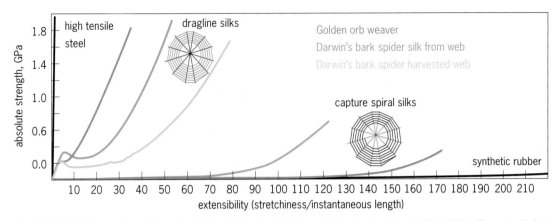

Fig. 2. Mechanical properties of the silks of Darwin's bark spider in comparison to the "standard" of spider silk research, that is, silks of the golden orb spider *Nephila*. The silks of both spiders are impressively strong and are comparable to synthetic materials like steel. However, spider silks are also very stretchy, such that they require much more energy to break, making them significantly tougher than steel. Darwin's bark spider dragline silk exhibits even more stretchiness than the dragline silk of other orb spiders like *Nephila*, which helps to explain why it is so incredibly tough. The highly elastic but much weaker capture spiral silks contain adhesive glues and are similar to rubber in their performance.

spider. The dragline silk of Darwin's bark spider is one of the strongest dragline silks spun by any species of spiders, being up to two times more elastic than typical dragline silk, which results in extraordinary toughness (Fig. 2).

Silk diversity and biomimetic fibers. Each spider produces a "tool kit" of different kinds of silks, with some female spiders producing up to eight distinct types of silk that are used for such functions as lifelines, egg protection, and web construction. Some of these silks are stiff, whereas others are pliable; some are strong and some are relatively weak; and some are highly elastic, whereas others do not stretch much at all. This variation also occurs for a single type of silk compared across spider species. Given that there are well over 41,000 known spider species on Earth, with many more to be discovered, and that each one of them can make several types of silk, nature has produced a virtual goldmine of probably more than 200,000 different silk fibers. Scientists have barely begun to sample this variation, with most silk research focusing on one silk type from a few spider species.

Much research on spider silk is ultimately driven by the desire to utilize these amazing materials for human benefit, whether using the spider silks directly or replicating the material properties of the silks in synthetic "biomimetic" materials. Spider silk and biomimetic fibers have many potential uses for humankind. By combining incredible strength and elasticity in a lightweight and durable fiber, spider silks or their synthetics could find use in many fields and industries, including those related to fabrics (lightweight superfabrics and tough ropes), medicine (bandages and ligaments), military equipment (ballistics), and robotics (sensors, activators, and artificial muscles). However, natural harvesting of spider silk is not feasible at commercial scales because spiders produce silk in relatively low quantities and spiders cannot be easily farmed (in contrast to silkworms). Unfortunately, biomimetic fibers so far do not approach spider silk's strength and elasticity. In part, this results from a lack of understanding

of the detailed mechanism of silk spinning, as well as limited comparative work on the protein building blocks of different kinds of spider silks. Future progress likely lies in using the variation of natural spider silks to understand how structural and molecular differences in silks determine which silks are relatively weak and which perform the best. For such applications, Darwin's bark spider silk is important as representing the toughest known natural fiber; then, even if it is impossible to fully replicate the silk properties in synthetics, it might be possible to still achieve something exceptional with a product that is only half as good as the original.

Scientific significance of discovering Darwin's bark spider. Why is the discovery of *Caerostris darwini* or Darwin's bark spider important? The scientific significance lies not so much in Darwin's bark spider building the "biggest" web, nor in it spinning the "toughest" silk, because these are both simply extremes in the enormous diversity of spider webs and silks. Instead, the discovery of Darwin's bark spider is significant for at least three reasons. First, it has been hypothesized that fundamental changes in the silk properties of Darwin's bark spider enabled a novel ecology, that is, the spinning of webs across large bodies of water. Second, the correlation between the extreme silk toughness and the unique habitat of these spiders allows scientists to use the knowledge of the natural history of spiders to predict something about the biomechanical properties of their silks. Such "bioprospecting" can speed discovery of exceptional biomaterials in nature, providing a new tool in the race to inventory biodiversity under the threats of habitat loss and extinction. Third, discovering the molecular basis for the extreme toughness of Darwin's bark spider silk could help unravel key molecular features and building blocks for biomimetic silks. Comparative studies of silks that range from relatively weak to relatively tough are critical for understanding how the structural composition of spider silks relates to their material properties, and hence also to the development of biologically inspired synthetic fibers. Therefore, for future research, Darwin's bark spider holds many promises.

For background information *see* ARACHNIDA; ARANEAE; ELASTICITY; ENGINEERING DESIGN; MANU-FACTURED FIBER; MATERIALS SCIENCE AND ENGINEER-ING; NATURAL FIBER; PREDATOR-PREY INTERACTIONS; SILK; SPIDER SILK; STRENGTH OF MATERIALS in the McGraw-Hill Encyclopedia of Science & Technology.

Ingi Agnarsson; Matjaž Kuntner; Todd A. Blackledge

Bibliography.   I. Agnarsson, M. Kuntner, and T. A. Blackledge, Bioprospecting finds the toughest bio-logical material: Extraordinary silk from a giant river-ine orb, *PLoS One*, 5(9):1–9, 2010; T. A. Blackledge, M. Kuntner, and I. Agnarsson, The form and function of spider orb webs: Evolution from silk to ecosys-tems, *Adv. Insect Physiol.*, 40, 2011, in press; M. Kuntner and I. Agnarsson, Web gigantism in Darwin's bark spider, a new species from Madagascar (Aranei-dae: *Caerostris*), *J. Arachnol.*, 38:346–356, 2010; D. Porter and F. Vollrath, Silk as a biomimetic ideal for structural polymers, *Adv. Mater.*, 21:487–492, 2009.

Denotational semantics

Computer programs are complex, structured assem-blies, as are buildings and television sets. However, programs are also linguistic assemblies, as are epic poems on the scale of Homer's *Iliad* or *Odyssey*. Even more so than with a building, television, or epic poem, a program must match its "blueprint" or specification exactly. For example, flight-control software or medical software must perform exactly as described; otherwise, someone might be harmed. Because of its linguistic aspect, a program's specifi-cation often looks like a mathematical formula, and a program's semantics (meaning) must be mathemati-cal in nature to provably match the specification.

Programs are written in a language, just as poems are, and the semantics of a computer program are derived from the semantics of the language used to write it. Because computer languages are structurally simpler than human languages, techniques from linguistics and symbolic logic can state precisely what a language's constructions mean and there-fore what a program written in the language means. This motivates the study of programming-language semantics.

The first attempts at stating precisely a program-ming language's semantics employed machine op-erations and computer hardware to describe what programs compute, but this approach was too detail-laden to be useful for proofs of mathematical cor-rectness. In the mid-1960s, in Oxford, England, Christopher Strachey, himself a computer hardware designer and also a language designer, adapted techniques from mathematical logic to define the semantics of computer language. His approach, called denotational semantics, established a mid-dle ground between computer-hardware detail and mathematical abstraction and permitted precise, yet intuitive, definitions of language constructions.

Denotational semantics is the standard starting point for stating what computer languages mean.

Semantics of arithmetic. Here is a simple example of denotational semantics. The first programming language that people learn is arithmetic. It has a syn-tax (spelling laws) as well as semantics. First, we must state precisely the syntax, that is, how to write grammatically correct arithmetic:

A numeral, N, such as 0 or 1 or 2 or ..., is an arithmetic expression.

If E_1 and E_2 are arithmetic expressions, then so are $(E_1 + E_2)$ and also $(E_1 \times E_2)$.

For example, 2, $(4 + 2)$, and $(2 \times (4 + 2))$ are all grammatically correct arithmetic; they are "pro-grams" in the language of arithmetic. The syntax definition is often written in equational form, as a Chomsky-style grammar law:

$$E ::= N \mid (E_1 + E_2) \mid (E_1 \times E_2)$$

We compute on arithmetic using laws for addition and multiplication. Thus, a hand-held calculator is a computer that understands the arithmetic language. Like a spoken language, the words and phrases of arithmetic have meaning, and the meaning of an arithmetic expression is formed from the meanings of its subexpressions. This approach underlies its denotational semantics, which looks like this:

$$\boldsymbol{E} : \text{Expression} \dashrightarrow \text{number}$$
$$\boldsymbol{E}[N] = N$$
$$\boldsymbol{E}[(E1 + E2)] = \text{plus } (\boldsymbol{E}[E1], \boldsymbol{E}[E2])$$
$$\boldsymbol{E}[(E1 \times E2)] = \text{times } (\boldsymbol{E}[E1], \boldsymbol{E}[E2])$$

The first line states that $\boldsymbol{E}$ is the name of a function that converts arithmetic expressions to their mean-ings, which are numbers. (You can read $\boldsymbol{E}[.]$ as "the meaning of.") The second line says that the meaning of a numeral, N, is just the corresponding number, *N*. The next line says that the meaning of E1 + E2 is the addition of the numerical meaning of E_1 to the numerical meaning of E_2. Multiplication is defined similarly. Here is how we determine the meaning of the program, $(2 \times (4 + 2))$:

$$\boldsymbol{E}[(2 \times (4 + 2))] = \text{times } (\boldsymbol{E}[2], \boldsymbol{E}[(4 + 2)])$$
$$\boldsymbol{E}[2] = two$$
$$\boldsymbol{E}[(4 + 2)] = \text{plus } (\boldsymbol{E}[4], \boldsymbol{E}[2])$$
$$\boldsymbol{E}[4] = four$$

The equations expose the phrase structure within the expression. The expression's meaning follows the structure, and when we solve the equation family, we deduce that the program's meaning is twelve:

$$\boldsymbol{E}[(2 \times (4 + 2))] = \text{times } (\boldsymbol{E}[2], \boldsymbol{E}[4 + 2]) =$$
times (*two*, plus $(\boldsymbol{E}[4], \boldsymbol{E}[2])$) = times (*two*, plus (*four*, *two*)) = times (*two*, *six*) = *twelve*

If there were a crucial correctness property of this program, it would be stated in terms of the structure of the program. (As an example, all numbers com-puted while totaling the meaning of this program are even-valued.) Because the program decomposes into its phrases, and each phrase has its own meaning, the correctness property is associated directly with each phrase and each meaning; this is the induction principle of mathematics and logic. Importantly, we reason about the program, not about the computer. In this fashion, denotational semantics treats computer programs as intellectual artifacts deserving of independent study, akin to Homer's epic poems.

Semantics of languages that maintain storage. Denotational semantics provides a useful blend of mathematical, logical, linguistic, and computational features, letting it tackle modern computer languages, which work with storage devices, databases, and networks. The following example is for the reader who has written computer programs.

Here is a small, Basic-like language that maintains primary storage. Programs are commands, and their syntax is summarized by these two equations, where

$$C ::= I = E \mid C_1; C_2 \mid \text{if } E : C_1 \text{ else } C_2 \mid \text{while } E : C$$
$$E ::= N \mid (E_1 + E_2) \mid (E_1 \times E_2) \mid I$$

N stands for a numeral, as before, and I stands for a variable name (a word of lowercase letters). The syntax of expressions, E, remains the same, but it also has variable lookup, I. Commands, C, are assignments, sequencing, conditionals, and loops. Here is an example program:

$$x = 0; \text{ if } x : y = (x + 1) \text{ else } y = 1$$

The meaning of expressions is provided by the function,

$$E[.] : \text{Expression} \; --> (\text{Storage} \; --> \text{Number})$$

The meaning of an expression is a function that examines the value of storage to compute a number. Storage is itself a function, mapping variable names to their numerical meanings:

$$\text{Storage} = \text{Variable} \; --> \text{Number}$$

Here is the meaning of expressions:

$$E[N](s) = N$$
$$E[(E_1 + E_2)](s) = \text{plus } (E[E_1](s), E[E_2](s))$$
$$E[(E_1 \times E_2)](s) = \text{times } (E[E_1](s), E[E_2](s))$$
$$E[I](s) = s(I)$$

Notice that the meaning of a variable, I is a lookup of the meaning of I saved in storage, s.

The meaning of a command is defined by a new function, $C[.]$, which indicates that a command updates the incoming storage to outgoing storage:

$$C : \text{Command} \; --> (\text{Storage} \; --> \text{Storage})$$

$$C[I = E](s) = s', \text{ where } s'(i) = (i == i) \; ? \; E[E](s) \mid s(i)$$
$$C[C_1; C_2] = C[C_2](C[C_1](s))$$
$$C[\text{if } E : C_1 \text{ else } C_2](s) = (E[E](s)! = 0) \; ?$$
$$\qquad\qquad\qquad\qquad C[C_1](s) \mid C[C_2](s)$$
$$C[\text{while } E : C](s) = w(s), \text{ where } w(s) =$$
$$\qquad\qquad (E[E](s) \; ! = 0)? \; w(C[C](s)) \mid s$$

The meaning of an assignment, I = E, is a function that converts incoming storage, s, to outgoing storage, s', where s' looks up variables, i, such that when i is I, then the meaning of E is retrieved (otherwise, the existing meaning of i in s is retrieved). That is, the meaning of assignment is to build updated storage for future lookups.

In a similar way, the meaning of two commands in sequence, $C_1; C_2$, is to construct the meaning of C_1 followed by the meaning of C_2. This is a composition of the two commands' update functions. A conditional command, if $E: C_1$ else C_2, is a conditional updating function, and a loop, while $E: C$, is a recursively defined updating function.

The meaning of a program written in this language is an updating function on the storage. Techniques from mathematics, such as reasoning by function extensionality and induction, are used to analyze programs' meanings. Here is a small example, which analyzes the program, $x = 0; \; x = x + 1$, and proves that it has the same meaning as the program, $x = 1$.

$$C[x = 0; x = x + 1](s) = C[x = x + 1](C[x = 0](s))$$
$$= C[x = x + 1](s_0),$$
$$\quad \text{where } s_0(i) = ((i == x) \; ? \; E[0](s) \mid s(i))$$
$$\quad \text{that is, } s_0(i) = (i == x) \; ? \; 0 \mid s(i)$$
$$= s_1,$$
$$\quad \text{where } s_1(i) = ((i == x) \; ? \; E[x + 1](s_0) \mid s_0(i))$$
$$\quad \text{that is, } s_1(i) = (i == x) \; ? \; \text{plus } (E[x](s_0), 1) \mid s_0(i)$$
$$\qquad = (i == x) \; ? \; \text{plus } (s_0(x), 1) \mid s_0(i)$$
$$\qquad = (i == x) \; ? \; \text{plus } (0, 1) \mid s_0(i)$$
$$\qquad = (i == x) \; ? \; 1 \mid s_0(i)$$
$$\qquad = (i == x) \; ? \; 1 \mid s(i)$$

However, s_1 equals exactly $C[x = 1](s)$. By function extensionality, $C[x = 0; x = x + 1]$ equals $C[x = 1]$.

Applications to correctness, implementation, analysis. The previous sections show how a computer language and its programs can be analyzed in a mathematically precise way. Humans can do such an analysis, but more important, a computer can itself be programmed to understand a language's syntax and its denotational semantics and use them to analyze other programs. Such automated program analysis can be used to prove that a program correctly computes the function stated by a mathematical specification (blueprint), analyze a computer program for erroneous or inefficient instructions that can be corrected or removed, and survey a computer program and extract properties that help explain the program's operation.

These applications appear in modern program verifiers, language compilers, and program-flow analyzers. They are central to the advancement of software engineering from a handcraft to a rigorous discipline by which huge, complex designs can be blueprinted, annotated with correctness properties, mapped into computer programs, and rigorously validated by either humans or computers for conformance to the designs. It is as if the *Iliad* itself can be designed, read, understood, and appreciated by the computer.

Denotational semantics plays a key historical and practical role in this process.

For background information *see* COMPUTER PROGRAMMING; LINGUISTICS; LOGIC; PROGRAMMING LANGUAGES; SOFTWARE; SOFTWARE ENGINEERING in the McGraw-Hill Encyclopedia of Science & Technology.
David Schmidt

Bibliography.   F. Nielson, H. R. Nielson, and C. Hankin, *Principles of Program Analysis*, Springer, New York, 1999; D. Schmidt, *Denotational Semantics: A Methodology for Language Development*, Allyn & Bacon, Boston, 1986; J. Stoy, *Denotational Semantics: The Scott-Strachey Approach to Programming-Language Theory*, MIT Press, Cambridge, MA, 1977; C. Strachey, Fundamental concepts in programming languages, *J. Higher-Order Symbolic Comput.*, 13:11–49, 2000.

Development and resilience

In psychology, resilience refers to positive adaptation among individuals at risk for maladjustment. It is characterized either by positive adjustment following experiences of adversity or by recovery from initial maladjustment. Resilience research is the study of the processes that foster positive adaptation. The goal of this research is to enhance the lives of individuals at risk by informing interventions and influencing social policy.

Resilience has been studied for more than five decades among children and adults exposed to a wide range of adversities, including parental mental illness, maltreatment, chronic illness, socioeconomic disadvantage, community violence, and catastrophic life events. In general, there are three classes of protective factors that help people to thrive in the face of significant adversities. These include support from close family and friends, support from the community, and personal traits such as intelligence, optimism, and emotional self-regulation.

Importance of positive relationships. Research has demonstrated unequivocally that, of these three protective factors, the strongest predictor of resilient adaptation is the quality of an individual's close, personal relationships. Stress and adversity can stifle brain development, weaken the biological systems needed for coping, and promote the expression of genes associated with psychopathology. Conversely, positive relationships can significantly buffer an individual from these negative effects of adversity and can provide the social control and emotional security needed to sustain positive adaptation. This is equally true for children and adults.

Close, supportive relationships buffer people from stress through several mechanisms. Positive relationships trigger the release of oxytocin, which is a hormone that reduces stress and anxiety. As a result, people who have strong social ties often perceive stressful situations as less stressful than do others who lack strong social ties. In addition, positive relationships help individuals to combat adversity through collective problem-solving and through the sharing of other emotional, intellectual, and physical resources. Supportive relationships also enhance the individual's capacity for future coping by engendering personal strengths and adaptive coping skills. For example, family and friends nurture cognitive development, encourage persistence, and foster emotional regulation, as well as high self-efficacy and the adoption of mature defense mechanisms.

Support from the wider social community can provide many of the same protective processes that help at-risk individuals to function well in the face of adversity. For children, quality day care, good schools, and safe, supportive, neighborhoods have the potential to offset risk by providing the support, structure, and stability needed to buffer stress and instill valuable psychological assets. Schools, in particular, have tremendous potential to meet these needs through formal and informal interventions. In much the same way, adults can derive strength from their wider social environment through community organizations, employers, religious organizations, and volunteer groups. Early community interventions aimed at fostering resilience in children can have profound positive effects on parents as well.

On the other hand, research clearly demonstrates that human beings flounder in the absence of close, positive relationships. The literature on maltreatment clearly documents the deleterious effects of chronic abuse and neglect. At the same time, research shows that the psychological experiences of loneliness and social isolation are themselves chronic stressors, and both are strongly associated with low emotional well-being, cognitive problems, psychopathology, and poor physical health. Thus, two primary goals of resilience research must be to focus on eradicating maltreatment and to focus on fostering positive social connections.

Role of personal attributes: findings and pitfalls. Research demonstrates that people who flourish in spite of adversity often exhibit adaptive personal strengths. For example, manifestly resilient individuals frequently show high intelligence; they also are often skilled at learning, planning, reasoning, and creative problem-solving, and they are often adept at abstract thought, communication, and introspection. Furthermore, resilient individuals tend to exhibit an easygoing temperament, and they are often extroverted, adaptive, persistent, attentive, and optimistic. Moreover, resilient people generally show a high sense of self-efficacy, self-esteem, and internal locus of control; they believe in their ability to affect change on the environment and, as a result, have a high sense of self-worth.

Some scholars maintain erroneously that resilience is a trait, dependent solely on inborn personal strengths. This individualistic view of resilience is problematic. It is crucial to reiterate that many of the personal strengths associated with resilience are nourished by human connections. As noted previously, relationships have a vast influence on the development of personal strengths, including intelligence, perseverance, emotional regulation, and self-efficacy. Human adaptation is always determined by a combination of what an individual brings to the situation and the quality and dependability of relationships from which he or she can draw strength.

Over the past few decades, pioneers of resilience research have emphasized the dangers of suggesting that any individual (child or adult) can be invincible to major life stresses. In the face of continuing adversities, even those blessed with the most salutary of personal traits (such as high intelligence) will inevitably succumb, unless they are bolstered by support from friends, family, and communities. Research shows that there is no single personal attribute that magically confers invulnerability to people battling adversity. It is critical that scholars keep in mind not only the successes but also the missteps of earlier works on resilience, strictly avoiding suggestions that resilience is somehow an innate feature.

From the standpoint of fostering resilience, a singular focus on developing personal strengths (in the absence of attention to positive relationships) is of limited value. For example, the benefits of an intervention designed to help at-risk individuals to develop coping skills, such as the capacity to reframe negative, self-defeating messages (for example, "I'll never get past this") into more realistic ones (for example, "This is a temporary setback that I am working to overcome"), will inevitably erode over time if these individuals continue to battle significant adversities without the love and support of others. By all means, interventionists can and should help people to develop efficacious coping skills. However, there has to be attention, simultaneously, to ensuring the presence of supportive, reliable relationships in order to bolster people's capacity to cope with current and future adversities.

Some scholars also suggest that resilience is "common," arguing that most people show positive adaptation in the face of major stressors. This view is extremely problematic as well because labels of "resilience" are conferred on the basis of measures of adaptation spanning multiple domains; people can function very well in some realms, whereas they struggle considerably in others. Moreover, resilient adaptation is not static. Individuals can seemingly rebound after a major stressor, but they begin to show significant vulnerabilities weeks or months later.

Most importantly, claims that resilience is innate or common facilitate "blaming the victim" and threaten the allocation of resources for interventions to help vulnerable populations, such as victims of disasters or those affected by dire, long-term poverty. To reiterate: the ultimate goal of resilience researchers is to understand, and communicate to policy makers, what we can do to help distressed individuals function better, and not simply to marvel at how many people function well. If, in fact, most people are resilient, then our charge, as scientists, is to carefully investigate why, and to apply these findings in ways that can most expediently improve the quality of life of those who do not show apparently resilient adaptation.

Outlook. Resilience is positive adaptation to stressful life circumstances. The desire to belong is a basic human need, and strong connections with others fuel resilient adaptation across the life span. Personal strengths can help in battling adversity; at the same time, these personal strengths themselves will falter when people are bereft of close, supportive relationships. Therefore, even when personal attributes are targeted in interventions, we must never lose sight of the undeniable fact that no man, woman, or child is an island. Attention to the quality of our relationships is indispensible if any intervention benefits are to endure over time. The crucial take-away message is the undeniable value of promoting positive relationships across every stage of development. In the future, researchers must prioritize better understanding of potential threats to positive relationships, while recommending ways to eliminate these threats as expediently and efficiently as possible.

For background information *see* BRAIN; CENTRAL NERVOUS SYSTEM; DEVELOPMENTAL PSYCHOLOGY; EMOTION; PERSONALITY THEORY; PSYCHOANALYSIS; PSYCHOLOGY; PSYCHOTHERAPY; STRESS (PSYCHOLOGY) in the McGraw-Hill Encyclopedia of Science & Technology.   Suniya S. Luthar; Jeremy D. Rothstein

Bibliography. S. Achor, Social investment, pp. 171–198, in *The Happiness Advantage: The Seven Principles of Positive Psychology That Fuel Success and Performance at Work*, Crown Publishing, New York, 2010; J. T. Cacioppo and W. Patrick, *Loneliness: Human Nature and the Need for Social Connection*, W. W. Norton, New York, 2008; N. A. Christakis and J. H. Fowler, *Connected: The Surprising Power of Our Social Networks and How They Shape Our Lives*, Little, Brown and Company, New York, 2009; S. S. Luthar, Resilience in development: A synthesis of research across five decades, pp. 739–795, in D. Cicchetti and D. J. Cohen (eds.), *Developmental Psychopathology*, vol. 3: *Risk, Disorder, and Adaptation*, 2d ed., John Wiley & Sons, Hoboken, NJ, 2006; S. S. Luthar and P. J. Brown, Maximizing resilience through diverse levels of inquiry: Prevailing paradigms, possibilities, and priorities for the future, *Dev. Psychopathol.*, 19:931–955, 2007; S. S. Luthar and L. B. Zelazo, Research on resilience: An integrative review, pp. 510–549, in S. S. Luthar (ed.), *Resilience and Vulnerability: Adaptation in the Context of Childhood Adversities*, Cambridge University Press, Cambridge/New York, 2003; J. W. Reich, A. J. Zautra, and J. S. Hall (eds.), *Handbook of Adult Resilience*, The Guilford Press, New York, 2010.

Dietary fructose and the physiology of body weight regulation

Consumption of dietary fructose has increased in conjunction with the rising intake of fructose-containing sugars, largely in the form of sugar-sweetened beverages. One systematic review of the relationship between sugar-sweetened beverage consumption and risk of weight gain has concluded that increased consumption of sugar-sweetened beverages is associated with weight gain, whereas another study has reported no such relationship. A recent meta-analysis of six sugar-sweetened beverage intervention studies showed dose-dependent increases in weight. In contrast, another meta-analysis concluded that long-term daily fructose consumption of up to 95% of normal intake is not associated with an increase in energy intake or body weight in healthy, normal-weight humans or overweight or obese humans. Despite this lack of consensus, there is a mechanistic basis for the hypothesis that consumption of fructose-containing sugars could be contributing to the obesity epidemic. The objectives herein are to review this mechanism, present evidence that does and does not support it, and discuss the reason that such evidence may be confounded. Additionally, the mechanism by which fructose may be affecting the distribution of weight gain will be discussed.

Role of leptin. Both short- and long-term studies have been conducted in which the effects of consuming fructose-sweetened beverages are compared to the effects of consuming isocaloric amounts of glucose-sweetened beverages (the term isocaloric means having equivalent calorie content/daily energy requirement). These studies, as well in vitro studies in isolated adipocytes, led to the hypothesis that fructose consumption could promote weight gain because it does not stimulate insulin secretion or leptin production by adipose tissue. Leptin production is regulated by insulin-mediated glucose metabolism. Ingestion of fructose does not result in meal-related increases of plasma glucose or insulin concentrations; therefore, meals accompanied with fructose-sweetened beverages result in reduced circulating leptin concentrations in comparison with glucose-sweetened beverages. Leptin acts, along with insulin, in the hypothalamus to regulate food intake and energy metabolism through neuropeptide systems, including neuropeptide-Y and melanocortins. Accordingly, leptin-deficient patients exhibit increased hunger and impaired satiety. Additionally, functional magnetic resonance imaging (fMRI) studies have shown that the areas of the brain associated with pleasure and reward are markedly activated when leptin-deficient patients are shown images of food, but this activation decreases to the level of normal subjects following 7 days of leptin administration. Leptin-responsive neurons also project to pathways that modulate signals to other body tissues that are involved in the regulation of energy expenditure and fat oxidation. Thus, leptin is a key regulator of energy homeostasis, and it has been hypothesized that a reduction of leptin production and circulating leptin concentrations during prolonged consumption of diets high in energy from fructose could lead to increased energy intake and/or decreased energy expenditure and weight gain.

This hypothesis was tested in a recent comprehensive comparison of the metabolic effects of glucose and fructose over an 8-week outpatient period, during which subjects consumed their usual diets ad libitum (that is, food is available at all times)

along with three servings per day of fructose- or glucose-sweetened beverages providing 25% of normal energy requirements. The results with regard to caloric intake or body weight gain did not support the hypothesis that decreases of leptin would lead to greater energy intake and weight gain in subjects consuming fructose. There were no significant differences in reported energy intake between the two groups, and the subjects consuming fructose gained 1.4 ± 0.3 kg (3.1 ± 0.7 lb), whereas the subjects consuming glucose gained 1.8 ± 0.5 kg (4 ± 1.1 lb). The similar weight gain occurred despite a decrease of energy expenditure in subjects consuming fructose, which was not observed in subjects consuming glucose.

Fructose malabsorption. These results can be possibly explained by the fact that excessive dietary intake of fructose as a monosaccharide can overwhelm the absorptive capacity of the small intestine, leading to fructose malabsorption and gastrointestinal distress; however, malabsorption was not quantitatively assessed in these subjects. A current study is assessing breath hydrogen concentrations, which are an index of the hydrogen derived from bacterial fermentation of nonabsorbed carbohydrate reaching the colon. These measures are taken every 30 minutes following consumption of the assigned sugar-sweetened beverage, containing 8% of the daily energy requirement. It has been found that hydrogen breath levels after consumption of fructose-sweetened beverages are 5–20 times higher at all time points compared to those after consumption of beverages sweetened with glucose or high-fructose corn syrup (containing 55% fructose). This suggests that fructose malabsorption could indeed be a confounding factor in intervention studies investigating the effects of fructose and glucose on energy intake and body weight gain. Fructose malabsorption would lead to less energy availability, and the gastrointestinal distress would reduce ad libitum consumption of other foods in subjects consuming fructose. Recently, it was reported that 10 subjects consuming an outpatient weight-maintaining diet plus 150 grams (5.3 oz) of glucose per day gained 1.7 ± 0.4 kg (3.75 ± 0.9 lb) [$P < 0.001$] in 4 weeks, whereas 10 subjects consuming similar diets plus 150 grams (5.3 oz) of fructose gained only 0.2 ± 0.6 kg (0.44 ± 1.3 lb) [$P = 0.4$]. It is possible that these results were confounded by fructose malabsorption; however, this issue was not addressed in the report.

Thus, the evidence to support the hypothesis that a reduction of leptin production and circulating leptin concentrations during prolonged consumption of high-fructose diets could lead to increased energy intake and/or decreased energy expenditure and weight gain is equivocal. Moreover, it is difficult to conduct a study that would definitively determine whether consumption of fructose with an ad libitum diet promotes body weight gain compared with consumption of glucose. It would require that subjects be provided and restricted to ad libitum consumption of a high-fructose or high-glucose diet that has been designed to achieve a comparable and controlled macronutrient distribution in all subjects, regardless of quantities consumed. Energy intake, energy expenditure, the sugar content of stools, and gastrointestinal discomfort would need to be monitored. It would also require that the intervention last at least 12 months because a difference in body weight change as small as 0.5 kg (1.1 lb) per year between groups would be a clinically relevant finding. The sample size and costs, as well as compliance and retention issues, involved in conducting such a study would likely prove prohibitive.

In an attempt to resolve the compliance and retention issues, rhesus monkeys were used to test the hypothesis that the consumption of fructose-sweetened beverages for 1 year would promote weight gain compared with glucose-sweetened beverages. Interestingly, the monkeys consuming fructose-sweetened beverages exhibited significantly increased body weight and reduced energy expenditure at 3 and 6 months of intervention, whereas monkeys consuming glucose did not. However, by the 12-month time point, body weight was increased and energy expenditure was decreased similarly in both groups. As in the previously discussed study in human subjects, energy intake was comparable between groups throughout the 12-month study period, and the potential for fructose malabsorption was not assessed.

Regional adipose distribution. Although it did not provide support for the hypothesis that fructose consumption promotes weight gain compared with glucose consumption, the aforementioned clinical intervention study did highlight some important metabolic differences between the effects of fructose and glucose consumption that were not confounded by differences in body weight gain. These differences include the observations that consumption of fructose-sweetened beverages at 25% of energy requirements increased de novo lipogenesis, produced unfavorable changes of lipids, and decreased glucose tolerance and insulin sensitivity in older, overweight or obese men and women, whereas consumption of glucose-sweetened beverages did not. Pertinent to the topic of fructose and the physiology of weight gain, it was also observed that consumption of fructose promoted visceral adipose tissue (VAT) deposition, and consumption of glucose resulted primarily in an increase of subcutaneous adipose tissue (SAT) deposition. The mechanistic explanation for this finding may involve depot-specific regulation of lipoprotein lipase (LPL), the rate-limiting enzyme involved in the uptake of triglyceride from the circulation storage in the adipose depots. LPL associated with SAT has been found to be significantly more sensitive to activation by insulin than LPL associated with VAT. This observation is pertinent because of the markedly different insulin profiles in subjects consuming glucose compared with fructose. Compared to when the complex carbohydrate baseline diet was consumed, consumption of glucose-sweetened beverages resulted in increased post-meal glucose and insulin peaks, whereas post-

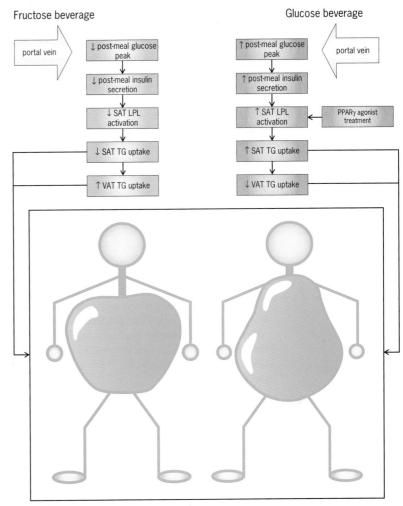

Mechanism for regional adipose distribution. Subcutaneous adipose tissue (SAT) is more sensitive to insulin-induced activation of lipoprotein lipase (LPL) activity than visceral adipose tissue (VAT). In subjects consuming fructose, reduced post-meal insulin exposure may lead to reduced activation of LPL in SAT, less triglyceride (TG) uptake in SAT, and thus increased TG uptake and accumulation in VAT. In subjects consuming glucose, increased post-meal insulin exposure may lead to greater activation of LPL specifically in SAT and thus increased TG uptake and accumulation in SAT. Similarly, SAT is more sensitive to activation of LPL by peroxisome proliferator-activated receptor-γ (PPARγ) agonists than VAT. Hence, treatment with PPARγ agonists leads to increased activation of LPL specifically in SAT and thus increased TG uptake and accumulation in SAT. (*Illustration by James Graham*)

meal glucose and insulin peaks were decreased during consumption of fructose-sweetened beverages. This led to the suggestion that activation of LPL in SAT was greater in subjects consuming glucose, and therefore uptake of triglyceride in SAT was increased. In contrast, activation of LPL in SAT was reduced in subjects consuming fructose, resulting in increased availability of triglyceride for uptake into VAT (see **illustration**).

To date, there are no other known reports suggesting that the macronutrient composition of the diet can influence adipose distribution in humans; therefore, these results require corroboration, and the suggestion as to the underlying mechanism is speculative. However, it is well documented that treatment with peroxisome proliferator-activated receptor-γ (PPARγ) agonists influences adipose distribution in humans (note that PPARs control a variety of target genes involved in lipid homeostasis). When investigating the mechanisms underlying the effect of PPARγ agonists to promote SAT accumulation, it has been observed that LPL in SAT from rats was significantly more activated by treatment with PPARγ agonists than LPL in VAT. Thus, it was concluded that LPL plays a key role in mediating the well-established effect of PPARγ agonists to specifically promote lipid accumulation in SAT. It is possible that the greater sensitivity of activation of LPL in SAT compared with VAT by both insulin and PPARγ agonists is involved in the effects of PPARγ agonists and the differential effects of fructose and glucose consumption on regional adipose distribution (see illustration).

Conclusions. The study comparing the metabolic effects of fructose and glucose in older, overweight or obese human subjects showed that fructose consumption promoted VAT accumulation and other adverse metabolic effects that were not apparent in subjects consuming glucose, despite comparable weight gains. However, this study and the aforementioned 12-month intervention in rhesus monkeys do not provide support for the hypothesis that increased consumption of fructose is a contributing factor in the obesity epidemic. Fructose malabsorption is a potentially confounding factor in the studies investigating this hypothesis.

For background information *see* ADIPOSE TISSUE; CARBOHYDRATE METABOLISM; ENERGY METABOLISM; FOOD; GLUCOSE; HUNGER; INSULIN; LEPTIN; LIPID; LIPID METABOLISM; MALNUTRITION; METABOLISM; NUTRITION; OBESITY; PEROXISOME PROLIFERATOR-ACTIVATED RECEPTOR; SUGAR in the McGraw-Hill Encyclopedia of Science & Technology.

Kimber L. Stanhope

Bibliography. I. S. Farooqi and S. O'Rahilly, Leptin: A pivotal regulator of human energy homeostasis, *Am. J. Clin. Nutr.*, 89(3):980S–984S, 2009; P. J. Havel, Dietary fructose: Implications for dysregulation of energy homeostasis and lipid/carbohydrate metabolism, *Nutr. Rev.*, 63(5):133–157, 2005; M. Laplante et al., Tissue-specific postprandial clearance is the major determinant of PPARγ-induced triglyceride lowering in the rat, *Am. J. Physiol. Regul. Integr. Comp. Physiol.*, 296(1):R57–R66, 2009; G. J. Morton et al., Central nervous system control of food intake and body weight, *Nature*, 443(7109):289–295, 2006; S. S. Rao et al., Ability of the normal human small intestine to absorb fructose: Evaluation by breath testing, *Clin. Gastroenterol. Hepatol.*, 5(8):959–963, 2007; K. L. Stanhope and P. J. Havel, Endocrine and metabolic effects of consuming beverages sweetened with fructose, glucose, sucrose, or high-fructose corn syrup, *Am. J. Clin. Nutr.*, 88(6):1733S–1737S, 2008; K. L. Stanhope et al., Consuming fructose-sweetened, not glucose-sweetened, beverages increases visceral adiposity and lipids and decreases insulin sensitivity in overweight/obese humans, *J. Clin. Invest.*, 119(5):1322–1334, 2009; K. L. Teff et al., Dietary fructose reduces circulating insulin and leptin, attenuates postprandial suppression of ghrelin, and increases triglycerides in women, *J. Clin. Endocrinol. Metab.*, 89(6):2963–2972, 2004.

Discovery of element 117

The discovery of new elements advances our understanding of the limits of the periodic table of the elements. The discovery of the new element with 117 protons described here is a significant advance in probing these limits. Each element is determined by the number of protons, given by symbol Z, inside its nucleus. The first element, with one proton, is hydrogen, and the last element found in nature is uranium, with 92 protons. Beginning in World War II, extensive searches were carried out to create new elements beyond uranium. The next two, with $Z = 93$ and 94, were discovered in the 1940s, and element 94, plutonium, played an important role as the fissile material in the first atomic bomb. By 1952, new elements through fermium ($Z = 100$) were discovered.

Nuclear shell model and magic numbers. The liquid drop model of the nucleus, formulated over 70 years ago, treats the atomic nucleus as a drop of incompressible fluid of nucleons (protons and neutrons). While the theory had many successes, it predicted that nuclei with greater than 100 protons could not exist but would promptly undergo fission. However, a nuclear shell model was developed by M. G. Mayer and J. H. D. Jensen to explain why nuclei with certain numbers of protons (Z) and neutrons (N) had especially stable spherical shapes. These so-called magic numbers correspond to closed (filled) proton and neutron shells with a gap in energy to the next available shell. In addition, nuclei like calcium-48 ($Z = 20$, $N = 28$) and lead-208 ($Z = 82$, $N = 126$), with magic numbers of both protons and neutrons, were especially stable. This is like the closed atomic electron shells that form the inert gases helium, neon, argon, and so on. *See* DOUBLY-MAGIC TIN-132.

This double-magic stability was a new quantum mechanical effect. It was then discovered that the binding energies of elements beyond lead ($Z = 82$) and uranium ($Z = 92$) were clearly dependent on the shell structure of the protons and neutrons and that there are new closed shells for deformed shapes.

Island of stability. These new shell structures give added stability to nuclei beyond $Z = 100$. However, as Z increased from 92 to 106, the half-lives for their radioactive decays decreased to seconds and milliseconds. Beginning around 40 years ago, theoretical calculations predicted the existence of the next spherical double magic closed shells for neutrons and protons. All models predicted that for neutrons it was $N = 184$, but different theories predicted $Z = 114$, 120, or 126 to be the next proton closed shell. Nuclei close to $N = 184$ and $Z = 114$, 120, or 126 were predicted to form an "island of stability" where the half-lives for their decay could be as long as 10–100 thousand years, and in some nuclei perhaps comparable even to the age of the Earth (4.5×10^9 years). **Figure 1** shows a plot of Z versus N for the known elements, with the possible islands of stability with $N = 184$ shown in the upper right corner.

Searches for heavy nuclei. The prediction of this island of stability stimulated major efforts to search for nuclei approaching the island. In Germany, the heaviest stable elements, lead ($Z = 82$) and bismuth ($Z = 83$), were bombarded with projectiles up to zinc ($Z = 30$) at low bombardment energies so that after their fusion there was enough energy left only to evaporate one neutron from the new compound nucleus (for example, $Z = 82 + Z = 30\ Z = 112$). New elements were discovered from $Z = 107$ to 112 in Germany. These elements have progressively shorter half-lives. The half life of $Z = 112$ is 0.7 milliseconds. Elements with $Z = 112$ and 113 (observed in Japan) are the essential limit for this technology because of the very low probability of forming them in fusion. **Figure 2** shows the cross sections (measures of the probability of fusion of the two nuclei), which sharply decrease with increasing Z in these reactions. All these nuclei had 165 neutrons or less, far from $N = 184$.

Then the group of the Flerov Laboratory of Nuclear Reactions at the Joint Institute for Nuclear Research (JINR) in Dubna, Russia, pioneered the bombardment of long-lived (with half-lives of hundreds to thousands of years) radioactive actinide targets of neptunium ($Z = 93$) to californium ($Z = 98$) with double-magic, neutron-rich calcium-48 ($Z = 20$) to discover the new elements with $Z =$ 113, 114, 115, 116, and 118. For example, fusion of plutonium ($Z = 94$) + calcium ($Z = 20$) makes element 114. The probabilities of fusion in these reactions are 100 to 500 times greater than in the reactions used in Germany (Fig. 2). The fact that these elements had much longer half lives and their neutron numbers ($N = 172$–177) were much closer to $N = 184$ provided new support for the importance of the island of stability. These experiments were a collaboration with Lawrence Livermore National Laboratory, which supplied the actinide targets obtained from the High Flux Isotope Reactor (HFIR) at Oak Ridge National Laboratory (ORNL). This reactor is a unique national resource that can provide milligram and higher quantities of radioactive actinide materials with $Z \geq 94$ for research. Since calcium-48 makes up less than 0.20% of natural calcium, the calcium-48 had to be significantly enriched in an electromagnetic separator to provide the necessary beam intensity for the experiments. Fortunately, the primary source of highly enriched calcium-48 is in Russia.

Element 117 was missing from the roster of newly discovered elements because there is no long-lived berkelium ($Z = 97$) isotope that can be made in the 10-20 mg quantities needed for a target. The longest-lived berkelium isotope that can be made in quantities of tens of milligrams is berkelium-249, which has a half-life of only 320 days.

Production of berkelium-249. In 2005, the Russian and Vanderbilt nuclear groups approached scientists at ORNL to form a collaboration to obtain the berkelium. Berkelium-249 can be produced in a high neutron flux reactor, but it is a very expensive process. Fortunately, berkelium-249 is a transitional nucleus

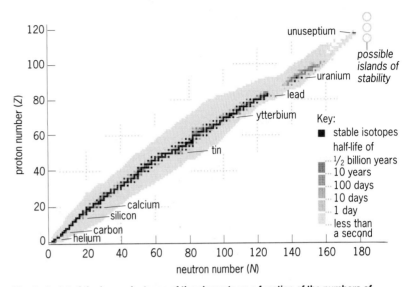

Fig. 1. A plot of the known isotopes of the elements as a function of the numbers of neutrons (N) and protons (Z). Some isotopes are stable while others will decay into more stable forms. The shadings on this chart indicate the relative stabilities of isotopes in terms of their half-lives. Possible islands of stability for $N = 184$ and $Z = 114, 120,$ or 126 are shown in the upper right hand corner. Ununseptium is a temporary name for element 117. (*Joint Institute for Nuclear Research, Nucleonica, Oak Ridge National Laboratory*)

formed in the production chain of californium-252 ($Z = 98$) for industrial uses. After three years, ORNL finally got a commercial order for californium. The Vanderbilt group arranged with ORNL to provide the additional funds to obtain the berkelium created in this process. The berkelium was produced at the HFIR.

To produce berkelium-249, targets of curium ($Z = 96$), americium ($Z = 95$), and plutonium ($Z = 94$) received intense neutron irradiation over 18 months. The samples had to be stored for 3–4 months to allow short-lived, highly radioactive

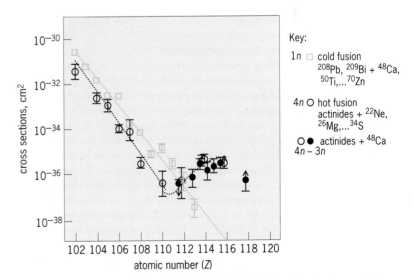

Fig. 2. Cross sections (measure of probabilities) for fusion in cold fusion (low excitation energy after fusion) reactions with stable targets and hot fusion (high excitation energy after fusion) reactions. In cold fusion, there is only enough energy to evaporate 1 neutron from the compound nucleus (1n), whereas in hot fusion, 3 neutrons (3n) or 4 neutrons (4n) are evaporated. The measured cross section for the "hot" fusion to form element 115 is 500 times greater than the extrapolation for a "cold" fusion reaction. In the key, the open and closed circles for actinide targets + ^{48}Ca refer to the open (4n) and closed (3n) circles for elements 112-118.

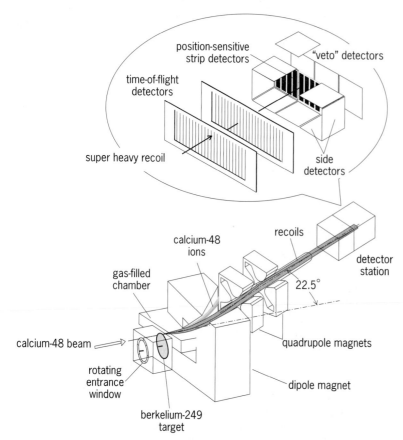

position-sensitive
strip detectors

"veto" detectors

time-of-flight
detectors

super heavy recoil

side
detectors

calcium-48
ions

recoils

gas-filled
chamber

detector
station

22.5°

calcium-48 beam

rotating
entrance
window

quadrupole magnets

dipole magnet

berkelium-249
target

Fig. 3. Schematic of the Dubna gas-filled recoil separator (consisting of a dipole magnet followed by a doublet of quadrupole magnets). Also shown is the schematic of the detector station at the separator focal plane.

isotopes like iodine–131 to decay away. This was followed by three months of continuous chemical separations in hot cells at ORNL to isolate 22 mg of ultrahigh-purity berkelium-249. This material was formed into a target of radioactive berkelium-249 in special facilities at the Research Institute of Atomic Reactors at Dimitrovgrad, Russia.

Discovery experiment. The experiment was carried out with the U400 cyclotron and gas-filled, magnetic separator dedicated to superheavy-element research at the Flerov Laboratory of Nuclear Reactions (JINR), Russia. The setup for the experiment is shown in **Fig. 3**. The cyclotron accelerated the calcium-48 beam to 252 million electronvolts (MeV) to irradiate the sample for three months. To keep the target from overheating, it was rotated at 1700 revolutions per minute (rpm), and the beam was wiggled up and down as the target rotated.

After the fusion of the calcium-48 and berkelium-249, and the evaporation of four neutrons, the $Z = 117$ evaporation residues (ER) that recoiled out of the target were passed through a gas-filled, magnetic separator where they were separated from other residues. Then they were passed through two time-of-flight detectors that identified them as coming from the separator. Next, they were implanted into a semiconductor detector array with 12 vertical position-sensitive strips. As shown in

Fig. 3, the position-sensitive strip detectors were surrounded by eight side detectors to detect subsequently emitted alpha particles that sometimes escaped from the strip detector array.

Following the detection of the residues, a chain of alpha decays transmuted the $Z = 117$ nuclei to 115, then to 113, and finally to 111, which then underwent spontaneous fission. Five such residues of element 117 and subsequent decays were observed. These observations demonstrated that the process had created the isotope of element 117 with 176 neutrons ($117 + 176 = 293$ atomic mass). The average of these five events is shown in **Fig. 4**.

Then the calcium-48 beam energy was decreased to 247 MeV to search for the reaction that creates a different isotope of element 117 with 177 neutrons. This involves the evaporation of three instead of four neutrons. The first experiment was repeated for four months at the new energy. One event was observed at this energy, as shown in Fig. 4. In contrast to the $4n$ channel with 3 alpha decays and fission, element 117 was observed to undergo six subsequent alpha decays before fission. The isotope formed in this $3n$ reaction and the isotope of element 116 that also has 177 neutrons are the most neutron-rich superheavy elements ever observed. They are the closest isotopes to $N = 184$ that have been discovered. Their half-lives are all much longer than those of other nearby isotopes. Thus, the discovery of element 117 with its long half-lives strongly supports the existence of an island of stability around $N = 184$.

Chemical properties of heavy elements. There are predictions that elements beyond $Z = 112$ will have different chemical properties from those expected because the high value of Z increases the Coulomb force on the electrons, and the larger Coulomb force changes the electron orbital structure. Thus, the discovery of the new element 117 and 9 new isotopes (of elements previously discovered) with longer half-lives not only strongly supports the predicted island of stability but also opens the door to studying the chemical behaviors of elements beyond the region where we now have knowledge.

In summary, the discovery of the new element 117 was a major international collaboration of two U.S. and two Russian National Laboratories and two U.S. universities. It required a tour de force that extended over five years and involved several unique facilities: facilities for the production and separation of berkelium-249, facilities for target fabrication, dedicated cyclotron and separator facilities, and powerful data analysis facilities in both Russia and the United States whose work culminated in the observation of six atoms of element 117 and their decay chains. This is the first new element to be discovered in 5 years.

For background information, *see* ALPHA PARTICLES; BERKELIUM; MAGIC NUMBERS; NUCLEAR FISSION; NUCLEAR REACTION; NUCLEAR STRUCTURE; PARTICLE ACCELERATOR; PARTICLE DETECTOR; RADIOACTIVITY; SUPERHEAVY ELEMENTS; TIME-OF-FLIGHT

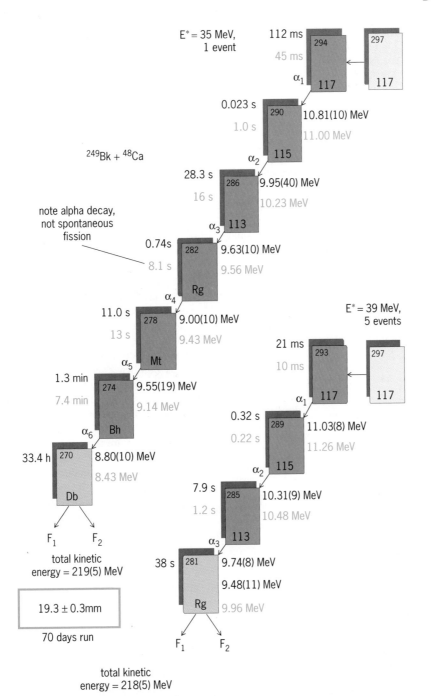

Fig. 4. Observed decay chains originating from the isotopes with mass number $A = 294$ (single event) and $A = 293$ (average of five events) of the new element $Z = 117$. The lifetimes (τ, related to the half-life $T_{1/2}$ by the equation $\tau = T_{1/2}/\ln 2$) and alpha-particle energies are shown, with measured and theoretically predicted values in black and color, respectively. The last isotope in each chain undergoes spontaneous fission into fragments F_1 and F_2. E^* is the internal excitation energy; at 39 MeV, 4 neutrons are more easily evaporated because of the extra internal energy. The 19.3 ± 0.3 mm indicates the positions in the detector strip of all the events from evaporation residues to fission in the long chain. The two different measured alpha-particle energies for the decay of the isotope 113-285 may arise from fine structure in the alpha decay (that is, one alpha decay goes to the ground state and the other goes to an excited state in the same nucleus) or from statistical fluctuations in the measurement of the energies.

SPECTROMETERS; TRANSURANIUM ELEMENTS in the McGraw-Hill Encyclopedia of Science & Technology.

Yuri Ts. Oganessian; Joseph H. Hamilton; Vladimir K. Utyonkov

Bibliography. J. Glanz, Scientists discover heavy new element, *The New York Times*, p. A18, April 7, 2010; Yu. Oganessian, Heaviest nuclei from ^{48}Ca-induced reactions, *J. Phys. G: Nucl. Part. Phys.*, 34:R165–R242, 2007; Yu.Ts. Oganessian, Superheavy elements, *Phys. World*, 17(7):25–29, July 2004; Yu.Ts.Oganessian et al., Eleven new heaviest isotopes of elements $Z = 105$ to $Z = 117$ identified among the products of ^{249}Bk + ^{48}Ca reactions, *Phys. Rev.*, C83:054315 (14 pp.), 2011; Yu.Ts.

Oganessian, et al., Synthesis of a new element with atomic number Z = 117, *Phys. Rev. Lett.*, 104:142502 (4 pp.), 2010; Yu. Ts. Oganessian, V. K. Utyonkov, and K. J. Moody, Voyage to the superheavy island, *Sci. Amer.*, 282(1): 45–49, January 2000; L. Schenkman, Discovery of 'missing' element 117 hints at stable isotopes to come, *Science*, 328:290–291, 2010.

Doubly magic tin-132

The year 2011 marks the centenary of the discovery of the atomic nucleus. In 1911, Ernest Rutherford, with his colleagues Hans Geiger and Ernest Marsden, demonstrated that at the heart of the atom lies the positively charged nucleus. In the intervening hundred years, much has been learned about the structure of the nucleus itself and the variations in the nature of different species of nuclei as the number of nucleons (neutrons or protons) is varied. For more than 50 years, the nuclear shell model has stood as a cornerstone of nuclear structure physics. Just as the inertness of noble gases and the volatility of alkali metals can be explained in terms of electrons moving in a well-defined structure of orbits, nucleons in nuclei occupy quantum shells. The shell model, developed by Maria Goeppert-Mayer, J. Hans D. Jensen, and others describes individual nucleons moving in an average potential generated by all the other nucleons. This model explains the existence of particularly well-bound nuclei with "magic" numbers of protons (Z) or neutrons (N). The standard magic numbers are 2, 8, 20, 28, 50, 82, and 126. However, recent experiments have demonstrated that the structure of shells in exotic light nuclei is markedly different to that for stable nuclear species. For example, the neutron magic numbers 20 and 28 disappear altogether for the most neutron-rich nuclei, and new shell gaps seem to emerge. No such experimental evidence has been observed in the heavier nuclei, and indeed such effects are expected only in isotopes beyond current experimental reach. It is therefore important to test the magicity of neutron-rich nuclei in the region around the nuclide tin-132 (^{132}Sn), whose proton number, $Z = 50$, and neutron number, $N = 82$, are both standard magic numbers, to study the evolution of shell structure away from stability.

Magicity. The nature of magic nuclei is expressed through a number of observables. Two of the clearest indications are the energy of the first collective electric quadrupole (2^+) state in nuclei with even N and even Z, and the amount of energy required to remove nucleons. The lowest-energy (ground) state of a magic nucleus is spherical. To be excited to the first 2^+ state, the nucleus first has to deform. The filled nuclear shells act to resist this deformation. An example of this can be seen in the case of the lead isotopes (**Fig. 1***a*). Having 82 protons, lead is a magic element as seen by the large energy of the first 2^+ energy, around 1 MeV, whereas it is closer to 500 keV for typical nonmagic nuclei. As the neutron number reaches the shell gap at magic number 126 for lead-208 (^{208}Pb), making it a particularly well bound doubly magic nucleus, the energy of the first 2^+ state jumps up by almost a factor of 4. A very similar observation can be made for the tin isotopes.

Nucleon separation energies are analogous to ionization energies in atoms: alkali metals are good electron donors, whereas noble gases show large discontinuities in their ionization energies when compared to their neighboring elements. When looking at nucleon separation energies, it is typical to use S_{2n}, the two-neutron separation energy. The lead and tin isotopes both show a drop in the two-neutron separation energy just past the shell closure (Fig. 1*b*). Simply put, it is much easier to remove two neutrons from ^{134}Sn (neutron number, $N = 84$) and produce ^{132}Sn than it is to remove two neutrons from the neutron doubly magic ^{132}Sn ($N = 82$) itself.

Studying the nature of the single-neutron (or proton) states just beyond a doubly magic nucleus provides a critical test of the shell closure. The spectroscopic factor, which describes the overlap (or similarity) between the initial nucleus with mass number

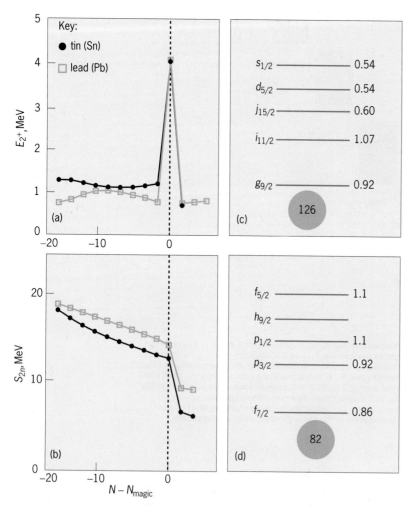

Fig. 1. Shell closures reveal themselves as discontinuities in (*a*) the energies of the first electric quadrupole 2^+ state and (*b*) the two-neutron separation energies, S_{2n}. In both cases, the observables are plotted against the number of neutrons beyond the shell closure, $N - N_{magic}$. The best indicator of magic nature lies in the single-particle states outside the closed shell. The single-particle states (*c*) in ^{209}Pb, above the magic number $N = 126$, and (*d*) in ^{133}Sn, above the magic number $N = 82$, are shown, together with the spectroscopic factors.

A and the final nucleus, *A* + 1, is a measure of the purity of the final state. In midshell nuclei it is common for the single-particle strength for a specific orbital to be fragmented through the spectrum of the nucleus, resulting in low spectroscopic factors. However, for a good magic nucleus *A*, the single-particle strengths in the *A* + 1 nucleus should each be concentrated in one state, resulting in high spectroscopic factors (close to 1). Additionally, characterization of these states outside double shell closures is essential to predicting the properties of thousands of currently unmeasured nuclei. These include those involved in the production of heavy elements in the astrophysical rapid neutron capture process, commonly known as the *r*-process, that is believed to happen in explosive cosmic environments such as supernovae. *See* ORIGIN OF THE HEAVY ELEMENTS.

Transfer reactions in inverse kinematics. Transfer reactions involve the addition (or removal) of one or more nucleons onto (from) a nucleus. One example is the (d,p) reaction where a neutron is transferred from a deuteron (^{2}H, composed of one proton and one neutron) onto a heavier nucleus. Traditionally, transfer reactions have been performed in "normal kinematics" using light beams—such as beams of deuterons for the (d,p) reaction—and solid targets. These methods proved very powerful for studying the structure of nuclei, and were used extensively in the 1960s and 1970s. With the advent of radioactive ion beams, these techniques have experienced a resurgence. As targets cannot be made of short-lived isotopes (meaning having half-lives typically in the range of milliseconds to a few months), the technique needed to be modified to use radioactive ions beams in "inverse kinematics." In the case of (d,p) this means that a beam of short-lived nuclei interacts with a target that contains deuterons (typically deuterated plastic).

The reaction Q-value (the difference in mass energy between the initial and final states) can be reconstructed from the measured angle and energy of protons emitted from a (d,p) reaction. When the mass of the heavy nucleus is known, this Q-value provides the energy of each state. This leads to the determination of the mass of the recoiling nucleus when the ground state is populated. When the reaction is performed at energies in the region above the Coulomb barrier (the energy required to overcome the electrostatic repulsion of the target and beam nuclei), the angular distribution of emitted protons is sensitive to the transferred orbital angular momentum. Additionally, by comparing the intensities of measured angular distributions to those produced in reaction calculations, spectroscopic factors can be extracted.

Neutron-addition transfer reactions, such as the (d,p) reaction, selectively populate single-particle states. When employed at energies just above the Coulomb barrier, low angular momentum transfers are preferred, and states with very negative reaction Q-values will not be populated. In practical terms, only the low-energy, low-spin, single-particle states will result from a (d,p) reaction on an even-even

nucleus (zero spin). This selectivity means that the single-particle nature of the individual states populated can be measured.

Structure of ^{133}Sn. Individual nuclear orbitals are labeled by orbital angular momentum, ℓ (with the letters *s, p, d, f*, and so forth, for $\ell = 0, 1, 2, 3, \ldots$); total angular momentum, *j* (1/2, 3/2, 5/2, and so forth, from coupling the intrinsic spin = 1/2 to ℓ); and number, *n*, denoting that this is the first/second/third, and so forth, state with that particular ℓ and *j*. (The total angular momentum is written as a subscript to the letter that represents the orbital angular momentum, and *n* precedes this letter. Thus, the orbital with $\ell = 3$, $j = 7/2$, and $n = 2$ is labeled $2f_{7/2}$.) The single-neutron states expected for ^{133}Sn, which is at the beginning of the 82–126 neutron shell, are $2f_{7/2}, 3p_{3/2}, 1h_{9/2}, 3p_{1/2}, 2f_{5/2}$, and $1i_{13/2}$. Previous to the study described here, candidates for all but the $3p_{1/2}$, and $1i_{13/2}$ states had been observed in ^{133}Sn through gamma rays following either beta decay or fission.

The aims of the experiment using the ^{132}Sn(d,p) reaction in inverse kinematics were to measure the previous unobserved $3p_{1/2}$ state, to confirm angular momentum assignments for the other *p*- and *f*-wave states, and to extract spectroscopic factors. The ^{132}Sn ions were mass separated from other fission fragments produced in the proton-induced fission of natural uranium, and accelerated to 630 MeV by the tandem accelerator of the Holifield Radioactive Ion Beam Facility at Oak Ridge National Laboratory. The beam reacted with a 160-μg/cm^2 deuterated polyethylene target. A large array of silicon detectors provided both energy and angle information for the reaction protons. The reaction Q-value was measured on an event-by-event basis, as shown in **Fig. 2**. As required by conservation of energy, the Q-value reduces as the excitation energy of the residual nucleus increases. The ground state and two previously observed states at 854 and 2005 keV are apparent as peaks in Q-value at approximately 0.15, −0.71, and −1.86 MeV, respectively. Additionally, a peak at approximately −1.22 MeV in Q-value is clearly

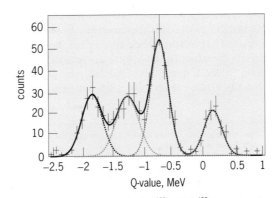

Fig. 2. Q-value spectrum for the ^{132}Sn$(d,p)^{133}$Sn reaction in inverse kinematics. The peaks in Q-value correspond, from right to left, to the ground state in ^{133}Sn, and excited states at 854 keV, 1363-keV, and 2005-keV. The black line is a four-Gaussian fit to the data. Note, as required by conservation of energy, the Q-value reduces as the excitation energy increases.

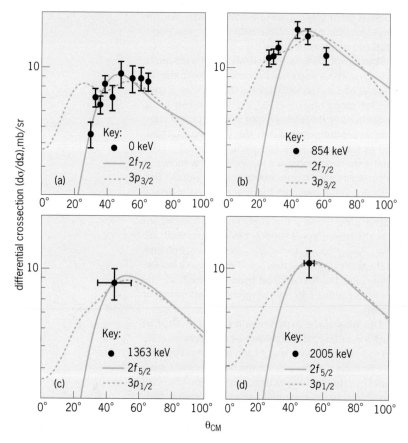

Fig. 3. Angular distributions, expressed as differential cross sections ($d\sigma/d\Omega$), of protons in the center of mass resulting from the ^{132}Sn(d,p)^{133}Sn reaction for the two lowest states populated, and cross-section measurements, also expressed as differential cross sections, for the two highest states. DWBA calculations are shown as solid lines for all $\ell = 3$ transfers and as broken lines for all $\ell = 1$ transfers. (*a*) Ground state. (*b*) 854 keV state. (*c*) 1363-keV state. (*d*) 2005-keV state.

sections were extracted for the 1363 and 2005 keV states owing to the reduced sensitive area of the silicon detectors for the low-energy protons emerging from the population of these states. Again, reaction calculations were scaled to the data and spectroscopic factors extracted. In all cases, the spectroscopic factors are compatible with 1.

The spectroscopic factors for the lowest-energy states in ^{209}Pb were extracted in a consistent manner to the DWBA results shown in Fig. 3 for ^{133}Sn. A comparison is made in panels Figs. 1*c* and 1*d*, where the single-neutron states outside the double-shell closure at ^{208}Pb and ^{132}Sn are shown, with the spectroscopic factors extracted using the method described here. In the ^{209}Pb case, very high spectroscopic factors are found for the lowest two states, but the single-neutron strength of the higher excited states appears to be more fragmented leading to spectroscopic factors close to 0.5. For ^{133}Sn the spectroscopic factors remain compatible with unity for all the states measured. The ^{208}Pb core of ^{209}Pb is well known for its magic nature. The persistence of high spectroscopic factors for the single-particle states in ^{133}Sn reflects the exceptionally robust magic nature of ^{132}Sn, confirming the doubly magic status of this short-lived [half-life ($t_{1/2}$) = 40 s], neutron-rich nucleus.

For background information *see* EXOTIC NUCLEI; MAGIC NUMBERS; NUCLEAR REACTION; NUCLEAR STRUCTURE; RADIOACTIVE BEAMS in the McGraw-Hill Encyclopedia of Science & Technology.

Kate L. Jones

Bibliography. J. H. D. Jensen, Glimpses at the history of nuclear structure theory, pp. 40–50, in Nobel Foundation, *Nobel Lectures in Physics (1963–1970)*, World Scientific, Singapore, 1998; K. L. Jones et al., The magic nature of ^{132}Sn explored through the single-particle states of ^{133}Sn, *Nature*, 465:454–457, 2010; M. G. Mayer, The shell model, pp. 20–37, in Nobel Foundation, *Nobel Lectures in Physics (1963–1970)*, World Scientific, Singapore, 1998; M.G. Mayer and J. H. D. Jensen, *Elementary Theory of Nuclear Shell Structure*, Wiley, New York, 1955.

discernible as a newly discovered state, with an excitation energy of 1363 ± 31 keV, interpolated from the known energies of the other three observed states.

The angular distributions of the ground and 854 keV states are compatible with $\ell = 3$ and $\ell = 1$ angular momentum transfer, respectively, as shown in **Fig. 3**. The angular distributions have been analyzed using both the traditional distorted-wave Born approximation (DWBA) and the more modern adiabatic wave approximation (ADWA). By scaling the calculations to the data, spectroscopic factors were extracted for both theoretical approaches, as summarized in the **table**. The DWBA calculations are shown in Fig. 3. Similar quality fits to the data were achieved with the ADWA calculations. Angle-integrated cross

Spectroscopic factors of the four single-particle states populated in the ^{132}Sn(d,p)^{133}Sn reaction

Energy (E_x)	$n\ell_j$	DWBA*	ADWA*
0	$2f_{7/2}$	0.86 ± 0.07	1.00 ± 0.08
854	$3p_{3/2}$	0.92 ± 0.07	0.92 ± 0.07
1363	($3p_{1/2}$)	1.1 ± 0.2	1.2 ± 0.2
2005	($2f_{5/2}$)	1.1 ± 0.2	1.2 ± 0.3

ADWA, adiabatic wave approximation; DWBA, distorted-wave Born approximation.

E-taxonomy

E-taxonomy (also termed electronic taxonomy and Web taxonomy) is an emerging area within the field of biodiversity informatics relating to the transformation of taxonomy into an Internet-based science (e-science). Taxonomy, here defined as the science and practice of the classification of biological organisms, is well suited to the technological opportunities offered by the Internet and stands to benefit through the resolution of challenges facing both practitioners and users of taxonomy. The term cybertaxonomy is another synonym for e-taxonomy, although it is regarded by some as a broader term incorporating the general use of information technology (IT) in taxonomic research.

In the mid-eighteenth century, the founding father of modern taxonomy, Carolus Linnaeus, sought to catalogue, describe, and order the natural world through the innovation of binomial nomenclature. By coining dual Latin names (for example, *Homo sapiens*), Linnaeus created a simple and efficient mechanism for communicating biological information that imparts both the uniqueness of species through the species name (*sapiens*) and the broader group to which they belong in the genus name (*Homo*). Two Linnaean works, *Species Plantarum* (1753) and *Systema Naturae* (1758), represent the official starting date for modern nomenclature. The binomial system has been a cornerstone of taxonomy ever since, and it is the primary means by which species are named and by which information about those species is retrieved.

Since the time of Linnaeus, almost 1.8 million organisms have been named and described. However, taxonomists face substantial challenges in their research. The often chaotic proliferation of names since the 1750s has resulted in many organisms being described more than once, creating numerous synonyms. International codes of nomenclature have thus been established to mitigate this problem and regulate the use of available names.

Taxonomists today are obliged to consider all names from the Linnaean (eighteenth century) starting point in their research (in contrast to most other fields of science, which rarely need to consult older literature). However, names are often published in obscure books and journals that are hard to access, creating an obstacle to research. Even now, many outlets for taxonomic research are not readily available outside major institutions. Moreover, complete listings of names and places of publication are available for only some groups of organisms and, as a result, published names are readily overlooked.

Nomenclatural ambiguity and the fragmented nature of taxonomic information are also problematic for the users of taxonomy, who may not be experts in taxonomic methods themselves. Moreover, taxonomists frequently differ in opinion regarding the delimitation of species (or higher taxonomic groups). Although competing hypotheses are to be expected in science, users of taxonomic information are not always qualified to judge between them.

Today, it has become widely accepted that the Internet offers an ideal environment within which to address many of the challenges facing taxonomy. Increasingly, taxonomic institutions and international consortia, such as the European Distributed Institute of Taxonomy (EDIT), are investing in Web-based solutions to the challenges of taxonomy. A wide diversity of taxonomic resources are being digitized, integrated, and served online. Importantly, increasing effort is being invested in the delivery of e-taxonomy in the strict sense in which taxonomic research is both conducted and disseminated online. Currently, a range of initiatives have been implemented.

E-taxonomic initiatives. The most fundamental types of taxonomic information, specifically names, nomenclatural literature, and specimen data, have been prioritized for online delivery. One of the most notable online initiatives is the International Plant Names Index (IPNI), an effectively complete index to all seed plant names, their authors, and their places of first publication. Checklists that distinguish the accepted name for each species from the various synonyms are a logical development of such names indices. For example, the World Checklist of Monocotyledons builds upon the IPNI to provide a synonymized names backbone for this group of 70,000 flowering plant species. The Catalogue of Life annually amalgamates such checklists for all organisms into a single checklist; the 2011 edition contains some 1.3 million species. Of course, specimens are the raw material for taxonomic research. As such, numerous museums, herbaria, and natural history institutions are actively building online databases for specimen data, often including images of the specimens themselves. The Global Biodiversity Information Facility (GBIF) provides a mechanism to search across such data sets shared by participating organizations and is the foremost access point for precise species occurrence information.

Tools that support e-taxonomy are numerous and wide ranging. Some tools support applications development by facilitating data exchange, such as globally unique identifiers (GUIDs) and data standards, including those administered by the Taxonomic Databases Working Group. Others assist the taxonomist in practical ways, such as software for the construction of interactive identification keys that can be deployed online. The Biodiversity Heritage Library (BHL) is a rapidly expanding online resource that provides free, online access to legacy literature for taxonomists and other users of biodiversity information.

The advent of several taxonomic Web journals, such as *Zootaxa* and *Phytotaxa*, represents an important step in the advancement of e-taxonomy. Because these journals are principally published online, they offer great flexibility to authors in the content that can be published, and the costs of publication are reduced. They also allow publishing of papers of any length and incorporate copious illustrations and potentially other media. Some Web journals (for example, *ZooKeys* and *PhytoKeys*) offer enhancements through XML (Extensible Markup Language) tagging that permit linkages to other online resources such as BHL, GBIF, or the Encyclopedia of Life (EoL). At present, the codes of nomenclature demand that all new species must be published in hard-copy publications. To satisfy this requirement, taxonomic Web journals ensure that sufficient copies of each publication are lodged in appropriate libraries. The amendment of the codes to permit online publication is a subject of current debate.

Although innovative, taxonomic Web journals in effect remain true to the traditional model of paper publication. However, new kinds of online resources are now revolutionizing the way biodiversity information is published and displayed, while also broadening user and contributor participation. Typically, these Web sites are structured around taxon pages, reached by searching or browsing for a particular species, genus, or higher group, which can deliver an array of rich content. The content may be generated in a variety of ways. One model (implemented most famously in Wikipedia, which contains many taxon pages, but also in Wikispecies) permits direct content contribution and editing by anybody regardless of expertise. EoL exploits an alternative model in which data from a wide variety of suppliers is aggregated into taxon pages, sometimes in combination with expert content contributed specifically for EoL. Aggregated content comes from a variety of sources, and the authoritativeness can vary. Content that has not been reviewed by experts is flagged.

E-taxonomy in the strict sense. Although the aforementioned initiatives are revolutionizing access to taxonomic information resources, they do not tackle the core of the e-taxonomic vision, that is, to convert the practice and dissemination of taxonomy into a Web-based science (e-science). This approach implies that revisionary taxonomy, comprising the description, naming, and illustration of taxa, will take place in an online environment, bringing with it all the opportunities of flexibility, community building, and linkages to other data sources. With access to a shared Web site, the expert community for a particular group of organisms can work collectively on revisionary taxonomy as a "team sport." In this way, a consensus or "unitary" taxonomy might be reached, in which an agreed community view of the taxonomy of the groups is displayed online, with alternative hypotheses visible only to those users who care to dig deeper. The ultimate goal is to provide a "one stop shop" for all biodiversity information for all groups of organisms.

Progress towards this strict view of e-taxonomy has been achieved in a number of initiatives, including the Creating a Taxonomic E-Science (CATE) project and the EDIT Cyberplatform for Taxonomy. These projects have produced a number of proof-of-concept taxonomic portals (for example, CATE Sphingidae, CATE Araceae, and Palmweb), providing content such as full nomenclature, descriptions, maps, images, glossaries, and keys for groups comprising several thousand species. The CATE sites also provide advanced editing tools that facilitate direct engagement by expert communities. The data included in these sites are structured around sophisticated models designed for taxonomic data. A less formal approach is taken in so-called "Scratchpads" and other social networking applications that employ a modular, open-access content management system enabling communities of researchers to share, manage, and publish taxonomic data online.

Benefits of e-taxonomy. The benefits of e-taxonomy promise to be real and long lasting. When fully implemented, e-taxonomy will increase the ease and speed of taxonomy because of the availability of new tools and efficiencies in publishing online. It will strengthen the global community of taxonomists and draw in the participation of new types of contributors with varying degrees of expertise. Sophisticated approaches to peer review will ensure that quality of content is maintained.

Data linkages will be exploited so that taxonomists can benefit from existing resources in their work and so that their output can be repurposed for other users.

From the user perspective, taxonomic outputs will become defragmented and more widely visible, dramatically increasing access to anyone with an Internet connection. Connections to other areas of science, such as genomics, will be more easily made as well. Overall, e-taxonomy will benefit the profile of taxonomy itself, which is increasingly threatened as a result of changing focuses of universities, institutions, and funding bodies, even though taxonomy is more relevant than ever because of the heightening global biodiversity crisis.

Outstanding issues. E-taxonomy is a rapidly evolving field, and many issues remain to be resolved. First, the requirement that new taxonomic names are published in hard copy persists in the codes of nomenclature. Amendments to the codes are currently under consideration that would permit online publication when certain conditions are satisfied (for example, the place of electronic publication is permanent, or the name is registered in a centralized place). Second, the use of published material in e-taxonomy is complicated by issues of copyright and licensing that remain controversial. Third, the credit that researchers receive for their publication is generally determined by the perceived impact of the journals they publish in, which can have a direct affect on career progression and rewards. Web sites and Web pages do not have a formal "impact factor" as many journals do, although there are other means of tracking the influence of Web pages (for example, by counting page views). Sophisticated approaches are required to ensure that researchers are appropriately credited for online publication. Finally, aspects of e-taxonomy have been criticized by some researchers. For example, "unitary" taxonomy (that is, a single taxonomy that is subscribed to by everyone) is regarded as authoritarian by some authors, transforming taxonomy from a hypothesis-driven science to a service industry, provoking a lively debate in the literature. Despite these issues, however, the current rate of development of the field and the growing global dependence on the Internet will guarantee a central role for e-taxonomy in the future.

For background information *see* BIODIVERSITY; CLASSIFICATION, BIOLOGICAL; DATABASE MANAGEMENT SYSTEM; DEOXYRIBONUCLEIC ACID (DNA); INTERNET; NUMERICAL TAXONOMY; SCIENTIFIC METHODS; TAXONOMIC CATEGORIES; TAXONOMY; TYPE METHOD; WORLD WIDE WEB in the McGraw-Hill Encyclopedia of Science & Technology.   William J. Baker

Bibliography. H. C. J. Godfray, Challenges for taxonomy: The discipline will have to reinvent itself if it is to survive and flourish, *Nature*, 417:17–19, 2002; H. C. J. Godfray and S. Knapp, Taxonomy for the twenty-first century: Introduction, *Philos. Trans. R. Soc. Lond. Ser. B Biol. Sci.*, 359:559–569, 2004; H. C. J. Godfray et al., The Web and the structure of taxonomy, *Syst. Biol.*, 56:943–955, 2007; S. J. Mayo et al., Alpha e-taxonomy: Responses from systematics community to the biodiversity crisis, *Kew Bull.*, 63:1-16, 2008; Q. D. Wheeler (ed.), *The New Taxonomy* (Systematics Association Special Volume 76), CRC Press, Boca Raton, FL, 2008.

Electric airplane

Growing interest in the use of renewable energy sources has led to rapid development of fuel cell technologies based on hydrogen due to its abundance and its non-polluting combustion products. Hydrogen is expected to be a fuel source in the medium term and for use in commercial aircraft applications. There are several potential advantages in using such a power source, ranging from environmental and economic to performance and operability.

Early examples derived from hydrogen-based fuel cell designs used in automotive systems are expected to facilitate the early introduction of proton exchange membrane (PEM) fuel cell systems in aeronautic applications in the near-term. Ford's global fleet of 30 Focus hydrogen fuel cell cars has accumulated more than 1.9 million kilometers (1.2 million miles) in real world testing. Daimler's fleet of 60 A-Class hydrogen fuel cell cars operated in a tour around the world has accumulated more than 3.2 million kilometers (2 million miles). Tens of fuel cell buses have been in service in many municipalities for several years.

Fuel cells could become the main power source for small general aviation aircraft or could replace several internal subsystems on transport aircraft [emergency power, ram air turbines (RAT), cabin power, auxiliary power units (APU), anti-icing systems, landing gear retraction, and so forth], to obtain all-electric or more-electric air vehicles.

The replacement of combustion engines or APU with fuel cell-powered electric motors can guarantee a massive reduction in air pollution, since the only emission produced by a fuel cell is water (although it is open to discussion if water vapor emitted at high altitude can produce a long-period greenhouse effect). Air pollution is not the only kind of pollution that can be reduced (if not eliminated) through fuel-cell-driven power systems; noise is indeed considered an important form of pollution produced by aircraft, and it is particularly important for airplanes taking off from airports located in urban areas or during night time when noise abatement regulations are even more stringent.

Nevertheless, the available fuel cell technologies need substantial improvement to meet safe operational requirements in terms of efficiency, reliability, performance, mass/volume, cost, and lifetime under flight conditions at altitude and under high and low ambient temperatures in the air and on the ground.

Hydrogen has the advantage of being obtainable from a variety of sources, so that it is less prone to market fluctuations. However, other forms of energy must be used to produce the hydrogen used as fuel. Widespread production, distribution, and use of hydrogen will require many innovations and investments in efficient and environmentally-acceptable production systems (such as wind energy, solar power), transportation systems, storage systems and usage devices.

Fuel cell system specific energy, defined as the energy output per unit weight, currently is the greatest issue affecting aerospace applications. It is about 500 Wh/kg, but it is estimated that specific energy would increase to 10 kWh/kg in 10-15 years, and to 20 kWh/kg in 20-30 years. This performance will enable an all-electric flight of a large commercial aircraft. However, even though hydrogen contains an amount of energy per unit mass three times higher than kerosene, the significantly lower density of hydrogen requires the adoption of pressurized or cryogenic fuel tanks; either solution means extra weight and volume.

As mentioned above, hydrogen can provide some advantages in performance too. For example, an all-electric or more-electric aircraft should exhibit higher reliability due to replacement of mechanical systems with electrical ones. Higher reliability implies also lower maintenance costs. Moreover electric propulsion is much less sensitive to altitude than combustion engines and thus greater altitudes can be reached with lesser power demands.

Several general aviation and motor-glider electric vehicles have been produced in recent years to attain a more environmentally friendly vehicle, including Antares 20E and 23E, Arcus E, Cri-Cri Cristaline, Electraflyer, Fascination E, Lambada HKS, PC-Aero Elektra One, Silent, Taurus Electro, and Yuneec E430. In all these vehicles, the power was supplied by a battery system, which introduced a strong limitation on flight endurance because of their high weight. Battery specific energy, in fact, is actually limited to 150-200 Wh/kg and is expected to increase in the middle term to a maximum value of 300 Wh/kg, one order of magnitude lower than fuel cell specific energies expected at that time.

Several examples are also available in which power is supplied by photovoltaic cells bonded on the wings. Among others, are the Solar Impulse aircraft (aiming at a world tour powered by solar cells but also supported by lithium batteries), the Solar Challenger, the Sunseeker motor-glider, the Pathfinder and Helios UAV, and the Heliplat project with the SESA UAV flight.

Different studies have been undertaken in recent years about the use of fuel cells in transport aeronautics. Boeing and Airbus are involved in feasibility studies of innovative APU for large transport aircraft in order to lower noise and emissions. Boeing Research & Technology Centre (Madrid) successfully flew the first fuel-cell-powered motor-glider in April 2008. The Boeing prototype flew for about 20 min at approximately 120 km/h (75 mph). DLR also flew in July 2009 with the motor-glider Antares-H2 powered by fuel cells.

In this framework, the European Commission has selected *ENFICA-FC* (ENvironmentally Friendly Inter City Aircraft powered by Fuel Cells) as one of the co-founded projects in the Aeronautics and Space priority of the 6th Framework Programme. The ENFICA-FC consortium, coordinated by G. Romeo, consists of nine partners. The main objective was to develop a fuel cell-based power system for more/all electric aircraft belonging to different categories and to validate the feasibility of a fuel-cell-propelled aircraft by converting an existing combustion engine two-seater airplane.

One study was carried out to provide a preliminary design for intercity aircraft power systems that could be supplied by fuel cell technologies; the safety, certification, and maintenance concepts were also defined. Parametric sizing of different aircraft categories (from two-seater aircraft to 32-passenger commuters) led to a better understanding of the practical meaning of the transition from kerosene to hydrogen in transport airplanes.

In another study, a two-seater electric motor-driven airplane, powered by fuel cell technology, was assembled and tested. The fuel cell, the hydrogen storage, and electric and electronic system were installed in a light sport aircraft RAPID 200, which was flight and performance tested as proof of the functionality and future applicability of this system for intercity aircraft.

Design of a conversion aircraft. After the selection of the proper aircraft for conversion, a complete but limited mission profile definition was selected to show the feasibility of a new-concept propulsion system. The requirements were chosen to guarantee that the mission could be flown while keeping the total weight at around 5500 N (1236 lb), that is, the maximum total weight at which original RAPID200 was tested.

An extensive computational fluid dynamics (CFD) analysis was performed to define power requested for flight, concerning not only the overall aircraft, but also critical components such as the engine cowl, which must guarantee proper air cooling of the various systems installed in the engine bay and provide a passive safety system for prevention of hydrogen accumulation. Predicted requested powers were found to be about 38 kW for the climbing phase and 18–20 kW for the cruising phase.

A particular architecture was adopted for the power system in order to achieve a safe and flyable aircraft for the prescribed mission: A battery system was added as a secondary power source to increase the rate of climb during the most power-demanding phases [takeoff and climbing to 500 m (1650 ft) altitude]. While fuel cells always provide up to maximum power output (20 kW) for normal flight (cruise and descending), the battery was designed to supply 20 kW for 18 min. During roll-out tests, a fuel-cell-only takeoff was also simulated to demonstrate the capability of a fuel-cell-only complete mission.

Having two completely separate power sources has a strong impact on flight safety, which is the main driver for all decisions taken during design. The battery is designed so that it can work as an emergency power source in case of failure of the fuel cell, allowing the pilot to land safely.

Fig. 1. Engine bay installments of the fuel cells.

The introduction of the second power source requires a more complex electronic control system. It is necessary indeed that the fuel cell always be selected automatically as the main power supplier in order to minimize battery use, which is activated only when requested power exceeds fuel cell maximum output. At the same time, the controller needs to be able to instantly draw power from the battery to replace the fuel cell in case of malfunction.

The conventional power system (the internal combustion engine, ICE) is very different from the fuel-cell-powered airplane both with regard to the number of items aboard and the volumes of those items, and aircraft balance must be maintained, keeping in mind safety constraints that are particularly important when operating with high-pressure hydrogen.

Some structural components were carefully redesigned, such as the engine mount as support for many different subsystems (**Fig. 1**) and a special lightweight support plate for hydrogen tanks (**Fig. 2**).

A new propeller was also designed, manufactured, and tested, since weight and available power of the converted aircraft strongly differ from the conventional one.

System testing. Several experimental test activities have been performed at different levels of increasing integration.

Individual subsystem testing. The fuel cell was carefully tested for endurance at its maximum power output. The system was continuously tested for more than six hours with no degradation of performance during

Fig. 2. Installment of the high-pressure hydrogen tank.

the experiment. Several long tests were performed to prove reliability of the fuel cell system.

Tests of the battery system were concerned mainly with the safety of the system during charge and discharge. Particular attention was paid to temperatures and minimum voltage during discharge. In fact, for safety reasons, the battery system was not provided with an automatic cut-off, so that the pilot was able to draw all the energy accumulated, eventually damaging the battery, in order to safely land during an emergency.

The motor, power electronics, and vehicle controller were simultaneously tested, mainly with regard to the temperatures that can be reached during a full duty cycle, and especially at the very beginning of the mission when airplane speed, and so fresh air mass flow, is very low. Temperature behaviors of critical components during the tests showed that the maximum temperature reached in the inverter was 78°C or 172°F (the maximum allowable is 120°C or 248°F), and in the motor it was 80°C or 176°F (the maximum allowable is 180°C or 356°F). Moreover, the vehicle controller was tested against its ability to immediately switch from the main to the secondary power source, and back, without any interruption or unexpected change of motor operation.

The hydrogen storage system [350 bar (approximately 5000 psi) is the working pressure for this application] was tested against maximum working pressure (438 bar or 6350 psi) and burst pressure (984 bar or 14,000 psi).

Semi-integrated system testing. The entire fuel cell system was completely installed in the final configuration on a fuselage mock-up, and the motor/power electronic block was linked to a bench brake. The main goal of this testing stage was to investigate and tune the communication between systems, above all the vehicle controller and fuel cell. Moreover, the fuel cell system is extremely complex and favorable strategies needed to be defined to pilot it during normal and abnormal operations that may occur during flight. Extensive testing hence was devoted to software related issues and tuning.

Integrated system testing. The final and most extensive campaign test was the one devoted to the complete aircraft (**Fig. 3**). The ground and flight tests were performed at Reggio Emilia airport (in northern Italy) with the goal to validate the design and installation of the complete converted aircraft.

This stage investigated mainly the behavior of output power when connected to the real load (the propeller), behavior of the propeller, handling of partial system failures, temperatures with the real cooling system, and finally aircraft performance in takeoff and cruise. Great attention was paid to correct handling of the two onboard power sources, testing several scenarios in which different failures were simulated.

Having the complete system installed on board allowed checking of the real efficiency of the cooling systems. In order to investigate this, temperatures were observed during high-speed roll-outs performed to test theoretical data about take-off dis-

Fig. 3. Rapid200FC during flight test. (*Copyright © Giulio Romeo, POLITO***)**

tances and speeds. The behavior of the cooling systems was very satisfactory, keeping temperatures below admissible limits.

After this extensive test campaign, the aircraft finally flew at Reggio Emilia airport. Six flights were performed, starting from an initial two-minute maiden flight and ending with world speed record attempts for electric aircraft powered by fuel cells according to the draft FAI (Fédération Aéronautique Internationale, The World Air Sports Federation) sporting code Class C, Aeroplane.

The flight path was chosen so that the aircraft was always able to land at the airport or at a close airfield, gliding with no available power. The main results obtained during flights were:

1. A maximum endurance of about 39 min was achieved, the limiting factor being the water consumption used for the stack humidification; the hydrogen measured pressure was 100 bar (1450 psi) at the end of that particular flight, in which a drop of 5.9 bar (86 psi) was recorded for each minute of flight.

2. A maximum average speed (according to the FAI definition) of 135 km/h (84 mph) was recorded during the best world record attempt, the speed measured during two continuous 3km (1.8 mi) long runs and with an altitude variation of less than 100 M (330 ft) between the start and the finish points.

3. A maximum speed of 158 km/h (98 mph) was reached during level flight, with a top speed of 180 km/h (112 mph), which was measured during several diving and pull-up maneuvre tests. The total Global Positioning System (GPS) horizontal path length for a flight (taxi + roll out + climb + horizontal flight + landing) was 76.5 km (47.5 mi) [**Fig. 4**].

Conclusions. The extensive experimental campaign carried out, as well as the theoretical estimations, have proven that fuel cell technologies represent a promising innovation in aeronautics as a key enabling technology for all-electric, zero-emission, low-noise aircraft.

The positive handling qualities and satisfactory engine performance led the team to consider these successful flights as a good starting point for further long-endurance, high-speed flights. At the moment,

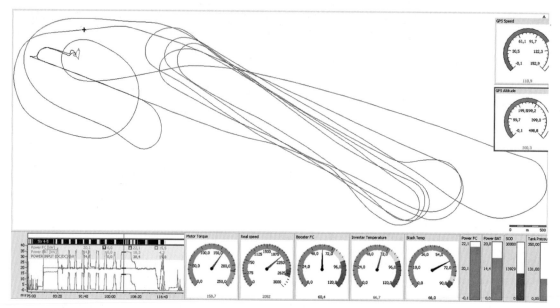

Fig. 4. Flight data and path for the fifh mission of the Rapid200FC flight tests.

for general aviation aircraft, fuel cells and the related technologies seem to need improvement from the gravimetric efficiency point of view. The current aircraft does not achieve the same performance as the original aircraft; a midrange technology development would be sufficient to obtain acceptable performance.

The real strength of the "all-electric aircraft" concept does not lie in an improvement in performance, but rather in the environmentally friendly use of the aircraft itself. Such an aircraft could be used in airports surrounded by urban centers, during the night, and in environments that are restricted because of excessive pollution risk.

The results obtained during these projects can be considered a further step towards introducing zero-emission flight.

The author acknowledges the important contribution made by the European Commission funding the ENFICA-FC, EC 6th FP programme—Contract No. AST5-CT-2006-030779 (2006–2010). The author also acknowledges the contribution of the Consortium partners.

For background information *see* AIRCRAFT DESIGN; AIRPLANE; BATTERY; FLIGHT CHARACTERISTICS; FUEL CELL; GENERAL AVIATION; HYDROGEN; PROPELLER in the McGraw-Hill Encyclopedia of Science & Technology.   Giulio Romeo

Bibliography.   S. Braid and A. Orsillo, Technical Demonstrator Report, Technical Report D7.3-2 ENFICA-FC, 2008; B. Glover, Fuel cells opportunity, *Proceedings of AIAA/AAAF Aircraft Noise and Emissions Reduction Symposium*, Monterey, CA, May 2005; V. Hiebel, Airbus fuel cell systems for aeronautic applications. a clean way from kerosene to energy, *Proceedings of 25th ICAS Congress*, Hamburg, Germany, Sept. 2006; P. Pistecky et al., Analysis, installation and mission performance of the fuel cell all-electric intercity transport aircraft, Technical Report D4.4 ENFICA-FC, 2009; G. Romeo et al., Engineering method for air-cooling design of two-seat propeller driven aircraft powered by fuel cells, *J. Aero. Eng.*, 24:79–88, 2011; G. Romeo and F. Borello, Design and realisation of a two-seater aircraft powered by fuel cell electric propulsion, *Aeronaut. J.*, 114(1155):281–297, May 2010; G. Romeo, F. Borello, and G. Correa, ENFICA-FC: Design, realization and flight test of all electric 2-seat aircraft powered by fuel cells, *27th Int. Conf. of the Aeronautical Sciences (ICAS 2010)*, Nice, France, Sept. 19–24, 2010; G. Romeo, F. Borello, and G. Correa, Set-up and test flights of an all-electric aeroplane powered by fuel cells, *AIAA J. Aircraft*, 49(4):1331–1341, 2011; G. Romeo, I. Moraglio, and C. Novarese, ENFICA-FC: Preliminary survey & design of 2-seat aircraft powered by fuel cells electric propulsion, *7th AIAA Aviation Technology, Integration & Operations Conference (ATIO)*, September 2007, Belfast, Northern Ireland, AIAA-2007-7754; G. Romeo, M. Pacino, and F. Borello, Solar cells and fuel cell power system for long endurance autonomous flight of solar powered UAS, *Proc. 4th Int. Conf., "UAV World,"* Frankfurt, Germany, November 2010; Z. Shavit, Fuel cell propulsion for all-electric and more-electric aircraft and the ENFICA-FC project, *Proceedings of the 48th Israel Annual Conference on Aerospace Sciences, (IACAS 48)*, Haifa, Israel, Feb. 2008, IACAS, Tel-Aviv, 2008.

Enceladus

Until the *Cassini-Huygens* mission began exploring the Saturn system in 2004, the diminutive moon Enceladus was thought to be just another small icy moon of Saturn, perhaps a source of particles for the tenuous E ring discovered in 1966. Since *Cassini's* first close flyby of Enceladus in 2005, it has become clear that the textbooks need to be rewritten. Rather than having a bit part in Saturn's cosmic drama, with the main role played by the distant and fascinating moon Titan, Enceladus has stolen the show in the region near to Saturn and co-stars in the overall system. Orbiting at 3.95 Saturn radii (a distance of 238,000 km or 148,000 mi), and with a diameter of 500 km (310 mi), Enceladus produces huge plumes of ice grains and neutral water molecules from its south polar region. The plumes emanate from relatively warm surface fissures known as "tiger stripes," and the plumes' existence and composition, as well as the heat associated with the stripes, hint strongly of a subsurface ocean there. The water ionizes and interacts with Saturn's rapidly rotating magnetosphere, and escaping ice grains feed Saturn's E ring. The exact process causing the plumes' emission, and its longevity, are not yet understood. What is clear is that Enceladus is not only unexpectedly active geologically at present, it also joins other outer-planet moons, including Europa and Titan, in potentially harboring the ingredients for life.

Exploration history. William Herschel discovered Enceladus in 1789. Almost 200 years later, the *Voyager 1* and *2* spacecraft flew through the Saturn system and took close-up images of Saturn's moons. The observations left many questions, particularly about Titan, but also about the icy moons including Enceladus.

Telescopic observations had shown that Titan's atmosphere, with surface pressure 1.5 times that of Earth's, was the only significant atmosphere of a solar-system moon, and contained methane and nitrogen. The relatively large Titan (5150 km, or 3200 mi, in diameter), orbiting at 20 Saturn radii, was hidden by orange smog in the visible imagery, and Titan's atmosphere was thought to be like that of the early Earth. It was thought that this dense atmosphere would provide a significant source of ions in the outer magnetosphere, ionized from an expected neutral torus a little like that of Io. Titan was a fascinating target for *Cassini-Huygens*, and many discoveries from the close flybys of Titan since 2005 have revealed the complexity of the chemistry at work all the way from the ionosphere, sampled by the *Cassini* orbiter, to the surface, visited by the *Huygens* probe. Imaging and radar measurements of the surface and in-situ measurements have shown a methane-driven world with processes analogous to those of Earth. Titan was also well placed to be used in planning *Cassini's* tour of the Saturn system, as the flybys could be used to tweak the orbital "petals"; there had been more than 75 Titan flybys as of May 2011. One of the additional prime targets in the tour was Enceladus.

Voyager's imagery of Enceladus, unimpeded by smog, had revealed a relatively young, icy, high-albedo surface with little evidence of the early bombardment characteristic of the solar system since its birth 4.6×10^9 years ago. This was something of a surprise, given the highly and spectacularly cratered surface of nearby Mimas, orbiting at 3.1 Saturn radii and only 20% smaller than Enceladus. Based on expected tidal forces (related to its greater distance from Saturn), Enceladus was expected to be less geologically active than Mimas. Some activity was inferred because Enceladus was found (by telescopic observations in the early 1980s) to be located at the heart of Saturn's E ring, but any activity was thought to be only occasional and perhaps at several surface locations, enough to feed particles to the E ring. Also, an extensive neutral OH cloud centered on Enceladus, discovered by the *Hubble Space Telescope* in the early 1990s, seemed to be linked with a source there. Before *Cassini's* arrival, then, Enceladus was thought to be an occasionally active icy moon with a likely role in augmenting the E ring and neutral clouds.

Cassini observations. The *Cassini* mission has revolutionized our view of Enceladus, initially based on three close flybys during the prime mission. Before Saturn orbit insertion in 2004, *Cassini's* ultraviolet instrument detected oxygen emission, adding to the neutral OH. The early flybys in February and March 2005 revealed spectacular tectonic activity in the equatorial region, as well as older, cratered terrain near the north pole. A key finding was a significant

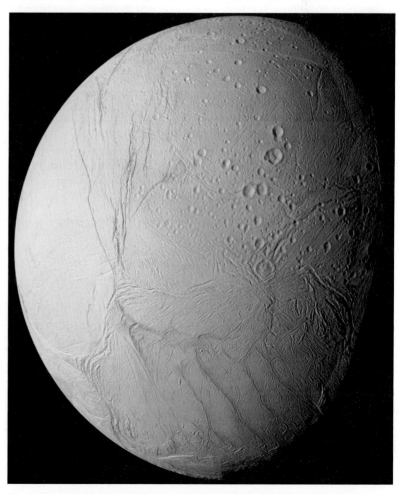

Fig. 1. Mosaic of 21 narrow-angle camera images (taken using visible, infrared, and ultraviolet filters) of Enceladus on July 14, 2005. The Southern Hemisphere region is smoother and younger, in contrast to the older, cratered Northern terrain. Signs of tectonic activity are present, and four pronounced "tiger stripes" are seen in the south polar region. (*Courtesy of NASA/JPL/Space Science Institute*)

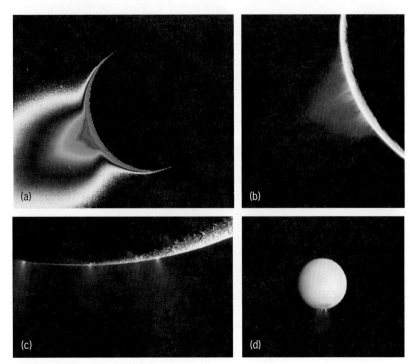

Fig. 2. Different views of the Enceladus plumes. (*a*) Image of back-lit (by the Sun) Enceladus, revealing the large scale of the plumes. (*b*) Imagery revealing eight different sources associated with the tiger-stripe features. (*c*) At least 30 differently sized jets from the tiger stripes, spraying plumes of water vapor, organics, and ice grains away from Enceladus's south polar region. (*d*) The Sun is behind Enceladus and illuminates the plumes, revealing their large-scale structure; Enceladus itself is lit by reflected light from Saturn. (*Courtesy of NASA/JPL/Space Science Institute*)

deflection of Saturn's magnetic field near to Enceladus. This discovery meant that Enceladus must be a significant source of ionized material, possibly an atmosphere—though this was not present at low latitudes according to ultraviolet stellar occultation measurements. The mission profile was revised to provide closer flybys of Enceladus, including changing

Fig. 3. Back-lit view of Saturn's E ring from September 15, 2006, showing wisps of bright, icy material from Enceladus joining the E ring. The structure of the wisps contains information about the direction of grain emission, the sweeping aspect of Enceladus itself, and the interaction of the ice grains with Saturn's magnetosphere. Dione is also visible in the image as a bright spot. (*Courtesy of NASA/JPL/Space Science Institute*)

the planned 1000-km (600-mi) flyby in July 2005 to 173 km (107 mi) and adding a 25-km (16-mi) flyby later in the mission. In total there were 13 flybys up to May 2011, providing a much closer look, with 10 more planned before the end of the mission in 2017.

Mass loading and deflection of the plasma near the south pole of Enceladus was confirmed by the Cassini Plasma Spectrometer (CAPS) and magnetometer results from July 2005, showing that the main region of ionized material was over the south pole. Ultraviolet stellar occultation measurements found evidence for gas over the south pole also, and provided speed estimates for the escaping gas.

Tiger stripes. Stunning visual imagery obtained by *Cassini* in July 2005 revealed the large plumes of material and remarkable surface features called "tiger stripes" in the south polar region (**Fig. 1**). The latter appeared as blue-tinged features on the otherwise lightly cratered, bright, and young-looking surface of Enceladus. The blue color of the stripes was interpreted as being due to ice grains, and later imagery and other data confirmed the tiger stripes as the sources of the plumes.

Infrared imagery confirmed that ice grains smaller than 1 mm (0.04 in.) were present in the tiger stripes, and showed the presence of carbon dioxide and organics as well as water. The tiger-stripe region was initially found to be a relatively warm "hot spot," with a temperature of about 85 K ($-188°$C, or $-307°$F) compared to its surroundings (75–80 K), but more detailed measurements of the stripe structure have revealed temperatures in localized regions up to 180 K ($-93°$C or $-136°$F), implying a source of heat there greater than 5 gigawatts, maintained by tidal forces partly due to orbital resonance with the moon Dione. This large heat flux has implications for the interior structure of Enceladus and may require a global or local subsurface ocean.

Plume emissions. Further encounters have allowed measurement of the plume environment in different ways, using in-situ plasma, field, and dust measurements, and multiwavelength remote sensing (**Fig. 2**). In-situ plasma measurements imply a mass addition rate to the plasma flow of approximately 300 kg/s. This rate is approximately one-third that of Io, the moon of Jupiter that is acknowledged to be the most volcanically active object in the solar system, its sulfurous emissions driven by tidal forces that produce a subsurface magma ocean there.

In addition to copious gas emission at a speed of about 500 m/s (1640 ft/s), some of which becomes plasma, about 10% of the plumes by mass emerges as ice grains at about 0.3–0.5 times the speed of the gas. Much of the dust ejected returns to the surface, accounting for the young-looking surface, as most of the dust particles move at less than the escape speed. The neutral ice particles that do escape, a few percent of those emitted, apparently populate Saturn's E ring (**Fig. 3**); their size (mostly in the micrometer range) is consistent with those forming the E ring. *Cassini's* Cosmic Dust Analyzer (CDA) confirmed ice grain emission from the south polar region, and established the south polar plume as the probable source of the E ring. Furthermore, some of these ice particles were found to contain sodium, possible evidence for a subsurface ocean.

Overall, the gas and ice emission was found to have significant time variability. A spot in Saturn's aurora at the magnetic footprint of Enceladus has been found, and its intensity can be used to probe this variability.

Plume ionosphere. The concentration of charged particles in the plume is sufficient that a "plume ionosphere" forms, with a region of stagnant plasma flow immersed in the rapidly moving plasma environment of Saturn's magnetosphere, which itself corotates with Saturn fairly closely at this distance. Positive and negative ions were found in this region within the plumes. The positive ions appeared as water-group ions (mass 16–19) and heavier ions. The negative ions appeared as clear multiples of the water or OH mass, with clusters of up to 100. This is further evidence of Enceladus as a source of water and is consistent with the idea of a subsurface ocean.

Cassini established that Enceladus is the main source of water in Saturn's inner magnetosphere, with an additional source being the rings. The inner magnetospheric region, which includes hydrogen ions mainly from Saturn's ionosphere, is dominated by water-based neutrals (O, OH) and permeated by the corotating magnetosphere. Enceladus, the rings, and the associated neutrals are the main source of heavy ions in the inner magnetosphere, and these populate the outer magnetosphere also. Some of the complex chemistry at Titan appears to involve particles, oxygen in particular, originally from Enceladus.

The Ion and Neutral Mass Spectrometer (INMS) aboard *Cassini* confirmed that most of the neutral gas is over the south polar region of Enceladus. Detailed composition measurements from the INMS show that, as well as water, carbon dioxide, methane, ammonia, argon-40, and a variety of organics are present in the neutral gas in smaller quantities (**Fig. 4**). CAPS measurements also indicate nitrogen, which may be from ammonia, and that Enceladus, rather than Titan, appears to be the dominant nitrogen source at Saturn. Ammonia is important because it can act as an "antifreeze," allowing liquid water at temperatures down to 176 K ($-97°$C or $-143°$F).

In addition to the population of neutral ice particles, charged nanograins were found by CAPS. The detailed timing of the variable densities of negative and positively charged grains was used to trace their trajectories back to particular sources within the tiger-stripe regions. In addition, the fact that the trajectories were different between the charged species implied separation with respect to each other and the neutral plume. Saturn's magnetic field effectively acts as a huge mass spectrometer for these particles. Ice grain–plasma interactions play a role in Saturn's inner magnetosphere.

Causes of geologic activity. Cryovolcanism (involving water ice and other low-boiling-point compounds rather than molten rock), driven by differences in the cold temperatures on the surface, is

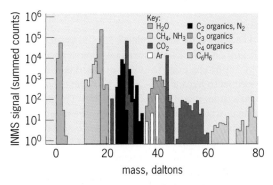

Fig. 4. Measurement of the neutral gas composition of the plume above Enceladus's south pole by *Cassini's* Ion and Neutral Mass Spectrometer (INMS), taken during the flyby of October 9, 2008. The geometry of this flyby allowed better measurements of minor species than previous flybys. In addition to carbon dioxide, methane, and hydrocarbons up to benzene, the composition includes ammonia, which acts to keep the moon's cold subsurface ocean in a liquid state. (*Courtesy of J. H. Waite et al., Liquid water on Enceladus from observations of ammonia and 40Ar in the plume, Nature, 460:487–490, 2009*)

an important process at Enceladus. Heating is produced by tidal forces (radioactive sources are not strong enough). This is thought to drive the variable activity of Enceladus, making it the most geologically active icy satellite.

One of the significant puzzles about Enceladus is why it is so active compared to Mimas, where the tidal heating should be higher. The answer may lie in the mechanical structure of the tiger stripes themselves, where diurnal shear may produce frictional heating at depth. An ocean is produced under the stripes as ice melts due to the huge stress (**Fig. 5**).

Effects of activity. Enceladus has significant effects on its environment. It is confirmed as the source of E-ring replenishment, a source for the neutral particles that dominate the inner regions of the Saturn system, and a source for plasma in the inner magnetosphere. In addition, it has an associated magnetic and plasma structure called an "Alfvén" wing which carries electrical current along the magnetic field to

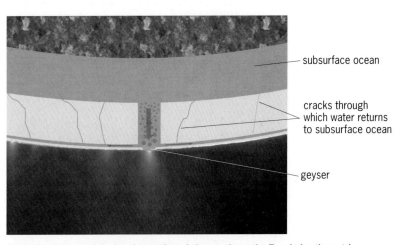

— subsurface ocean

— cracks through which water returns to subsurface ocean

— geyser

Fig. 5. Model to explain the observation of plumes above the Enceladus tiger stripe fissures, and of sodium salts and organics in the plumes. Slightly bubbly seawater from a subsurface ocean travels through a passage in the ice crust to feed a geyser, returning to the ocean via small cracks. (*NASA/JPL/SSI*)

Saturn, where it forms an activity-dependent spot in Saturn's aurora.

Life in the outer solar system? Several aspects of the *Cassini* discoveries at Enceladus make it an astrobiologically interesting location—that is, it may have the potential to harbor life—along with the other outer-solar-system moons Europa and Titan. The presence of heat, water in a subsurface salty ocean, and complex molecules including organics are all key ingredients. We do not yet know if all of the biogenic elements (carbon, hydrogen, nitrogen, oxygen, phosphorus, and sulfur) are present, because phosphorus has not been detected. Another significant requirement for life, however, is time for it to develop. More exploration well beyond *Cassini* is required to answer whether there has been enough time for life to develop at Enceladus. Meanwhile, although its instruments will never be able to answer all the questions, *Cassini* will make further exciting discoveries until the scheduled end of its mission in 2017.

For background information *See* ALBEDO; ASTRO-BIOLOGY; ATMOSPHERE; HUBBLE SPACE TELESCOPE; MAGNETOSPHERE; MASS SPECTROMETRY; PLASMA (PHYSICS); SATURN; SPACE PROBE.   Andrew J. Coates

Bibliography. O. Grasset et al., *Satellites of the Outer Solar System*, Space Sciences Series of ISSI, vol. 35, Springer, New York, 2010; J. R. Spencer et al., Enceladus: An active cryovolcanic satellite, in M. K. Dougherty, L. W. Esposito, and S. M. Krimigis (eds.), *Saturn from Cassini-Huygens*, pp. 683–724, Springer, Dordrecht, the Netherlands, 2009.

Endophyte-associated ergot alkaloids

Tall fescue (*Lolium arundinaceum*) occupies nearly 35 million acres (12 million hectares) in the United States and is the most abundant and economically important forage grass of the transition zone east of the Mississippi River. It is a hardy and persistent forage with excellent nutritive value. These characteristics led to early and rapid adoption of tall fescue as a primary forage crop for cattle, sheep, and horses during the 1940s. Although nutritional analyses predicted that tall fescue would support good body weight gains in grazing livestock, it failed to produce gains on par with similar forages (for example, orchard-

grass). In fact, tall fescue induced a form of ergotism (intoxication), now known as fescue toxicosis, in these animals. Fescue toxicosis is actually made up of three metabolic syndromes: "fescue foot," "summer slump," and "fat necrosis."

During the late 1970s, the presence of an endophytic (grows inside a plant as a symbiont) fungus (*Neotyphodium coenophialum*) in tall fescue was linked to poor body weight gains. Subsequently, it was determined that the endophyte produced ergot alkaloids (for example, ergovaline and lysergic acid), which constitute a class of chemicals that contain ergotamine and the hallucinogenic drug of abuse, LSD (lysergic acid diethylamide). Likewise, these alkaloids (see **illustration**) are of the same class as those responsible for severe outbreaks of ergotism (for example, St. Anthony's Fire) in humans during the Middle Ages. However, in human cases of ergotism, the fungus responsible for production of the ergot alkaloids (for example, ergotamine) was the ergot (*Claviceps purpurea*) of rye (*Secale cereale*).

Vasculature. The circulatory system, through its network of blood vessels, provides tissue level removal of metabolic waste products and delivery of necessary nutrients, oxygen, hormones, growth factors, and immune components needed to maintain health and productivity. Blood vessels (arteries, veins, and capillaries) are composed of tissues arranged in three concentric layers (tunicae) of tissue. Beginning at the lumen (blood side of the vessel), the intima (first layer) is composed of endothelial cells (lining the lumen), subendothelial connective tissue, and an internal elastic lamina. The internal elastic lamina forms the separation of the intima from the media (middle layer). The media consists of smooth muscle cells, elastic fibers, and the external elastic lamina. The external elastic lamina separates the media from the adventitia (outer layer). The adventitia is a poorly defined layer of connective tissue containing nerves and small blood vessels needed to support the media of larger blood vessels. Generally, these layers are best defined in larger arteries and less defined in smaller vessels and veins. Capillaries, the smallest blood vessels, lack the smooth muscle cells found in the media of other blood vessels, but have a small number of pericytes that constrict similar to smooth muscle cells to reduce the diameter of the vessel.

ergotamine   ergovaline   lysergic acid

Chemical structures of the ergopeptines (ergotamine and ergovaline) and lysergic acid.

Fescue toxicosis: component syndromes and symptoms/outcomes	
Syndromes	Symptoms
Fescue foot	• Lameness • Fetlock/hoof swelling • Dry gangrene of extremities • Loss of tail tips, hooves, and ear tips
Summer slump	• Reduced grazing • Extended time in shade and water (mud holes, ponds, and streams) • Reduced weight gains • Rough hair • Elevated rectal temperature • Increased respiration rate and salivation • Reduced serum prolactin and milk production • Dystocia (difficult birth) in pregnant mare • Death of foal and mare
Fat necrosis	• Hard fat deposits around internal organs • Reduced weight gains • Poor health and appearance

Compromise of the circulatory system with fescue toxicosis is well documented and explains many of the symptoms and outcomes (see **table**). For example, dry gangrene of extremities occurs as a result of the additive effects of alkaloid-induced vasoconstriction with constriction due to cold weather (a heat conservation mechanism), leading to cessation of blood flow to extremities and subsequent death of the tissue. Likewise, alkaloid-induced vasoconstriction can impede heat transfer from internal organs to the skin for dissipation and thus induces increased rectal temperatures (representative of increased internal body temperature). As a consequence of the temperature increase, behavioral adaptations (for example, reduced grazing, and standing in the shade and water) are induced. An additional complication occurs because of alkaloid-induced cell proliferation and death, resulting in altered vascular morphology.

Ergot alkaloids and vascular morphology. During the 1960s and 1970s, prior to the linkage of endophyte-infected tall fescue to fescue toxicosis, researchers described gross changes in vascular morphology in cattle consuming tall fescue or alcohol extracts of the plant. Some of the reported morphological changes included blood vessel congestion, perivascular hemorrhage, blood vessel distension, thickened walls, and small luminal diameters in peripheral blood vessels (for example, in the foot). The noted changes may occur as a result of hypertrophy and/or hyperplasia of smooth muscle cells and/or death of endothelial cells.

Hypertrophy (an increase in size of the smooth muscle cells) would reduce luminal size and blood flow. This is analogous to a smaller internal diameter pipe not being able to carry as much water as a larger one. However, this mode has yet to be empirically proven. Hyperplasia (an increase in the number of smooth muscle cells) would have an effect similar to hypertrophy, that is, thickening of the media while reducing luminal size and blood flow. Evidence for hyperplasia was demonstrated in the mid-1990s, when it was reported that ergovaline was able to stimulate quiescent smooth muscle cell growth in a petri dish. These first two modes can provide explanation of how ergot alkaloids contribute to blood vessel congestion, thickened walls, and small luminal diameters. However, compromise of the endothelial cell layer would be required for hemorrhage or leaking. Endothelial cell death has been shown by treating bovine (cattle) pedal vein endothelial cells grown in a petri dish with ergovaline and monitoring cell viability. Ergovaline induced endothelial cell death in this in vitro system and thus provided evidence for how ergot alkaloids may induce perivascular hemorrhage, as these cells normally prevent blood components from escaping into perivascular tissues.

Ergot alkaloids, biogenic amine receptors, and vasoconstriction. Although ergot alkaloids may induce morphological changes in the vasculature, the primary mechanism by which these alkaloids induce circulatory compromise is via vasoconstriction (contraction of smooth muscle cells resulting in decreased luminal diameter and blood flow). Both in vivo (in the animal) and in vitro (tissue culture) research studies have clearly demonstrated the contractile potential of a number of ergot alkaloids (for example, lysergic acid and ergovaline). Recent studies, using Doppler ultrasound, have conclusively demonstrated that ergovaline causes constriction of luminal diameter and reduces blood flow through the caudal (tail) vein and artery in cattle consuming endophyte-infected tall fescue. Doppler ultrasound has improved our ability to study the circulatory compromise of fescue toxicosis in real time and should provide a good picture of an animal's recovery following removal of alkaloid exposure during future studies. The tail vessel findings are representative of what happens to blood flow in a number of other cattle and sheep tissues, including leg skin, adrenal glands, rib skin, cerebellum, duodenum (small intestine), and colon, upon consumption of ergot alkaloids by these animals. Blood flow through these latter tissues was determined by tracking radioactive beads through the blood vessels, which is a more difficult and costly method compared to Doppler ultrasound.

In vitro contractility studies have been used extensively to investigate the vascular activity of drugs and toxicants. In these studies, a ring of blood vessels is suspended between a fixed point and a transducer that converts the mechanical contraction exerted by the vessels in response to the addition of ergot alkaloids to an electrical signal that is recorded by a computer as a contractile force. Studies conducted in this manner have conclusively demonstrated that the ergopeptines (for example, ergotamine and ergovaline) are more effective as vasoconstrictors (that is, they are more toxic) than lysergic acid. Additionally, bioaccumulation in vascular tissue has been shown for ergovaline, but not for lysergic acid, using these types of studies. Furthermore, these studies have provided researchers with potential cell surface targets for treating fescue toxicosis. From these in vitro studies, it has been demonstrated using receptor antagonists that ergot alkaloid-induced vasoconstriction is mediated through activation of biogenic amine receptors (serotonin and α-adrenergic receptors). These receptors act as on/off switches for muscle contraction in the blood vessels. When turned on, these switches cause reduced blood flow as a consequence of the reduced luminal diameter occurring via constriction.

The findings of relative potency and bioaccumulation have clearly demonstrated that lysergic acid, although a potential hallucinogen, is not a major player in circulatory compromise of peripheral tissues, whereas the ergopeptines are of great concern. The latter indicates a need to focus research efforts on reducing exposure to ergopeptines, decreasing sensitivity of the animal's tissues to ergopeptines, and/or enhancing clearance of the ergopeptines in order to reduce or eliminate fescue toxicosis.

Potential solutions to fescue toxicosis. During the last two decades, significant progress has been made in developing novel endophyte-infected tall fescue cultivars that retain hardiness and persistence while eliminating the ergot alkaloids and thus fescue toxicosis. However, these new cultivars have not yet significantly replaced the toxic endophyte-infected tall fescue varieties in the field. Locations with rough and steep topographies are not conducive to renovation (replacement) with the new tall fescue varieties. Additionally, problems with the old variety recovering from stray plants and existing seed banks continue to decrease the effectiveness of replacement with novel endophytes. Efforts to dilute the amounts of ergot alkaloids that are consumed or absorbed by livestock have also been tried through a variety of methods: interseeding pastures with clover, using supplemental feeds while animals are on pasture, and using binding agents to inhibit ergot alkaloid absorption. However, none of these have been entirely successful.

Problems with replacing toxic cultivars of tall fescue and/or decreasing ergot alkaloid intake and absorption have led to treatments designed to block alkaloid interactions at the vascular biogenic amine receptors. This approach has been partially successful in reducing the severity of fescue toxicosis when drugs (receptor antagonists) such as metoclopramide, sulpiride, and domperidone are used. In fact, the effectiveness of domperidone in treating fescue toxicosis in pregnant mares has led many veterinarians in the Bluegrass region of Kentucky to use the drug to avoid costly reproductive problems (see table). However, no drug has been successful in completely reversing fescue toxicosis.

Expansion of livestock production to meet world protein demand without expanding land use will require that fescue toxicosis be overcome using new and innovative approaches. Future efforts to address fescue toxicosis should continue to exploit efforts currently showing promise and should explore modifying animal metabolism of the alkaloids. This latter approach may be accomplished through genetic selection of animals naturally resistant to the alkaloids, modification of rumen microbes, or modification of the metabolism of animals through the use of nutrients and nutraceuticals.

For background information *see* ALKALOID; BLOOD VESSELS; ENDOPHYTIC FUNGI; ERGOT AND ERGOTISM; FESCUE; FORAGE CROPS; FUNGAL BIOTECHNOLOGY; FUNGAL ECOLOGY; FUNGAL INFECTIONS; FUNGI; GRASS CROPS; MYCOLOGY; MYCOTOXIN in the McGraw-Hill Encyclopedia of Science & Technology.

James R. Strickland

Bibliography. D. L. Cross, L. M. Redmond, and J. R. Strickland, Equine fescue toxicosis: Symptoms and solutions, *J. Anim. Sci.*, 73:899–908, 1995; F. J. Schoen, Chapter 11: Blood vessels, pp. 467–516, in R. S. Cotran, V. Kumar, and S. L. Robbins (eds.), *Pathological Basis of Disease*, 5th ed., Saunders, Philadelphia, 1994; J. R. Strickland, G. E. Aiken, and J. L. Klotz, Ergot alkaloid-induced blood vessel dysfunction contributes to fescue toxicosis, *Forage Grazinglands*, November 4, 2009; J. R. Strickland et al., Board-invited review: St. Anthony's Fire in livestock: Causes, mechanisms, and potential solutions, *J. Anim. Sci.*, 89:1603–1626, 2011; J. R. Strickland et al., Physiological basis of fescue toxicosis, in H. A. Fribourg, D. B. Hannaway, and C. P. West (eds.), *Tall Fescue for the Twenty-first Century*, American Society of Agronomy, Madison, WI, 2009.

Ethical leadership

Ethical leadership deals with the virtuousness of leaders and the nature of their behavior. The twenty-first century began with a series of high-profile ethical scandals, including the bankruptcies of an energy company (Enron) and a telecommunications giant (WorldCom), and the global financial collapse of 2007. Many of the leaders of the companies involved in the ensuing financial turmoil were convicted on criminal charges. In addition, despotic and corrupt heads of state lead many nations. Unethical leaders abound, so it is little wonder that the past decade has led to increased interest in ethical leadership on the part of scholars.

The study of leadership has focused primarily on leader effectiveness, with relatively little attention given to good, or ethical, leadership. It is particularly important that leaders focus on their ethical responsibilities because of the power, status, and privilege afforded them. Misbehavior by a powerful leader, such as a head of state or the chief executive officer (CEO) of a large corporation, can have an enormous negative impact on a wide range and number of constituents.

Ethical theories and leadership. The two general approaches to ethical leadership concentrate either on the leader's conduct or on the leader's character. Teleological theories focus on whether a leader's behavior will lead to desirable ends or purposes. The ends may benefit the individual (self-interests, also referred to as ethical egoism), or they may benefit a group or collective. A business leader, for example, might strive to increase profits in order to benefit the company and its shareholders, and to allow wages (including the leader's own salary) to be raised. A danger of this approach is that the positive ends may be seen as justifying the means. In other words, the leader may engage in questionable behavior in an effort to do good, such as lying to customers in an effort to increase profits.

Utilitarianism is a teleological theory that argues that leaders should act to create the greatest good for the greatest number of people. Government leaders, for example, are often concerned with trying to use limited financial resources in order to maximize the benefits for the greatest number of constituents. For example, a politician may weigh whether tax breaks for corporations to create jobs or government entitlement programs such as welfare or food stamps would have the greatest impact on alleviating poverty.

Altruism involves a leader who ignores self-interests in an effort to promote the best interests of others. The theory of servant leadership suggests an altruistic base, as the leader's role is to forgo personal gains in an effort to serve the followers and promote their development and well-being. Mohandas Gandhi and Mother Teresa would be prototypes of altruistic servant leaders.

Whereas teleological theories of ethical leadership focus on the outcomes or consequences of the leader's actions, deontological theories focus on whether the leader's action itself is good. The deontological approach argues that it is the leader's obligation to do the right thing, that is, actions that do no harm to others. In other words, the ethical leader has a duty to be respectful of others, to be fair, and to be honest. Honesty, integrity, and fairness are at the heart of the deontological approach, and these are among the qualities we admire most in our leaders.

Dating back to the ancient Greek philosophers Plato and Aristotle, a second set of theories focuses on the character of the leader. These typically are referred to as virtue-based theories. Virtues are characteristics that are valued because they promote individual and collective well-being. Importantly, virtues can be acquired, developed, and honed. Core, or cardinal, virtues include courage–fortitude, temperance, justice, and prudence (or wisdom). Other important virtues are truthfulness, humility, and benevolence–generosity. Virtue theories focus more on what leaders are (as persons) rather than on how they behave. This approach involves choosing or developing leaders who are considered good people.

Ethics and leadership theories. Theories of leadership emerging from management and psychology rarely considered leader ethics in their formulations. The focus was primarily on leader effectiveness. However, since the 1970s, a number of theories have incorporated leader ethics in effective leadership.

The first of these was the aforementioned theory of servant leadership. In essence, servant leadership emphasizes that leaders must be attuned to the needs of followers and must care for those needs, while nurturing follower personal development. The altruistic servant leader does not focus on personal gains, but on the welfare of the followers and other constituents.

In the late 1970s, James MacGregor Burns developed a theory of transformational leadership whereby leaders attempt to effect change by inspiring followers to higher standards of moral responsibility. Transformational leaders serve as positive and ethical role models for followers; these leaders are concerned with the needs and personal development of followers.

Building on these theories, more recent leadership theories have begun to consider ethics as core elements. Most notable is authentic leadership theory, which consists of four key positive psychological characteristics related to ethical leadership. These are self-awareness, an internalized moral perspective, balanced processing, and relational transparency. Self-awareness allows the leader to know personal strengths and limitations (a requirement for the leader's ethical development), whereas the internalized moral perspective serves as a moral compass to help guide the leader's ethical behavior. Balanced processing helps a leader to be objective in analyzing potentially unethical situations, and relational transparency relates to straightforward and fair dealings with others. These dimensions of authentic leadership have a connection and similarity to the moral virtues espoused by Aristotle.

Ethics can be developed in leaders through increasing moral awareness, helping leaders recognize ethical dilemmas, and reinforcing a culture that values ethical behavior. Often, unethical leader behavior occurs because the leader does not realize the ethical problem, he or she holds to a false ethical rule (for example, bribery in certain cultures is acceptable), or the leader rationalizes that unethical behavior is actually ethical.

Moving forward, it is hard to imagine that new theories of leadership would not incorporate ethics into their formulations. There also have been several recent attempts to measure ethical leadership, for research purposes as well as for identifying and selecting ethical leaders. Nearly all of these instruments require others (most often followers of leaders, but also peers and supervisors) to evaluate or rate a target leader's ethicality or moral character. A problem with this is that others must make inferences about a leader's ethical nature and moral character from their observations of that behavior. We know that leaders often engage in impression management, and try to put their best face forward; thus, it may be difficult to know the whole person, and it may be impossible for others to know the true motivations of a leader.

The practice of ethical leadership. It is important for leaders to behave ethically in carrying out their roles, but leaders also play an important part in setting an ethical climate for their groups, organizations, or nations. Leaders serve as role models for followers, so it is imperative that leaders behave ethically in order to prevent the spread of unethical behavior in their collectives. By virtue of their power and position, leaders also have an obligation to prevent negative and destructive misbehavior in their organizations or collectives; this can be accomplished, for example, by combating harassment, discrimination, incivility, aggression, and bullying by members.

In order for a leader to be ethically effective, there are certain key components that govern moral or ethical action; these are represented by four psychological processes. The first is moral recognition, that is, being able to recognize and identify ethical problems. All too often, leaders have a sort of moral blind spot, which renders them unaware of the ethical implications of certain courses of action. Therefore, it is important that leaders increase their sensitivity to the presence of ethical issues—recognizing these issues when they see them or being able to anticipate when ethical problems may present themselves.

A second key component of ethical action involves ethical judgment, or knowing the objectively right and wrong courses of action in certain situations. Theories and research on moral reasoning suggest that leaders can develop their ethical decision-making abilities and help set an ethical climate in their groups and organizations.

A third process is ethical motivation, which means that a leader needs to be motivated to behave in an ethical manner, despite the backlash that may ensue. In addition, leaders need to ensure that followers are encouraged to behave in ethical ways, and to discourage their unethical actions.

The final process is the actual implementation of ethical actions. This involves having the courage to follow through and take the objectively right course. Often, the leader may suffer because the ethical course may conflict with the agenda of the organization, or there may be costs associated with acting ethically.

Leadership experts argue that leaders must be aware of and accept the ethical imperative that goes along with the leadership role. Leaders have a responsibility to behave ethically and to avoid falling into the trap of unethical behavior caused by personal gains, greed, fears, and ego. They also should take steps to promote an ethical climate in the groups and collectives that they lead and should work to build their personal integrity and character. In sum, ethical leaders should learn to master the psychological processes that are required for effective ethical action.

For background information *see* DEVELOPMENTAL PSYCHOLOGY; MOTIVATION; PERSONALITY THEORY; PSYCHOLOGY; SOCIOBIOLOGY in the McGraw-Hill Encyclopedia of Science & Technology.

Ronald E. Riggio

Bibliography. J. M. Burns, *Leadership*, Harper & Row, New York, 1978; J. B. Ciulla (ed.), *Ethics: The Heart of Leadership*, Praeger, Westport, CT, 2004; C. E. Johnson, *Meeting the Ethical Challenges of Leadership: Casting Light or Shadow*, 3d ed., Sage, Los Angeles, 2009; T. L. Price, *Leadership Ethics: An Introduction*, Cambridge University Press, New York, 2008.

Evolution of phagotrophy

Phagotrophy describes the process by which unicellular organisms derive their food from engulfing and digesting other cells (**Fig. 1**). The process is thought to be ancient, but establishing exactly when phagotrophy emerged in the history of life is difficult and a source of ongoing debate. Opinions range from very early origins to a relatively late emergence in the evolution of life, well after the appearance of fully formed eukaryote cells (that is, cells with a nucleus and mitochondria).

Phagotrophic predators in the Proterozoic (2.5–0.542 BYA). Cellular life-forms are divided into three distinct groups: Eukaryotae, Archaea, and Bacteria. Among modern organisms, all cells capable of phagotrophy are eukaryotic. Eukaryotes include organisms as diverse as animals, fungi, and plants, but they also include a diverse array of unicellular organisms such as diatoms and amebas. Their genetic material is housed in a membrane-bounded nucleus, distinguishing eukaryotes from the prokaryotic groups, Bacteria and Archaea, which lack this structure. The earliest definitive appearance of eukaryotes in the fossil record is in the Proterozoic, around 1.8 BYA (billion years ago). However, there are indications that this group could have evolved in the Archean (3.8–2.5 BYA), with controversial hydrocarbon biomarker data suggesting a date as early as 2.7 BYA. The recent characterization of large microfossils from 3.2-billion-year-old rocks from South Africa has led some researchers to suggest that an even earlier origin is plausible (**Fig. 2**).

A difficulty with these early fossils is relating them to modern eukaryotic groups. Evidence for diversification leading to recognizably modern eukaryote groups (called the crown group) occurs first around 800–720 MYA (million years ago), indicating that a billion years may separate the origin of eukaryotes from the radiation of modern groups. Eukaryote fossils that predate this radiation are part of the stem group (**Fig. 3**). Stem lineages are all extinct.

A key prerequisite for phagotrophy is the capacity for the cell membrane to be flexible so that food particles can be engulfed. There is no direct evidence of this in the early fossil record; however, there are certainly indications that early eukaryotes were capable of modifying cell shape, which is necessary although not sufficient for engulfment.

Comparisons of living eukaryote lineages indicate that key parts of the molecular machinery for phagocytosis (engulfment of one cell by another) are found across all eukaryotes, and there are examples of phagotrophs in all major eukaryote groups. However, opinions vary considerably regarding when this capacity evolved in eukaryote evolution.

When did phagotrophy evolve? Establishing the timing of the emergence of phagotrophy is not trivial. One possibility is that phagotrophy emerged extremely early in the evolution of life on Earth. Proponents of this view point out that, if phagotrophy only evolved in eukaryotes, there would have been a very long "Garden of Eden" period (perhaps as long

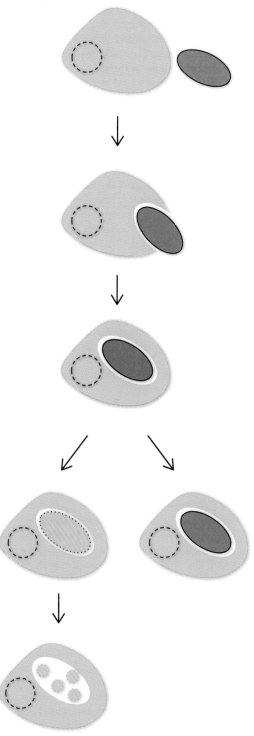

Fig. 1. Schematic of the process of phagotrophy. A bacterial cell (gray) is engulfed by a larger eukaryotic cell (tinted). The eukaryote cell nucleus is shown as a dashed circle. Following engulfment, two possible outcomes are shown. The outcome on the left depicts the engulfed cell being degraded by and digested by the engulfing cell. In the right-hand scenario, the engulfed cell resists degradation and persists within the host cell. Such digestion-resistant bacteria are frequently found in phagotrophic amebae, and this process is thought to have been crucial to the evolution of a stable association between the bacterial ancestor of mitochondria and the early eukaryote that engulfed it.

as 3 billion years) when there were no predators on Earth. As direct fossil evidence of phagotrophy is lacking, one group of researchers recently took a modeling approach to assess this possibility. They generated a simple system with only basic physical, biological, and ecological properties to answer whether such a long predator-free period might be plausible. Two key properties in their model were that an engulfing cell must be larger than its prey and that there is a predation advantage. Their simulations revealed that predation emerged from a broad range of basic initial conditions, arguing against a "Garden of Eden" scenario for early life.

However, other researchers suggest that the reason for the long predator-free period can be explained by the many complex cellular structures that first had to evolve. In order to engulf prey, a cell membrane needs to have the capacity to change shape, the prey must be encapsulated [forming a vacuole inside the cell (Fig. 1)], and there must have been complex mechanisms for controlled degradation

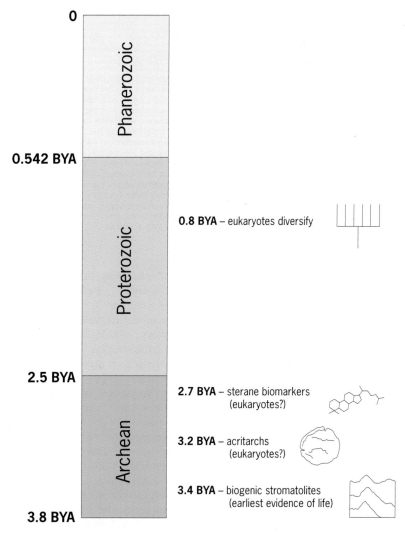

Fig. 2. Geological timeline showing evidence suggestive of eukaryote cells in the fossil record. It is uncertain whether acritarch microfossils from 3.2 BYA (billion years ago) and sterane biomarkers from 2.7 BYA indicate the presence of eukaryote cells in these early periods. Earlier dates (>1 BYA) for the diversification of major eukaryote groups in the fossil record have also been proposed.

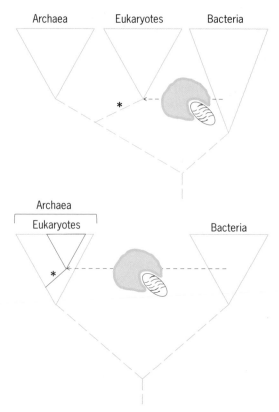

Fig. 3. Two possible relationships between Archaea and Eukaryotae. The top graphic shows Archaea and Eukaryotae as sister groups, with each having evolved separately from a common ancestor. In the bottom graphic, eukaryotes have evolved from within the Archaea. Triangles indicate crown groups, which trace to the modern radiation of each group from their respective common ancestor. Stems (dashed lines) connect groups, but species traceable to each stem are now extinct and only represented by fossils. In both scenarios, the bacterial ancestor of the mitochondrion enters the eukaryote lineage via engulfment. An asterisk indicates the evolution of phagocytosis along the eukaryote stem, prior to the engulfment event and prior to the radiation of modern eukaryote groups about 800–720 MYA.

and transport of nutrients out of the vacuole, plus mechanisms to remove waste. Engulfment and controlled digestion are complex, coordinated processes; therefore, it is argued that the large number of requisite cellular features precludes an early origin.

Although there is no doubt that phagotrophy is complex, recent investigations show that engulfment and digestion are not unique to eukaryotes. The bacterium *Gemmata obscuriglobus* has a number of cellular structures that look remarkably similar to those found in eukaryote cells, including internal membranes that resemble the eukaryote nucleus. Careful experimentation revealed that these bacteria engulf and degrade proteins. While this is a far cry from engulfing something as large as another cell, the important insight is that the process of ingesting food through internalization of a section of membrane is more widespread than previously thought. Moreover, it suggests that ingestion is not as difficult to evolve as an analysis of modern eukaryote cell biology might indicate.

Mitochondria in eukaryote evolution. Evidence for past phagocytoses comes from chloroplasts (cell plastids) and mitochondria, which are organelles found in a wide range of eukaryotes. Chloroplasts are responsible for photosynthesis in plants and algae, and they probably evolved following engulfment (but not ingestion) of a photosynthetic bacterium, likely a cyanobacterium. Mitochondria have a role in energy generation. A combination of cell biological, genetic, and evolutionary analyses indicate that mitochondria have features in common with bacterial cells, including bacteria-like cell division. Moreover, mitochondria carry DNA. This DNA displays similarities to the genomes of bacteria, showing that mitochondria evolved from a group of free-living bacteria, most probably the α-proteobacteria.

Until recently, mitochondrial capture was thought to be a late event in the history of eukaryotes. Evolutionary phylogenetic trees and cell biological studies led researchers to conclude that mitochondria were absent from a number of eukaryotes. These mitochondria-free groups, dubbed Archezoa, carried all other hallmark eukaryote features and were thus thought to be missing links in the evolution of eukaryotes. Archezoa were therefore considered living proof that the earliest eukaryote groups never possessed mitochondria.

However, more recent research has led to rejection of this model. Reanalysis of the genetic and cell biological evidence reveals that modern Archezoa do in fact possess mitochondria (or related organelles), and it is now accepted that all modern eukaryotes either carry or once carried mitochondria, or organelles that derive from them such as hydrogenosomes and mitosomes.

Mitochondria before phagocytosis? Following the realization that Archezoa are not "living fossils," a range of models have emerged for the origin of eukaryotes. One line of reasoning is that, if all eukaryote lineages evolved from a mitochondrion-bearing ancestor, engulfment of the mitochondrion was the defining event in the evolution of the eukaryote cell. Perhaps then the cell that did the engulfing was something else. The attention of some researchers has therefore turned to the third domain of life, the Archaea. This makes sense for multiple reasons. Chief among these is the observation that eukaryote genomes carry genes similar to those in Bacteria (many of which migrated from the mitochondrion to the cell nucleus) and genes with similarity to archaeal genes. There is ongoing debate regarding exactly how Eukaryotae and Archaea are related, but this similarity suggests two possibilities. Eukaryotae and Archaea could be sister groups that share a common ancestry (top panel of Fig. 3), or eukaryotes could in fact be an unusual group within the diversity of the Archaea (bottom panel of Fig. 3). With Archezoa losing their missing-link status, one popular suggestion is that the two partner cells in this symbiosis were an archaeon and an α-proteobacterium. This scenario fits well with emerging genetic data, but creates a problem for cell biology because there is no known mechanism by which an archaeon can internalize a bacterial cell. If phagocytosis is excluded because it is specific to eukaryotes, proponents of the archaeal–bacterial model require that there must have been an alternative process, which has since disappeared from modern archaeal lineages.

A straightforward resolution. The confusion surrounding the origin of eukaryotes is simply resolved by applying the same stem and crown framework that paleontologists have applied in placing early fossils in the context of eukaryote origins. Doing so reveals that phagocytosis can still explain the origin of mitochondria, even if eukaryotes evolved from Archaea (Fig. 3). The reasoning is as follows: if rudimentary phagocytosis and mitochondria can both be placed in the last eukaryotic common ancestor (LECA), both features evolved in the stem and the order in which they evolved is invisible. However, because these events postdate the divergence of the eukaryote lineage from a possible archaeal ancestor, no new mechanisms are required as long as phagocytosis evolves on the stem before engulfment of the mitochondrial ancestor. The alternative model is rather convoluted and cannot be readily tested; it requires that an archaeal-specific mechanism evolved that enabled engulfment of the mitochondrial ancestor, but that the fledgling eukaryote lineage and the archaeal lineage that it evolved from both subsequently lost this process, which was then replaced by phagocytosis in eukaryotes. As there is no evidence for this more complex series of events, it makes most sense to stick with the simpler phagocytic scenario.

Although the origin of eukaryotes is difficult to pinpoint in geological time, analyses of fossils, genes, cells, and computer modeling are all contributing to an increasingly clear picture of the early origins of phagotrophy in the stem lineage.

For background information *see* ARCHAEA; ARCHEAN; BACTERIA; EUKARYOTAE; FOSSIL; GENE; GEOLOGIC TIME SCALE; MACROEVOLUTION; MITO-CHONDRIA; ORGANIC EVOLUTION; PHAGOCYTOSIS; PROTEROZOIC in the McGraw-Hill Encyclopedia of Science & Technology.

Nadja Neumann; Anthony M. Poole

Bibliography. C. J. Cox et al., The archaebacterial origin of eukaryotes, *Proc. Natl. Acad. Sci. USA*, 105:20356–20361, 2008; S. de Nooijer, B. R. Holland, and D. Penny, The emergence of predators in early life: There was no Garden of Eden, *PLoS One*, 4:e5507, 2009; E. J. Javaux, C. P. Marshall, and A. Bekker, Organic-walled microfossils in 3.2-billion-year-old shallow-marine siliciclastic deposits, *Nature*, 463:934–938, 2010; T. G. Lonhienne et al., Endocytosis-like protein uptake in the bacterium *Gemmata obscuriglobus*, *Proc. Natl. Acad. Sci. USA*, 107:12883–12888, 2010; N. Neumann and A. M. Poole, Reconciling an archaeal origin of eukaryotes with engulfment: A biologically plausible update of the Eocyte hypothesis, *Res. Microbiol.*, 162:71–76, 2011; M. van der Giezen and J. Tovar, Degenerate mitochondria, *EMBO Rep.*, 6:525–530, 2005.

Evolution of theropod dinosaurs

Theropods (suborder Theropoda) are among the most evocative dinosaurs and include such famous predators as *Tyrannosaurus* and *Velociraptor*. All carnivorous dinosaurs were theropods, although some theropods were secondarily omnivorous or herbivorous. Their origins lie among the earliest dinosaurs, and they persisted for more than 160 million years until the end of the Cretaceous. In fact, because birds evolved from theropods, they still exist today. (Hereafter, "theropods" will refer only to nonavian forms.)

Evolutionary history. The earliest theropods lived in the early Late Triassic of Argentina, being already represented by small forms [*Eodromaeus*, 5 kg (11 lb)] and medium-sized forms [*Herrerasaurus*, 200 kg (440 lb)]. Just 10 million years later, theropods had reached nearly every part of Pangaea and had begun to diversify. Hundreds of species are known during the Mesozoic, representing numerous distinct lineages (**Fig. 1**).

Coelophysoids (Coelophysoidea) were predominantly small [approximately 20 kg (44 lb)], long-necked, low-skulled forms. *Dilophosaurus* was the largest of these, reaching 6 m (20 ft) in length and 200 kg (440 lb) in weight. Adult and juvenile *Coelophysis* specimens occur by the dozens in bone beds from North America and southern Africa. Common in the Late Triassic and Early Jurassic, coelophysoids were extinct by the Middle Jurassic.

Ceratosaurs (Ceratosauria) were larger animals, with shortened arms, pronounced skull growths, and stockier proportions. They lived from Early Jurassic times through end-Cretaceous. The large abelisaurids were dominant carnivores on southern continents in the Cretaceous. They lived alongside smaller, slender noasaurids, which had unusually procumbent teeth.

The remaining theropods form the group Tetanurae (named for tail modifications that made the base flexible and the end stiffer). Megalosaurs, allosaurs, and coelurosaurs are included in this group.

Megalosaurs (Megalosauroidea) include megalosaurids and spinosaurids. These two groups consisted of large predators with long skulls and four-fingered, three-clawed hands. Megalosaurids were less specialized and were common in the Jurassic. Spinosaurs were mostly Cretaceous forms. These sail-backed giants may have been piscivores (fish-eaters), with conical teeth in long, crocodile-like snouts.

The diverse allosaurs (Allosauroidea) were large predators that appeared in the Middle Jurassic. Metriacanthosaurids were common in eastern Asia. *Allosaurus* is the quintessential carnivore of the North American Jurassic and is among the best-known theropods. The Cretaceous carcharodontosaurids were true giants, sporting low back sails, sharklike teeth, and short, robust arms.

The most birdlike theropods form the Coelurosauria, the most diverse group of tetanurans. Many species were covered with filaments or feathers, and some were among the smallest

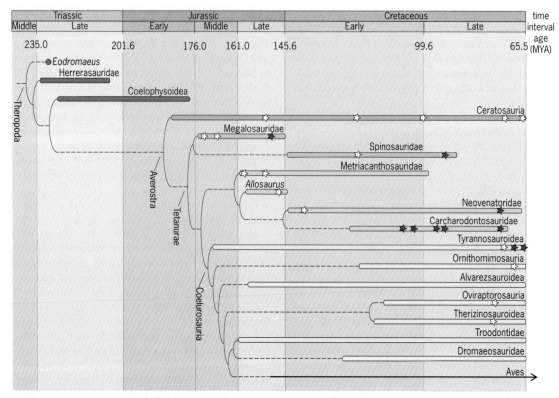

Fig. 1. Evolutionary relationships of major theropod groups, showing their estimated times of origin and durations. Megalosauridae and Spinosauridae are the megalosauroids. Metriacanthosauridae, Allosaurus, Neovenatoridae, and Carcharodontosauridae are the allosauroids. Together these groups are the tentanurans. Light colored bars show the coelurosaurs. White stars indicate large theropods [450–900 kg (1000–2000 lb)]; gray stars show giant theropods [more than 900 kg (2000 lb)]. Dates are million of years ago (MYA).

theropods. This group includes tyrannosauroids, ornithomimosaurs, alvarezsaurids, therizinosaurs, oviraptorosaurs, troodontids, and dromaeosaurids.

Tyrannosauroids include *Tyrannosaurus rex* and its relatives, the last lineage of giant predatory dinosaurs. Their origins are in the Jurassic and began with small, feathered forms such as *Guanlong*.

Ornithomimosaurs were ostrichlike "bird-mimics," having beaks, long legs, and small heads. Most species were omnivorous or herbivorous, as shown by fossilized stomach contents. The bizarre, short-armed alvarezsauroids were once thought to be flightless birds, but are now considered to be more primitive.

The bulky therizinosaurs were large herbivorous forms with long necks and giant hand claws. They were related to the beaked, crested oviraptorosaurs.

Troodontids and dromaeosaurids ("raptors") were closely related forms that had a large claw on the inner toe for use in tackling prey. These theropods, including *Velociraptor*, had feathers and were the closest relatives of birds.

Biogeography. Theropods were globally distributed, with fossils known from every continent (**Fig. 2**). Early in the Mesozoic, theropods were generally similar and were widely spread across Pangaea [a former supercontinent composed of all the continental crust of the Earth, and later fragmented by drift into Laurasia and Gondwana]. By the end of the Mesozoic, when Laurasia (comprising the combined continents of North America, Europe, and Asia) and Gondwana (comprising the combined landmasses of Africa, Antarctica, Australia, India, and South America) were separated, theropods (and other dinosaurs) developed greater geographic distinctions.

For example, *Coelophysis* and similar species are known from the early Mesozoic of North America, southern Africa, Europe, and China. In the Late Jurassic, *Allosaurus* is known from North America and Portugal, and close relatives were present in China. By the Late Cretaceous, nearly all Laurasian theropods were coelurosaurs, with tyrannosaurs as apex predators. By contrast, Gondwanan faunas were dominated by large abelisaurids and late-surviving allosauroids.

In the future, the interrelationships of individual species will prove important in understanding how dinosaurs responded to continental fragmentation and associated climate and habitat changes.

Body size. Adult theropods ranged from less than 1 kg (2.2 lb) [*Microraptor*] to approximately 4000 kg (8800 lb) [*Tyrannosaurus, Spinosaurus,* and *Giganotosaurus*] in mass and from 1 to 14 m (3.3 to 46 ft) in length. This impressive array is interesting because both large and small forms evolved throughout the Mesozoic and in different groups. Giant (multiton) theropods arose in several lineages: tyrannosaurids, carcharodontosaurids, spinosaurids, and megalosaurids (Fig. 1). Likewise, very small

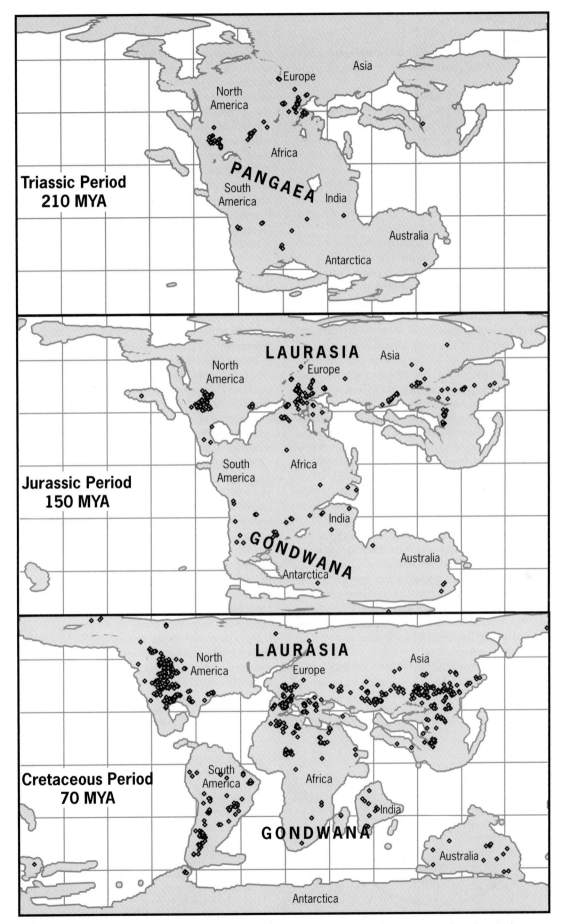

Fig. 2. Geographic distributions of theropods through geologic time. The locations of the theropod fossils from each Mesozoic time (Triassic, Jurassic, and Cretaceous) are shown as diamonds superimposed on maps of the continents at the dates indicated. Paleolandmass names are indicated in bold type; modern landmass names are in light type. (*Base maps plotted using software copyright © John Alroy and Chris Scotese; data obtained from the Paleobiology Database*)

Triassic Period
210 MYA

PANGAEA

North America
Europe
Asia
Africa
India
South America
Australia
Antarctica

Jurassic Period
150 MYA

LAURASIA

North America
Europe
Asia
South America
Africa
India
Australia

GONDWANA

Antarctica

Cretaceous Period
70 MYA

LAURASIA

North America
Europe
Asia
South America
Africa
India

GONDWANA

Australia

Antarctica

theropods evolved among coelophysids, noasaurids, compsognathids, alvarezsaurids, troodontids, and dromaeosaurids.

Theropod size evolution seems to conform to Cope's rule, that is, the observation that lineages tend to get larger over time. Most giant theropods were the product of tens of millions of years of gradual increase from much smaller ancestors. Giant theropods appeared largely in succession, with one lineage replacing another in the role of "largest predator." Interestingly, a concerted trend toward small size occurred among coelurosaurs as they became more birdlike.

Locomotion. The most primitive theropods were bipeds, with grasping hands that served poorly for support and locomotion. There were no quadrupedal theropods. Hind limb and hip bones included a hingelike ankle with little rotation, digitigrade (walking on the toes) feet, and upright posture.

Several modifications occurred in the skeleton and muscles during the course of theropod evolution. At the hip, the head of the thigh bone pointed more directly into the socket. The foot was reduced from four main toes to three, and the ankle bones became more extensively fixed to the shin.

Other variations suggest significant differences in speed, stamina, and agility among species, based on studies of similar correlations in living tetrapods. Several lineages achieved a cursorial limb in which the toes were extremely long; this would have been effective for either high speeds or wide ranges. Other forms had robust limbs and short toes that would have been better suited to weight bearing.

Computer modeling has allowed significant advances in testing ideas about dinosaur locomotion. Such models incorporate muscle reconstructions (based on scars left on fossil bones and inferences from living relatives), bone sizes and strengths, and joint articulations. One recent study rejected the notion that *Tyrannosaurus* was a fast runner because to do so would have required more musculature than was anatomically possible.

Footprints provide additional data, provided that one can identify distinctive theropod traits in the tracks, infer the general size of the track maker, and interpret the original sediment and setting. Typically, speed and behavior are studied; unsurprisingly, most theropods moving across mud suitable for track preservation were not going very fast, but this does not indicate they were generally slow. An unusually extensive track site in Lesotho shows one theropod using several stride types to cross a stream bed.

Growth and reproduction. Histology (microscopic bone analysis) allows detailed study of theropod growth rates and bone deposition. The identification of medullary bone, a specialized tissue produced by egg-laying individuals, now provides a way to identify female dinosaurs. Growth rates were rapid, with most theropods (and dinosaurs) experiencing a "growth spurt" and reaching adult size in fewer than 20 years. Although theropod eggs ranged in size, none were larger than a grapefruit, so giant theropods required exceptionally high growth rates to reach adulthood. Recent work has indicated that tyrannosaurs were fully grown in about 15 years.

Theropod reproduction was similar to that of other dinosaurs: they laid eggs with hard shells (resembling the eggs of birds, but not the leathery eggs of crocodilians). Theropods frequently laid eggs in pairs within nest structures. This suggests that they retained two functioning oviducts (only one remains in birds). Some theropods incubated their eggs, as shown by discoveries of oviraptorids entombed atop their nests. Many species were too large for this and may have covered their nests with vegetation, guarded them nearby, or simply abandoned the eggs after they were laid.

Arms, hands, and wings. All theropods had a grasping hand (**Fig. 3**). Over time, it became modified to include fewer fingers (from 5 to 3, 2, or 1) and some thumb opposability. The three largest fingers were equivalent to the human thumb and first two fingers. More birdlike theropods typically had less flexibility in the hand than more primitive forms.

Most theropod hands bore strong, curved claws, with the largest on the thumb, but unusual modifications also evolved. Tyrannosaurids h two-fingered hands on very short arms. The short arms of

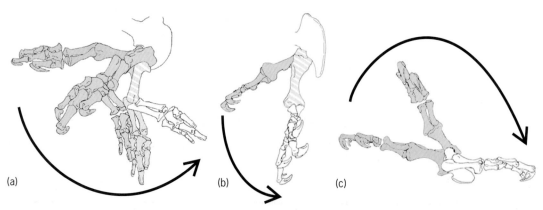

(a)   (b)   (c)

Fig. 3. Reconstructed arm movements in the carcharodontosaurid theropod *Acrocanthosaurus*, showing extensive shoulder mobility but limited wrist motion. Similar shades and patterns show corresponding positions as the arm is moved forward in the three views. The views are from the right side (*a*), front (*b*), and below (*c*). (*Modified from original images by P. Senter; copyright © Journal of Zoology, John Wiley and Sons*)

abelisaurids retained four fingers and very stocky proportions. Alvarezsaurids had exceptionally short and stubby hands with 1–3 fingers. However, hand functions are not well understood in any of these creatures.

The theropod forelimb (Fig. 3) could not move in the same manner as its human equivalent. In most theropods, a half-moon-shaped bone permitted only side-to-side motion of the wrist. The forearm bones could not cross, so the hands could not be held palms-down (as often seen in popular depictions). Thus, the typical posture of the theropod arm would be with the hands facing in (as if applauding). The elbow could open and close, but most motion would have been at the shoulder. These are the origins of the motions that a bird uses to flap its wings and fold them closed.

Origin of birds. The origin of birds from theropod dinosaurs has been well studied since the 1970s and 1980s, resulting in a detailed framework for understanding this important evolutionary transition. New discoveries have demonstrated that many classic "bird" features actually arose in their theropod ancestors long before the origin of flight.

Feathers are one dramatic example, which are now known from several coelurosaurian species. These range from simple filaments in more primitive forms (*Sinosauropteryx* and *Guanlong*) to downy and vaned feathers in more derived species (*Caudipteryx* and *Sinornithosaurus*). Their presence in the large therizinosaur *Beipiaosaurus* underscores the fact that feathers were not originally flight-related. Even the most primitive feathers retain evidence of having been distinctly colored.

Other "bird" features known in theropods include fused clavicles (furculae or wishbones), pneumatic outpockets of the respiratory tract into the skeleton, and a modified wrist.

For background information *see* ANIMAL EVOLUTION; AVES; BIOGEOGRAPHY; DINOSAURIA; EXTINCTION (BIOLOGY); FEATHER; FOSSIL; MESOZOIC; PALEOGEOGRAPHY; PALEONTOLOGY; SAURISCHIA in the McGraw-Hill Encyclopedia of Science & Technology.
Matthew T. Carrano

Bibliography. M. T. Carrano, Body-size evolution in the Dinosauria, pp. 225–268, in M. T. Carrano et al. (eds.), *Amniote Paleobiology: Perspectives on the Evolution of Mammals, Birds, and Reptiles*, University of Chicago Press, Chicago, 2006; D. E. Fastovsky and D. B. Weishampel, *Dinosaurs: A Concise Natural History*, Cambridge University Press, Cambridge, U.K., 2009; T. R. Holtz, Jr., *Dinosaurs: The Most Complete, Up-to-Date Encyclopedia for Dinosaur Lovers of All Ages*, Random House, New York, 2007; J. R. Hutchinson and V. Allen, The evolutionary continuum of limb function from early theropods to birds, *Naturwissenschaften*, 96:423–448, 2008; M. C. Langer, M. D. Ezcurra, and J. S. Bittencourt, The origin and early evolution of dinosaurs, *Biol. Rev.*, 85:55–110, 2009; K. M. Middleton and S. M. Gatesy, Theropod forelimb design and evolution, *Zool. J. Linn. Soc.*, 128:149–187, 2000; P. Senter and J. H. Robins, Range of motion in the forelimb of the theropod dinosaur *Acrocanthosaurus atokensis*, and implications for predatory behavior, *J. Zool.*, 266:307–318, 2005; D. B. Weishampel, P. Dodson, and H. Osmólska, *The Dinosauria*, 2d ed., University of California Press, Berkeley, 2004.

First-order reversal curves (FORCs)

Magnetic properties can be used to determine the magnetic composition, grain size, concentration, and even the origin of a material or its magnetization, as is done in environmental and rock magnetic studies of natural materials, or to assess how synthetic materials can be exploited as media for magnetic recording or in other industries. Many of the fundamental bulk magnetic properties of a material, such as the saturation magnetization, saturation remanent magnetization, and coercivity (the field needed to demagnetize a sample), are obtained from magnetic hysteresis studies, which generally focus on the major hysteresis loop (**Fig. 1**). Bulk magnetic properties, however, only tell part of the story because magnetic materials are comprised of numerous microscopic magnetic domains, which are regions (for example, a submicrometer grain of magnetite) with uniform magnetization. These domains may interact with each other and often have variable magnetic properties even when the material is compositionally homogeneous. Hence rather than a single (bulk) coercivity, a material will have a distribution of coercivities and that distribution will be influenced by how each micromagnetic domain interacts with neighboring domains.

First-order reversal curves (FORCs) are partial hysteresis curves that provide a means of characterizing these micromagnetic properties. A FORC is essentially a single curve or path along or inside the major hysteresis loop (**Figs. 1, 2**). It is obtained by first placing a sample in a magnetic field large enough to reach its saturation magnetization. Usually the applied fields are about 1 Tesla or larger, which is at or near the limit of most laboratory electromagnets. The field is then decreased to some value lower than the saturating field. This new field is referred to as the reversal field (B_r; Fig. 1), which may be the same or opposite sign as the saturating field. The applied field (B_a) is then progressively increased in small steps while measuring the moment (M) of the sample.

To produce a FORC diagram for a sample, multiple FORCs are collected with the first FORC starting at a reversal field just below the saturating field and each additional FORC starting at a reversal field slightly lower than that of the previous FORC, until the reversal field has been decreased to the negative saturating field. The FORC distribution (ρ) is then computed at each point at which a moment measurement was made using a mixed partial derivative given in Eq. (1). This equation indicates that the

$$\rho(B_a, B_r) = -\frac{1}{2}\frac{\partial^2 M(B_a, B_r)}{\partial B_r \partial B_a} \qquad (1)$$

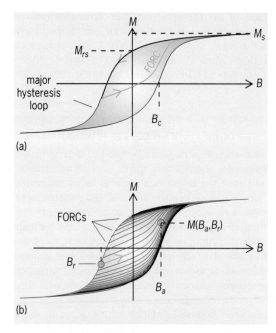

Fig. 1. Magnetic hysteresis. (a) A hysteresis loop is shown with some commonly determined magnetic properties, including the saturation magnetization (M_s), saturation remanent magnetization (M_{rs}), and coercivity (B_c). The colored curve is a single FORC. **(b)** Multiple moment measurements for multiple FORCs are required to generate a FORC diagram. The point $M(B_r, B_a)$ represents a single moment measurement made at an applied field of B_a along a FORC that started at a reversal field B_r.

size of the FORC distribution is related to the change of the moment along a FORC and the change in that change relative to adjacent FORCs. Hence, samples with purely reversible magnetizations, such as paramagnetic materials, would yield zero FORC distributions because, no matter what reversal field is used, all FORCs for a sample would follow the same path and that path would coincide with the major hysteresis loop. Samples with irreversible magnetizations will yield nonzero FORC distributions, with the shape and position of the distribution depending on the coercivity of the domains and the interactions of those domains with other domains that have reversible and/or irreversible magnetizations.

Besides being simply zero or non-zero, FORC distributions can have a variety of modes that may be circular, elongated, or complex shapes and, may even have negative regions. These distributions can be displayed in plots of the applied field versus the reversal field (B_a versus B_r) [Fig. 2c] but are more typically transformed such that the axes correspond to coercivity (B_c) and interaction (also referred to as bias, B_b) using Eqs. (2). Plots created using this transfor-

$$B_c = (B_a + B_r)/2$$
$$B_b = (B_a - B_r)/2 \tag{2}$$

mation are what are referred to as FORC diagrams (Fig. 2e).

The interpretation and modeling of FORC distributions is still developing, with a focus on qualitative description of FORC diagrams combined with quantitative modeling based mainly upon the Preisach

model, which presumes hysteresis of a macroscopic system can be decomposed into a number of tiny subsystems referred to as hysterons. In such models, each hysteron has a switching field and an interaction field. The hysterons are thus analogous to micromagnetic domains that have a coercivity and that may interact with each other.

Both experiments and modeling have clarified the origin of features observed in FORC diagrams and have aided in their interpretation. For example, a material consisting of noninteracting domains with a single coercivity of β would produce a virtual spike on a FORC diagram that would lie on the B_c axis and have its peak at β (**Fig. 3a–c**). Because FORC observations are collected from discrete observations that have noise and that require smoothing, the spike would be smeared somewhat along and across the peak value. If instead the material consisting of noninteracting domains with a normal distribution of coercivities with a mode of β, the resulting FORC distribution would have a very narrow elliptical shape with the long axis of the ellipse occurring along the B_c axis and the peak value of the FORC distribution at β. If these domains were close enough to interact, then the FORC distribution would be widened perpendicular to the B_c axis and the peak value might be offset from the B_c axis, particularly in the interactions produce a mean field across the sample. Introducing a second population of domains with a different coercivity that interacts strongly with the first distribution can produce elongated ridges and troughs, which are typically in the lower half of the FORC diagram and at about a $45°$ angle to the B_c axis (Figs. 2 and 3d–f). Interactions are not limited to those between particles with single domains. Some materials contain ferromagnetic particles that are sufficiently large (larger than about 1 μm) that each particle contains multiple domains separated by domain walls. The domain walls can move and the domains can be reorganized when the sample is placed in a magnetic field. FORC diagrams for materials with multidomain particles generally have low coercivity distributions that are highly elongated perpendicular to the B_c axis. Such behavior results from the many domains within each grain interacting with each other very strongly as well as interactions between grains.

Although FORC distributions are most commonly displayed in in FORC diagrams (that is, B_c-versus-B_b space), additional insights can be gained from different perspectives. For example, even the location of the FORC paths relative to the major hysteresis loop can be informative because those paths that follow the upper portion of the hysteresis loop display reversible magnetization behavior, whereas those that pass through the hysteresis loop display irreversible behavior. The relationship of the FORC distribution relative to the hysteresis loop can be even further studied by transforming the FORC distribution into B-versus-M space, as is done for a chondrule from the Bjurbole meteorite in Fig. 2d. Chondrules are millimeter-size spherules of rock thought to be formed from melting of primordial material in the accretionary disk of the solar nebula. The magnetic

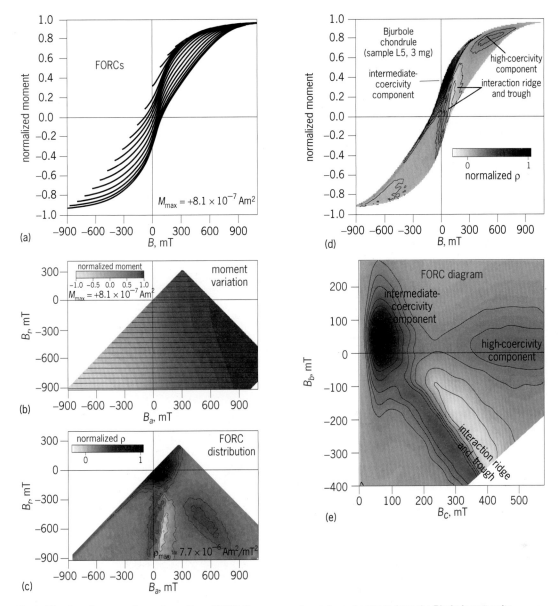

Fig. 2. The steps in measuring and creating a FORC diagram are shown for a chondrule from the Bjurbole meteorite. (a) The FORCs along which moment measurements are collected are the curves. (b) The FORCs (every fourth one is shown) are the horizontal lines. The measured moment is gray-scale contoured in a plot of the applied field versus the reversal field (B_a-versus-B_r). These moment data are used to compute the FORC distribution (ρ) as given by Eq. (1). (c) The FORC distribution is shown in B_a-versus.-B_r space. (d) The same FORC distribution is plotted in B-versus-M space, showing the relationship of the distribution relative to the hysteresis curve. (e) The FORC diagram is produced by transforming the FORC distribution [using Eq. (2)] into coercivity versus bias (= interaction) [B_c-versus-B_b] space. (*From G. Acton et al., Micromagnetic coercivity distributions and interactions in chondrules with implications for paleointensities of the early solar system, J. Geophys. Res., 112: B03S90, 2007, doi:10.1029/2006JB004655*)

minerals in the Bjurbole chondrule are mainly iron-nickel alloys (such as tetrataenite, taenite, and kamacite). Even though an intermediate coercivity component (30–100 millitesla [mT]) dominates the FORC distribution, a very high coercivity component (400–900 mT) exists owing to the presence of tetrataenite. Interactions of the two components produce the conspicuous interaction ridge and trough.

To date, FORC analysis has largely focused on magnetic systems, although the methodology can be applied to other systems that display memory or irreversible behavior, including thermal, electric, elastic, economic, and biologic systems. For example, studies of metal-insulator transitions have

shown that metallic domains can exist within certain materials even below the temperatures at which the materials act as insulators. Similar to magnetic domains, these metallic domains interact with each other, resulting in hysteresis in metal-insulator systems. Examples of applications in magnetic systems include numerous studies on the properties of exchange spring magnets and their uses as magnetic recording media and a growing number of studies on rock magnetic properties. In the latter, the quantification of micromagnetic coercivity distributions and interactions has allowed geoscientists to investigate dissolution processes in sediment, the formation and alteration of oceanic basalts, and the

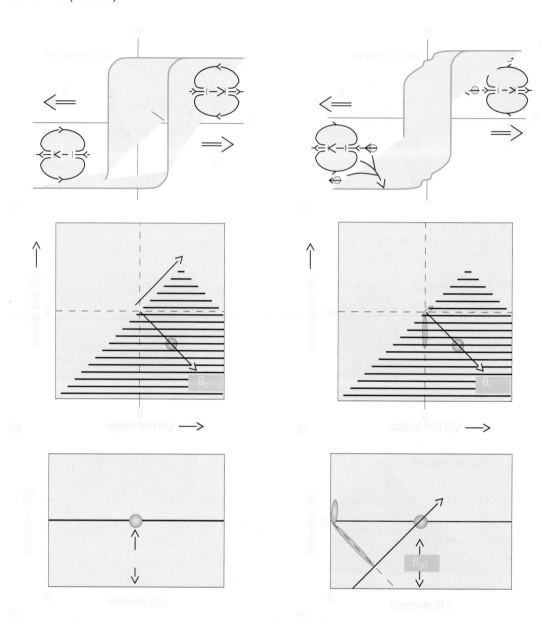

Fig. 3. Schematic models illustrating the FORC distributions for a population of non-interacting magnetic particles, all of which have a coercivity of B_{SD} (a–c), and for two populations of particles with different coercivities that interact (d–f). Parts a and d show the hysteresis loops for the two models with insets that illustrate the orientation of the magnetic moments for the particles at points along the hysteresis curve. For each inset, the applied field B (that is, the field of the electromagnet) is shown as an arrow with double lines. For a magnetic particle, the coercivity of the particle, the applied field, and the field of neighboring magnetic particles are what determine when the moment of the particle will flip or switch. In panels b and e, the resulting FORC distribution is shown relative to the applied field (B_a) and reversal field (B_r). In parts c and f, the FORC distribution is transformed into the standard coordinate system of FORC diagrams. In part f, interactions between the two populations of particles with different coercivities result in the formation of the interaction ridge.

size of ancient magnetic fields generated by Earth and by the early solar system.

For background information *See* DOMAIN (ELECTRICITY AND MAGNETISM); HYSTERESIS; MAGNETIC MATERIALS; MAGNETIC RECORDING; MAGNETISM; MAGNETIZATION; METEORITE; ROCK MAGNETISM in the McGraw-Hill Encyclopedia of Science & Technology.   Gary Acton

Bibliography. G. Acton et al., Micromagnetic coercivity distributions and interactions in chondrules with implications for paleointensities of the early solar system, *J. Geophys. Res.*,112: B03S90, 2007; L. Alonso, T. R. F. Peixoto, and D. R. Cornejo, Magnetic interaction in exchange-biased bilayers: a first-order reversal curve analysis, *J. Phys. D: Appl. Phys.*,43:465001 (8 pp.), 2010; J. E. Davies et al., Anisotropy dependence of irreversible switching in Fe/SmCo and FeNi/FePt exchange spring magnet films, *Appl. Phys. Lett.*, 86:262503 (3 pp.), 2005; C. R. Pike, A. P. Roberts, and K. L. Verosub, Characterizing interactions in fine magnetic particle systems using first order reversal curves, *J. Appl. Phys.*,85:6660–6667, 1999; J. G. Ramírez et al., First-order reversal curve measurements of the metal-insulator transition in VO$_2$: Signatures of persistent metallic domains, *Phys. Rev. B*, 79:235110, 2009.

Fossil record of vertebrate responses to global warming

In the twenty-first century, the Earth is expected to experience a significant [2–4°C (3.6–7.2°F)] rise in global mean surface air temperature (SAT) in consequence of the large volume of greenhouse gases, particularly CO_2, emitted principally by modern industrial and industrializing nations. This rise will be milder in the tropics but more severe toward the North Pole, and will probably be accompanied by other disruptive climatic changes. The net increase in SAT observed in the last century [0.74°C (1.3°F)] has already been implicated in morphological and phenological (periodic life cycle) changes and range shifts by various species. Prediction of ecosystem resilience to projected climate changes and the extent of species responses can be made on the basis of ecological and physiological models. However, the magnitude of warming is unparalleled in recorded (scientific) history, and uncertainties about thresholds and state transitions remain.

An alternative is to turn to the geologic record to examine the influence that rapid global warming had on life in the past. Climate has varied significantly throughout Earth's history, and episodes of global warming are particularly well documented in the record of the last 65 million years. For instance, after the last glacial maximum, about 21,000 years ago, the temperature rose by 4–7°C (7.2–12.6°F) over the course of about 7000 years. The record of this system change is excellent because it is so recent, but interpretation of the consequences of climatic change is complicated in some areas by the fact that the dispersal of *Homo sapiens* into these areas preceded or roughly coincided with the climate change.

Paleocene–Eocene Thermal Maximum. An earlier episode of significant global warming is the Paleocene–Eocene Thermal Maximum (PETM), whose beginning, approximately 56.3 million years ago (MYA), marks the boundary between those two epochs. This event, which was the most severe of a number of similar events known as hyperthermals, was initially recognized by two isotopic anomalies in deep-sea sediments. The first, known as the carbon isotope excursion (CIE), was a very significant but short-term decrease in the ratio of heavy to light stable carbon isotopes ($^{13}C/^{12}C$) [**Fig. 1**]. The CIE was global in extent, indicating a major perturbation of the carbon cycle. Vast quantities of carbon (thousands of gigatons, or trillions of tons) were added to the atmosphere–ocean system from uncertain sources. It has been suggested that isotopically light (^{12}C-rich) carbon from deep-sea methane reservoirs (clathrates) or from volcanic sources flooded the ocean and atmosphere; alternatively, isotopically light carbon may have been released by large-scale conflagrations of biomass or surficial carbon deposits such as peat, or from a combination of sources in sequence. There is no evidence for an extraterrestrial impact at this time.

The second isotopic anomaly was seen in ratios of stable oxygen isotopes ($^{18}O/^{16}O$) [Fig. 1]. These ratios, as preserved in various geological objects, allow estimation of past temperatures in particular places, with higher ratios being associated with lower temperatures. Studies of deep-sea sediment cores, complemented on land by the analysis of fossil leaf margins and other proxy methods, indicate that the anomaly was associated with a hyperthermal event, specifically an increase in mean annual temperature of 5–9°C (9–16.2°F) over 10,000 years or less.

The basic interpretation of these anomalies is as follows: Large amounts of carbon were rapidly released into the atmosphere (and turned into CO_2, if it began as some other form); then, increased CO_2 concentrations drove global warming. Negative feedbacks may have brought temperatures back down again. For instance, CO_2 ultimately participates in the weathering of silicate rocks, and weathering rates are thought to increase in response to higher atmospheric CO_2 concentrations and higher temperatures; peat and other surficial carbon reservoirs may also have been replenished. The duration of the entire PETM event is estimated by various methods to be 150,000–200,000 years.

Endothermic (metabolically thermoregulated) vertebrates. The PETM had profound consequences for the biota of the time. The best terrestrial record of the PETM is found in North America in the state of Wyoming, where sediments of the Fort Union and Willwood Formations have been prospected for nearly 150 years. Fossil collections of recent decades have been more systematic (with detailed stratigraphic data) and provide an unparalleled view of terrestrial community structure over time.

Standing mammal species diversity in Wyoming increased suddenly at the PETM (**Fig. 2**), when established Paleocene lineages were joined during the PETM by a host of newcomers. Extinction at this time, at least at the generic level, scarcely rose above background rates. This increase in diversity is even more pronounced when one normalizes fossil collections for sample size using a statistical technique

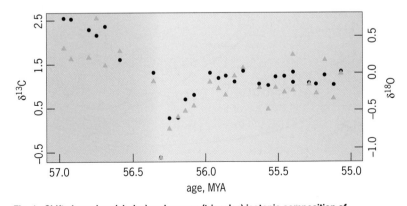

Fig. 1. Shifts in carbon (circles) and oxygen (triangles) isotopic composition of the ocean during the PETM, as seen at Ocean Drilling Program Site 690 near Antarctica. For practical reasons, isotopic ratios are given in delta notation, where $\delta^{18}O = [(^{18}O/^{16}O_{sample} - {}^{18}O/^{16}O_{standard})/(^{18}O/^{16}O_{standard})] \times 1000$, with units of per mill (‰). All data are from tests (shells) of the planktonic foraminifer *Subbotina eocaenica*. Relative temperatures, as reflected in the oxygen isotope ratios, are indicated by the background shades of the graph, with warmer temperatures being indicated by brighter (lighter) shades. Age is given in MYA (million years ago).

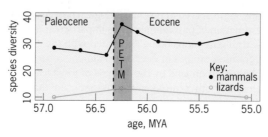

Fig. 2. Species diversity of mammals and lizards across the PETM in Wyoming. Species diversity is statistically corrected for sample size (that is, rarefied). Because of this correction, the increase in lizard diversity near the PETM is subdued with respect to the actual number of known immigrants. Age is given in MYA (million years ago).

called rarefaction. At one time, it was thought that many of these immigrant species came from the south; however, it is now recognized that most of them dispersed into North America (perhaps directly from Asia) across high-latitude corridors made newly habitable by increased temperatures. Intercontinental immigrants quickly formed a numerically significant proportion of the biota.

Earliest Eocene species are, on average, larger than late Paleocene species, which probably reflects the larger average size of immigrants and the small size of the few species that went extinct at the boundary. Curiously, individual body size for a number of

lineages during the PETM is markedly smaller (by up to 50%) than in preceding or succeeding periods. It is not yet clear whether this indicates temporary dwarfing of individual species or rather the temporary ecological replacement of individual species by smaller, close relatives. This circumstance may reflect a number of selective pressures associated with higher temperatures. Decreasing body size is already observed among some extant species.

Bird communities before, during, and after the PETM are virtually unstudied.

Ectothermic vertebrates. Ectotherms, that is, animals that regulate their body temperature by differential exposure to environmental heat sources and sinks, are particularly sensitive to changes in ambient temperature, and they were also dramatically affected by the PETM. The best-studied ectothermic vertebrate groups in this time period are Testudinata (Chelonia) [turtles] and Squamata (lizards, snakes, and relatives); the latter, with about 8500 recognized living species, is the largest such tetrapod group. Recent advances, including methods to reconstruct skeletons from many isolated pieces deriving from different individuals, have provided considerable new data on the spatiotemporal distribution of small squamates.

Even in the late Paleocene, the Earth was considerably warmer, and ectothermic vertebrates consequently thrived at high latitude. Ten species of lizards are recorded in an individual locality in the latest Paleocene of Wyoming; as many as seven lineages of turtles are known in Wyoming. [By way of comparison, in all of Wyoming today, there live seven species of lizards (most belonging to a single subtropical-to-temperate group of Iguanidae) and four species of turtles (softshells, snappers, pond turtles, and box turtles).] In the earliest Eocene, in the middle of the PETM, 18 squamate species, including two snakes, have been documented at a single locality (Castle Gardens). New turtle lineages are also recorded at this time. As with mammals, the difference in faunal composition reflects primarily the addition of new species to an existing fauna of long-established Paleocene lineages, rather than wholesale faunal replacement. (This pattern may also have been repeated in Europe, although stratigraphic resolution is poorer there.)

Thus, many reptile species also immigrated into Wyoming near the Paleocene–Eocene boundary, perhaps in conjunction with the PETM, enriching the existing fauna. One such immigrant lizard during the PETM is *Suzanniwana patriciana* (**Fig. 3**), which is a primitive relative of the extant basilisks. However, where might the immigrants have come from? The phylogenetic (evolutionary) relationships of the immigrant species provide a strong indication. The living relatives of most of the squamate immigrants occur today in the Neotropics (tropics of the Americas). These include the basilisks (Corytophaninae), anoles (*Anolis*), and galliwasps (Diploglossinae). Similarly, one PETM turtle species is related to the living Central American river turtle (*Dermatemys mawii*). Most immigrant squamates appear to

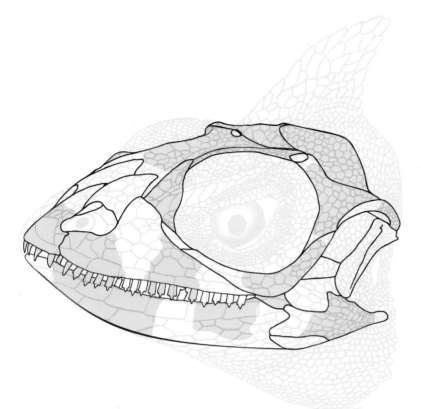

Fig. 3. Reconstruction of the head of *Suzanniwana patriciana*, one of the immigrant lizards in Wyoming, during the PETM. Known portions of the cranial skeleton are tinted. This species is a primitive relative of the extant basilisks (*Basiliscus, Corytophanes,* and *Laemanctus*) of the Neotropics. The scaly covering overlies the reconstructed cranial skeleton. The midline projecting blade on the top of the head is speculative but based on the living relatives and other fossil material; the blade, if present, was not supported by bone.

have moved northward in response to global warming, and so did some turtles.

Amphibian and freshwater fish communities before, during, and after the PETM are virtually unstudied. Invertebrate ectotherms, at least arthropods, are notably smaller during the PETM, as indicated by the trace fossils that have been left.

Prospectus. So far, few deleterious effects on terrestrial vertebrate diversity can be ascribed to the PETM, and this is noteworthy. The proximate cause of the PETM and of anthropogenic global warming is the same: higher greenhouse gas concentrations in the atmosphere. Furthermore, population-level processes of birth, death, and dispersal among many animals are rapid with respect to both the PETM and modern global warming.

However, caution is warranted. The rate of temperature increase during the PETM was as much as two orders of magnitude lower than (that is, 2% of) the projected rate of temperature increase for this century. Thus, the PETM does not offer an exact analog of modern global warming. Moreover, anthropogenic habitat alteration was clearly not coincident with the PETM. Finally, a detailed understanding of the PETM is only now being developed for the terrestrial realm in one small area of Wyoming. Species diversity is a complex concept; local patterns of diversity do not necessarily mirror regional or global patterns. Hence, in addition to better environmental and ecosystemic characterizations of the PETM in individual localities, it will also be important to evaluate the geographic extent of this pattern.

For background information *see* BIODIVERSITY; BIOGEOGRAPHY; CLIMATE HISTORY; CLIMATE MODELING; FOSSIL; GLOBAL CLIMATE CHANGE; GREENHOUSE EFFECT; ISOTOPE; PALEOCLIMATOLOGY; PALEOECOLOGY; PALEONTOLOGY; VERTEBRATA in the McGraw-Hill Encyclopedia of Science & Technology.

Krister T. Smith

Bibliography. W. C. Clyde and P. D. Gingerich, Mammalian community response to the latest Paleocene thermal maximum: An isotaphonomic study in the northern Bighorn Basin, Wyoming, *Geology*, 26:1011–1014, 1998; P. D. Gingerich, Mammalian responses to climate change at the Paleocene-Eocene boundary: Polecat Bench record in the northern Bighorn Basin, Wyoming, pp. 463–478, in S. L. Wing et al. (eds.), *Causes and Consequences of Globally Warm Climates in the Early Paleogene* (GSA Special Paper 369), Geological Society of America, Boulder, CO, 2003; J. H. Hutchison, Turtles across the Paleocene/Eocene epoch boundary in west-central North America, pp. 401–408, in M.-P. Aubry, S. G. Lucas, and W. A. Berggren (eds.), *Late Paleocene-Early Eocene Climatic and Biotic Events in the Marine and Terrestrial Records*, Columbia University Press, New York, 1998; F. A. McInerney and S. L. Wing, The Paleocene-Eocene Thermal Maximum: A perturbation of carbon cycle, climate, and biosphere with implications for the future, *Annu. Rev. Earth Planet. Sci.*, 39:489–516, 2011; K. T. Smith, A new lizard assemblage from the earliest Eocene (zone Wa0) of the Bighorn Basin, Wyoming, USA: Biogeography during the warmest interval of the Cenozoic, *J. Syst. Palaeontol.*, 7:299–358, 2009.

From forest log to products

Two basic necessities—logs and production means—are essential to produce wood products. However, minimization of the environmental loads for a specific wood product is also important. In addition, the costs, both monetary and environmental, of today's wood production methods are higher than necessary. In fact, approximately 10% of the raw materials and production resources are lost as waste. The cost of the waste alone, which is added to the price paid by today's world wood consumers, equals about 10 billion euros (€) per year [about $14 billion (U.S. dollars) per year].

Compared to most engineering materials, wood is a biological and short-cycle renewable material. This is a positive feature because it takes only 10–100 years from seedling to log. However, this has a negative aspect as wood's individual characteristics depend on inheritance and environmental factors.

In the preindustrial era, wood products were better optimized because the carpenter and logger were closely affiliated, if not being the same person. Today, though, a new approach is needed to retain efficient wood production. Automatic wood traceability is one such approach that enables the feedback of information from product user to harvesting and wood production operations. Since 2000, full-scale traceability solutions have been demonstrated in two large European projects: LINESET and Indisputable Key (IK). LINESET was a European Union–financed project conducted between 2000 and 2003; its aim was to facilitate a secure and automated system for the forestry-wood production chain. The main objective of Indisputable Key (2006–2010) was to initiate and stimulate an industrial breakthrough in traceability systems for biological raw materials, specifically wood, leading to substantial economical and environmental improvements in the wood processing chain.

Wood production waste problem. The world production of sawn wood in 2008 was 404 million m^3, using 899 million m^3 of roundwood (that is, cylindrical pieces of timber used without being squared by sawing or hewing). With present wood production strategies, great amounts of waste are generated (**Fig. 1**). For example, at 10% waste, a level estimated by LINESET and IK, the lost revenue is 8–13 billion € per year ($11.2–18.2 billion per year), depending on waste value calculations. In addition, the environmental load per m^3 of wood products produced is mainly related to the excessive (about 10%) energy used. Therefore, the magnitude of this waste cost should serve as a true driving force for needed development and should help to institute a swift change to more efficient and environmentally healthy production technologies.

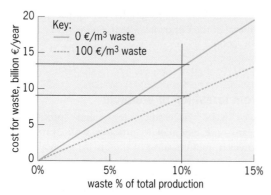

Fig. 1. Simplified correlation of waste % and world cost in billons of euros (€) based on the average cost for roundwood (50 €/m³; $70/m³) and the average price for sawn wood (200 €/m³; $280/m³). At 10% waste, a level estimated by LINESET and Indisputable Key, the lost revenue is 8–13 billion € per year ($11.2–18.2 billion per year), depending on waste value calculations. Note that this value can involve a recycling cost in some cases. Consequently, two alternatives are indicated in the figure: the value of waste is 0 € ($0) [solid line] or 100 € ($140) [dashed line]. The most probable cost (in billions of € per year) for waste should be found somewhere between these lines.

Fig. 2. A harvesting machine. (*Rottne Industri AB*)

The forestry–wood production chain. The automated production of wood, prevailing in the Nordic countries, starts in the forest when the harvester (**Fig. 2**) fells, delimbs, and cuts the tree into logs (saw logs and pulp logs). The logs are transported

from the felling site by forwarders to forest roadside piles, where the logs are picked up by timber trucks for transport to sawmills and pulp mills for further processing into products. This forestry–wood production chain, from harvesting of logs to final cutting of boards at the end of the sawmill process, is schematized in **Fig. 3**. Note that any information for the optimization of individual production steps (for example, positioning the log before it is sawn to optimize volume yield) ends up in a local memory or is discarded. The value of this information is limited to aggregated statistics because it is not associated with individual items as logs or boards.

Five basic stakeholders are involved in the conversion of forest raw materials into wood products. These are the forest owner, the supplier of raw materials, the sawmill operation, the wood production utility, and the final product buyer (**Fig. 4**). At times, some of these can be combined together (for example, forest owner and raw material supplier, or sawmill and wood production), but mostly they are split into groups of actors (for example, the raw material supplier can be split into a forest management company, a harvesting company, and a transporting company). Regardless of the number of actors involved, they all have different short-term goals, including the maximization of profits. However, a long-term relationship with buyers is essential. In addition, the product's final value needs to be considered. Too high of a price would make the buyer more likely to purchase products made from materials other than wood.

In general, the price of the raw materials of wood is based on regional price lists. The price depends on wood species, dimensions, and quality (and also on the proportions of these in some cases). Other factors such as harvesting costs and transport distances (costs) are taken into consideration before a buyer signs the sales contract. Furthermore, market fluctuations with regard to wood production play a large role. Extreme variations from one year to the next can occur. As such, a sawmill needs to have enough long-term financial strength and proper means to react swiftly to minimize any resulting economical consequences or it will soon find itself being placed out of the market.

Chain optimization. Since the start of wood production industrialization in the early eighteenth century, technological advances have dramatically increased wood productivity (cubic meters of wood per sawmill employer). As a recent example, between 1953 and 1995, during the present electrical motor epoch, sawn wood productivity in Sweden increased more than sixfold and is presently well over 1 m³ per person-hour. In one sense, though, wood production today is similar to what it was more than 100 years ago. Generally, a board for a specific product is still graded according to its final production stage and is selected from a flow of boards with similar dimensions. However, to maximize yield and efficiency and to minimize downgrading and production of waste, chain optimization should be implemented in all stages of the process,

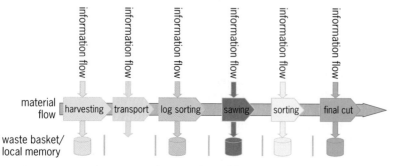

Fig. 3. Local information from various production steps along the forestry–wood production chain is wasted today.

Fig. 4. Players involved in transactions and wood transfer operations from the initial forest raw material to the wood product.

including selection from specific types of trees, cutting of suitable logs at harvest, and final processing of the desired board.

Currently, in 2011, chain optimization is not in full use, mainly because the necessary tools do not exist yet (that is, proven equipment and reliable methods are not yet commercially available). To obtain most of the environmental benefits that chain optimization makes possible, international incentives could speed up the introduction and breakthrough of new technologies. One example is controlling the illegal use of wood (which results in increased CO_2 emissions and the resultant environmental effects) by legislation requiring securely identifiable saw logs. Since 2000, a number of large research projects have highlighted the need to change the prevailing wood production philosophy into chain optimization and also have demonstrated practical solutions to speed up the change. The latest of these projects, Indisputable Key, has introduced wood traceability solutions for the complete production chain from forest to final product.

Many aspects in the forestry–wood production chain are inefficient. In the first process of the chain (felling the tree), the initial choice is either to fell or not to fell. Consequently, the first opportunity for an error can occur at the start of the chain. Then, when logs from the felling site are hauled to the sawmill, they must be transported along the most efficient routes and they must end up in the correct pile for which they were destined. Moreover, when a log arrives at the first sawmill process (the log sorting into saw classes), it is not always evident if it should undergo further processing (or be used for pulp production). Other errors at the sawmill stage include the sorting of logs into the wrong saw class, which means that the logs will not be sawn into the products that they were optimized for in the log sorting station; faulty optimization of sawing settings (for example, rotation and positioning), leading to lower yields; and boards that have been wrongly stacked, reducing the value for both the sawmill and customer. Thus, depending on the needed quality and dimensions, the final outcome can vary greatly, affecting the final value of the wood product. However, it is not known until the end of the production chain (the final cut) if processing of the log was beneficial or not. Of course, any bad decisions result in excessive costs. Any mistake has a cost that the final wood user must pay at the end of the chain.

Therefore, to avoid the shortcomings inherent in the present wood production methodology (that is,

final production-stage selection), a chain optimization or closed-loop control system is ideal (**Fig. 5**). With this type of system, feedback information is available from within the sawmill process and from within the whole forestry–wood production chain process, and at magnitudes faster and more detailed than today.

Components and systems needed for chain optimization. In order to change the current wood production methodology into chain optimization, means must be provided to save, process, and utilize present information. This information is available at most of the production steps. To utilize this information along the production chain, it is necessary for this information to be uniquely addressable, that is, securely connected to the individual logs and boards generated along the process (Fig. 5).

The main components needed for chain optimization include (1) code marking and reading systems for logs and boards (**Fig. 6**), based on radio-frequency identification (RFID) technology and ink-jet printing/optical code detection technology; (2) reliable connections between codes and items; (3) standardized communication between components and between actors; and (4) software to optimize the use of wood.

Log coding technology. The main requirements for log coding technology are code readability, functionality, and minimization of costs. Two quite different technologies have been investigated: the Forest RFID-transponder (tag) system (**Fig. 7**) and SIMPLE (Saw Integrated Micro Printer of Log Ends) [**Fig. 8**], with the latter being an ink-marking technology. These two technologies have been developed because the cost of an RFID-transponder is considerably higher than that of ink marking the log. The present drawback with SIMPLE, though, is a lower amount of codes (up to 1 million) and a lower readability (presently about 60%) compared to RFID (which has a code readability greater than 95%).

Board coding technology. The main requirements for code marking of boards (**Fig. 9**) are similar to those for logs. However, the final system chosen was

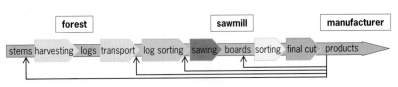

Fig. 5. A chain optimization or closed-loop control system for wood production.

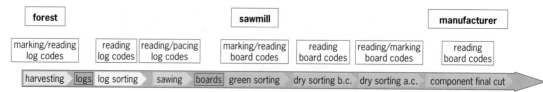

Fig. 6. Marking/reading positions in the forestry–wood production chain from logs to boards. Forest: harvesting machine. Sawmill: log sorting, saw line, green sorting, final sorting before cut, and final sorting after cut. Product manufacturer: component optimization. Pacing (shown above the sawing step) is a method to secure a correct connection between a log and its boards.

based on ink-marking technology, with a code readability of about 85% during industrial production.

Standardized communication between components and software. Because wood production involves many players who need to communicate with each other to reap the full benefits of chain optimization, a communication standard is essential. One of the main wood traceability objectives has been to develop, implement, and make such a standard commonly available.

The information pertaining to an object (log, board, and so on) is referred to as the IAD (Individual Associated Data). The IAD contains a unique identifier attached to the object (for example, an RFID-marked log or a data matrix–marked board). From the "birth" of the log (by the harvester operation) or board (by the sawing operation), information can be

Fig. 7. Forest RFID-transponder (tag) system. (a) Forest RFID-transponder (8 × 1 × 0.5 cm). (b) Automatic tag applicator. (c) Applied RFID-transponder in log. (d) Sawmill reader.

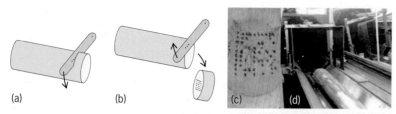

Fig. 8. SIMPLE code marking of logs (an ink-marking technology). (a) Saw sword cuts into the wood and prints at the same time on the top end of the log. (b) Saw sword returns to home position, and the top part of the coded log falls to the ground. (c) Data matrix (10 × 10) code. (d) Code reader in log sorting station.

Fig. 9. Requirements for code marking of boards. (a) Code printer for boards. (b) Board code printing; inset: data matrix (8 × 18) code. (c) Code reading. (Markem-Imaje)

tied to the specific IAD of the log or board, and new information can be added along the production process. The information of the IAD is then contained in a database.

The information exchange is either internal (available only to the company) or external (business-to-business). The system architecture can handle both "push" (to provide data when available) and "pull" (to provide data on demand). The architecture is based on XML (Extensible Markup Language) messages as carriers of information.

The final architecture and communication standard has been reached through testing and discussion. It allows for the exchange of traceable information, including the efficient storage, retrieval, and sharing of information associated with individual objects (for example, logs and boards) and processes (for example, transportation, sorting, cutting, and drying). The identification code is also used for identifying different process steps, for example, a board in the green sorting step versus the final sorting before cutting operation (Fig. 6).

A traceability system thus makes information for individual items available from different stages of the forestry–wood production chain. This information can be incorporated in automated systems to improve the utilization of raw material and production resources (for example, to optimally adjust processing parameters for individual boards, depending on their properties measured earlier in the process; or to sort similar boards together for downstream bulk processing).

Increased profitability. An example of chain optimization profitability can be seen in how a selected raw material can increase yield (with regard to volume and total revenue). In this example, a higher internode length (that is, the distance between branch whorls) means a higher added value for the final manufacturer (as a result of higher productivity and fewer costly glue operations). Hence, if it were possible to select logs with higher internode length and process these into boards for a specific product (for example, furniture blanks for bed constructions), lots of energy, time, and production utilities could be spared. Also, the waste would be less.

Another study verified an end-manufacturer's profitability by analyzing the suitability of the logs used (**Fig. 10**). Logs that were determined to be "suitable" (as opposed to "not suitable") showed significantly increased yields, resulting in increased productivity and revenue.

Future steps. To increase the implementation speed for wood chain optimization components

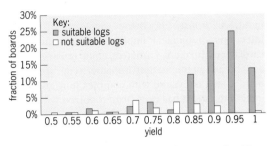

Fig. 10. Yields obtained from different log categories. The yield from "suitable" logs is on average 88%, whereas the yield from "unsuitable" logs is on average 73%.

and systems and to allow a quicker reduction of the current excessive levels of waste and resource consumption, the following objectives should be beneficial: (1) lowering the cost and size of RFID tags; (2) increasing the reliable connection between the log and tag (based on directional antennas and enhanced software solutions); (3) enhancing automatic tag applicator functions (including reduction of production time for application and having increased magazine capacity of more than 500 tags); (4) enhancing code readability and ruggedness; (5) extending implementations and demonstrations in forestry–wood production chains (from small systems with a few production steps to full chains incorporating all steps from the forest to final components/products); (6) decentralizing systems based on scalable IAD storage and transfer (for all sizes of industries); (7) enhancing systems for pacing (based on printing technology); and (8) developing innovative and smart RFID solutions that can deliver identifiable information (based on semipassive solutions with low-cost energy converters instead of expensive batteries).

For background information *see* DATABASE MANAGEMENT SYSTEM; FOREST AND FORESTRY; FOREST ENGINEERING; FOREST TIMBER RESOURCES; INFORMATION SYSTEMS ENGINEERING; LOGGING; LUMBER; WOOD ANATOMY; WOOD PROCESSING; WOOD PRODUCTS; WOOD PROPERTIES in the McGraw-Hill Encyclopedia of Science & Technology.   Richard Uusijärvi

Bibliography. J. L. Bowyer, R. Shmulsky, and J. G. Haygreen, *Forest Products and Wood Science: An Introduction*, 5th ed., Blackwell, Ames, IA, 2007; Swedish Forest Agency, *Swedish Statistical Yearbook of Forestry 2010*, Jönköping, Sweden, 2010; R. Uusijärvi, *Indisputable Key Final Report*, SP Technical Research Institute of Sweden, Stockholm, Sweden, 2003; R. Uusijärvi, *Linking Raw Material Characteristics with Industrial Needs for Environmentally Sustainable Efficient Transformation Processes (LINESET)*, SP Technical Research Institute of Sweden, Stockholm, Sweden, 2003.

Fungal biofilms

Microbiologists have historically studied homogeneous planktonic (that is, free-floating) cells in pure culture. However, the link between heterogeneous sessile (that is, surface-attached) cells, microbial pathogenesis, and human infection is now widely accepted. It is apparent that a wide range of bacteria and fungi are able to alternate between planktonic growth and sessile multicellular communities, commonly referred to as biofilms. In fact, up to 80% of all microorganisms in the environment exist in biofilm communities.

Biofilms are defined as highly structured communities of microorganisms that are either surface-associated or attached to one another, and are enclosed within a self-produced protective extracellular matrix (ECM). The advantages to an organism of forming a biofilm include protection from the environment, resistance to physical and chemical stress, metabolic cooperation, and a community-based regulation of gene expression. In recent years, there has been an increased appreciation of the role that fungal biofilms play in human disease because microbes growing within biofilms exhibit unique phenotypic characteristics compared to their planktonic counterpart cells, particularly increased resistance to antimicrobial agents. In addition to providing safe sanctuary for microorganisms, biofilms may also act as reservoirs for persistent sources of infection in a patient. As such, they adversely affect the health of an increasing number of individuals, including patients with HIV infection, cancer, and transplants; patients requiring surgery or intensive care; and newborn infants.

Clinical significance. Fungi represent a significant burden of infection to the hospital population. The use of broad-spectrum antibiotics, parenteral (intravenous) nutrition, indwelling catheters, the presence of immunosuppression, and disruption of mucosal barriers as a result of surgery, chemotherapy, and radiotherapy are among the most important predisposing factors for invasive fungal infection.

Candida bloodstream infection is the third most common cause of nosocomial bacteremia in patients requiring intensive care and the most common etiologic agent of fungi-related biofilm infection. *Candida albicans*, a normal commensal of human mucosal surfaces and an opportunistic pathogen in immunocompromised patients, is most frequently associated with biofilm formation (**Fig. 1***a*). Indwelling medical devices, such as intravascular catheters, can become colonized with *Candida* species, allowing the development of adherent biofilm structures from which cells can then detach and cause an acute fungemia or disseminated infection. Recently, it has been shown that the cells that detach from the biofilm have a greater association with mortality than equivalent planktonic yeasts. These implant-associated infections are inherently difficult to resolve and may require both long-term antifungal therapy and the physical removal of the implant to control the infection. In addition to *C. albicans*, other *Candida* species associated with biofilm formation and catheter-related bloodstream or device-related infections include *C. glabrata*, *C. parapsilosis*, *C. krusei*, and *C. tropicalis*.

Biofilm-related infections resulting from other yeasts and filamentous fungi have also been described, including *Aspergillus* species, *Cryptococcus*, *Saccharomyces*, *Blastoschizomyces*,

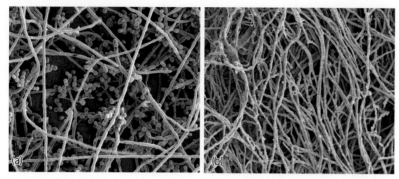

Fig. 1. Scanning electron micrographs of fungal biofilms: (*a*) *Candida albicans* and (*b*) *Aspergillus fumigatus* biofilms. *Candida albicans* biofilms are characterized by the presence of both budding yeasts and hyphae within a complex three-dimensional matrix. *Aspergillus fumigatus* biofilms are composed of long hyphae that are intertwined to form a coherent structure.

Malassezia, *Trichosporon*, *Pneumocystis*, and *Coccidioides*. *Cryptococcus neoformans* is an encapsulated opportunistic yeast that causes life-threatening meningoencephalitis in immunocompromised individuals. This organism has been shown to colonize and subsequently form biofilms on ventricular shunts, peritoneal dialysis fistulas, and cardiac valves. Different *Trichosporon* species, also opportunistic yeasts, can cause disseminated life threatening infections associated with medical implants, including catheters, breast implants, and cardiac grafts. *Blastoschizomyces capitatus* has been associated with catheter-related fungemia, *Malassezia pachydermatis* has been isolated from patients undergoing parenteral nutrition, *Saccharomyces cerevisiae* has been detected from dentures of stomatitis patients, and recurrent meningitis has been associated with a *Coccidioides immitis* biofilm at the tip of a ventriculoperitoneal shunt tubing.

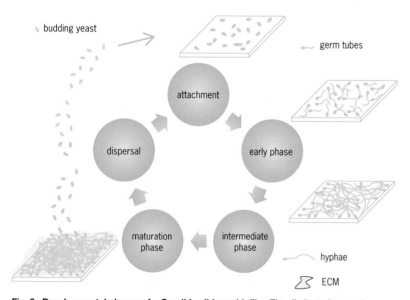

Fig. 2. Developmental phases of a *Candida albicans* biofilm. The distinct phases of biofilm development are demonstrated: initial adhesion, followed by germ tube formation, hyphal elongation, and colonization with some initial extracellular matrix (ECM) production. Maturation of the biofilm is characterized by further ECM encasing the entire structure, which has a three-dimensional structure with channels to allow water and nutrients to penetrate. Budding yeast cells can then disperse from this structure and start the cycle again.

There are also limited reports of the filamentous mold, *A. fumigatus*, being involved in biofilm infections of medical implant devices, including cardiac pacemakers, catheters, joint replacements, and breast augmentation implants (Fig. 1*b*). This mold is also associated with complex lung and sinus infections, and it has even been reported within the urinary tract.

Developmental characteristics. Initial attachment is mediated by electrostatic interactions and specific adhesins that enable direct binding to the surface or to a proteinaceous conditioning film. After attachment, microcolonies are formed and the fungi then multiply, with initial deposition of the ECM. A wide variety of factors contribute to the initial surface attachment, including forces from the flow of the surrounding medium (urine, blood, and saliva), pH, temperature, the presence of antimicrobial agents, and host immune factors. Hyphal growth is an important factor for some fungal species, although not essential. Microscopic analysis has demonstrated that fungal biofilm formation can be separated into five distinct developmental phases (**Fig. 2**): (1) arrival at an appropriate substratum and adhesion (irreversible), (2) early phase (colonization), (3) intermediate phase (hyphal elongation), (4) maturation (ECM production), and (5) dispersal.

Mature biofilms are entirely enclosed in the ECM, which consists of proteins, chitins, DNA, and carbohydrates. The ECM covers the biofilm and is thought to act as a protective barrier, preventing penetration of host immune factors and antifungal drugs, as well as impeding physical disruption of underlying cells. The architecture of the mature biofilm is highly ordered to enable the perfusion of nutrients and the expulsion of waste products and exhibits spatial heterogeneity with both microcolonies and water channels being present, which are characteristics common to bacterial and fungal biofilms.

These features are governed by defined genetic pathways. For example, in *C. albicans*, a gene encoding a major regulator of hyphal development (*EFG1*) has been shown to be involved in regulation of the morphological transition and with ability to form coherent biofilm structures on polystyrene, polyurethane, and glass. Also, members of the *ALS* gene family are expressed on both inanimate and biological substrates, and may facilitate coaggregation with other yeasts and bacteria. Moreover, the transcriptional regulator Bcr1p activates cell-surface protein and adhesin genes required for biofilm formation. The implementation of powerful techniques such as DNA microarray and proteomic analyses to identify global patterns of gene and protein expression during the biofilm lifestyle has enabled researchers to compare planktonic and biofilm cells grown in different conditions [for example, varying levels of nutrient flow, aerobiosis (life existing in air or oxygen), and glucose concentration]. This has demonstrated transcriptional correlation of culture conditions between the biofilms, indicating that similar and specific transcriptional events occur during biofilm formation independent of the growth conditions. The different phases of biofilm development have also been examined in biofilms grown on both denture acrylic and catheter substrates, and it was shown that the transcriptional profiles are both phase- and material-specific.

Pathogenesis of fungal biofilms. The high density of cells present within a biofilm represents a challenge to the host through both direct and indirect interaction with sessile cells and associated products, ultimately leading to inflammation and further host cell damage. It has been shown that both *Candida* biofilm diversity and quantity contribute to high levels of inflammation. It has also been shown that epithelial surfaces are protected from the development of mucosal infections through specific ("toll-like") receptor signaling and neutrophil infiltration. The ECM plays a key role in protecting microbial biofilms from host immune responses, with antibodies failing to penetrate biofilms because of this coating.

The ECM is also important with respect to resistance to antifungal drugs. Recent studies have shown that the matrix binds key antifungal drugs, including the azole and polyene families, and reduces their effects. For example, azoles are up to 1000 times less effective against fungal biofilms compared to planktonic cells in a wide variety of in vitro and in vivo biofilm models. In contrast, liposomal amphotericin B formulations and the echinocandins retain activity against *C. albicans* biofilms. Other key factors responsible for the increased resistance of biofilms to antifungal agents include the structural complexity and increased cell density of biofilms and the differential gene expression of biofilms compared to planktonic cells. Efflux pump-encoded genes (*MDR* and *CDR* genes) have been described as a resistance mechanism against azoles in planktonic cells; however, studies in *Candida* and *Aspergillus* have shown that efflux pumps are differentially regulated in biofilms and are likely to play a role during the early phases of biofilms until maturation, when the ECM plays the predominant role. Another mechanism of resistance is that of persister cells; in this case, a small subset of yeast cells in a *C. albicans* biofilm are highly resistant to polyene drugs following adhesion-independent upregulation of efflux pumps and cell membrane composition.

It is likely that cell-cell communication or quorum sensing (the ability of microorganisms to communicate and coordinate their behavior via the secretion of signaling molecules in a population-dependent manner) plays an important role in pathogenesis and resistance. Farnesol was the first quorum sensing molecule described in fungi and has been found to reduce biofilm development in a concentration-dependent manner and modulate hyphal-related gene expression. The reciprocal molecule, tyrosol, has been found to promote germ tube formation, suggesting that *C. albicans* morphology is under a complex regulatory system of positive (tyrosol) and negative (farnesol) control. Biofilms produce significantly more tyrosol than planktonic cell cultures, and the early stages of biofilm development are enhanced following exposure to tyrosol.

Conclusions. Fungal biofilms are beginning to be featured more prominently in clinical medicine. The diversity of fungal species playing a role in surface-attached infections is increasing, and our ability to treat these infections is hindered as a result of intrinsic antifungal resistance. Scientific progress is continuously being made to understand the biological processes of fungal biofilms so that we can create and develop therapeutic strategies against fungal biofilms.

For background information *see* BIOFILM; CELL ADHESION; FUNGAL BIOTECHNOLOGY; FUNGAL ECOLOGY; FUNGI; HOSPITAL INFECTIONS; INFECTION; MEDICAL MYCOLOGY; MYCOLOGY; YEAST in the McGraw-Hill Encyclopedia of Science & Technology.

Gordon Ramage; Craig Williams

Bibliography. H. R. Ashbee and E. M. Bignell, *Yeast Handbook: Pathogenic Yeasts*, Springer-Verlag, Berlin, 2010; J. R. Blankenship and A. P. Mitchell, How to build a biofilm: A fungal perspective, *Curr. Opin. Microbiol.*, 9:588–594, 2006; R. Calderone, *Candida and Candidiasis*, ASM Press, Washington, D.C., 2002; J. S. Fikel and A. P. Mitchell, Genetic control of *Candida albicans* biofilm development, *Nat. Rev. Microbiol.*, 9:109–118, 2011; F. Odds, *Candida and Candidosis*, 2d ed., Bailliere Tindall, London, 1988; G. Ramage, J. P. Martinez, and J. L. Lopez-Ribot, *Candida* biofilms on implanted biomaterials: A clinically significant problem, *FEMS Yeast Res.*, 6:979–986, 2006; G. Ramage et al., Our current understanding of fungal biofilms, *Crit. Rev. Microbiol.*, 35:340–355, 2009.

Fungal diversity in GenBank

Genetic databases are indispensable in systematics, phylogenetic studies, and identification of fungal taxa of different ranks on the basis of gene sequence comparisons. GenBank, located at the National Center for Biotechnology Information (NCBI) in the United States, and its international database partners, the European Molecular Biology Laboratory (EMBL, United Kingdom) and the DNA Data Bank of Japan (DDBJ, Japan), house the majority of DNA sequences produced by researchers investigating fungal diversity. Each of these databases is updated daily, but at the expense of interchanging the new DNA sequences submitted to any one of them. According to recent information, the size of GenBank exceeds 195 gigabytes and the number of sequences therein is doubled every 18 months. The total number of nucleotide fungal sequences in GenBank is approximately 3.0 million. GenBank also contains data on the number of nucleotide sequences for any specific taxa. For example, *Saccharomyces pastorianus* has 978,121 sequences, *Saccharomyces cerevisiae* has 57,038 sequences, and *Moniliophthora perniciosa* has 53,740 sequences.

Analysis of online fungal data sources. Fungal diversity data obtained from several online sources have been analyzed. These sources include GenBank (NCBI), the Dictionary of the Fungi [Centre

TABLE 1. Fungal diversity of Ascomycota in GenBank

Kingdom	Phylum	Class	Count of genera	Count of species
Fungi	Ascomycota	*Arthoniomycetes*	42	164
Fungi	Ascomycota	*Dothideomycetes*	423	2597
Fungi	Ascomycota	*Eurotiomycetes*	171	1541
Fungi	Ascomycota	*Incertae sedis* (uncertain placement)	172	381
Fungi	Ascomycota	*Laboulbeniomycetes*	11	17
Fungi	Ascomycota	*Lecanoromycetes*	435	3283
Fungi	Ascomycota	*Leotiomycetes*	245	1271
Fungi	Ascomycota	*Lichinomycetes*	22	40
Fungi	Ascomycota	*Orbiliomycetes*	13	137
Fungi	Ascomycota	*Pezizomycetes*	146	693
Fungi	Ascomycota	*Saccharomycetes*	77	1002
Fungi	Ascomycota	*Sordariomycetes*	615	4110
Fungi	Ascomycota	*Taphrinomycetes*	4	51
Total			**2376**	**15,287**

for Agricultural Bioscience International (CABI) and Assembling the Fungal Tree of Life (AFTOL)], Index Fungorum, StrainInfo, and FungalDC. The analysis is possible because of the previously developed format of the specialized FungalDC database, which incorporates three features: (1) a modern classification scheme of fungi; (2) fungal species diversity as deposited in GenBank; and (3) fungal culture species diversity found in collections of the world. In the database, the names of all fungal taxa (from ranks of kingdom to genus) are included. This includes all species names in GenBank and the catalogs of the world's 386 fungal collections. The classification of higher taxa is given in compliance with the 10th edition of the *Dictionary of the Fungi*.

Fungal taxonomic diversity in GenBank. The number of fungal taxa presently in GenBank has been estimated, including taxa from Chromista and Protista, which are maintained traditionally by fungal culture collections and herbaria. Special attention must be given to the fact that DNA sequences are not always deposited with the correct species name. Presently, out of the more than 69,110 fungal taxa entered in GenBank, only 23,254 have definitive species names, equating to 3700 genera. The remaining entries are represented at the level of higher taxa (for example, families, orders, and classes). A considerable number (more than 40,000) do not have any rank at

all. They are designated as fungal isolates extracted from natural environments or have other nonspecific designations.

When considering the total diversity in GenBank, with regard to fungal higher taxa, it is noted that Ascomycota has the greatest number of sequences of known genera and species (63.6% genera, 58.1% species) [**Table 1**], followed by Basidiomycota (29.2% genera, 37.8% species) [**Table 2**]. Other phyla are shown in **Table 3** (Zygomycota: 3.0% genera, 1.6% species; remaining taxa: <5% genera, <3% species).

Deficiencies of fungal genetic data in GenBank. By analyzing the dynamics of fungal data obtained since 1993, the initial stage of this period is characterized primarily by the deposition of putatively identified taxon sequences in GenBank (**Fig. 1**). However, in the last few years, the number of unidentified samples has increased dramatically, partially the result of more journals requiring sequence accession numbers for publication. It also reflects the increase in environmental projects that catalog DNA diversity. Numerous names are derived from environmental sequences obtained from BLAST (Basic Local Alignment Search Tool, a bioinformatics tool used to find regions of similarity between biological sequences). Unfortunately, there are a number of deficiencies associated with this process: (1) Incomplete fungal

TABLE 2. Fungal diversity of Basidiomycota in GenBank

Kingdom	Phylum	Class	Count of genera	Count of species
Fungi	Basidiomycota	*Agaricomycetes*	846	8227
Fungi	Basidiomycota	*Agaricostilbomycetes*	8	28
Fungi	Basidiomycota	*Atractiellomycetes*	5	9
Fungi	Basidiomycota	*Classiculomycetes*	2	2
Fungi	Basidiomycota	*Cryptomycocolacomycetes*	2	2
Fungi	Basidiomycota	*Cystobasidiomycetes*	7	7
Fungi	Basidiomycota	*Dacrymycetes*	11	36
Fungi	Basidiomycota	*Exobasidiomycetes*	35	193
Fungi	Basidiomycota	*Incertae sedis* (uncertain placement)	6	19
Fungi	Basidiomycota	*Microbotryomycetes*	21	201
Fungi	Basidiomycota	*Pucciniomycetes*	70	538
Fungi	Basidiomycota	*Tremellomycetes*	39	425
Fungi	Basidiomycota	*Ustilaginomycetes*	37	261
Total			**1089**	**9948**

TABLE 3. Minor components of GenBank fungal diversity

Kingdom	Phylum	Class	Count of genera	Count of species
Fungi	Blastocladiomycota	*Blastocladiomycetes*	6	14
Fungi	Chytridiomycota	*Chytridiomycetes*	40	88
Fungi	Chytridiomycota	*Monoblepharidomycetes*	3	8
Fungi	Glomeromycota	*Glomeromycetes*	12	132
Fungi	Neocallimastigomycota	*Neocallimastigomycetes*	6	12
Fungi	Zygomycota	*Incertae sedis* (uncertain placement)	111	420
Chromista	Oomycota	*Oomycetes*	4	34
Protista	Microspora	*Microsporea*	73	183
Protista	Amoebozoa	*Myxogastria*	2	26

taxonomic diversity in GenBank. (2) GenBank is not an authorized nomenclature and taxonomy system. (3) The amount of new incorrect records significantly exceeds the number of corrected records. (4) Erroneous sequences (about 20%) are entered. Moreover, the assignment of a certain species name to an organism depends on the researcher's qualifications and cannot exclude the effect of human error: for example, the author can simply misidentify the species by relying on other diagnostic characteristics; the author can follow a certain specific classification scheme, resulting in a conception of a particular taxon that is different from other conceptions; or the living culture or herbarium specimens subjected to sequencing are contaminated, resulting in error-filled data and false conclusions.

Importance of genetic data on type specimens for fungal systematics. However, in spite of the aforementioned limitations, molecular genetic data banks are undoubtedly important for fungal systematics, and GenBank fulfills an extremely important archival function, maintaining a diversity of sequences isolated from different organisms and affording an opportunity of quick data access for comparative analysis. In this connection, it is absolutely essential to deposit molecular genetic data on type specimens, ex-type cultures (cultivations derived from some kind of type material), and herbarium samples for different fungal species in GenBank.

Type specimens and authentic samples are maintained in biological resource centers and fungal

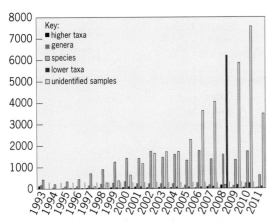

Fig. 1. Dynamics of the data accumulated on different fungal taxa in GenBank.

herbaria. It should be pointed out that this task must be performed not only by curators of collections and herbaria but also by those researchers who use type specimens and authentic samples in their work. The study of type specimens is of great importance for generating new species and variants, particularly if they are maintained in regional collections with limited resources. In such cases, it is necessary to conserve them in duplicated collections and herbaria to facilitate additional molecular studies and to deposit obtained sequences in available databases.

Fungal diversity in world culture collections and other data sets. By comparing the fungal genera present in GenBank with the Index Fungorum database, it is possible to estimate the number of individual taxa identified by molecular genetic methods. The data reveal those taxa that have the greatest number of "unknowns" and those that have the most sequences deposited in GenBank. Because of incomplete submission data, it is impossible to determine how many of these were obtained from type cultures or specimens. The necessity to designate type sample use in different areas of mycological research has been frequently noted by various authors, and this also has been advocated by open letters written by members of the scientific community. In several journals, numerous authors cite the numbers of taxa studied with molecular genetic methods; however, they do not often specify whether these taxa are isolated from types or not, nor do they give other designations of the studied samples. Often only GenBank accession numbers are indicated. Such disregard to the type designation results in a level of uncertainty that limits the taxonomic value for comparative molecular genetic research. To be more definitive on the status of studied samples, it is imperative to carry out labor-intensive verifications of their numbers by comparison of relevant collection and herbarium catalogs, and this is possible only if they are openly accessible.

The possibility of providing an interface between the EMBL genetic database and the database on fungal cultures maintained in the collections of CABRI (Common Access to Biological Resources and Information) has been successfully demonstrated by StrainInfo (the network of Biological Resource Centers). The FungalDC database affords the opportunity to obtain online data about the fungal cultures of certain species in different culture collections of the

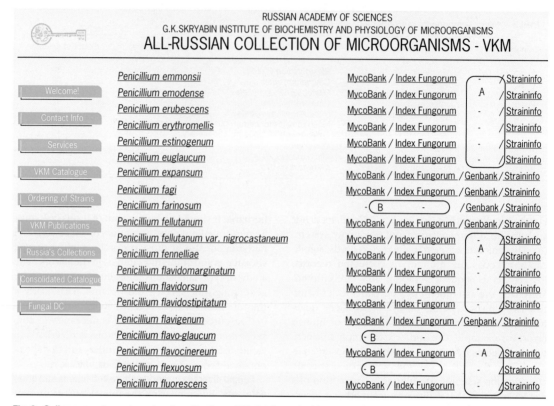

RUSSIAN ACADEMY OF SCIENCES
G.K.SKRYABIN INSTITUTE OF BIOCHEMISTRY AND PHYSIOLOGY OF MICROORGANISMS
ALL-RUSSIAN COLLECTION OF MICROORGANISMS - VKM

Fig. 2. Online computer-screen view from the FungalDC database. In this case, there is an absence of information (blank fields) about different taxa in GenBank ("A") and MycoBank/Index Fungorum ("B").

world. This information is presented in comparison with a fungal diversity in GenBank (**Fig. 2**). The availability of this information allows genetic researchers to locate the necessary living material (including extype and authentic samples) for different systematic groups of fungi. A sample ratio of fungal diversity in the culture collections and in GenBank is represented in **Fig. 3**. These data indicate that the diversity of culture collections has not been sufficiently used yet for studies of the fungal genetic diversity.

Perspectives. Some of the major challenges facing fungal databases include the need for integration of data between resources, standardized formats of data entry, updating and interpreting additional sequences, and developing systems that integrate molecular and conventional information. As examples, the present format of the GenBank database introduced by the International Nucleotide Sequence Database Collaboration (INSDC) for data entry does not include a special field for indicating the type status for a strain, that is, whether the specimen was a type or ex-type for a given species. GenBank is now establishing a curated, nonredundant sequence database (RefSeq Targeted Loci) that will facilitate flagging of extype and authoritative sequences in similarity searches. Likewise, because GenBank acts as both an archive and a database, the freedom to change submitted data is restricted by the willingness of the original submitters to allow this. Efforts are now in progress to allow for third-party notes to be added to the database without explicit changes to the original data. Another way has also been sug-

gested: instead of correcting the records that have been deposited in GenBank, the relevant materials could be placed in FungalDC. This would allow real-time tracing of the accumulated data on nomenclature types for different species. For this purpose, it would be possible to use both FungalDC and

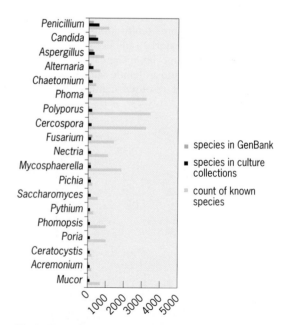

Fig. 3. Comparison of fungal diversity (as of May 2011) in GenBank, world culture collections, and Index Fungorum (count of known species).

MycoBank formats by means of addition of online links with GenBank on the basis of accession numbers for fungal type cultures. Therefore, it would be possible to pool the information about sequences generating from types, the comments of different researchers, and the information concerning other properties of type samples (morphology, physiology, diagnostic characteristics, illustrations, and so forth) into one database. A similar data organization scheme already exists for StrainInfo, which consolidates data on fungal cultures relating to one common strain and maintained in different collections. Again, it is important for access to be open in this database, including the information on each species.

The central problem connected with fungal diversity databases is the lack of precise data indicating which strains have ex-type status. As mentioned previously, the present format of the GenBank database introduced by INSDC for data entry does not include a special field for indicating whether the organism (source of consequence) is a type specimen or an ex-type culture for the given species. Some questions are included about the sources of sequences, but they can be ignored by submitters.

The integration of resources can help to collate fragmentary data that are available in different databases, simplifying access and ensuring their simultaneous use. One of the most important trends in developing these integrated databases is the organization of cooperation when making special formats to introduce comments and analytical notes concerning different records deposited in GenBank. One possibility is to allow corrections of erroneous fungal sequences to be made by third parties unaffiliated with GenBank.

For background information *see* BIODIVERSITY; CLASSIFICATION, BIOLOGICAL; DATABASE MANAGEMENT SYSTEM; FUNGAL GENOMICS; FUNGI; MYCOLOGY; PHYLOGENY; SYSTEMATICS; TAXONOMIC CATEGORIES; TAXONOMY; TYPE METHOD in the McGraw-Hill Encyclopedia of Science & Technology.

Svetlana M. Ozerskaya

Bibliography. D. A. Benson et al., GenBank, *Nucleic Acids Res.*, 38(suppl. 1):D46–D51, 2010; M. I. Bidartonto et al., Preserving accuracy in GenBank, *Science*, 319:1616, 2008; P. D. Bridge et al., On the unreliability of published DNA sequences, *New Phytol.*, 160:43–48, 2003; P. D. Bridge, B. M. Spooner, and P. J. Roberts, The impact of molecular data in fungal systematics, *Adv. Bot. Res.*, 42:33–67, 2005; P. W. Crous et al., MycoBank: An online initiative to launch mycology into the 21st century, *Stud. Mycol.*, 50:19–22, 2004; D. L. Hawksworth, Fungal diversity and its implications for genetic resource collections, *Stud. Mycol.*, 50:9–18, 2004; D. L. Hawksworth, "Misidentifications" in fungal DNA sequence databanks, *New Phytol.*, 161:13–15, 2004; D. S. Hibbett et al., A higher-level phylogenetic classification of the Fungi, *Mycol. Res.*, 111:509–547, 2007; P. M. Kirk et al. (eds.), *Dictionary of the Fungi*, 10th ed., CABI, Wallingford, U.K., 2008; S. Miyazaki et al., DNA data bank of Japan (DDBJ) in XML, *Nucleic Acids Res.*, 31(1):13–17, 2003; R. H. Nilsson et al., Taxonomic reliability of DNA sequences in public sequence databases: A fungal perspective, *PLoS ONE*, 1(1):e59, 2006; S. M. Ozerskaya, G. A. Kochkina, and N. E. Ivanushkina, Fungal diversity in GenBank: Problems and possible solutions, *Inoculum*, 61(4):1–4, 2010; S. M. Ozerskaya et al., FungalDC: A database on fungal diversity in culture collections of the world, *Inoculum*, 61(3):1–5, 2010; E. Pennisi, Proposal to "wikify" GenBank meets stiff resistance, *Science*, 319:1598–1599, 2008; M. Ryberg et al., An outlook on the fungal internal transcribed spacer sequences in GenBank and the introduction of a web-based tool for the exploration of fungal diversity, *New Phytol.*, 181:471–477, 2009; S. L. Salzberg, Genome re-annotation: A wiki solution?, *Genome Biol.*, 8:102, 2007; R. Vilgalys, Taxonomic misidentification in public DNA databases, *New Phytol.*, 160:4–5, 2003.

Harvesting waste energy

Batteries now represent the dominant energy source for electronic devices. However, in addition to environmental concerns, the use of batteries can be troublesome because of their limited energy-storage capacity and finite lifespan, thus necessitating periodic replacement or recharging. For this reason, alternative power sources have been sought to replace batteries, which would allow the devices to function over extended periods of time. A promising approach to this challenge is known as energy harvesting: collecting waste energy from the environment and converting it to usable electricity. The field of energy harvesting has become of growing interest in the past few years as a potential approach for a wide variety of self-powered devices. The waste energy sources available for harvesting are primarily of four forms: light, thermal gradients, motion or vibration, and radio-frequency (RF) electromagnetic radiation (**Fig. 1**). All have received attention as alternative power supplies offering different degrees of usefulness depending on the application. Most fields using this energy harvesting technology rely on

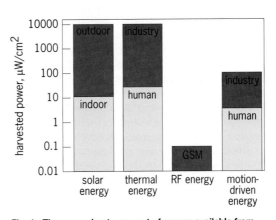

Fig. 1. The approximate amount of energy available from four energy sources. (*Adapted from R. J. M. Vullers et al., Micropower energy harvesting, Solid-State Electron., 53:684–693, 2009*)

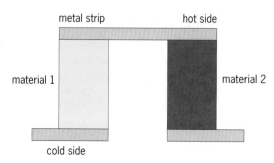

Fig. 2. Schematic of a thermocouple.

sensors located in remote, dangerous, or sensitive areas that require maintenance-free power at low voltages.

Solar energy. The best-known alternative energy source is solar or light energy. Solar energy systems are one of the most commonly considered energy-harvesting solutions. Energy from solar cells results from the photovoltaic effect, which is the conversion of incoming light into electricity. They offer excellent power density in direct sunlight, but indoors, the solar energy available is significantly reduced compared to the outdoor environment. *See* TRANSPORTING LIGHT-HARVESTING MATERIALS.

Thermoelectric energy. Temperature gradients existing in nature and the human body have the ability to generate electricity. The simplest thermoelectric generator is a thermocouple consisting of two legs made of different materials and a metal strip (**Fig. 2**). The temperature difference between the bottom and the top of the legs results in an electrical current. This phenomenon, known as the Seebeck effect, is the basis of thermoelectric generation. For best performance, the legs typically are made of heavily doped *p*-type and *n*-type semiconductors. In practice, a thermoelectric energy harvester is a thermopile, which is usually formed by many thermocouples connected electrically in series and thermally in parallel. A good example of commercial realization of thermoelectric generators is the thermoelectric wristwatch, which demonstrates the conversion of body heat into electrical power to drive the watch. The key advantages of thermoelectric energy harvesters are that they are solid-state devices, reliable, and scalable, making them ideal for small portable applications. Their scalability to small sizes through micromachining has enabled the possibility of harvesting very small amounts of heat for low-power applications, such as wireless sensor networks, medical devices, and mobile devices. Recently, microthermoelectric devices utilizing thin thermoelectric materials have received a lot of attention for powering microfabricated devices. Power produced from the microthermoelectric generators is low, typically in the range of microwatts to milliwatts. Therefore, most research has focused on optimization of thin-film and nanostructured thermoelectric materials and their geometries so as to produce sufficient power and voltage from temperature differences as low as 5–10°C (9–18°F). One of the

challenges is to maintain the temperature gradient between the hot and cold regions at such a small scale.

RF energy. One of the energy-harvesting methods for passively powered devices is to acquire power from propagating RF signals carried by electromagnetic fields, such as television signals, signals from cell phone towers, and signals of wireless radio networks (such as GSM and Wi-Fi). RF energy can be harvested and converted into usable dc voltage through a power-conversion circuit integrated with a receiving antenna. Recently, an ultrahigh-frequency (UHF) antenna connected to a four-stage charge pump power-harvesting circuit has been developed, which can harvest a power of 60 μW from a television station 4 km (2.5 mi) away. Other commercial products based on RF remote-powering technology have also been developed to provide power-over-distance for low-power applications. However, the available power level in the RF energy-harvesting system is an issue, because the propagation of RF energy decreases rapidly as distance from the source is increased. The main advancement that will allow this technology to generate sufficient power is the development of receivers with broad ranges of frequencies, and highly sensitive and efficient power-conversion systems.

Motion-driven energy. One of the most effective methods of implementing an energy-harvesting system is to exploit ambient motion or mechanical vibration. Motion-driven or vibration energy harvesting involves conversion of mechanical energy to electrical energy using an electromechanical transducer. Mechanical-to-electrical energy conversion falls into two types of mechanisms: the direct application of force and the use of inertial forces acting on a proof mass (**Fig. 3**). In the case of a direct-force generator, the force $f(t)$ acts directly on a mass m supported on a suspension with a spring constant k and a damping element c, resulting in the displacement $z(t)$. Energy is converted from mechanical to electrical form if a suitable transduction mechanism is implemented as a damping element. The operating principle of an inertial generator is based on a proof mass m supported on a suspension with a spring constant k. The inertia of a proof mass causes the mass to move relative to the frame with relative displacement $z(t)$ when the frame with displacement $y(t)$ experiences acceleration. The energy of the inertial generator is

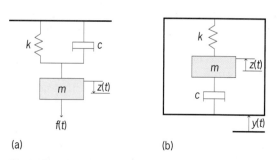

Fig. 3. Generic models of (a) direct-force generator and (b) inertial generator.

converted via a damping element that opposes the motion of the proof mass in the frame.

A variety of motion-powered energy harvesters have been demonstrated, particularly at the microscale. The development of devices able to convert kinetic energy from vibrations, forces, or displacements into electric output has advanced rapidly during the past few years because such energy can be found in numerous applications, including industrial machines, transportation equipment, household goods, civil engineering structures, and portable and wearable electronics. The amount of energy generated by the external mechanical excitations depends specifically on the amplitude of the vibration or force, the frequency of the source, and the efficiency of the transducer. Several transduction methods can be used for energy harvesters, including electromagnetic, electrostatic, and piezoelectric transductions (**Fig. 4**). The choice of transduction method depends mainly on the application, because there is no clear evidence as to which transduction method is preferable.

Electromagnetic transduction. Electromagnetic power conversion is the generation of electric current $i(t)$ in a conductor within a magnetic field B based on Faraday's law (Fig. 4a). The electricity is generated either by the relative motion $z(t)$ of the magnet and the coil or by a change in the magnetic flux. The amount of power generated depends on the velocity of the relative motion, the number of turns of the coil, and the strength of the magnetic field. Electromagnetic generators are widely used for the large-scale generation of power because of their extremely efficient mechanical-to-electrical energy conversion. The use of this transduction mechanism in small-scale energy-harvesting applications is also well researched. Commercially available generators capable of producing output powers in the milliwatt range have been developed by at least three companies. Attempts to miniaturize the generator using MEMS (micro-electro-mechanical systems) technology, however, reduce efficiency levels considerably.

Electrostatic transduction. Electrostatic generators consist of two conductive plates that are electrically isolated via air, vacuum, or an insulator (Fig. 4b). The movement or vibration of one movable conductive plate $z(t)$ causes a voltage change across the plates and results in a current flow $i(t)$ in an external circuit. These generators require a precharge voltage via an external voltage source such as a battery for their operation. The charging of the plates creates equal and opposite charges q on them, resulting in stored charge when the voltage source is disconnected. In general, the power density of electrostatic generators is increased by the reduction of plate surface area or plate spacing. For this reason, electrostatic generators are efficient at smaller scales and are well suited to MEMS fabrication. However, there are two main challenges for practical realization of electrostatic generators: The generators require an input charge and extensive power circuitry to operate, and they produce low amounts of current because of high output impedance.

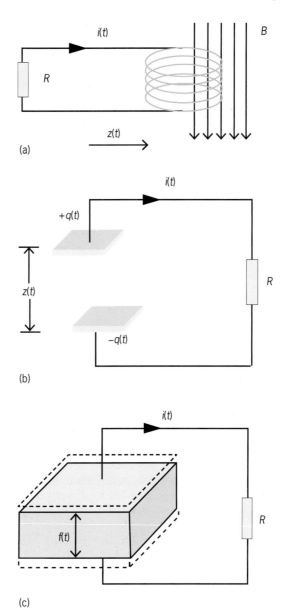

(a)

(b)

(c)

Fig. 4. Principles of operation of (*a*) electromagnetic, (*b*) electrostatic, and (*c*) piezoelectric transduction.

Piezoelectric transduction. Piezoelectric generators utilize the direct piezoelectric effect of a piezoelectric material to transform mechanical strain $f(t)$ into electrical charge $i(t)$[Fig. 4c]. This transduction principle offers the simplest approach, whereby kinetic energy from structural vibrations or displacements is converted directly into electric output with a relatively high power density and no requirement for having complex geometries and numerous additional components. The most often used piezoelectric materials for energy harvesting are aluminum nitride (AlN) and ferroelectric lead zirconate titanate (PZT). One major advantage is that piezoelectric transduction is particularly well suited for miniaturization using microfabrication techniques, because several processes are available for depositing piezoelectric thin and thick films. The piezoelectric materials must be strained directly, and thus their mechanical properties limit the performance, efficiency, and lifetime.

For background information *See* ANTENNA (ELECTROMAGNETISM); ELECTROMAGNETIC INDUCTION; ELECTROSTATICS; FARADAY'S LAW OF INDUCTION; GENERATOR; MICRO-ELECTRO-MECHANICAL SYSTEMS (MEMS); PIEZOELECTRICITY; SEEBECK EFFECT; THERMOCOUPLE; THERMOELECTRICITY; TRANSDUCER in the McGraw-Hill Encyclopedia of Science & Technology.   Don Isarakorn; Danick Briand; Nico de Rooij

Bibliography. S. R. Anton and H. A. Sodano, A review of power harvesting using piezoelectric materials (2003–2006), *Smart Mater. Struct.*, 16:R1–R21, 2007; P. D. Mitcheson et al., Energy harvesting from human and machine motion for wireless electronic devices, *Proc. IEEE*, 96:1457–1486, 2008; S. Priya and D. J. Inman (eds.), *Energy Harvesting Technologies*, Springer, New York, 2009; R. J. M. Vullers et al., Micropower energy harvesting, *Solid-State Electron.*, 53:684–693, 2009.

HD Radio technology

HD Radio™ technology allows broadcasters to upgrade AM and FM radio from analog to digital. The HD Radio digital carriers are broadcast in the current AM and FM bands alongside the existing analog signal, and are therefore termed in-band. Because the digital signal is within the U.S. Federal Communications Commission (FCC) spectral emissions mask established for each AM and FM station, it is also termed on-channel. This design gives HD Radio technology its generic technical name: in-band, on-channel digital broadcasting, or IBOC.

The HD Radio system supports three principal services. The Main Program Service (MPS) enables the simulcast of analog programming in a digital format. The Station Identification Service (SIS) provides data information about the broadcast station. The Advanced Application Service (AAS) offers broadcasters the ability to transmit auxiliary data services to listeners.

The system is comprised of four basic functions: (1) the modem, which modulates and demodulates the signal; (2) a physical layer, which specifies where the digital information is placed in the spectrum and how much throughput is dedicated to specific services; (3) HDC audio compression technology, which compresses, encodes, and decodes the audio signal; and (4) a transport layer, which supports the transmission of compressed audio and data and defines how the physical layer is used. The physical layer design specifies the placement of bits in a station's allocated spectrum to enable the broadcast modes, as well as allowing for a high level of flexibility for broadcasters to control which types of services they broadcast, how much throughput to dedicate to each, and where to locate them in the physical spectrum.

These functions are illustrated by the layered protocol stack in **Fig. 1**. Source material (audio or data) moves down the protocol stack from layer 5 to layer 1 at the transmitter, is broadcast over the air, and is passed back up the protocol stack from layer 1 to layer 5 at the receiver. At the transmitter, layer 5 receives audio or data program content from the broadcaster. Layer 4 provides content-specific source encoding (such as audio compression), as well as station identification and control capabilities. Layer 3 ensures robust and efficient transfer of layer 4 data. Layer 2 provides limited error detection, addressing, and multiplexing. Layer 1 receives the formatted content from layer 2 and creates an IBOC waveform for over-the-air transmission.

FM system. The FM HD radio system includes three modes of operation. In all modes, the digital signal is modulated using orthogonal frequency-division multiplexing (OFDM). In the hybrid mode (**Fig. 2***a*), the existing analog program is simulcast using low-power digital carriers on either side of the analog signal. Without any additional spectrum, stations are able to offer a transition period when listeners can receive both analog and digital signals. The use of redundant digital carriers on either side of the analog signal creates frequency diversity and promotes greater robustness for the digital signal in the presence of multipath or adjacent-channel interference. The hybrid mode supports 99 kbps of audio or data information. The extended hybrid mode is created by extending the OFDM subcarriers inward, closer to the analog signal. This allows the broadcast of up to 148 kbps of digital information. In the all-digital mode (Fig. 2*b*), the analog signal is terminated and lower-power secondary digital carriers are added in the spectrum previously

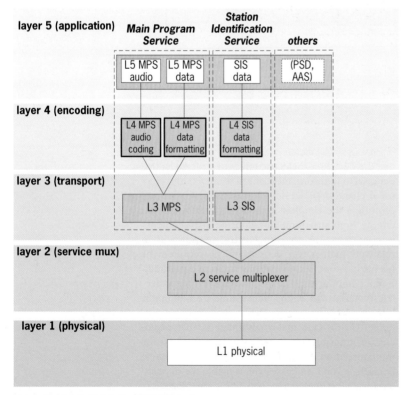

Fig. 1. Protocol stack in the HD Radio system.

occupied by the analog signal. Using the all-digital mode, broadcasters can transmit up to 254 kbps of digital information.

In the hybrid and extended hybrid modes, the system uses time diversity and blending between the analog and digital signals to provide rapid tuning and a backup in case of loss of the digital signal. In the all-digital mode, a secondary digital signal is used for tuning and backup. The hybrid digital carriers are set in power and frequency to minimize interference to and from the existing host analog signal while at the same time providing robust digital coverage. In the hybrid and extended hybrid modes, the digital signal operates at -20 dB relative to the existing analog signal. Stations recently were authorized to increase their power to -14 dBc. In cases where stations are able to demonstrate sufficient geographic separation to avoid interference to adjacent channel stations, they may be authorized to increase power to -10 dBc.

AM system. The AM system uses quadrature amplitude modulation (QAM) on each OFDM subcarrier and offers a hybrid and an all-digital mode. In the hybrid mode (**Fig. 3a**), the digital subcarriers are located in primary and secondary sidebands on either side of the host analog signal, as well as underneath the host signal in tertiary sidebands. The hybrid mode supports 36 kbps of digital information. For stations that experience increased risk of interference from the digital signal into existing analog operations, a reduced-bandwidth version of the hybrid mode eliminates the secondary and tertiary carriers and offers 20 kbps of digital information throughput. In the all-digital mode (Fig. 3b), the analog signal is replaced with higher-power primary digital sidebands in the center of the channel, thereby reducing adjacent-channel interference.

Benefits of technology. The HD Radio system facilitates an efficient upgrade to digital broadcasting by using the existing broadcasting infrastructure and minimizing implementation costs. Analog and digital broadcasts can exist side-by-side so broadcasters can upgrade to digital in their normal equipment replacement cycles. Broadcasters' existing studios, towers, and antenna equipment are maintained, thereby limiting the capital investment required to upgrade to digital operation. Consumers purchasing an HD Radio receiver are able to receive the upgraded digital programming and features provided by HD Radio technology while still receiving existing analog channels. Existing analog receivers remain functional during the hybrid mode of operation.

HD Radio technology provides consumers with substantial advantages over analog. HD Radio broadcasting provides consumers with CD-quality sound in the FM band and FM-like quality sound in the AM band. In addition, HD Radio signals are less prone to multipath interference and other channel impairments than existing analog signals, leading to higher fidelity audio. Through multicasting capability, a digital FM broadcaster is able to offer multiple audio streams over its existing FM frequency, providing consumers with new, diverse, targeted content not

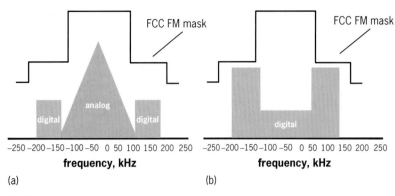

Fig. 2. FM waveforms. (*a*) Hybrid mode. (*b*) All-digital mode.

available on analog AM and FM. These additional channels are commonly referred to as HD2 or HD3. AM broadcasters can also expand their programming options, with FM-quality digital audio enabling them to competitively program music formats again. HD radio technology also provides consumers with new services and interactivity, such as Program Service Data (PSD), which enables stations to display artist name, song title, and station call letters; scrolling data displaying stock quotes, sports scores, and other real-time information on the screen of the radio receiver; tagging, which allows listeners to tag songs they hear on the radio for future purchase; Artist Experience, which allows broadcasters to transmit album art and other images for display on the screen of the receiver; and traffic and navigation updates for receivers.

Regulatory background. Early versions of the HD Radio system were developed in the mid-1990s and tested on the air starting in 1998. The final system design was completed in the early 2000s. In 2001, the National Radio Systems Committee (NRSC), a standards setting organization sponsored by the National Association of Broadcasters and the Consumer Electronics Association, completed an evaluation and issued an endorsement of the FM system. The NRSC completed a similar evaluation and endorsement of the AM system in 2002. Also in 2002, the FCC selected the HD Radio system as the sole upgrade for digital AM and FM broadcasting in the United States and provided interim authorization for AM and FM broadcasters to commence digital operations. In 2005, the NRSC adopted a formal industry standard for IBOC technology known as NRSC-5.

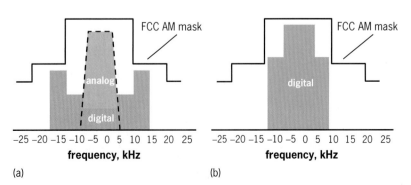

Fig. 3. AM waveforms. (*a*) Hybrid mode. (*b*) All-digital mode.

Subsequently, in 2008, the FCC established the final AM and FM digital broadcasting rules and authorized nighttime digital AM broadcasting, multicasting, and datacasting. In 2010, the FCC authorized the increase in digital power levels allowing broadcasters to operate at up to −10 dBc.

The International Telecommunication Union (ITU) endorsed the AM HD Radio system as a recommended system for digital broadcasting in the broadcasting bands below 30 MHz starting in 2001. In 2002, the FM HD Radio system (referred to as Digital System C) was recognized as a recommended system for digital sound broadcasting in the very high frequency (VHF, 30–300 MHz) band. The HD Radio system has been authorized in Mexico, the Philippines, Thailand, and Panama. Numerous other countries are actively testing the technology.

The system has been on-air commercially since 2002, and consumer receiver products have been available since 2004. As of July 2011, there were more than 2,100 radio stations broadcasting using HD Radio technology and more than 1,200 new multicast streams available. Some stations offer up to four digital streams. There are more than 100 receiver models available in OEM automobile, aftermarket automobile, portable, and home HiFi formats.

For background information *see* AMPLITUDE-MODULATION RADIO; DATA COMPRESSION; FREQUENCY-MODULATION RADIO; MODULATION; RADIO; RADIO BROADCASTING; RADIO SPECTRUM ALLOCATION in the McGraw-Hill Encyclopedia of Science & Technology.   Albert D. Shuldiner

Herschel Space Observatory

The *Herschel Space Observatory* is a satellite carrying out astronomical observations in the far infrared and submillimeter part of the electromagnetic spectrum, at wavelengths between 57 and 670 μm (roughly 100–1000 times longer than those of visible light). It is one of the major missions in the European Space Agency (ESA) science program, and it has been built with substantial additional participation from NASA.

Herschel carries a passively cooled 3.5-m (11.5-ft)–diameter telescope, the largest astronomical telescope yet launched, and three scientific instruments. *Herschel*'s wavelength range is one of the least explored portions of the spectrum, and it is mostly inaccessible from the ground due to absorption by the Earth's atmosphere. The main scientific goals of the mission are to investigate star formation in the Milky Way and in nearby galaxies, and to study the formation and evolution of galaxies over cosmic time.

Around half of the electromagnetic energy that comes to us from the universe is in the far infrared and submillimeter region of the spectrum. This is because much of the radiation produced at shorter wavelengths from luminous young stars, and from the accretion of material onto protostars and onto black holes at the centers of galaxies, is absorbed by surrounding clouds of gas and small grains of solid material, known as dust. Stars form within such clouds, and from the initial stages until the emergence of a fully formed star, no visible light emerges due to obscuration by the dust particles. However, the embedded young stellar objects heat the gas and dust in the surrounding cloud to temperatures that are a few tens of degrees above absolute zero, and the absorbed energy is reradiated at longer wavelengths. Emission from the dust grains produces a broad continuum of wavelengths, and radiation from the gas is in the form of spectral lines characteristic of the chemical composition and the physical conditions. This so-called reprocessing of stellar and accretion energy means that, as galaxies have evolved, most of their energy output during the most active periods of star formation has been in the far infrared. Likewise, in studying the obscured star formation that is going on today in our own or nearby galaxies, it is essential to make observations at these wavelengths. The heavy elements making up the dust grains that are responsible for the obscuration and reprocessing of shorter-wavelength light have been produced by earlier generations of stars. Toward the end of their lives, stars eject large quantities of material into interstellar space, including atoms such as oxygen, carbon, and silicon (which are the building blocks of planets and life). So to investigate the final stages of stellar evolution, we also need to observe in the far infrared and submillimeter region of the spectrum.

Spacecraft and instruments. The *Herschel* spacecraft (**Fig. 1**) comprises the telescope; a cryostat, containing the instruments' focal plane units (FPUs), and a tank of superfluid liquid helium that cools them to ultralow temperatures; the service module (SVM), containing electronics and spacecraft control and telemetry systems; and the sunshield to protect the telescope and cryostat from solar illumination. Electric power is provided by solar panels on the opposite side of the sunshield.

The telescope is of a Cassegrain design, with radiation being collected by the primary mirror and directed by the secondary mirror through a hole in the center of the primary to the instruments below. Both mirrors are made from silicon carbide and are coated with a reflective aluminum layer. Protected by the sunshade, the telescope and the outer vessel of the cryostat are passively cooled (by their own thermal radiation into space) to an operating temperature of approximately 85 K (−188°C or −307°F). The emissivity of the telescope in the operational wavelength range is less than 2%, giving a very low level of telescope thermal emission (which is the ultimate limit to the sensitivity of photometric observations).

All three *Herschel* instruments require cryogenic temperatures for their operation. The FPUs are mounted on an optical bench inside the cryostat, and their warm electronics units are located in the service module. The 2370-L (626-gal)-capacity tank contains superfluid helium at a temperature of 1.6 K. As it boils off, the helium gas flows through a network of pipes to cool the instrument enclosures to a temperature of approximately 5 K. For internal parts

(a)   (b)

Fig. 1. *Herschel* satellite. (*a*) Artist's impression. (*b*) Assembled satellite under test. (*ESA*)

of the instruments that require even colder temperatures, additional direct thermal connections to the liquid helium are provided by high-conductivity copper straps.

The Heterodyne Instrument for the Far Infrared (HIFI) is a heterodyne receiver that operates as a nonimaging (single-pixel) high-resolution spectrometer. It mixes the incoming telescope beam with a local oscillator to generate a band of intermediate frequencies (IF) that contain both the amplitude and phase information of the input beam. The spectral content of the IF band is then analyzed onboard to produce a high-resolution spectrum with resolving power, $\lambda/\Delta\lambda$, of up to 10^6, where λ is the wavelength. HIFI is designed to detect and measure the detailed shapes of spectral features from atoms and molecules. Its advantages over such receivers operating on ground-based telescopes are its high sensitivity, uniform calibration, and unfettered access to the complete spectral range, allowing it, for instance, to detect astrophysically important water (H_2O) lines, which cannot be observed from Earth.

The Photodetector Array Camera and Spectrometer (PACS) works in the 55–210-μm range. The camera uses arrays of silicon bolometric detectors cooled to a temperature of 0.3 K by an internal helium-3 refrigerator. The spectrometer is a grating design, optimized for maximum sensitivity when observing a single spectral line of known wavelength. Its detectors are two arrays of gallium-doped germanium photoconductors working at a temperature of around 2 K.

The spectral resolving power varies between 1500 and 4000 across the band.

The Spectral and Photometric Imaging Receiver (SPIRE) complements PACS by covering the longer-wavelength part of the *Herschel* range. It has a three-band camera and an imaging Fourier transform spectrometer (FTS), both of which use germanium bolometer arrays at 0.3 K. Three camera arrays provide simultaneous imaging of the same field of view in spectral bands centered on approximately 250, 350, and 500 μm. The FTS is based on a Michelson interferometer, and it produces a full spectrum over its entire wavelength range in each observation. Two bolometer arrays are sensitive in overlapping bands covering the whole 194–672-μm range simultaneously. SPIRE's internal helium-3 cooler cools all five detector arrays to 0.3 K, and is identical in all key respects to the PACS cooler.

Launch and operations. The 7.5-m (25-ft)-tall, 3.4-ton *Herschel* satellite was launched from the Spaceport of the European Space Agency (ESA) in Kourou, French Guiana, on May 14, 2009, on board an *Ariane 5* rocket. It will operate until late 2012 or early 2013, the lifetime being limited by the eventual exhaustion of the liquid helium used to cool the scientific instruments. Cryogenic satellites such as *Herschel* are best operated far from the Earth, which would otherwise pose thermal problems, as it would act as a large thermal radiator in the vicinity of the satellite. For this reason, *Herschel* is located near a position known as L2, the second Lagrange point of

Fig. 2. Composite 250/350/500–μm image of a 4 × 4° area of sky, part of the Herschel-ATLAS Key Project. (*ESA; H-ATLAS Consortium*)

Fig. 3. Composite 70/160/250–μm image of the Rosette Nebula, showing a 2 × 2° area of sky, part of the Herschel-HOBYS Key Project. (*ESA; SPIRE and PACS Consortia*)

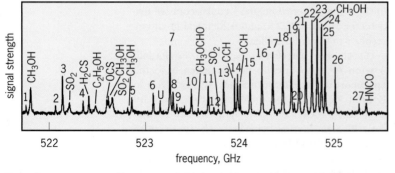

Fig. 4. Small portion of a HIFI spectrum (signal strength versus frequency in GHz) of the Orion Nebula. Spectral lines from numerous interstellar molecules are seen. The numbered lines are from methanol (CH$_3$OH). "U" indicates a currently unidentified spectral line. (*Adapted from S. Wang et al., Herschel observations of EXtra-Ordinary Sources (HEXOS): Methanol as a probe of physical conditions in Orion KL, Astron. Astrophys., 527:A95, 2011, reproduced with permission © ESO*)

the Sun-Earth system. L2 is approximately 1.5 million kilometers (1 million miles) further out from the Sun than the Earth, with the Sun, the Earth, and L2 in a straight line. A satellite near L2 orbits the Sun at the same angular rate as does the Earth, always keeping the same position with respect to Earth and Sun, so that this location provides a highly stable thermal environment.

After 6 months of commissioning, performance verification, and science demonstration, *Herschel* has been in routine science operation since early 2010. At the time of publication, mid-2011, the spacecraft and its instruments were performing as planned, with scientific capabilities and data quality matching or exceeding prelaunch expectations. The first half of *Herschel*'s routine operation phase (roughly 18 months) was devoted to key projects designed to exploit *Herschel*'s unique capabilities to address important science areas in a comprehensive manner and to produce well-characterized and uniform datasets of high archival value. The second half of the mission is being largely used for smaller and more focused investigations, many of which are following up on discoveries made during the key project observations.

Scientific results. The range of astrophysics research that *Herschel* covers is far too extensive to review here, but a few examples are given to illustrate the data quality, the breadth of topics, and the results being obtained. **Figure 2** shows a 4 × 4° area of extragalactic sky (about 60 times the area of the full moon) made in a 16-h observation with SPIRE. The single image reveals more than 7000 galaxies, together with some wispy structures resulting from thin foreground clouds of dust within our own galaxy. The brightness of the depicted galaxies depends on their dust content, and the observed color (relative brightness at the three wavelengths) depends on the distance. Through statistical analysis of the properties of such large samples of galaxies, astronomers are able to reconstruct the star formation history of the universe and to understand the origin and evolution of the galaxies that populate the cosmos today.

Figure 3 illustrates the combined power of PACS and SPIRE surveys of our own galaxy to reveal the structure of the interstellar medium, the nature of the clouds from which stars form, and the physical processes and evolutionary sequences involved. It shows a 2 × 2° image of the Rosette Nebula, about 5000 light-years from Earth, and combines images taken at 70, 160, and 250 μm. Radiation from a group of massive young stars to the right of the image (but not visible to *Herschel*) is seen to be heating and eroding the surface layers of the cloud. Numerous bright compact sources are apparent in the cloud; these are dusty cocoons hiding protostars. This and similar images have been analyzed to reveal the manner in which star formation is associated with the triggering influence of nearby stars and with filamentary structure in the parent clouds, and to investigate the relationship between the mass distribution of star-forming cores and the mass distribution that we observe for fully formed stars.

The spectroscopic power of HIFI is illustrated in **Fig.** 4, which shows a small portion of the spectrum of the Orion Nebula, a nearby region of active star formation about 1500 light-years away. The spectrum shows features from several organic molecules, including methanol (CH_3OH), ethanol (C_2H_5OH), methyl formate (CH_3OCHO), and iso-cyanic acid (HNCO), as well as simpler molecules, such as sulfur dioxide (SO_2) and carbonyl sulphide (OCS). In the complete spectral survey, more than 100,000 spectral features have been detected. The presence and intensities of such lines provide information on chemical composition, temperature, and density, while the detailed line profiles can be used to infer kinematic properties of the gas, such as rotation, inflow, or outflow.

Herschel's combination of angular resolution and wavelength coverage, and its sensitive and versatile instruments, are bringing about a revolution in our understanding of the obscured universe. All *Herschel* data will be maintained by ESA in an archive that is expected to be used as a resource by the worldwide astronomical community for decades to come.

For background information *See* ASTRONOMICAL SPECTROSCOPY; BOLOMETER; GALAXY FORMATION AND EVOLUTION; HETERODYNE PRINCIPLE; INFRARED ASTRONOMY; LIQUID HELIUM; MOLECULAR CLOUD; NEBULA; ORION NEBULA; PROTOSTAR; SUBMILLIMETER ASTRONOMY; TELESCOPE in the McGraw-Hill Encyclopedia of Science & Technology.   Matt J. Griffin

Bibliography. Th. de Graauw et al., The *Herschel*-Heterodyne Instrument for the Far-Infrared (HIFI), *Astron. Astrophys*, 518:L6 (7 pp.), 2010; M. J. Griffin et al., The *Herschel*-SPIRE instrument and its in-flight performance, *Astron. Astrophys.*, 518:L3 (7 pp.), 2010; G. L. Pilbratt et al., *Herschel* Space Observatory. An ESA facility for far-infrared and sub-millimeter astronomy, *Astron. Astrophys.*, 518:L1 (6 pp.), 2010; A. Poglitsch et al., The Photodetector Array Camera and Spectrometer (PACS) on the *Herschel* Space Observatory, *Astron. Astrophys.*, 518:L2 (12 pp.), 2010.

HIV and bone marrow transplantation

This article chronicles a novel treatment that saved a patient from two different diseases, one of which was thought to be incurable. The patient was infected with human immunodeficiency virus (HIV) and also suffered from acute myeloid leukemia (AML). Based on research known about HIV's interactions with T-cells, doctors tailored the patient's treatment to not only be an effective cancer treatment for AML but also a treatment to cure the patient's HIV infection. HIV infects T-cells, which are specialized immune cells that are depleted as the course of HIV infection progresses in the patient. The patient received a bone marrow transplantation from a donor who had a natural resistance to HIV infection. Bone marrow transplantation is also part of the protocol to treat leukemia patients in addition to chemotherapy and radiation treatment.

HIV patient background. Timothy Ray Brown is the first known person to be cured of HIV infection, according to his doctors. In the media, he is referred to as the "Berlin patient." Brown is a U.S. citizen living in Germany, who was infected with HIV during the 1990s. From approximately 2003 to 2007, he began a standard drug regimen called highly active antiretroviral therapy (HAART). HAART is the use of a cocktail of at least three antiretroviral drugs to aggressively decrease the viral load of a patient infected with HIV. HAART usually consists of treating a patient with two nucleoside reverse transcriptase inhibitors and one protease inhibitor, or two nucleoside reverse transcriptase inhibitors and one nonnucleoside reverse transcriptase inhibitor. These drug inhibitors target the HIV reverse transcriptase or a viral protease that is active during the maturation of HIV particles.

Brown's regimen consisted of 600 mg efavirenz (an HIV nonnucleoside reverse transcriptase inhibitor), 200 mg emtricitabine (an HIV nucleoside reverse transcriptase inhibitor), and 300 mg tenofovir (an HIV nucleoside reverse transcriptase inhibitor) per day. During this time, Brown did not experience any opportunistic infections associated with acquired immune deficiency syndrome (AIDS). Brown's HIV drug regimen kept his plasma HIV-1 RNA viral loads undetectable, and he remained healthy.

Acute myeloid leukemia: a curse or a blessing? However, in 2007, he was diagnosed with AML at the age of 40. AML is a type of cancer in which the bone marrow makes abnormal myeloblasts (a type of white blood cell), red blood cells, or platelets. At the time of his AML diagnosis, Brown's plasma HIV-1 RNA loads were not detectable. AML is the most common type of acute leukemia in adults and gets worse quickly if not treated. The standard treatment for AML is chemotherapy to destroy most of the patient's blood cells (a process called conditioning) and then infusing the patient with stem cells from the blood or bone marrow of a human leukocyte antigen (HLA) matching donor. These antigens are present on all nucleated cells of the body, allowing the body's immune system to recognize self or foreign cells. HLA typing, also called tissue typing, plays an important role in the compatibility of tissue, graft, stem cell, and organ transplants. The new stem cells will repopulate the immune system and kill any remaining leukemia cells that survived chemotherapy.

Timothy Ray Brown was treated with two initial courses of chemotherapy; at a later point, a final course of chemotherapy was instituted. The first course of chemotherapy treatment caused harsh side effects. He experienced liver toxicity and kidney failure. His regimen of HAART therapy was discontinued. Brown's viral load shot up to 6.9×10^6 copies of HIV-1 RNA per mL of plasma. Once his viral load stabilized, however, chemotherapy was resumed. For a period of 3 months after the chemotherapy treatments, his HIV-1 RNA load was not detectable.

A special bone marrow donor. Unfortunately, 7 months later, his leukemia returned. Brown's

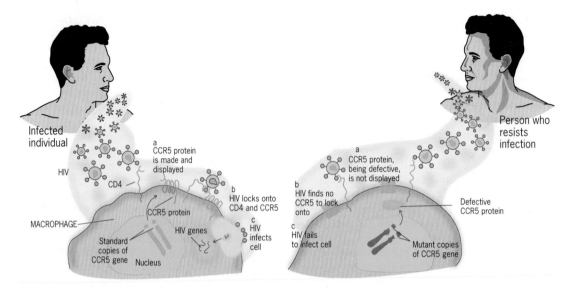

Fig. 1. HIV enters cells by fusing with a cellular coreceptor such as CCR5. Individuals with a defective CCR5 gene (CCR5-Δ32) are resistant to HIV infection because HIV cannot lock onto and enter cells. (*Adapted from http://www.wellesley.edu/Chemistry/Chem101/hiv/resist.gif*)

physician, Dr. Gero Hütter, a hematologist at the Charité University Hospital in Berlin, Germany, came up with the idea of finding a matching HLA-type donor who had stem cells resistant to HIV-1 infection. Because of the work of previous researchers, it was known to Hütter that persons who had two gene copies (making them homozygous) of a defective CCR5 gene, called CCR5-Δ32, were highly resistant to infection by HIV-1. CCR5 is a chemokine (chemoattractant cytokine) receptor. It is a coreceptor for HIV-1 entry into target cells. The CCR5-Δ32 gene codes for a truncated or shorter gene product (a protein that is missing 32 amino acids). HIV-1 cannot gain entry into cells because it is unable to use the truncated CCR5 coreceptor (**Fig. 1**). About 1% of Caucasians carry the homozygous CCR5-Δ32 gene, but it is not present in persons of African descent. Matching HLA-type donors were screened for homozygosity for the CCR5-Δ32 allele; it was not easy to find a donor match with the double mutation.

Following through with this idea and after receiving the consent of the Berlin patient and a clinical internal review board, Hütter initiated a conditioning regimen for Brown that included four chemotherapeutic agents: amsacrine, fludarabine, cytarabine, and cyclophosphamide. The conditioning was followed by a full-body dose of irradiation. Subsequently, Brown received a stem cell transplantation from an HLA-identical donor who was homozygous for the CCR5-Δ32 allele; this was administered in an attempt to repopulate his immune system. In addition, the patient was put on cyclosporine to prevent rejection of the stem cell transplant. He was monitored for rejection and with regard to his HIV status. His viral loads were reduced to undetectable levels after the introduction of the HIV-1-resistant stem cells, and HAART was discontinued. However, nearly a year after the transplantation of CCR5-Δ32 donor cells, Brown's leukemia relapsed again.

The procedure was repeated with a chemotherapy conditioning regimen of cytarabine and gemtuzumab and a single dose of whole-body irradiation. Subsequently, Brown received a second transplant of CCR5-Δ32 stem cells from the same previous donor. Since then, his AML has been in remission for more than 3 years. He is HIV-free and is not receiving antiretroviral therapy. His anti-HIV antibody levels have declined, suggesting that there are no HIV-1 particles to stimulate antibody production. This is why Hütter and a team of doctors believe that Timothy Ray Brown is cured.

A stunning but risky medical breakthrough. The course of Brown's treatment was not easy, and his recovery has been long. At 5.5, 24, and 29 months after he received the first transplantation with the CCR5-Δ32 stem cells, he underwent a colonoscopy, more than 13 colon biopsies, rectal biopsies, and a liver biopsy because doctors thought he was suffering from graft-versus-host disease (GVHD). GVHD is a common complication of bone marrow transplantation. Symptoms include liver damage, rash, and intestinal problems. Some of the biopsy specimens were collected for research purposes. Immunosuppressive drugs to prevent rejection of the new stem cells were discontinued at 38 months after the first transplantation with the CCR5-Δ32 stem cells. Symptoms of GVHD have not recurred since immunosuppressive drugs were discontinued.

Seventeen months after the initial transplantation with the CCR5-Δ32 stem cells, Timothy Ray Brown experienced neurological symptoms that included an unsteady gait (difficulty in walking), coordination problems, changes in his personality, and problems in speech, vision, and memory. In addition, he has needed to continue extensive speech and physical therapy. He also underwent magnetic resonance imaging (MRI) tests, and cerebral spinal fluid has been collected repeatedly. A brain biopsy was performed to determine the cause of the brain abnormalities. It was determined that he was suffering from astrogliosis. Astrogliosis is the proliferation of astrocytes (neuroglial cells) in the brain as a result of the destruction of nearby neurons. Viral infections were ruled out as the cause of the astrogliosis. Doctors concluded that the astrogliosis was the result of the chemotherapy treatments (**Fig. 2**).

Is Timothy Ray Brown cured? Doctors needed to prove that Brown did not harbor any viral reservoirs in body tissues. HIV-1 reservoirs would allow the virus to rebound in the body after HAART was discontinued, and disease progression would occur. To determine that no HIV-1 reservoirs were present in Brown's body, repeated blood samples as well as colon and rectal biopsy tissues were collected and analyzed. All evidence suggests that the donor cells repopulated Brown's immune system. The new donor cells attacked and killed the patient's remaining white blood cells, killing off many of the remaining cells that were a reservoir source of HIV. As the donor HIV-resistant cells repopulated the patient, HIV had no susceptible cells to infect. Early during the course of an HIV infection, the virus infects cells containing the CCR5 coreceptor. Later during the course of HIV infection, the virus infects cells containing CXCR4 coreceptors. The donor cells in Timothy Ray Brown are susceptible to HIV infection through the CXCR4 coreceptor, but so far the Berlin patient still remains HIV-free. His T-cell count has increased to 800 per μL plasma (a normal T-cell count range is 500–1300 cells per μL blood). Of course,

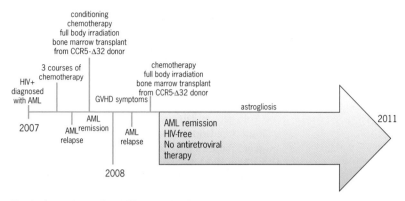

Fig. 2. General overview of Timothy Ray Brown's medical treatment for AML and HIV.

additional proof that this strategy works would be to reproduce this result in additional patients.

The future: CCR5 deletion gene therapy. The amazing medical breakthrough in this case suggests that the development of ways to create CCR5-deficient T-cells may cure more HIV-infected individuals. It is not practical or feasible to submit relatively healthy HIV-infected individuals to massive chemotherapy and whole-body irradiation. Also, as the homozygous CCR5-Δ32 phenotype occurs at very low frequency in the population (1%–3%), it is very difficult to find a matching donor for patients.

However, researchers are exploring the possibility of removing the cells of patients and genetically engineering them to resist HIV infection. A clinical trial in which a DNA-binding nuclease will be used to

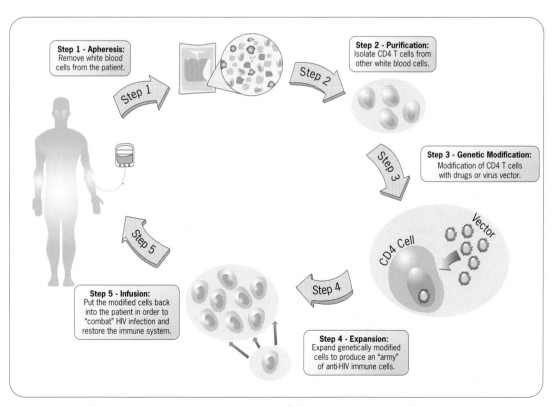

Fig. 3. The development of gene therapies aimed to block HIV–CCR5 interaction may be an effective therapy for patients with drug-resistant strains of HIV-1 or a potential cure if used in combination with mild chemotherapy conditioning. (*Adapted from http://www.virxsys.com/media/MOA.jpg*)

disrupt the CCR5 gene in 12 patients was approved for clinical trials. The CD4+ T-cells were removed from the blood of these patients and treated with the zinc-finger nuclease (known as the SB-728-T agent; a zinc-finger nuclease is an engineered DNA-binding protein that facilitates targeted editing of the genome). The zinc-finger nuclease bound to the CCR5 gene and excised a portion of it, thereby creating a population of T-cells that had a disrupted CCR5 gene; these T-cells were then injected back into the patients. Results to date are encouraging. The modified T-cells rapidly replaced the T-cells depleted by HIV, leading to a strong antiviral effect. Perhaps a combination of modified T-cells in conjunction with mild chemotherapy may open up this treatment to many more HIV-infected individuals (**Fig. 3**).

For background information *see* ACQUIRED IMMUNE DEFICIENCY SYNDROME (AIDS); CANCER (MEDICINE); CELLULAR IMMUNOLOGY; CHEMOTHERAPY AND OTHER ANTINEOPLASTIC DRUGS; CLINICAL IMMUNOLOGY; CLINICAL PATHOLOGY; GENE; GENETIC ENGINEERING; IMMUNOLOGY; IMMUNOTHERAPY; STEM CELLS; TRANSPLANTATION BIOLOGY in the McGraw-Hill Encyclopedia of Science & Technology.

Teri Shors

Bibliography. K. Allers et al., Evidence for the cure of HIV infection by CCR5 Δ32/Δ32 stem cell transplantation, *Blood*, 117(10):2791–2799, 2011; G. Hütter and E. Thiel, Allogeneic transplantation of CCR5-deficient progenitor cells in a patient with HIV infection: An update after 3 years and the search for patient no. 2, *AIDS*, 25(2):273–274, 2011; G. Hütter et al., Long-term control of HIV by *CCR5* Delta32/Delta32 stem-cell transplantation, *N. Engl. J. Med.*, 360(7):692–698, 2009; G. Hütter et al., Transplantation of selected transgenic blood stem-cells—A future treatment for HIV/AIDS?, *J. Int. AIDS Soc.*, 12:10, 2009; M. S. Mueller and J. R. Bogner, Treatment with CCR5 antagonists: Which patient may have a benefit?, *Eur. J. Med. Res.*, 12:441–452, 2007.

Howieson's Poort

The Howieson's Poort is a stone tool industry of southern Africa (**Fig. 1**). It is considered a horizon marker within the Middle Stone Age, which refers to the archeological phase in sub-Saharan Africa roughly equivalent in time to the Eurasian Middle Paleolithic between approximately 300,000 and 50,000 years ago. The industry lasted for roughly 5–8 thousand years between approximately 66,000–58,000 years ago. The name Howieson's Poort derives from a pass in the mountains of the Eastern Cape Province of South Africa. "Poort" is Afrikaans for "gateway" or "portal," and in this case it was named after Mr. Howison (misspelled in the literature as Howieson), who lived in the area in the early 1900s. It is also the geographical area where the stone tool industry was first identified.

The industry is characterized by blade technology with backed (retouched) blades and tools. Backed tools include different geometric shapes such as segments (sometimes called crescents or lunates) and trapezes. Segments are sometimes (but not always) the principal backed tool class, or the "type fossil," associated with the Howieson's Poort (**Fig. 2a**). In addition to the Howieson's Poort being a rather advanced blade-based technology, the geometrically shaped backed tools are also often of microlithic dimensions [smaller than about 20 × 10 mm (0.8 × 0.4 in.)]. Thus, before its discovery sandwiched between other Middle Stone Age industries at the Klasies River site (see **table**), and before the advent of recently developed dating techniques such as optically stimulated luminescence, it was assumed to have been much younger, possibly contemporaneous with the Eurasian Upper Paleolithic (less than 40,000 years ago). During the late 1980s, the world's attention turned towards sub-Saharan Africa when the results of DNA analyses suggested an origin for *Homo sapiens* in this region more than 100,000 years ago. The Middle Stone Age became central to discussions about the origins of anatomically, cognitively, and culturally modern humans; that is, humans who looked, thought, and behaved in ways similar to us. The Howieson's Poort Industry was at the forefront of this debate. It is during this phase that there is a culmination of a range of behaviors signifying levels of cognitive and cultural complexity and variability similar to those of recent hunter-gatherers.

Paleoenvironmental setting. Human behavior and culture unfold against a certain setting. This does not imply a simplistic view of the ecosystem in which we find our subsistence or shelter from the elements. Rather, it requires a multilayered interpretation of the environment as a milieu that would include physical, geographical, ecological, social, economic, cultural, and ritual aspects. For the first time, it is possible to use numerical dating techniques that are free from assumptions about ecological or behavioral patterns. However, detailed correlation between paleoenvironmental conditions and the occurrence of the Howieson's Poort Industry remains mostly hypothetical because continuous and well-dated paleoenvironmental records across the range of southern African biomes are not yet available. Processes of environmental change during the Middle Stone Age in southern Africa were complex, and they cannot be viewed as congruent over time and space; hence, most blanket explanations fall short of adequately addressing contextual variables in human behavior against the setting of environmental change and complexity. It is only by developing detailed localized chronologies for the Howieson's Poort and associated paleoenvironmental databases that the potential relationship between ecosystems, the physical environment, and past human behavior can be explored.

Multidisciplinary studies with archeobotany, archeozoology, and geoscience foci are necessary. At sites with good organic preservation where such studies are part of the research design, results can be used to generate, test, and restrain reconstructions of past ecosystems. For example, investigating hunting behavior at the important site of Sibudu

[KwaZulu-Natal, South Africa (Fig. 1)] from multiple angles showed that (1) segments were integral parts of flexible, composite, and complex hunting technologies (Fig. 2*b, c*); (2) bone points and small crystal quartz segments herald the earliest known use of bows and arrows (Fig. 2*d, e*); (3) species that prefer closed (forested) niches such as the blue duiker and bushpig predominate hunted prey during the Howieson's Poort; and (4) there is circumstantial evidence for the use of snares and traps during the same phase at the site. These innovations and modifications in hunting behavior coincide with a local ecosystem that included a mosaic of habitats. Yet, certain records indicate that maximum summer temperatures were cooler than those of the present, that humidity was greater than usual, and that Sibudu was surrounded by evergreen forests. These data seem to corroborate interpretations of the introduction of the Howieson's Poort as a cultural–social adaptation to changing ecological conditions. It does not imply, however, that it was the only contributing factor; instead, it could have been one of many influences within a multifaceted decision-making process.

Howieson's Poort and change. The most apparent characteristic of the Howieson's Poort is that it represents a definitive change in the approach to lithic technology and its applications long before similar changes took place elsewhere. First, the manufacturing technology changes from the traditional Middle Stone Age flake-based method to a distinct blade-based technology, often produced with soft hammers. Second, there are marked differences in the general and specific sourcing and application of raw materials in comparison with pre- and post–Howieson's Poort phases. Third, there is compelling evidence for an increase in the complexity of hunting and hafting technologies regarding diversity in weapons and prey-capturing systems, the use of raw materials, adhesive recipes, hafting materials, and hafting configurations (Fig. 2). Not only are the temporal boundaries of the industry defined by technological change, but it is also increasingly shown to be a relatively short, dynamic phase within which subtle technological, material, and functional changes occur.

Innovations or changes in subsistence technologies coinciding with ecological challenges imply that new subsistence strategies became appropriate, and that resources could be more efficiently exploited with novel technologies. In addition to potential subsistence and social implications, the increased technological efficiency and economic productivity caused by such innovations could also have led to a substantial increase in the carrying capacity of the environment for human populations. This might have resulted in growing human population numbers and densities. On the other hand, swelling groups and numbers of groups on the landscape would have increased interpersonal and intergroup competition and interaction, stimulating the need to cope with changing physical and social boundaries, and causing florescence in sociocultural

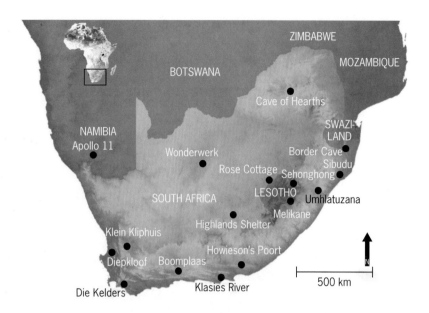

Fig. 1. Map of southern Africa with the location of some Howieson's Poort sites.

behavior. One of the innovations associated with the Howieson's Poort is bow-and-arrow technology (Fig. 2*d, e*). Regardless of disagreement on the place and timing of the origins of this technology, there seems to be consensus that it was used exclusively by *Homo sapiens*. Such complex technologies may have been a key strategic innovation, enabling our species to overcome obstacles, and driving Late Pleistocene human dispersal into western Eurasia after approximately 50,000 years ago.

Developments in cognitive modeling. Symbolic behavior, considered by some as indicative of so-called modern behavior, was expressed during the Howieson's Poort in the form of marine shell beads from Sibudu, engraved ostrich eggshell fragments from Diepkloof, and engraved ochre from Klein Kliphuis (Fig. 1). Similar artifacts, though, have been found associated with older contexts from approximately 100,000 years ago elsewhere in Africa. Increasingly, signs of symbolic behavior are also being found in association with hominin species other than *Homo sapiens*, so its value to distinguish our species in the archeological record is decreasing.

Evidence for complex cognition, however, does not have to hinge on the presence of unambiguous symbolic objects. In-depth information about the application of materials, resulting from the technological repertoire of the Howieson's Poort, provides opportunities for reinforced cognitive modeling. Taking into account neuroscientific principles and applying stringent bridging arguments, certain types of data can be interpreted as behavioral proxies for complex cognition during the Middle Stone Age, informing on types of mental architecture. For example, a series of functional and experimental projects focused on backed tool components from Howieson's Poort sites have resulted in a body of data providing direct and indirect evidence for compound-adhesive manufacture (Fig. 2*a, b*), and

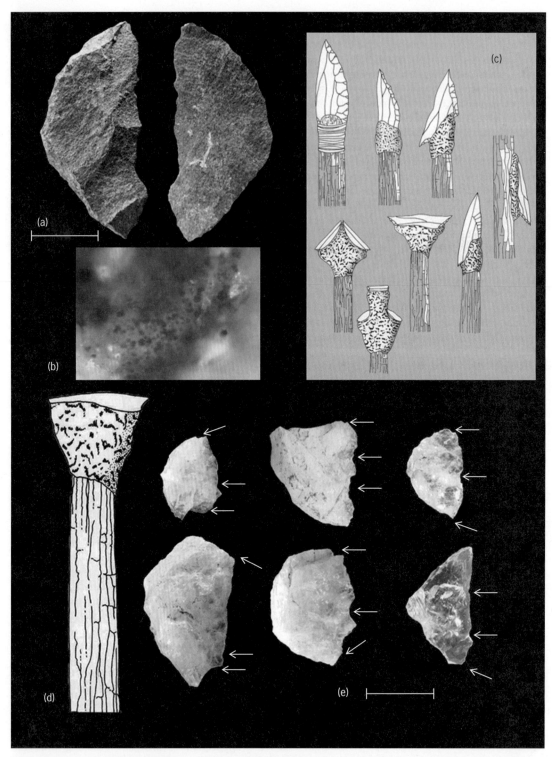

Fig. 2. Hafting and hunting technologies associated with backed pieces from the Howieson's Poort [scale bars = 1 cm (0.4 in.)]: (*a*) hornfels segment from Sibudu with ochre-loaded adhesive along the backed portion; (*b*) micrograph (200×) of ochre-loaded tree gum adhesive recorded on a Howieson's Poort segment from Sibudu; (*c*) potential hafting configurations of segments (*after D. Nuzhnyj, Development of microlithic projectile weapons in the Stone Age, Anthropol. Préhist., 111:95–101, 2000*); (*d*) hypothetical reconstruction of transversely hafted arrowhead; (*e*) macrofractures (*arrows*) on quartz backed tools from Sibudu and Umhlatuzana indicating the use of the tools as transversely hafted arrowheads. (*Copyright* © *Marlize Lombard*)

the hafting of similarly shaped tools in different configurations (Fig. 2*c*). It has been shown that these activities indicate advanced levels of multitasking, mental fluidity, working memory capacity, and the use of abstraction and recursion. On a neuroscien-

tific level and considered together, these traits represent complex cognition, or the type of cognition attributed to people who think like us.

Conclusions. There can be little doubt that there is strong support for behavioral complexity and

Stone tool phases and associated ages*	
Cultural phase	Approximate age range
Final Middle Stone Age (informal designation based on Sibudu sequence)	30,000–40,000 years ago
Late Middle Stone Age (informal designation based on Sibudu sequence)	45,000–50,000 years ago
Post–Howieson's Poort (also referred to as MSA 3)	47,000–58,000 years ago
Howieson's Poort Industry	58,000–66,000 years ago
Still Bay Industry	70,000–77,000 years ago
Mossel Bay Industry (also referred to as MSA 2b)	85,000–105,000 years ago
Klasies River substage (also referred to as MSA 2a)	105,000–115,000 years ago
MSA 1 (informal)	Suggested age: 130,000–195,000 years ago [oxygen isotope stage 6 (OIS6)]

*Although the Middle Stone Age (MSA) cultural sequence of southern Africa is more complex, with contextual variations, this table provides an approximation of stone tool phases with their roughly associated ages.

high-level thought processes during the Howieson's Poort, but the industry is no longer viewed as an anomaly within the Middle Stone Age sequence of sub-Saharan Africa. Rather, it is considered a phase within an ever-changing behavioral dynamic. Based on shifts in the debate of the origins of modern human behavior, pressure to find evidence for modern or symbolic behavior is less intense. Consequently, current research aims to focus more tightly on variation within the phase, assessing the levels of complexity in cognition and behavior, and on hunter-gatherer/forager behavioral model building. Change and diversity as observed in the archeological record throughout the Howieson's Poort, and the immediately pre- and post-dating phases of the cultural sequence in southern Africa (see table), indicate plasticity in decision-making processes. These processes were probably not set in motion by a single factor, but evolved during the interplay between aspects of the social, cultural, economical, and physical environments against which they unfolded. Early evidence of such flexibility represents an important evolutionary step towards the supple responsiveness associated with modern-day cultures.

For background information *see* ANTHROPOLOGY; ARCHEOLOGICAL CHRONOLOGY; ARCHEOLOGY; COGNITION; DATING METHODS; EARLY MODERN HUMANS; PALEOLITHIC; PHYSICAL ANTHROPOLOGY; PREHISTORIC TECHNOLOGY; SOCIOBIOLOGY in the McGraw-Hill Encyclopedia of Science & Technology.
Marlize Lombard

Bibliography. Z. Jacobs et al., Ages for the Middle Stone Age of southern Africa: Implications for human behaviour and dispersal, *Science*, 322:733–735, 2008; M. Lombard, The Howieson's Poort of South Africa amplified, *S. Afr. Archaeol. Bull.*, 189:4–12, 2009; M. Lombard and L. Phillipson, Indications of bow and stone-tipped arrow use 64,000 years ago in KwaZulu-Natal, South Africa, *Antiquity*, 84:1–14, 2010; J. J. Shea and M. L. Sisk, Complex projectile technology and *Homo sapiens* dispersal into western Eurasia, *PaleoAnthropology*, 2010:100–122, 2010; L. Wadley, Compound-adhesive manufacture as a behavioral proxy for complex cognition in the Middle Stone Age, *Curr. Anthropol.*, 51:S111–S119, 2010.

Hubble constant and dark energy

One great surprise in science was the discovery published in 1998 that the expansion of the universe has been speeding up over the past 5 billion years. This cosmic acceleration is attributed to the effects of a "dark energy" that provides a springy quality to empty space itself, producing faster expansion as time goes by. Astronomers are closing in on a precise measurement of the present rate of cosmic expansion (the Hubble constant), determining how the expansion has changed over the 13.7 billion years since the big bang, and using the results to learn more about the nature of dark energy.

A telescope is a no-nonsense time machine that gathers ancient light from distant exploding stars. We measure the apparent brightness of a supernova to compute its distance: that tells us how long the light has been in flight. We also measure the stretching of its light due to cosmic expansion, the redshift: that tells us how much the universe has expanded during that time. Putting these clues together for many supernovae, astronomers have built a persuasive case that the expansion of the universe has not been slowing down, as expected from the effects of gravitation, but actually speeding up.

Doing this measurement correctly is not easy. Though all type Ia supernovae reach nearly the same absolute brightness, type Ia supernovae are not identical, so we have had to learn how to tell which are extra bright and which are extra dim from other clues. The present state of the art allows astronomers to measure the distance to a single well-observed type Ia supernova with an uncertainty of only about 7% (**Fig. 1**).

As a result of all this effort to develop the tools to measure cosmic acceleration, we now have exquisite samples of nearby supernovae that have recently been put to work to improve knowledge of the Hubble constant.

Measurement of the Hubble constant, H. In 1929, Edwin Hubble made a plot of velocity against distance for 46 galaxies and showed that the velocity increases with distance in a simple way: the velocity V is proportional to the distance, $V = H \times$ (distance). The numerical value of the Hubble constant H was a very difficult problem from Hubble's time to the 1990s. Measuring astronomical distances has many pitfalls, most of which were discovered by falling into these traps of misunderstanding or systematic error. Astronomers are always confident, but seldom correct.

This picture changed in the 1990s with the use of the *Hubble Space Telescope* to find Cepheid

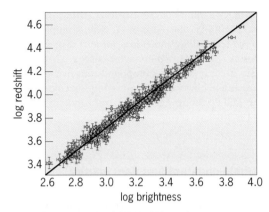

Fig. 1. Distance and redshift for 253 nearby supernovae. The line is the inverse square law. The small scatter about the line shows that the uncertainty in the distance of a single object is only 7%. The intercept of the line with the *y*-axis determines the Hubble constant once the distance to the supernova hosts has been determined. As described, the uncertainty in that measurement is now about 3%. *(From M. Hicken et al., Improved dark energy constraints from ~100 new CfA supernova type Ia light curves, Astrophys. J.,700:1097–1140, 2009)*

variable stars in more distant galaxies (including the host galaxies of type Ia supernovae). The Key Project quoted a value for the Hubble constant, in astronomers' units, of 72 km/s/Mpc $\pm$ 9 (statistical) $\pm$ 7 (systematic). However, there was still room for improvement and an important use for improvement.

Radio astronomers, using the technique of very long baseline radio interferometry (VLBI), have measured the motion of powerful radio sources that orbit the center of the galaxy NGC 4258. This technique permits a very precise distance to be determined to this galaxy, skipping over the nearby Large Magellanic Cloud, which formed the first step in previous calibration of the Hubble constant. This jump is helpful because the Large Magellanic Cloud has lower abundances of the heavy elements than our Galaxy or any of the galaxies that have been hosts to well-studied type Ia supernovae, while NGC 4258 is quite similar to those host galaxies. Uncertainties in the exact way that chemistry affects the properties of Cepheid variables are less important if you vault over the Large Magellanic Cloud to NGC 4258.

The new camera on the *Hubble Space Telescope*, installed in 2009, is sensitive in the infrared, while the camera used by the Key Project operated at shorter, visible wavelengths. Since interstellar dust absorbs infrared light less than it absorbs visible light, the new observations are less sensitive to imperfect knowledge of the amount of dust along the path from a Cepheid to us. What the new program on *Hubble Space Telescope* has done is to find and measure 600 Cepheids, in the infrared, in 8 galaxies that have been the hosts to very well-observed nearby supernovae of type Ia. This provides the calibration for Fig. 1 and a more precise measurement of the Hubble constant. The authors of the most recent study find a value of $H = 73.8 \pm 2.4$ km/s/Mpc, where the uncertainty includes the systematic error. The path forward for this technique is clear: continue to discover and study

well the supernovae that occur in the (rather limited) volume of the universe where it is also possible to measure distances with Cepheids and search for galaxies in addition to NGC 4258 to which radio observations can provide geometric distances. The present uncertainty of 3% can be reduced further.

Importance of precisely measuring H. Is it important to have a more precise measurement of the Hubble Constant? After all, our concept of an evolving universe whose expansion is governed by the properties of its dark matter and dark energy contents unfolding according to the equations of general relativity does not really depend much on whether the Hubble constant is 71 or 73. However, measurements of the subtle ripples in the cosmic microwave background, and the traces of the properties of the big bang that are imprinted in the clustering of galaxies, can be used more effectively if we have prior knowledge of the present expansion rate. One intriguing result is that the present data on the cosmic microwave background and clustering, given this new value for the Hubble constant, has strengthened the case for the presence of four relativistic particles in the early universe. That is interesting because we only know of three neutrino species. Could it be that an improved value of the Hubble constant will make the case for a new subatomic particle, as yet undetected by experiments on Earth? It is too early to say, but it would be a startling example of the way in which studying the largest object we know of—the universe—can help us understand the submicroscopic nature of the world.

Using infrared light to measure dark energy. What's more, the same advantages that arise from being more-or-less indifferent to the effects of interstellar dust accrue to the supernovae as well as to the Cepheids. In the quest to pin down dark energy, the effect of dust is to make a supernova seem farther away than it really is, and this looks like cosmic deceleration. By making observations of distant supernova in light that was emitted in the infrared, we could make more reliable determinations of the properties of dark energy. Because the distant galaxies have substantial redshifts, the light that is emitted in the infrared is shifted to even longer wavelengths. This makes these measurements, which are straightforward for supernovae in nearby galaxies, technically difficult for the distant supernovae. The same infrared camera that is on the *Hubble Space Telescope*, or the more powerful infrared systems being built for the *James Webb Space Telescope*, or more specialized infrared instruments proposed for the *WFIRST* satellite can do this job. The infrared will be the frontier for measurements of dark energy in the future.

Dark energy and the cosmological constant. One question we can ask and answer in a quantitative way is: could the dark energy be a modern version of the cosmological constant? This is the notorious constant that Einstein inserted into his equations for general relativity in 1916 to make the universe static. Once observations established that the universe was not static, but expanding, Einstein regarded the cosmological constant with disgust, and banished it from further consideration.

But not all his contemporaries agreed. Georges Lemaître, one of the inventors of expanding cosmological solutions to Einstein's equations, noted that the presence of a cosmological constant, which technically provides a negative pressure, could make the universe accelerate. Lemaître speculated that this might account for the expansion that was being observed by Hubble and others. In 1934, he said "Everything happens as though the energy in vacuo would be different from zero... we associate a pressure $p = -\rho c^2$ to the density of energy ρc^2 of vacuum. This is essentially the meaning of the cosmological constant Λ."

Today, we can use the measurements of supernova distances and redshifts to test whether the pressure p is or is not equal to -1 times the energy by observing, as best we can, the onset of cosmic acceleration. We call this number the equation-of-state index "w." If the dark energy is the cosmological constant, then we will observe $w = -1$. If the dark energy is something different, for example, a dark energy that is not constant, but changes with time, we might observe a value of w that is different from -1. For convenience, it is handy to discuss the quantity $(1 + w)$, which is zero for the cosmological constant.

How are we doing on pinning down the value of $(1 + w)$? Today, the state of the art has a sample of 580 type Ia supernovae, which constrain the value of $(1 + w)$ to be 0.015 ± 0.07 (statistical) ± 0.08 (systematic) [**Fig. 2**]. To state this simply, everything we know today is consistent with the dark energy behaving like the cosmological constant.

We do not, as yet, have any convincing explanation for the reason why the cosmological constant that is observed is so much smaller (by a factor of 10^{60}) than the estimates based on quantum mechan-ical ideas about how the vacuum ought to behave at the scale that corresponds to gravitational effects. Whether the true answer lies in a time-varying dark energy or modifications to general relativity, we do not yet know, but measurements underway today will help guide tomorrow's thoughts.

For background information *see* ACCELERATING UNIVERSE; CEPHEIDS; COSMIC BACKGROUND RADIATION; COSMOLOGY; DARK ENERGY; DARK MATTER; GALAXY, EXTERNAL; HUBBLE CONSTANT; HUBBLE SPACE TELESCOPE; INFRARED ASTRONOMY; MAGELLANIC CLOUDS; NEUTRINO; RADIO TELESCOPE; RELATIVITY; SUPERNOVA; UNIVERSE in the McGraw-Hill Encyclopedia of Science & Technology.

Robert P. Kirshner

Bibliography. W. L. Freedman and B. F. Madore, The Hubble constant, *Annu. Rev. Astron. Astrophysics*, 48:673–710, 2010; M. Hicken et al., Improved dark energy constraints from ~100 new CfA supernova type Ia light curves, *Astrophys. J.*, 700:1097–1140, 2009; G. Lemaître, Evolution of the expanding universe, *Proc. Natl. Acad. Sci.*, 20:12–17, 1934; K. S. Mandel, G. Narayan, and R. P. Kirshner, Type Ia supernova light curve inference: Hierarchical models in the optical and near-infrared, *Astrophys. J.*, 731:120, 2011; A. G. Riess et al., A 3% solution: Determination of the Hubble constant with the *Hubble Space Telescope* and Wide Field Camera 3, *Astrophys. J.*, 730:119, 2011; N. Suzuki et al., The Hubble Space Telescope Cluster Supernova Survey: V. Improving the dark energy constraints above $z > 1$ and building an early-type-hosted supernova sample, *Astrophys. J.*, in press.

Hubble Ultra Deep Field

The Hubble Ultra Deep Field (HUDF) represents the deepest image ever produced of the universe at visible wavelengths (that is, the image containing the faintest objects ever seen) [**Fig. 1**]. The first HUDF image was created by combining exposures obtained between September 2003 and January 2004 by the Advanced Camera for Surveys (ACS) onboard the *Hubble Space Telescope*. The total exposure time exceeded 300 h.

History of the Hubble Deep Fields. The *Hubble Space Telescope* (HST) had first observed a deep field in 1995. The so-called Hubble Deep Field (HDF) project was an exploration of what deep exposures (that is, exposures that would detect very faint objects) using Hubble's Wide Field and Planetary Camera 2 (WFPC-2) could reveal. The idea for the HDF project was not selected through peer review, as is customary in allocating time on the *HST* and all other major facilities, but was carried out using Director's Discretionary time by Robert Williams, then Director of the Space Telescope Science Institute (STScI), the organization operating *Hubble* on behalf of the National Aeronatîcs and Space Administration (NASA). Because of this noncompetitive selection process, the data were combined at STScI and the final images released to the scientific community in such a way

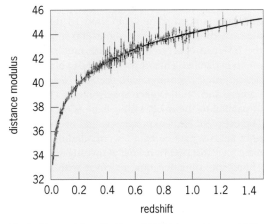

Fig. 2. The world's collection of 580 high- and low-redshift supernovae. In this plot, redshift is plotted linearly on the *x*-axis. The line shown is the fit of a model with cosmic acceleration corresponding to a universe that is 73% dark energy and 27% gravitating matter. (*From N. Suzuki et al., The Hubble Space Telescope Cluster Supernova Survey: V. Improving the dark energy constraints above z > 1 and building an early-type-hosted supernova sample, Astrophys. J., in press, and http://arxiv.org/abs/1105.3470*)

Fig. 1. The Hubble Ultra Deep Field. The image was obtained by combining data in 4 wavelength bands obtained by the *Hubble Space Telescope* Advanced Camera for Surveys. The bands used range from the blue (435 nm) to the near-infrared (900 nm).

as to provide a level playing field for all researchers. This model of public data release to the community was very successful and proved to have longer lasting effects than even the HDF data themselves.

Another long-lasting consequence of the HDF is the prominence it gave to the Lyman break technique to identify high-redshift galaxies. The technique is based on the idea that neutral hydrogen in the galaxy and in the intervening intergalactic medium will absorb all ionizing ultraviolet (UV) radiation below a certain limit (known as the Lyman limit) and also, as the redshift increases, increasingly dim the spectrum at wavelengths between the Lyman limit (91.2 nm) and the Lyman alpha line (121.6 nm). This absorption due to hydrogen introduces a clear discontinuity in the spectral energy distribution, and this discontinuity can be used to identify galaxies at high redshift using broadband filters, with reasonable confidence and without the need for costly spectroscopy.

Once the ACS was installed on *Hubble* in 2002 during Servicing Mission 3B, it was natural for the new STScI Director, Steven Beckwith, to think about repeating a deep field project with the new, more capable instrument. The new project, which came to be known as the Hubble Ultra Deep Field, was targeted to the study of the galaxies responsible for the reionization of hydrogen. The idea was a consequence of the discovery a few years earlier that diffuse hydrogen in the universe was reionized by about 1 billion years after the big bang, at a redshift of 6 or so. This was based on obtaining spectra of redshift 6 quasars discovered with the Sloan Digital Sky Survey (SDSS). These spectra showed evidence that there was significant absorption by intervening diffuse neutral hydrogen for those quasars at redshift greater than 6, while for quasars at smaller redshift, hydrogen absorption was due to individual neutral hydrogen clouds and there was an absence of any diffuse neutral hydrogen. Hydrogen became

neutral after recombination, which took place about 380,000 years after the big bang, at a redshift of 1300, and caused the release of the cosmic microwave background (CMB) radiation. If most diffuse hydrogen was reionized at redshift 6, then there had to be sources—most likely galaxies—responsible for it, and one could theoretically predict their properties. When this was done, it was found that an Ultra Deep Field (UDF) survey with the ACS was capable of detecting the objects responsible for reionization. The UDF project was thus conceived to detect these objects.

Hubble Ultra Deep Field (HUDF) survey. The duration of the Hubble UDF project was chosen to be 400 orbits of the *HST* around the Earth. The *HST* orbits the Earth once every 96 minutes, and the orbits are a natural way to measure time for *HST* (each orbit being effectively one *HST* day). 400 orbits was the total amount of Director's Discretionary time that Beckwith could spare in two years, and it was about twice as long as was spent on the HDF. Spending twice as much observation time with a more sensitive instrument like the ACS promised to open up significant discovery space. The observations were very carefully planned, because it was realized that this field would become the main staple of high-redshift studies. Thus it had to be easily observable by other facilities.

The specific target chosen for the HUDF was within the southern field of the Great Observatories Origins Deep Survey (GOODS), a peer-review-approved program exploring a much wider field at shallower depth. The field was covered by very deep exposures using the *Chandra X-ray Observatory* telescope and the *Spitzer Space Telescope*, enabling multiwavelength studies. The field also had low galactic dust and no bright nearby stars or galaxies, and was ideally placed for follow-up by the terrestrial Atacama Large Millimeter Array (ALMA), a large international radio telescope operating at millimeter wavelengths that began operations in 2011. A field observed with the *HST* undergoes eclipses, and the instruments need some setup time, so that the effective exposure time was about 300 h. However, the *HST* can observe the sky with multiple instruments in parallel. Thus, it was decided to obtain deep near-infrared exposures with the *HST* Near-Infrared Camera and Multi-Object Spectrometer (NICMOS) in parallel to the main, visible-light ACS exposures. The deep NICMOS exposures would be in two distinct fields separated by a few arcminutes from the main UDF. NICMOS's principal investigator, Rodger Thompson, obtained time to cover with NICMOS, albeit at a shallower depth, the full UDF surveyed by the ACS so that some near-infrared data would be available even for the main field.

Normally when talking about the UDF one refers to the main UDF, observed the longest with the Advanced Camera. However, the UDF effectively consists of three fields, the main one and the two adjacent parallel fields. Each of these fields covers an area of the sky that is about 60 times smaller than the full Moon.

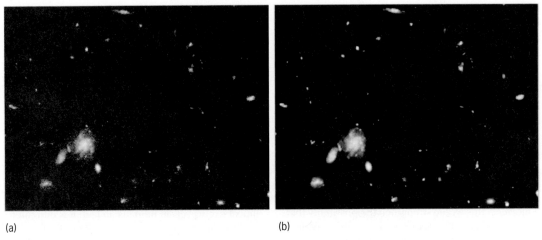

(a)   (b)

Fig. 2. Comparison of (*a*) the Great Observatories Origins Deep Survey and (*b*) the Ultra Deep Field. GOODS images were obtained over a larger area with an exposure time of about 5 h, against the 300 h spent in a single pointing on the UDF. The bright galaxies are detected in both images but there are many faint galaxies visible only in the UDF. The difference in background noise level is also very noticeable.

UDF images. The raw images obtained from the *HST* need to be processed to be scientifically useful, because they contain instrumental signatures due to imperfections in the detectors and optics. The ACS images, processed to remove the instrumental signatures, were released on March 9, 2004. The depth of the images is nicely illustrated by the comparison of a small area observed within the GOODS survey and the same area as seen in the UDF (**Fig. 2**). The bright objects appear about the same, but the background level in the GOODS image is much higher and several faint galaxies are lost in the noise. In contrast, the faintest light sources significantly detected in the UDF have a flux of about 4 nanojanskys or 4×10^{-35} Wm^{-2}Hz^{-1}. This sensitivity is sufficient to detect from Earth a 2-W incandescent tungsten night-light bulb at the distance of the Moon. One can also appreciate how faint this really is by looking at the equivalent photon rate, which is about 10 photons per minute over the whole *HST* primary mirror aperture.

UDF science. The release of the Ultra Deep Field images was quickly followed by the publication of several scientific papers based on these data. Many of the early papers focused on the highest-redshift galaxies visible in the images (redshift 6) and on the implications of their detection for the reionization of hydrogen. Several groups realized that the inferred emission of ionizing radiation from the detected galaxies was tantalizingly close to what would have been needed to reionize hydrogen but not quite sufficient, unless one assumed a very blue spectral energy distribution; such an energy distribution would be characteristic of extremely metal-poor or possibly primordial stars (the so-called population III), coupled to a very high fraction of ionizing radiation escaping from galaxies. In the years since the release of the UDF data, the debate about reionization has centered on assessing the contribution of galaxies below the detection threshold, on understanding the spectral energy distribution appropriate to as-

sume for galaxies at such high redshifts, and on the assumed or estimated values of the escape fraction. Most results to date depend on assumptions about these three hard-to-measure quantities.

In addition to the papers on reionization, the UDF was used to study low-surface-brightness features in intermediate-redshift galaxies, galaxy size evolution, massive evolved galaxies at intermediate and high redshift, substructure in galaxies, and the evolution of elliptical galaxies.

UDF follow-up programs. The initial UDF studies demonstrated that we were very close to identifying the sources of reionization but not quite there. Obtaining deeper data than the UDF or spectra for the redshift 6 galaxies was unlikely with available technology, so the focus moved to extending the search to higher redshifts, which also implied moving to near-infrared wavelengths.

The two NICMOS near-infrared parallel fields already observed as part of the UDF represented the deepest near-infrared data available at the time. The first follow-up of the UDF, dubbed UDF05, was proposed (and granted time through peer review) to exploit these deep infrared fields by obtaining visible ACS data of matching depth. In fact, even though redshift 7 galaxies would be detected only in the infrared spectrum, one needs deep visible data to exclude all the lower redshift "interlopers." The small field of view of NICMOS and its modest sensitivity turned out to be limiting factors for these searches and only a handful of galaxies at redshift 7 were found in the UDF, making it hard to draw major conclusions from them.

Luckily, the *HST* project was at work developing the Wide Field Camera 3 (WFC3). The WFC3 was installed on *Hubble* during Servicing Mission 4 in 2009 and was immediately put to use to carry out another peer-review-selected follow-up program: UDF09. The WFC3 instrument has two channels, the ultraviolet-visible channel (WFC3/UVIS) and the near-infrared channel (WFC3/IR). The near-infrared channel WFC3/IR has about 40 times the discovery efficiency of NICMOS thanks to its larger field of view and higher sensitivity. Even with partial data it was possible to identify 16 galaxies at redshift 7 and even five at redshift 8. More galaxies were identified once the full data set was in hand. Higher redshifts than these seemed to push the limits of UDF09, even though one candidate object at redshift 10 was identified. The galaxies identified thanks to the UDF09 survey showed that the ionizing luminosity density of galaxies at redshift 7 and 8 was even lower than those at 6 but, at the same time, gave hints that the luminosity function was very steep, that is, there were many faint galaxies for each bright one. If confirmed, this would strengthen the possibility that reionization could be achieved due to the contribution of faint galaxies. In 2011, another peer-review-selected follow-up program was approved which was aimed at pushing even further the depth of the UDF, to attempt the detection of more galaxies at redshift 9.

Hubble is not the only space telescope engaged in follow-up programs in the UDF region. As previously noted, the field had been selected for study because of the availability of deep *Spitzer* and *Chandra* space telescope observations. By mid-2011, *Chandra* exposures of this region were reaching a total of 1100 h. Combined with the *Hubble* data, the x-ray data have been fundamental to studying the population of faint quasars at high redshift.

Future of the UDF. It is likely that further attempts to push *Hubble* and its instrument will be carried out, but it remains unlikely that objects at redshift greater than 10 will be detected as they begin to redshift out of the near-infrared wavelengths where WFC3 is highly sensitive. However, the location of the field and the richness of observations already taken and in the public domain will preserve the importance of the UDF as the premier field for the study of high-redshift galaxies. Some of the questions posed by the study of the UDF may have to wait for the development of the *HST*'s successor, the *James Webb Space Telescope* (*JWST*). With its infrared sensitivity, the *JWST* will be able to push the study of high-redshift galaxies well beyond what we can do now, while at the same time building on the "shoulders of a giant".

For background information *see* ASTRONOMICAL SPECTROSCOPY; BIG BANG THEORY; CHANDRA X-RAY OBSERVATORY; COSMIC BACKGROUND RADIATION; GALAXY, EXTERNAL; HUBBLE SPACE TELESCOPE; INFRARED ASTRONOMY; QUASAR; REDSHIFT; SLOAN DIGITAL SKY SURVEY; SPITZER SPACE TELESCOPE; UNIVERSE in the McGraw-Hill Encyclopedia of Science & Technology.   Massimo Stiavelli

Bibliography. S. V. Beckwith et al., The Hubble Ultra Deep Field, *Astron. J.*,132:1729–1755, 2006; R. J. Bouwens et al., A candidate redshift $z \approx 10$ galaxy and rapid changes in that population at an age of 500 Myr, *Nature*, 469:504–507, 2011; M. Stiavelli, *From First Light to Reionization*, Wiley-VCH, Weinheim, Germany, 2009; M. Stiavelli et al., Observable Properties of Cosmological Reionization Sources, *Astrophys. J.*, 600:508–529, 2004; M. Trenti et al., The Galaxy Luminosity Function During the Reionization Epoch, *Astrophys. J.*, 714:L202–L207, 2010; R. E. Williams, et al., The Hubble Deep Field: Observations, Data Reduction, and Galaxy Photometry, *Astron. J.*,112:1335–1389, 1996.

Human susceptibility to Staphylococcus aureus

Staphylococcus aureus is a significant cause of human disease worldwide. Although it is sometimes carried by other animals, *S. aureus* causes disease almost exclusively in humans. In fact, it is difficult to study this pathogen using an animal model, because much higher numbers of the bacteria are required to infect laboratory animals. A recently released study has demonstrated that the bacteria appear to prefer human hosts over animal hosts because of the type of hemoglobin in human blood.

Background. *Staphylococcus aureus* is a member of the staphylococcal group of bacteria (genus *Staphylococcus*), which are gram-positive cocci typically found growing in clusters (**Fig. 1**). These bacteria are known to produce a wide range of virulence factors; the most prominent of these factors are hemolysins, which are proteins capable of destroying erythrocytes (red blood cells) in order to cannibalize the substances found within the cells. One of these substances is iron, which is held by the human hemoglobin in red blood cells. Iron is a growth-limiting factor for bacteria; for them to grow and reproduce, they must obtain iron from some source. When *S. aureus* infects humans, these

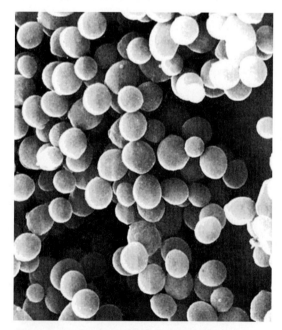

Fig. 1. An electron micrograph of *Staphylococcus aureus*. Note the characteristic spherical shape and arrangement of the cocci into grapelike clusters. (*Reprinted from M. K. Cowan and K. P. Talaro, Microbiology: A Systems Approach, 2d ed., p. 541, McGraw-Hill, New York, 2008*)

pathogens use human red blood cells as their source for this element.

Staphylococci are ubiquitous, meaning they are found nearly everywhere. Colonization of the skin by staphylococci is universal in human populations, although short-term *S. aureus* carriage is more common in the anterior nasopharynx than in other parts of the body. *Staphylococcus aureus* is persistently carried in the nasopharynx by about 30% of the normal adult population; these individuals are asymptomatic but may spread the bacteria to other, more susceptible individuals.

Diseases caused by *S. aureus* depend on the site that is infected and on the general condition of the patient. Surgical wound infections are frequently caused by *S. aureus*; it is among the most common causes of nosocomial (hospital-acquired) infections. Other infections caused by this species include toxic shock syndrome, impetigo (an acute, contagious, inflammatory skin disease), osteomyelitis (inflammation of bone tissue and bone marrow), scalded skin syndrome (**Fig. 2**), and pneumonia. Treatment for staphylococcal infections is problematic because resistance to a number of different antibiotics is common. In fact, penicillin resistance in staphylococci developed within a very short period after the introduction of penicillin in the 1940s. MRSA (methicillin-resistant *Staphylococcus aureus*) originated in hospital settings in the early 1960s, although in recent years it has moved into the general population with frequent outbreaks. MRSA is resistant to the penicillin group of antibiotics; infections caused by MRSA strains require treatment with vancomycin. However, vancomycin-resistant strains have subsequently been isolated in hospitals in Japan, demonstrating the need for further research on the mechanisms that are used by this pathogen to cause disease.

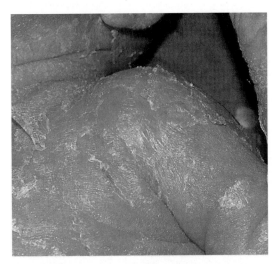

Fig. 2. Staphylococcal scalded skin syndrome in a baby. Note the appearance of the skin layers peeling as seen in a severe burn. This syndrome is caused by the production of an exotoxin by *S. aureus*. (*Reprinted from M. K. Cowan and K. P. Talaro, Microbiology: A Systems Approach, 2d ed., p. 545, McGraw-Hill, New York, 2008*)

Stealing iron. All bacteria require a source of iron for their metabolism and growth. Once inside the host, *S. aureus* follows a stepwise process to remove iron from the host red blood cells. First, the cell is lysed by the bacteria. Next, the hemoglobin binds to a hemoglobin receptor on the bacterial cell wall that is known as the IsdB receptor. After attachment, the heme is extracted and then carried into the bacterial cell, where the iron is freed and used by the pathogen to grow and reproduce.

A new discovery by researchers led by Eric P. Skaar at Vanderbilt University in Nashville, Tennessee, may explain why *S. aureus* appears so discriminatory in infecting humans as compared to other animals. These scientists examined differences in the ability of *S. aureus* to bind to and remove iron from different types of animal hemoglobin. In the process of doing this, it was discovered that *S. aureus* has a strong preference for binding to human hemoglobin as compared to hemoglobin of other animals.

In their study, the researchers analyzed how efficiently the IsdB receptor bound to human hemoglobin (hHb) and compared the result to its binding to mouse hemoglobin (mHb). Iron-starved *S. aureus* was incubated separately with either hHb or mHb before loose hemoglobin was washed away, and the amount of hemoglobin still attached to the bacterial cells was determined. It was found that hHb bound more efficiently, indicating a strong preference of the bacteria for human blood.

Next, the scientists tested the effect of the different types of hemoglobin on the reproduction of the bacteria. Each type of hemoglobin was added separately to iron-free media prior to inoculation with the bacterial cells. The reproductive rate of the bacteria was measured over time. It was found that there was a definite delay in the reproduction of *S. aureus* using the mouse hemoglobin compared to the cells grown using the human form.

Finally, the researchers compared the ability of different species of bacteria to use human hemoglobin as their sole source of iron. It was found that bacterial pathogens could be divided into three major groups: (1) those that preferred human hemoglobin to mouse hemoglobin; (2) those that used either form equally well; and (3) those that could not use human hemoglobin as a sole iron source at all. *Corynebacterium diphtheriae*, the bacterial species that causes diphtheria, shared the preference of *S. aureus*, whereas *Bacillus anthracis*, the cause of anthrax, used both forms. Interestingly enough, *Staphylococcus epidermidis*, a close relative of *S. aureus*, could not grow at all with either type of hemoglobin as a sole iron source. These results reinforce the role of *S. aureus* as a primary human pathogen, whereas *S. epidermidis* causes opportunistic rather than invasive infections.

The mechanisms by which *S. aureus* causes disease have been studied extensively using mice as an animal model. However, the mice are not a perfect model for determining how *S. aureus* interacts with its host because of the differences between humans and mice. Therefore, development of mice with human characteristics is a priority of the scientific community studying human disease. In one such study, the susceptibility to infection by *S. aureus* was measured in transgenic mice expressing equal amounts of human and mouse hemoglobin. These mice were found to be more susceptible to infection by *S. aureus* than normal mice expressing mouse hemoglobin only. Hence, these results emphasize the usefulness of transgenic mice with human characteristics in studying human infectious disease.

Genetic susceptibility. Why are some people colonized by *S. aureus* but do not develop disease? A possible answer may lie in the fact that variations exist among the hemoglobin genes found in different individuals, leading to variability in the structure of their hemoglobin proteins. These differences may play a role in the susceptibility of an individual to infection with *S. aureus*. As stated previously, *S. aureus* is carried in the nasopharynx by about 30% of the normal adult population, yet these individuals rarely suffer from invasive infection. Why, then, do so many people carry this pathogenic species with impunity, whereas others suffer from invasive infection when they are exposed? Genetic links have been found to predispose certain individuals to diseases such as breast cancer and diabetes; perhaps such a link is responsible in this case as well.

Current studies at Vanderbilt University are also examining the connection between hemoglobin structure and susceptibility to *S. aureus* infection. Using the patient gene bank at the university, the hemoglobin genes of individuals who were diagnosed with *S. aureus* infection are being analyzed and compared to those of uninfected patients. By establishing potential gene sequences that confer susceptibility or resistance to *S. aureus*, the hope is to be able to screen new patients for susceptibility to infection by this pathogen.

Conclusions. *Staphylococcus aureus* is a significant infectious cause of human suffering and death. Recent advances have shown that this species relies heavily on its ability to extract iron from human hemoglobin, and that it is much more likely to infect humans than other animals. The results discussed here also demonstrate that transgenic mice are a much more reliable animal model than wild-type mice, making them of great value in the study of this dangerous pathogen. Finally, variations in human hemoglobin genes among individuals may be significant in determining why some individuals are more susceptible to *S. aureus* infection. By studying these differences, it may be possible to prevent future infections from occurring in susceptible individuals.

For background information *see* ANTIBIOTIC; BACTERIA; DISEASE; DRUG RESISTANCE; GENE; GENETIC ENGINEERING; HEMOGLOBIN; HOSPITAL INFECTIONS; INFECTIOUS DISEASE; IRON; MEDICAL BACTERIOLOGY; PATHOGEN; PUBLIC HEALTH; STAPHYLOCOCCUS in the McGraw-Hill Encyclopedia of Science & Technology.

Marcia M. Pierce

Bibliography. M. K. Cowan and K. P. Talaro, *Microbiology: A Systems Approach*, 2d ed., McGraw-Hill, New York, 2008; N. Legrand et al., Humanized mice for modeling human infectious disease: Challenges, progress, and outlook, *Cell Host Microbe*, 6:5–9, 2009; S. K. Mazmanian et al., Passage of heme-iron across the envelope of *Staphylococcus aureus*, *Science*, 299:906–909, 2003; P. R. Murray, K. S. Rosenthal, and M. A. Pfaller, *Medical Microbiology*, 6th ed., Mosby, St. Louis, 2009; E. Nester et al., *Microbiology: A Human Perspective*, 6th ed., McGraw-Hill, New York, 2008; G. Pishchany et al., Specificity for human hemoglobin enhances *Staphylococcus aureus* infection, *Cell Host Microbe*, 8:544–550, 2010; L. D. Shultz, F. Ishikawa, and D. L. Greiner, Humanized mice in translational biomedical research, *Nat. Rev. Immunol.*, 7:118–130, 2007.

Ice-templated ceramics

The solidification or freezing of colloidal suspensions or colloids is commonly encountered in a variety of natural processes, such as the freezing of soils and the growth of sea ice in northern regions, as well as in everyday life and engineering situations, such as food engineering, materials science, cryobiology, filtration or water purification, and the removal of pollutants from waste. It is an amazingly common phenomenon in natural, physical, social, and technological environments. The associated costs (degradation of roads) or benefits (climate control, cryopreservation protocols, and tissue-engineering scaffolds) are of tremendous importance. Among the many applications of colloid freezing, its potential use as a processing route for bio-inspired porous materials is particularly innovative and exciting. This processing route is currently referred to as freeze casting or ice templating.

Principles of ice templating. The freezing of colloids can be used as a self-assembly process. In the natural occurrence of colloids freezing, such as the freezing of sea ice, pure ice crystals with a platelet morphology are formed, and the various impurities originally present in seawater (salt, biological organisms) are expelled from the forming ice and entrapped within the brine channels between the ice crystals. Using ceramic colloidal particles instead of biological impurities, advantage can be taken of this natural segregation principle, using ice as a natural and environmentally friendly templating agent, which is then removed by sublimation. The final result is a porous body with a complex and usually anisotropic porous structure generated during freezing, with the structure being a negative of the ice structure before drying (**Fig. 1**). This ceramic scaffold can then be used as the basis for a porous or dense composite, if infiltrated with a suitable second phase.

The process is often referred to as freeze casting, although ice templating should be more appropriate. Freeze casting has been developed as a processing route for yielding dense and near-net shape ceramic pieces. In freeze casting, highly concentrated ceramic suspensions are poured into a mold with the details that should replicated. A subsequent freezing stage ensures the consolidation of the piece, and water is removed by sublimation. Any growth of ice crystals leading to defect formation when the ice is removed should be avoided, since complete densification is sought. The growth of ice is therefore considered for its deleterious influence over the structure. In ice templating, advantage is taken of this phenomenon, and the ice is used as a templating agent for the porosity. It is therefore necessary to use a lower particle concentration in the ceramic suspension, so that the ice crystals have enough room to grow uniformly and achieve their templating effect (**Fig. 2**).

Advantages and limitations. The freezing route is an appealing process because of its simplicity, with no equivalent at the moment in terms of both process (environmentally friendly, with no solvent except water) and structures obtained (highly directional). Other techniques that structure bulk materials at submicrometer scales usually require unrealistic processing time, hindering further realistic development. Samples a few cm thick can be processed within a few minutes with the ice-templating approach. In addition, the equipment required is readily available and has been developed, tested, and used for years in various fields, including cryopreservation and food and materials engineering.

More importantly, the ice-templating process is extremely versatile. The vast majority of the biomimetic or bio-inspired processing routes developed so far often rely on highly specific interfacial compatibilities. The structural formation mechanisms involved here are based mostly on physical interactions between the solidification front and the particles, so any type of ceramic, metallic, or even polymer particles can be used (**Fig. 3**). In ceramic processing to date, a variety of materials have been ice-templated of both ionic (alumina, zirconia, silica) and covalent (silicon carbide, silicon nitride) nature. The range of potential applications derived from this approach is extremely wide, including filtration, catalysts support, scaffolds for biomaterials, heat guides, and wear-resistant materials for cutting tools, to name some.

Of all the solvents that have been considered for the ice-templating processes, water is the most commonly encountered and the most peculiar. Several specificities of water and ice must be taken into account. For example, under the usual temperature and pressure conditions, ice exhibits highly anisotropic crystal growth. The hexagonal structure of ice results in the formation of platelets or lamellae, because the ice growth kinetics in the basal plane is orders of magnitude larger than that occurring perpendicular to the basal plane. This anisotropy is the underlying reason for the peculiar morphology of the brine channels in sea ice and the lamellar morphology of the ice-templated ceramics.

As expected from the unique structure, some unique properties are exhibited by such materials (see **table**). In particular, the strong anisotropy can considerably enhance the functional properties in one direction when needed, such as mechanical properties or electrical or thermal conductivity. For example, a 400% increase in compressive strength of

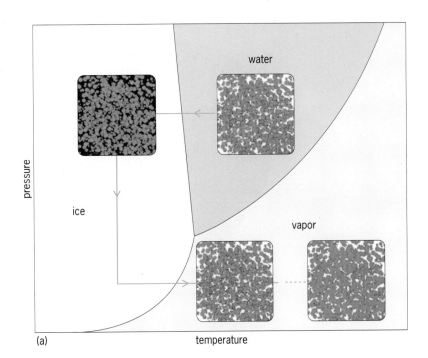

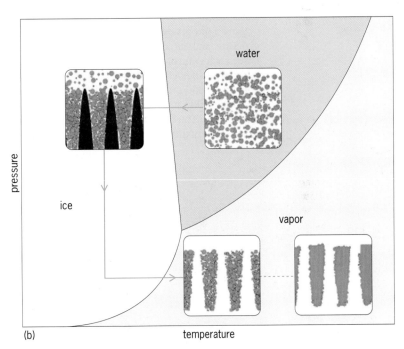

Fig. 1. Principles of (*a*) freeze-casting and (*b*) ice-templating processing routes, presented on the pressure-temperature phase diagram of water. The following steps must be followed: freezing, freeze-drying, and densification, usually at high temperature in the case of ceramics.

calcium phosphate scaffolds for bone-substitute applications was demonstrated, compared to materials obtained through other state-of-the-art techniques. Such anisotropy can be highly beneficial for a wide range of other applications involved with mass, gas, or species transport.

If infiltrated with a suitable second phase, these ceramic scaffolds can then be used to make porous or dense composites. Such composites take advantage of both polymer and ceramics qualities, ideally to achieve materials with high stiffness and high

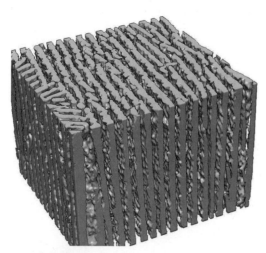

Fig. 2. Three-dimensional representation of a typical structure of an ice-templated ceramic, obtained by x-ray computer tomography. The typical thickness of a lamellar feature is 10–20 μm.

toughness. The porous scaffolds and dense composites obtained by this process exhibit a complex and highly hierarchic structure, defined at several length scales.

Many approaches can be modified to control the structure and its morphology at different levels. Three such control categories are (1) control of the starting powder characteristics and of the suspension formulation, (2) control of the freezing stage and sublimation stage, and (3) control of the densification stage. The first and third categories are common to most of the ceramic processing routes, so that

the available knowledge can be used rather straightforwardly. More interesting is control of the freezing stage that is specific to the process. Currently, the focus is on controlling the ice crystals (kinetics, morphology) and particle redistribution and the packing between the crystals (Fig. 3).

Several limitations are nevertheless associated with the process, some linked to the intrinsic nature of the process and some to its technological implementation. The most important limitation is the processing time. A steady freezing velocity is required to achieve a homogeneous and defect-free structure. The typical solidification velocities being used are in the range 5–50 micrometers/second (μm/s); a few minutes are therefore necessary to freeze pieces a few cm thick. Achieving homogeneity in pieces thicker than 3 or 4 cm can be tricky, in particular if a certain range of pore size is targeted. The lateral dimensions of the pieces are only constrained by the capabilities of the freezing setup, which can be scaled up rather easily. Related to the processing time is the pore size. The greater the freezing velocity, the smaller the pore dimensions. Processing pores larger than 100 μm can quickly become time-consuming, and homogeneity is more difficult to achieve. Ice templating in such a case is not the most appropriate choice, in comparison to extrusion techniques, for example. The greatest constraint on the process is imposed by the size of the particles being used. Ice templating works best with submicrometer particles. Using larger particles, it is more difficult to obtain well-defined pore morphologies, in particular when the particle size is in the same

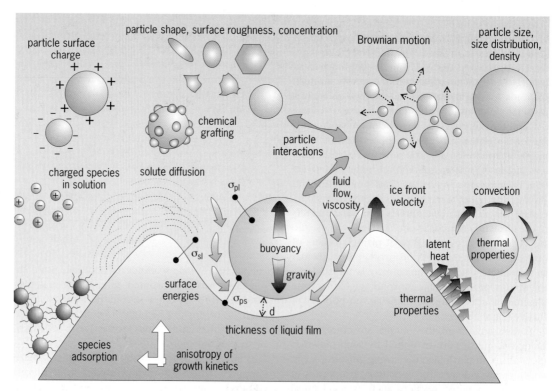

Fig. 3. Underlying mechanisms and parameters controlling the formation of the structure, schematic representation. The particles must be repelled by the growing crystals and concentrated between adjacent crystals for the ice-templating mechanism to operate.

Typical properties of ice-templated ceramics and derived composites

Property	Typical values
Porosity	40–95%
Pore dimension (cross-section)	0.2–200 μm
Compressive strength	up to 150 MPa
Strain to failure	up to 20%
Toughness (for composites)	up to 40 MPa$\sqrt{m}$
Tensile strength (for composites)	up to 300 MPa

size range of the dimensions of the ice crystals. Sedimentation effects can also become predominant if not properly controlled.

Perspectives. The phenomenon itself is simple to describe—the water-ice interface propagates through a colloidal suspension of particles. This simplicity is, nevertheless, only apparent, and the freezing of colloids is still a puzzling phenomenon with many unexplained features. A very large number of disparate parameters should be accounted for when trying to understand, model, and control the freezing of colloids (Fig. 3). Fundamental work is needed to progress in this understanding, with particular attention on the freezing stage, which is the core of the process. A solid experimental foundation on which further theoretical developments can be built and validated is, of course, a preliminary and strong requirement. If the various manifestations of colloids freezing can be observed in numerous occurrences in natural or technological situations, precise and quantitative observations are required for understanding the process. Ideally, in situ, real-time, and three-dimensional observations of both the crystal growth and colloid movement in the suspensions would be needed, with individual colloid tracking and concentration measurement to investigate the colloid redistribution during freezing. The corresponding space (submicrometer) and velocity (tens to hundreds of μm/s) scales severely restrict the choice of experimental techniques.

To date, the control of the structure is still empirical because of the large number of parameters involved in the process. Achieving predictable control of the structure is certainly an exciting and difficult challenge and will required an in-depth understanding of the crystal growth conditions and the interactions between the particles and the crystals.

Finally, little is known about the properties of such materials and their structure-to-property relationships (see table). Current properties can far exceed what could be expected from a simple mixture of their components. Several degrees of structural hierarchy have already been identified in ice-templated structures, throughout all the length scales, making the structure unique. A wide range of techniques and parameters can be used to control the structure and each of these degrees of hierarchy, which might be used either to promote a single property or to introduce several functionalities. A precise qualitative and quantitative description of the structure is therefore necessary, combined with a complete assessment of the functional properties. Once the structure-to-property relationships have been established, proper control and tailoring of the functional properties will be achieved for a wide spectrum of applications and will bring the ice-templating process to its full potential.

For background information *see* CERAMICS; COLLOID; COMPOSITE MATERIAL; POLYMER COMPOSITES in the McGraw-Hill Encyclopedia of Science & Technology.   Sylvain Deville

Bibliography. S. Deville, Freeze-casting of porous ceramics: A review of current achievements and issues, *Adv. Eng. Mater.*, 10:155–169, 2008; S. Deville et al., Freezing as a path to build complex composites, *Science*, 311:515–518, 2006; E. Munch et al., Tough, bio-inspired hybrid materials, *Science*, 322:1516–1520, 2008; V. F. Petrenko and R. W. Whitworth, *Physics of Ice*, Oxford University Press, New York, 1999; M. Scheffler and P. Colombo, *Cellular Ceramics: Structure, Manufacturing, Properties and Applications*, Wiley-VCH, Berlin, 2005.

Invasive species and their effects on native species

Many thousands of species, including animals, plants, fungi, bacteria, and viruses, have been transported to areas outside their native ranges and have successfully established new populations there. If one of these introduced species then causes significant negative ecological, economic, or human health impacts, it is categorized as an invasive species. The movement of species around the globe is not a new phenomenon, but human intervention has drastically increased the rate of movement, the distances that organisms can travel, and the types of species that are now transported around the globe. Invasive species are accidentally or intentionally introduced to new areas via many routes, including the horticultural trade, agriculture, aquaculture, agroforestry, the game animal trade, the pet trade, ship ballast, fishing bait, transport of untreated wood, and mud stuck to boots or tires. They also can be introduced as food items. The economic cost of invasive species is considerable. For example, the impact of introduced agricultural pests and weedy species on cropland, pastures, and forests costs billions of dollars annually in the United States. Invasive trees in South Africa have been calculated to reduce current river flow rates by up to 22%; controlling their spread will cost many millions of dollars, but allowing them to spread further would be costlier still, given their threat to critical water supplies. Invasive species also increase public health risks. In the Northeast United States, areas with invasive Japanese barberry bushes (*Berberis thunbergii*) had nine times more black-legged ticks (*Ixodes scapularis*, the species responsible for transmitting Lyme disease to humans) than areas without Japanese barberry.

Invasives also affect many native species through predation, disease, herbivory, competition for resources, and hybridization, resulting in genetic swamping of the native species. Entire ecosystems are affected by invasives that alter fire regimes, cause nitrogen enrichment, and remove native foundation species. The following case studies demonstrate both the diversity of invasive species and the range of effects that they can have on native species and ecosystems.

Predators. Cats and snakes are two important predator invasives.

Cats. Invasive cats (*Felis catus*) are responsible for 26% of all predator-related island bird extinctions worldwide and have been described as being by far the most dangerous of any introduced carnivore (**Fig. 1**). Cats are particularly damaging predators as they continue to hunt even when fed regularly by humans. Because cats are good at controlling rats aboard ships, they have traveled around the world; some have escaped to create feral populations, whereas others have been intentionally introduced to control mice and rats in new colonies. In addition to killing unwanted introduced rodents, they also kill great numbers of other native small mammals, birds, reptiles, and amphibians. Feral cats in the United States kill an estimated 240 million birds per year, and that estimate doubles with the inclusion of cats that are pets. Invasive predators on oceanic islands often severely affect native prey species, many of which are endemic species that occur nowhere else on Earth. For example, the cat of a lighthouse keeper was solely responsible for the extinction of the Stephens Island wren (*Xenicus lyalli*), the only flightless songbird in the world, in 1894, with 12 additional native bird species being extirpated from this island by cats within a few years. Feral cat populations have been controlled successfully with the

Fig. 1. Cats are one of the most effective introduced predators, preying on birds and other small animals. (*Photograph courtesy of Keri Van Camp*)

introduction of feline-specific viruses or toxins, followed by hunting and trapping once the population size is reduced. For example, the removal of cats from Long Cay Island in the British West Indies resulted in the recovery of the highly endangered iguana *Cyclura carinata*.

However, the consequences of removing invasive species such as cats can be much more complex than anticipated. Cats were introduced on Macquarie Island, Australia, in 1818. Initially, the cats had little effect on the Macquarie Island parakeet. Then, rabbits were introduced to the island in 1879. With this year-round food supply, the cat population grew rapidly, ultimately hunting the parakeet to extinction. By the 1960s, it became clear that the rabbits were destroying the maritime tall grassland and megaforb associations, and thus the *Myxoma* virus was introduced as a biocontrol for the rabbit population. Rabbit populations became much lower by the 1970s, and the cats switched to eating seabirds. A cat eradication effort began in 1985 to protect the seabirds, with the final cats being removed from the island in 2001. At this point, the remaining rabbits (that is, those that survived the virus and the cats) began to reproduce rapidly, and once again they began to devastate the unique plant communities on the island, transforming them into bare ground or grazed lawns. Clearly, the best solution is to prevent invasions from occurring in the first place.

Brown tree snakes and Burmese pythons. The brown tree snake (*Boiga irregularis*) was accidentally introduced to Guam after World War II, probably in the undercarriage of military aircraft. By the 1970s, the population size of the brown tree snake had reached a density of 100 per hectare (1 ha = 2.47 acres), as they had ample food supplies, no known diseases or parasites, and very few predators on Guam. The brown tree snake has devastated the native fauna on the island. Of the 18 native bird species, 12 went extinct and five others declined more than 90%. Of the three native bat species, two have gone extinct, whereas the third, the Mariana fruit bat (*Pteropus mariannus*), is now endangered. These species did not coevolve with predators such as the brown tree snake and thus had few defenses against such an efficient nocturnal, arboreal predator. The losses of these bird and bat species have cascading effects across the entire island because they played important roles in pollinating flowers, dispersing seeds, and controlling insect populations. Many other invasive species, including introduced rats, pigs, deer, birds, and skinks, have become established on Guam as well, further affecting the native ecosystems. Old-growth forests on the island are no longer regenerating and the fate of other native species is unclear. In addition to the effects of the brown tree snakes on native species, direct economic costs are estimated to run as high as $12 million (U.S. dollars) per year, primarily as a result of power outages when the snakes crawl into and short out the electrical transmission system. Brown tree snakes are continuing to disperse around the globe as stowaways on airplanes, with some traveling as far away as Spain. Interdiction has proven difficult.

Fig. 2. Invasive reptiles such as the Burmese python (*Python molurus bivittatus*) are often introduced via the pet trade. (*Photograph courtesy of Everglades National Park, Florida; National Park Service*)

Barriers have been erected, traps around ports and airports in Guam catch hundreds of snakes per year, and detector dogs are used on Guam and at destination ports. This vigilance will need to be maintained indefinitely if brown tree snakes are to be kept in check.

Invasions of nonnative snake species continue to occur around the globe. Burmese pythons (*Python molurus bivittatus*) are a recent invader in Florida (**Fig. 2**). They prey on many species, including endangered species such as the wood stork and Key Largo woodrat, as well as species such as white ibis, limpkins, and round-tailed muskrats that are of conservation concern. Burmese pythons now occupy thousands of square kilometers in southern Florida. In fact, although cold temperatures will eventually limit their spread northward, it is unclear whether measures will be able to prevent their spread to other suitable habitats nearby, such as the Florida Keys.

Disease and insect pests. Fungi and insects are notable invasives, often decimating native species.

Chestnut blight. Chestnut blight (*Cryphonectria parasitica*) is an invasive fungus from Asia that first arrived in North America on infected Japanese chestnut trees in the late 1800s. It was first detected on American chestnut trees (*Castanea dentata*) in 1904 at the New York Zoological Park in the Bronx. As the trees lacked any resistance or tolerance to the disease, it rapidly spread throughout the deciduous forests of the eastern United States. The fungus spores invade chestnut trees through any small break in the bark and then the fungal mycelium grows throughout the vascular tissue of the tree, killing the tissue and stopping the transport of water and nutrients. Billions of American chestnut trees were lost from the forests of the eastern United States and were replaced by other hardwood species, primarily oak trees. The chestnut was once a dominant forest species, amounting to 25% of the trees in some areas. These trees produced huge quantities of nuts every year, and therefore the loss of the chestnut had a very negative impact on other species that depended on those nuts for food, including bears, deer, elk, squirrels, raccoons, and turkeys, as well as the people who harvested the nuts. Chestnut trees now exist

only as small understory tree sprouts that grow from the old stumps. These sprouts eventually succumb again to the blight, and new sprouts emerge. Reproduction by seed is absent. At some point, these old stumps will lose their vigor and the sprouts will die out. There are several organizations that are working to develop blight-resistant strains of the American chestnut, but it seems highly unlikely that it will ever be able to return as a dominant tree species in the forests of the eastern United States.

Emerald ash borer. The emerald ash borer (*Agrilus planipennis*) is an invasive insect recently introduced from Asia that is decimating the native ash tree species in North America. It was first discovered in 2002, but had probably been introduced about a decade or more earlier from Asia into the Detroit area in untreated ash wood used in crates or packing materials. It is now found from Wisconsin to New York and as far south as Tennessee. Larvae feed on the cambium layer under the bark, and their feeding galleries disrupt the flow of water and nutrients in the tree. They remain in the bark and wood as larvae and pupae; then, when the adults emerge, they may disperse up to 0.8 km (0.5 mi) away. It also spreads by the shipping of infected nursery trees and the transport of firewood. Efforts are being made to identify early infestations to limit its spread (**Fig. 3**), but tens of millions of trees have already been killed. It is estimated that the emerald ash borer may

Fig. 3. Sticky trap boxes, usually purple in color (such as the one shown here), are hung in ash trees to track the spread of the emerald ash borer and assist in control efforts. (*Photograph courtesy of Keri Van Camp*)

result in the loss of all 18 native ash species in North America. Researchers are searching for natural enemies against the borer in its native range to use as biocontrol agents, as well as looking for trees that demonstrate genetic resistance to the borer. As with the American chestnut, the loss of North American ash trees is likely to have a cascade effect on other native species throughout the northern deciduous forests of the United States.

Ecosystem transformers. Ecosystems can be dramatically changed by invasive species.

Cheatgrass. Cheatgrass (*Bromus tectorum*) has invaded arid regions worldwide after accidental introductions as a seed contaminant in cereal crops. Between 1890 and 1935, cheatgrass transformed the Intermountain West region of the United States (that is, the area between the Rocky Mountains to the east and the Pacific mountain ranges to the west) into a virtual monoculture. It is now the dominant plant across at least 200,000 km^2 (77,220 mi^2). Cheatgrass is an annual that dies in the summer, creating a blanket of highly combustible fuel across the landscape. The resulting fires cover hundreds of thousands of hectares, moving quickly across the landscape and totally consuming all aboveground plant material. The loss of vegetation during the fires leaves the soil vulnerable to erosion, which then creates sedimentation problems in the surrounding creeks and rivers. Previously, these areas had a fire frequency range of 60–100 years, but now the fire frequency is 3–5 years. This increase in fire frequency has led to the complete loss of the shrub community. The loss of these shrubs affects the endangered sage grouse (*Centrocercus urophasianus*), as well as many other native animals and plants.

Earthworms. Earthworms have not inhabited many northern regions for 10,000 years, since the Pleistocene glaciation eliminated the previous earthworm fauna. Over the past few hundred years, the use of soil as ballast, the relocation of landfill, the nursery trade, reforestation projects, and the release of unused bait have resulted in the introduction of well over 100 different earthworm species. When earthworms invade, the leaf litter layer vanishes as the surface organic matter is mixed into the underlying mineral soil. The loss of the litter layer shifts the food web relationships of litter-dwelling and litter-feeding organisms, altering their abundance and composition. Declines in native ferns, woodland flowers, and tree species such as sugar maple have been linked to the presence of invasive earthworms, whereas the growth of invasive plants such as European barberry (*Berberis vulgaris*), stiltgrass (*Microstegium vimineum*), and garlic mustard (*Alliaria petiolata*) is facilitated. In addition to redistributing soil organic matter, earthworms also alter soil porosity and soil structure, increase the rate of nutrient cycling, create higher levels of nitrogen in the soil, raise the soil pH, and increase the rate of litter decomposition. These transformations, coupled with excessive deer herbivory and pressure from invasive plants, are having a profound influence on the regeneration of deciduous forests in the eastern United States.

Future conservation. As seen in these examples, invasive species can have devastating effects on native species and ecosystems. The challenge is to decide how to respond to these invasions. Some conservationists see the task of removing invaders as a very expensive and nearly impossible goal, and feel that we should accept that they are here to stay. Preserving native biodiversity within these novel ecosystems is seen as a more realistic and reachable goal than trying to restore the original ecosystem. These conservationists also see great value in preserving ecosystem function, regardless of the species composition. Others vehemently disagree and point to the numerous examples of successful eradication and control. They argue that the threat to global biodiversity is too great to not continue the fight against invasives. Conservationists on both sides of this issue agree that the best solution is to prevent further invasions and, whenever possible, eradicate invasives in areas before they become fully established.

For background information *see* BIODIVERSITY; BIOGEOGRAPHY; ECOLOGICAL COMMUNITIES; ECOLOGICAL COMPETITION; ECOLOGY; ECOSYSTEM; ENDANGERED SPECIES; EXTINCTION (BIOLOGY); INVASION ECOLOGY; POPULATION ECOLOGY; POPULATION VIABILITY; PREDATOR-PREY INTERACTIONS; RESTORATION ECOLOGY in the McGraw-Hill Encyclopedia of Science & Technology.   Margaret L. Ronsheim

Bibliography. A. M. Ellison et al., Loss of foundation species: Consequences for the structure and dynamics of forested ecosystems, *Front. Ecol. Environ.*, 3:479–486, 2005; S. Freinkel, *American Chestnut: The Life, Death and Rebirth of a Perfect Tree*, University of California Press, Berkeley, 2007; J. Gurevitch and D. K. Padilla, Are invasive species a major cause of extinctions?, *Trends Ecol. Evol.*, 19:470–474, 2004; J. A. McNeely et al. (eds.), *A Global Strategy on Invasive Alien Species*, IUCN, Gland, Switzerland, 2001; V. A. Nuzzo, J. C. Maerz, and B. Blossey, Earthworm invasion as the driving force behind plant invasion and community change in northeastern North American forests, *Conserv. Biol.*, 23:966–974, 2009; D. Pimentel, R. Zuniga, and D. Morrison, Update on the environmental and economic costs associated with alien-invasive species in the United States, *Ecol. Econ.*, 52:273–288, 2005; D. Simberloff and M. Rejmanek (eds.), *Encyclopedia of Biological Invasions*, University of California Press, Berkeley, 2011; D. Simberloff et al., Recognizing conservation success, *Science*, 332:419, 2011; G. Vince, Embracing invasives, *Science*, 331:1383–1384, 2011.

Iron catalysis

The industrial chemistry of the modern world is following a general trend of becoming more sustainable, economical, and environmentally friendly. Improvements in the field of catalysis are contributing to this trend. Typical catalytic processes that operate in a laboratory or industry utilize precious metals that are toxic, expensive, and rare. To solve this problem, chemists have been developing catalysts based on enzymes, organic molecules, main-group elements, and soluble complexes of the more abundant transition metals. Recently, soluble iron complexes have been discovered with desirable catalytic properties. These catalytic systems are of great importance, because iron is cheap, abundant, and relatively nontoxic.

Background. Chemical reactions occur because of an intrinsic reactivity of two or more reagents toward each other when mixed. A catalyzed chemical reaction, on the other hand, involves two or more inert (nonreactive) substances that have to be activated for the reaction to proceed by a third molecule (catalyst) that is regenerated after each cycle. This type of reaction is employed successfully by nature in a large variety of regulatory processes. High selectivity, mild conditions, and minimum waste production associated with reactions catalyzed by enzymes (protein-based catalysts) encouraged scientists to study these processes so they could be applied on a large scale in academic and industrial laboratories.

A historic example is the enzyme system required for biological nitrogen fixation; that is, the reduction of nitrogen (N_2) in the air to ammonia (NH_3), which is a building block for the synthesis of proteins and nucleic acids. This very challenging reaction needs to be catalyzed because the nitrogen molecule has a very strong and inert $N\equiv N$ triple bond to be broken. As the $N\equiv N$ triple bond breaks, three strong N–H bonds form to balance the energy and drive the reaction to completion. The protein catalyst that is used by nature in this case is nitrogenase, an enzyme with long chains of amino acids and iron–sulfur clusters. Nitrogen and hydrogen are thought to combine to form ammonia at iron sites at the surface of one of these clusters. The iron sites are thought to be in the +2 oxidation state [Fe(II)]. The direct use of enzymes is problematic because of the high complexity of their structures and their mode of action. For that reason, researchers focused mostly on the preparation and use of catalytic analogs with simple structures that have similar activating abilities.

In the early part of the twentieth century, Haber and Bosch revolutionized agriculture by developing a process for the production of ammonia. In this process, ammonia is synthesized by passing hot, pressurized hydrogen and nitrogen gas over a hot, solid iron catalyst, with iron in the zero oxidation state. No one has yet found a simple iron compound that can duplicate nature's feat of catalytic nitrogen reduction at room temperature. Ruthenium, the rare, expensive, and somewhat toxic element that sits below iron in the periodic table, is a better catalyst for this process, allowing the reaction to occur closer to room temperature. Some chemical plants actually use ruthenium as a catalyst; however, a "greener" solution would use an iron compound, yet to be discovered, which follows from nature's example.

In the past 50 years, the field of catalysis has grown substantially. Major discoveries were based on the observation that transition metals and their compounds, especially those of the platinum group of metals (PGM)—ruthenium (Ru), osmium (Os), rhodium (Rh), iridium (Ir), palladium (Pd), and platinum (Pt)—were able to activate otherwise inert chemical bonds. Those involved in the activation of hydrocarbons were used in the preparation of fuels, textiles, fragrances, and pharmaceuticals.

The use of catalytic reactions addresses some of the most important issues of our time, including waste and pollution reduction, safer and "greener" industrial processes, and a reduction in the costs to produce goods. Researchers involved in the development of important catalytic transformations were recognized in the past decade with the Nobel Prizes in Chemistry: in 2001 to W. S. Knowles and R. Noyori for the development of catalysts based on Rh and Ru for asymmetric hydrogenation, and to K. B. Sharpless for asymmetric oxidation catalysts based on Os; in 2005 to R. H. Grubbs, R. R. Schrock, and Y. Chauvin for the development of metathesis catalysts based on Ru or Mo; and recently, in 2010, to R. F. Heck, E. Negishi, and A. Suzuki for the development of Pd-catalyzed cross-coupling reactions. All of these catalysts are now used in the synthesis of fine chemicals.

Asymmetric hydrogenation, a particularly valuable reaction, refers to the addition of hydrogen to a carbon–carbon, carbon–oxygen, or carbon–nitrogen double bond, so that it results in a carbon with four different atoms or groups attached. If the substituents are different, then mirror-image molecules (left-handed or right-handed), referred to as enantiomers, are formed. Asymmetric hydrogenation refers to a process in which one enantiomer is synthesized preferentially.

The PGMs commonly used in these valuable catalytic transformations are expensive and toxic, placing limitations on their use in the production of bioconsumable goods, such as pharmaceuticals, foods, and fragrances. Also, because these metals are the rarest in the Earth's crust, their supply will be exhausted if they are used extensively. Only in some cases can these metals be recycled and reused.

These limitations may be resolved by the discovery of catalytic processes utilizing the first-row transition metals of the periodic table. Metals such as titanium, chromium, iron, cobalt, nickel, and copper are more abundant, inexpensive, and generally less toxic than the PGMs. The reactivities of these metals are more complicated than that of the precious metals because of the lower stability of their compounds and the tendency to cause uncontrollable reactions involving odd-electron (free-radical) compounds. Those drawbacks made them less attractive candidates as catalysts in the past. Recently, however, the importance of creating more sustainable, environmentally friendly, and inexpensive industrial processes has triggered an interest in the development of catalytic systems based on the first-row transition metals.

The most interesting candidate for catalysis is iron. It is an essential element in the chemistry of living organisms, catalyzing key chemical transformations. For this reason iron is less toxic than the PGMs, which are not essential to life. Iron is the most abundant transition metal in the Earth's crust. The technologies for mining and purifying iron have been known from ancient times and are optimized to perfection. This makes iron the cheapest and most available metal. The chemistry of complexes with iron–carbon bonds has been studied for more than a century. This study centered first on derivatives of iron pentacarbonyl, $Fe(CO)_5$, which is formed by passing carbon monoxide over finely divided iron. Here iron is in the zero oxidation state. Such complexes catalyze the water-gas-shift reaction, in which carbon monoxide from coal gasification reacts with basic water in the presence of this iron complex to produce hydrogen gas and carbon dioxide. Other carbonyl derivatives have been prepared that have two or more irons in carbonyl cluster complexes, such as $Fe_2(CO)_9$ and $Fe_4(CO)_{12}$. It was a surprise to scientists in the 1990s that enzymes (hydrogenases) in bacteria contain related iron carbonyl clusters. These enzymes catalyze the uptake or generation of hydrogen gas efficiently, allowing the bacteria to use hydrogen gas as their energy source. With this discovery in mind, some research is currently underway to replace platinum in hydrogen fuel cells with iron complexes.

Iron catalysis today. The properties of soluble (homogeneous) catalysts depend on the properties of the metal center and its attached organic framework (ligand). Research into changing the structure of the ligand or ligands attached to iron has led to recent breakthroughs in the field of iron catalysis.

T. Collins and co-workers at Carnegie Mellon University have prepared an iron(III) complex containing a tetradentate (four binding atoms) anionic ligand and two water molecules [TAML-Fe(III)]. This complex and its analogs are exceptional activators of hydrogen peroxide (H_2O_2) under ambient conditions in the presence of water and air (see **illustration**). The high solubility of these compounds in water makes it possible to use this complex in combination with H_2O_2 for water purification. This approach avoids the use of chlorinating compounds or other toxic bleaching reagents. This active system can catalyze the removal (via oxidation) of 75% of the colored impurities within hours from a contaminated water solution using very low concentrations of the iron catalyst.

The hydrogenation of carbon–carbon, carbon–oxygen, and carbon–nitrogen double bonds is one of the fundamental reactions in organic chemistry and is used extensively in industry. The catalytic reduction usually proceeds via the addition of a hydrogen molecule or its equivalent to a PGM catalyst to produce a hydride. This hydride then reacts with the substrate to give the hydrogenated product. Only recently has this reduction process been accomplished efficiently with iron complexes (see illustration). In 2004, P. J. Chirik's group at Cornell University and J. C. Peters' group at the California Institute of Technology reported that certain soluble iron complexes catalyze the addition of hydrogen to the carbon–carbon double bonds of olefins at rates comparable

Iron-based catalyst	Reaction	Applications
X = H or Cl, R = CH₃ or F, L = H₂O	Oxidation of organic compounds in water. H_2O_2 is used as an oxidizing agent.	Water purification Water oxidation
	Hydrogenation of a carbon-carbon double bond in an olefin to produce an alkane. H_2 is used as a reducing agent.	Fine chemical synthesis
	Hydrogenation of a carbon-oxygen double bond in a compound to produce an alcohol. H_2 is used as a reducing agent.	Fine chemical synthesis
	Asymmetric transfer hydrogenation of a compound containing a carbon-oxygen double bond to produce the left-handed isomer of an alcohol. 2-propanol used as a reducing agent.	Fragrance, pharmaceutical fine chemical synthesis

Iron-based catalysts.

to those using conventional rhodium catalysts. Since then, the groups of C. P. Casey at the University of Wisconsin, M. Beller at the University of Rostock, R. H. Morris at the University of Toronto, and D. Milstein at the Weizmann Institute of Science have all found catalysts for the hydrogenation of the carbon-oxygen double bond of a ketone to produce an alcohol (see illustration).

The transfer hydrogenation (TH) reaction is a subclass of reduction reactions in which a molecule of hydrogen is transferred to the substrate from a sacrificial reductant such as 2-propanol or formic acid. This type of hydrogenation is less favorable for large industries if the reductant cannot be easily regenerated; thus, direct hydrogenation is simpler and more cost-effective. Nevertheless, transfer hydrogenation is safer to use on small- to medium-scale reactions because high pressures are not required. Complexes prepared by the Morris and Casey groups, which were used for direct hydrogenation, are also active for TH of ketones and certain imines to produce alcohols and amines, respectively. Morris et al.'s catalysts are unique in allowing efficient asymmetric TH of ketones with activities comparable to catalysts based on PGMs such as ruthenium (see illustration). J. X. Gao's group at Xiamen University and M. Beller's group have reported that catalyst systems produced by mixing different iron carbonyl clusters and organic ligands also catalyze asymmetric TH. Although the structures of the catalysts that are formed in solution are unknown, the reactivity of these catalytic systems is excellent under mild conditions.

The high charge and stability of iron in the +3 oxidation state [Fe(III)] makes it an excellent Lewis acid. Salts such as iron(III) chloride were found to be very efficient catalysts for Lewis acid–assisted reactions, such as addition reactions (Michael addition, allylation and alkylation of carbonyl compounds, and substitution reactions), as well as cycloadditions. It is also worth noting that microwave-assisted catalytic reactions of these types usually give higher yields of the product in shorter reaction times.

An active field for catalyst development is the cross-coupling of organic molecules via carbon-X bond formation, where X is commonly carbon, oxygen, or nitrogen. Palladium and rhodium are the metals of choice for these transformations. Often a carbon–hydrogen bond in one of the reactants has to be broken at a metal center in a lower oxidation state [for example, Rh(I) or Pd(0)], a process called carbon–hydrogen bond activation. This results in a metal–carbon bond in which the metal is two oxidation states higher, Rh(III) or Pd(III). Nickel complexes are also useful in this regard, where the catalyst is proposed to shuttle between Ni(II) and Ni(IV). Iron poses a problem for this reaction because its common oxidation states are Fe(II) and Fe(III). Nevertheless, iron-catalyzed cross-coupling reactions are possible and are believed to occur via a radical mechanism. So far, iron-catalyzed cross-coupling reactions are not able to compete in reactivity with those catalyzed by the precious metals. However, the concept that iron is useful in these transformations is proven.

Outlook. Recent results in the field of catalysis suggest that many classical catalytic transformations can be performed using cheap, abundant, and less toxic iron-based catalysts. The principles of operation of these catalytic systems are currently being uncovered in order to optimize or better develop these new chemical transformations, which promise greener chemical processes.

For background information *see* ATOM ECONOMY; CATALYSIS; ENZYME; GREEN CHEMISTRY; HOMOGENEOUS CATALYSIS; IRON; LIGAND; NITROGEN FIXATION; TRANSITION ELEMENTS in the McGraw-Hill Encyclopedia of Science & Technology.

Alexandre A. Mikhailine; Robert H. Morris

Bibliography. R. M. Bullock, *Catalysis without Precious Metals*, Wiley-VCH, Weinheim, Germany, 2010; I. Ojima, *Catalytic Asymmetric Synthesis*, 2d ed., Wiley-VCH, Weinheim, Germany, 2000; B. Plietker, *Iron Catalysis in Organic Chemistry: Reactions and Applications*, Wiley-VCH, Weinheim, Germany, 2008.

Large Hadron Collider: first year

The Large Hadron Collider (LHC) is the largest and most complex scientific undertaking ever attempted. Its results will determine the future direction of high-energy physics. The LHC produced 7 TeV proton–proton collisions in 2010 and 2011, and it is expected to produce 14 TeV proton–proton collisions in 2014. Experiments at the LHC may discover new particles and forces associated with electroweak symmetry breaking, dark matter, or properties of the early universe.

LHC: the machine. The LHC is located in a circular tunnel 27 km (17 mi) in circumference. The tunnel is buried 50–175 m (164–574 ft) underground and straddles the Swiss and French borders on the outskirts of Geneva (**Fig. 1a**). It circulated the first beams on September 10, 2008; on March 30, 2010, proton–proton collisions were achieved at 7 TeV, an energy value 3.5 times higher than any previously achieved at a particle collider. Since then the LHC has been operating almost continuously, apart from technical shutdowns and a short run with heavy-ion collisions.

In accelerator-based high-energy physics experiments, the type and number of particles brought into collision and their center-of-mass energy characterize the interaction. In its main operating mode the LHC is colliding proton beams. A monoenergetic proton beam is equivalent to a wide-band parton beam, where parton refers to the quarks, antiquarks, and gluons that constitute the proton. An impressive technological challenge of the LHC is the 1232 superconducting dipoles operating at a temperature of 1.9 K that bend the two proton beams around the 27-km-circumference tunnel. At 7 TeV these magnets must produce a magnetic field of 8.33 teslas (T) at a current of around 11,700 A just to keep the proton beams circulating in the ring. The magnets have two side-by-side apertures (a dual-core or "two-in-one" design), one for each of the counterrotating

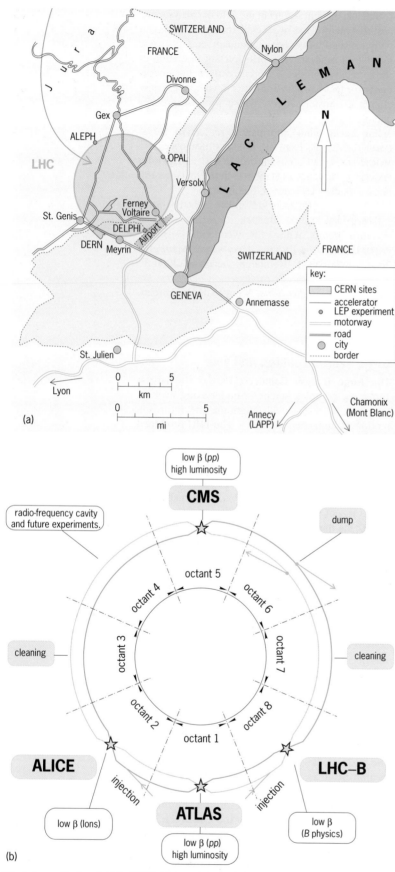

proton beams. Each dipole is 14.3 m (47 ft) long. The niobium–titanium coils create the magnetic fields to guide the two counterrotating proton beams in separate magnetic channels but within the same physical structure. The coils are surrounded by nonmagnetic "collars" of austenitic steel, which hold the coils in place against the strong magnetic forces that arise when the coils are at full field—the force loading each meter length of dipole is about 400 tons.

The LHC dipoles comprise 7600 km (4700 mi) of superconducting cable that weighs 1200 tons. Each cable is made up of 36 strands of superconducting wire, and each strand houses 6300 superconducting filaments of niobium–titanium alloy; thus the total length of the filaments is astronomical, more than 10 astronomical units. For the cooling of the machine, the LHC cryogenics need about 40,000 leak-tight pipe junctions, while 12 million liters (3 million gallons) of liquid nitrogen are vaporized during the initial cool-down of 31,000 tons of material; the total inventory of liquid helium is 700,000 liters (185,000 gallons).

The LHC run in 2010–2011 used up to 1400 bunches of protons each with up to 10^{11} protons. The two beams are focused and brought into collision at the four interaction points shown in Fig. 1b. At the interaction point, 10^{11} protons are squeezed down to a beam width of only 16 μm (micrometers). The collision rate is up to 10^9 per second. The LHC experiments apply fast real-time primary selection, known as triggering, on the raw analog and digital signals from the particle detectors; this selection records events that are deemed sufficiently interesting at a rate of about 300 Hz. The amount of data that the experiments collect is unprecedented. There exists no single computing site that could analyze the LHC data in its entirety. The challenge of distribution, storage, accessing, and analyzing the data of the LHC has already helped spawn and develop new technologies in computer science, referred to as grid computing.

Detectors. The LHC detectors are built to uncover new particles and new fundamental dynamics by reconstructing the identity and kinematics of particles produced in high-energy collisions, and comparing the reconstructed events to what is expected from the standard model of particle physics. The detectors identify the particles based on their interaction with matter, their decays, or their measured mass; they determine the momentum of the charged particles by measuring their curvature in a magnetic field once their mass is known, and they determine where inside the detector the particles originated. The LHC detectors have up to 10^8 individual channels of readout for a single collision event. The four main LHC detectors are ATLAS (A large ToroidaL ApparatuS; **Fig. 2**), CMS (Compact Muon Solenoid; **Fig. 3**), ALICE (A Large Ion Collider Experiment), and LHCb (the Large Hadron Collider beauty detector). ATLAS and CMS are multipurpose detectors designed to discover the so far elusive Higgs boson, the quantum of the quintessential field that is thought to be

Fig. 1. Large Hadron Collider (LHC). (*a*) Map of the geographic location, showing the scale of the machine. (*b*) Overall layout. "Low β" means that the beam is very narrow and "squeezed" in order to achieve the desired collision rate.

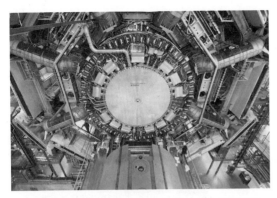

Fig. 2. The ATLAS multipurpose detector at the LHC. This is a view of the ATLAS cavern, side A, in 2007, while the End Cap Calorimeter was being moved. (*Photo courtesy of CERN*)

responsible for giving the observed masses to *W* and *Z* bosons as well as quarks and charged leptons.

Measurements of standard model processes. From its first 4 months of operation at 7 TeV, the LHC delivered enough data to the experiments to observe most standard model particles and their interactions. By the end of 2010, the experiments reproduced the standard model measurements that originally took physicists over four decades to perform in a multitude of experiments around the world. Measurements of standard model processes continued in 2011, covering precise properties of *W* and *Z* boson interactions as well as study of the top quark, the most massive of all known elementary particles. The importance of these measurements lies in gaining a complete understanding of the standard model at 7 TeV as the required foundation for the discovery program. *See* TOP QUARK AT THE TEVATRON.

Search for the Higgs boson. Although a number of elementary particles have no known reason to exist, the Higgs boson (or some equivalent physics mechanism) appears to be a prerequisite for the rest of the standard model to make sense. The Higgs mechanism is thought to be the simplest way for elementary particles to acquire mass, through a new kind of fundamental interaction that also implies the existence of a Higgs boson. The mass of this particle is not predicted, but it should be light enough to be produced in collisions at the LHC. At best, one high-energy collision in a billion will produce a Higgs boson, which will then decay immediately to a pair of lighter particles, such as photons, *W* or *Z* bosons, or heavy quarks. The ATLAS and CMS detectors were designed to be able to identify and measure such events with great accuracy, but it is still a challenge to discriminate a handful of Higgs boson decays from standard model processes that occur much more frequently in high-energy collisions. With data collected in the first half of 2011, the ATLAS and CMS experiments have already demonstrated the necessary sensitivity to discover a Higgs boson if its mass were within a certain range of values. The hunt for the Higgs continues with high expectations, as much larger data sets will be analyzed in the near future.

Beyond the standard model. Experiments at high-energy colliders have the potential to discover new particles produced, via Einstein's mass–energy relation, $E = mc^2$, from the kinetic energy of the colliding beams. These new particles may be too heavy to be produced at lower energies, or they may appear more readily at higher energies through a new mechanism, such as the decay of other heavy particles. High-energy collisions also re-create on a microscopic scale the extreme conditions that prevailed universally during the first instant of the big bang. Physicists expect that the laws of physics in these extreme conditions of the early universe may differ in fundamental ways from those that prevail today, explaining, for example, the excess of matter over antimatter to which we owe our own existence. Searches for new particles, new interactions, and deviations from accepted physical laws are the largest effort in the LHC experiments. Already in 2010, dozens of searches were conducted for new heavy particles of various types associated with theoretical ideas such as supersymmetry, string theory, and extra dimensions of space. Because no new particles were found in 2010, the experiments were able to set lower bounds on the masses of these hypothetical particles; these mass bounds range from 2 to 3 times the mass of the top quark to more than 20 times the mass of the top quark, depending on the particle. These searches will continue and expand in scope and sophistication as more data are collected.

For background information *see* ANTIMATTER; BIG BANG THEORY; ELEMENTARY PARTICLE; HIGGS BOSON; INTERMEDIATE VECTOR BOSON; PARTICLE ACCELERATOR; QUARKS; STANDARD MODEL; SUPERSTRING THEORY; SUPERSYMMETRY in the McGraw-Hill Encyclopedia of Science and Technology.

Maria Spiropulu

Bibliography. G. Kane and A. Pierce, *Perspectives on LHC Physics*, World Scientific, Singapore, 2008; The CERN Large Hadron Collider: Accelerator and experiments, Special section, *J. Instrum.*, 3:S08001–S08007, 2008.

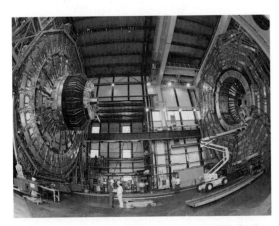

Fig. 3. The CMS multipurpose detector at the LHC. On the left are the 15-m-diameter (50-ft) end-cap "disks" installed with muon chambers. On the right is part of the CMS barrel with the solenoid and the hadron calorimeter visible, as well as the muon barrel wheel. This picture was taken before "closing" the detector for the first time at the assembly for the 2006 magnet commissioning and cosmic data run. (*Photo courtesy of CERN*)

LEAFY: a master regulator of flower development

Flowers represent a powerful reproductive structure, considered to be largely responsible for the evolutionary success of flowering plants or angiosperms. The flower is usually composed of two reproductive organs, the carpels and the stamens, surrounded by protective and decorative organs, the sepals and the petals. Variations in the presence, number, color, and shape of these organs, and their associated scents, have driven coevolution with pollinators, resulting in accelerated speciation in both plants and insects.

Flowering is dependent on growth conditions such as day length and winter cold. Once flowering is triggered, the stem apex, or apical meristem, that is, the plant structure that produces new organs, becomes able to produce flowers in addition to leaves. Many floral mutants or variants exist in gardens or research laboratories. The study of these mutants or variants has been instrumental for understanding the genetic control of flower development. Among the many genes controlling the induction of flowering, the development of flower buds, and the nature of the floral organs, one gene, called *LEAFY* (*LFY*), plays a central role. This gene was discovered as a result of *lfy* loss-of-function mutants in the model plants, snapdragon and *Arabidopsis*, in which flowers are replaced by leafy structures (leaves, shoots, or shoot/flower intermediates). Conversely, the engineered constitutive expression of *LFY* in *Arabidopsis* was shown to convert inflorescences into solitary flowers. Therefore, *LFY* appears necessary and sufficient to confer a floral identity to meristematic structures.

Genetic control of floral identity by LFY in model species. In *Arabidopsis*, *LFY* gene expression is regulated by day length, winter cold, and even endogenous signals such as hormones. Because it perceives the influence of various pathways, *LFY* has been called a floral integrator. Its expression level slowly increases in each young leaf until it is high enough in the newly formed primordium to turn it into a flower. After conferring a floral identity to a nascent bud on the flanks of the shoot tip, LFY protein also contributes to its patterning, by inducing the ABCE genes. These genes are expressed in different locations on the developing flower bud, and the combination of their activities specifies the nature of the floral organs (**Fig. 1**). The most externally expressed A genes specify sepals in the outer whorl and, in combination with B genes, specify petals in the second whorl. The C genes are expressed in the center of the flower; they control the development of carpels in the fourth whorl and, together with B genes, the development of stamens in the third whorl. The E activity is present and required in all four floral whorls. LFY is expressed uniformly in the young flower, but it interacts with various coregulators expressed in specific floral regions that help to locally induce the ABCE genes. A recent genome-wide study suggests that LFY likely regulates several

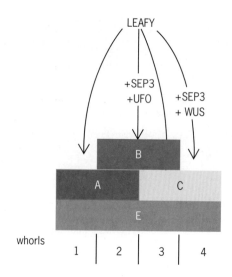

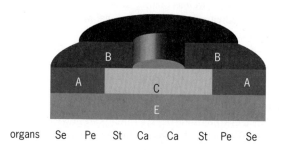

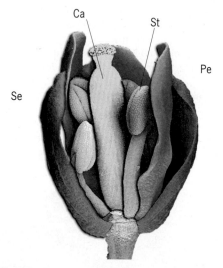

Fig. 1. Floral regulatory network controlled by *LFY*. LFY controls the expression of several A, B, C, and E genes in association with known coregulators, including WUSCHEL (WUS), UNUSUAL FLORAL ORGANS (UFO), or SEPALLATA3 (SEP3). The ABCE model for floral organ identity is shown together with a flower and its four types of organs: sepals (Se), petals (Pe), stamens (St), and carpels (Ca).

hundred genes, either positively or negatively, including some of its coregulators such as SEPALLATA3 (**Fig. 2**), thereby forming feed-forward loops (a motif commonly found in gene regulatory networks). LFY is also involved in a series of feedback loops with some of its target genes that help to turn the progressive increase of LFY level in a sharp developmental transition.

Molecular mechanisms responsible for LFY function in angiosperms. LFY exists only in land plants and shows no similarity to other proteins. Its biochemical characterization reveals that it acts as a transcription factor, and structural analysis of its DNA-binding domain (LFY-DBD) shows that it contains a helix-turn-helix motif (a common motif interacting with nucleic acids) buried in a unique seven-helix fold (Fig. 2). The LFY-DBD binds DNA as a dimer and interacts with both the major and minor DNA grooves, reminiscent of homeobox transcription factors (which help to control gene expression during plant and animal morphogenesis and development). The interaction surface with DNA is relatively large (19 base pairs), allowing LFY to have a high degree of DNA-binding specificity. A mathematical model has recently been constructed that captures this DNA-binding specificity and provides an accurate prediction of LFY-binding sites in plant genomes.

The function of LFY is well established in the model plant *Arabidopsis*, but whether it can be generalized to other angiosperms is still under investigation. To understand how flowers have appeared and evolved, it would be interesting to gain insight into the origin and the evolution of the floral network orchestrated by LFY. Among angiosperms, the LFY-DBD sequences are extremely similar, suggesting that binding specificity of LFY-DNA is also relatively conserved. A mathematical DNA-binding model for *Arabidopsis* based on this assumption and applied to other angiosperms revealed that the link between LFY and the C gene *AGAMOUS* preceded the divergence between monocots and eudicots. Applied to other target genes and more evolutionary basal species, a similar strategy might help to pinpoint the origin of the floral network from direct inspection of genomic sequences.

Vegetative function for LFY? Genes closely related to *LFY* are present in all other land plants, including nonflowering plants such as gymnosperms, ferns, and mosses. This raises two main questions: What is the function of *LFY* in these plants? How did *LFY* acquire its floral function?

Gymnosperms constitute a group of seed plants that were predominant during the Mesozoic (or Secondary) era. Extant species include conifers and cycads. They are usually characterized by independent male and female cones, which lack the flower-specific organs (sepals and petals). In most gymnosperms, *LFY* possesses a paralog called *NEEDLY* (*NLY*), which was probably lost in angiosperms. Both are expressed in reproductive structures, in overlapping domains that sometimes become exclusive as organs differentiate. The expression data suggest that *LFY* and *NLY* could regulate preflowering gene networks involving some MADS-box genes (transcription factors that play prominent roles in plant development) related to the B and C genes from angiosperms. These networks would specify male or female reproductive organ identity. Based on this assumption, the flower could have been created following modifications of LFY properties that allowed it to acquire NLY function, resulting in

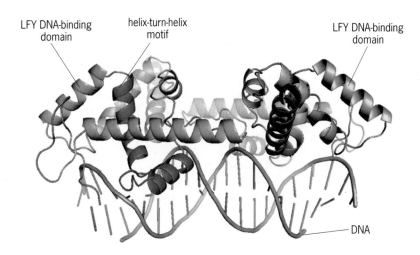

Fig. 2. Structure of the LFY DNA-binding domain on its target DNA. Two LFY DNA-binding domains associate in a dimer to bind DNA. One helix-turn-helix motif is labeled; the DNA is shown on the bottom.

the assembly of male and female structures on the same axis.

In mosses and ferns, which predate seed plants during evolution, *LFY*-related genes are also present; however, functional data are available only in the moss *Physcomitrella patens*. In *lfy* mutants of this plant, cell division is arrested after fertilization in the diploid structure called the sporophyte. How *LFY* from moss controls cell division is unknown, but understanding this function is of particular interest as it could reflect its ancestral role. In angiosperms, most investigations on *LFY* function have been focused on flowers. Nevertheless, in addition to controlling floral development, *LFY* is also involved in other processes: it controls the architecture of inflorescences in cereals (in rice, for example) and the development of compound leaves in some legumes (such as pea). These two processes depend on whether a specific group of cells continues dividing to form new stems or leaf lobes. *LFY* might thus bear two functions in angiosperms: controlling cell division (its putative ancestral role) and converting meristems into flowers.

It is likely that the novel techniques [for example, chromatin immunoprecipitation sequencing (ChIP-seq)] available to discover target genes of transcription factors will not only help to clarify the mechanisms of known *LFY* functions but might also reveal novel processes involving this versatile transcription factor. The combination of such genome-wide techniques (applied to the growing body of genome sequences available) will help elucidate how this master regulator works in model species and how it acquired its multifaceted function.

For background information *see* APICAL MERISTEM; CELL DIVISION; FLOWER; GENE; GENETICS; INFLORESCENCE; MAGNOLIOPHYTA; PLANT DEVELOPMENT; PLANT EVOLUTION; PLANT GROWTH; PLANT ORGANS; PLANT REPRODUCTION; TRANSCRIPTION in the McGraw-Hill Encyclopedia of Science & Technology.   Gabrielle Tichtinsky; Gilles Vachon; François Parcy

Bibliography. M. W. Frohlich and M. W. Chase, After a dozen years of progress the origin of angiosperms is still a great mystery, *Nature*, 450:1184–1189, 2007; C. Liu, Z. Thong, and H. Yu, Coming into bloom: The specification of floral meristems, *Development*, 136:3379–3391, 2009; E. Moyroud et al., LEAFY blossoms, *Trends Plant Sci.*, 15:346–352, 2010; D. Wagner, Flower morphogenesis: Timing is key, *Dev. Cell*, 16:621–622, 2009.

Malware

Malware, a contraction of the phrase "malicious software," is computer code that intentionally performs actions that are harmful. There are many different types of malware, and, as we become more reliant on information systems in our day-to-day lives, the danger posed by malware will continue to grow. An estimate from the June 2011 *Consumer Reports* put the economic cost of malware to consumers at $2.3 billion in 2010. The Stuxnet worm in 2010 demonstrated that malware can harm not only computer systems and networks, but also real-world devices and, potentially, people. It is possible that one day malware will be responsible for the death of a human being (and perhaps many human beings). An ongoing race pits those who create malware against those who defend against it. Each time the defenders invent a new way to counteract a particular class of malware, the malware creators adapt their code to circumvent the new defenses and the cycle continues.

Motivations for malware. Historically there have been two major motivations for people who develop malware: not-for-profit and for-profit. Those who develop malware with no profit motive do so for a variety of reasons. They may be seeking to demonstrate their security knowledge or skills, publicize a vulnerability they have discovered, or cause mischief just for the fun of it. All three of these motivations probably contributed to the creation in 1988 of the famous Morris worm, which was one of the first pieces of malware to spread widely on the Internet.

For-profit malware developers are seeking financial gain. Based on how lucrative malware can be and the low probability of getting caught, it is not surprising that making money is currently the main motivation for most malware authors. A good example is Zeus. Zeus is a package that allows an individual to remotely control a compromised host and spy on any unsuspecting user of the host. Zeus is particularly adept at stealing passwords and online banking credentials. The author of Zeus develops and sells the malware to criminals for a few thousand dollars. The criminals then use a variety of techniques to install Zeus on computers worldwide, take control of them, and use the information that Zeus harvests to steal money. Most experts agree that the number of hosts worldwide infected by Zeus easily number in the millions.

A relatively recent development is malware developed for use in furthering political, economic, or military goals and sometimes with the support of nation states or non-state actors (such as organized crime). One example is the Stuxnet worm, which was mentioned earlier. This malware is purported to have been developed with the backing of at least one government and to have targeted Iran's uranium enrichment equipment in 2010. Another example is the distributed denial of service (DDoS) attacks on PostFinance and PayPal in 2010 by a group of hackers in retaliation for the freezing of accounts used by the controversial WikiLeaks website.

Types of malware. Although there are many different types of malware, we can divide malware into two broad classes. Parasitic malware is code embedded inside another (apparently useful and benign) program. Examples of parasitic malware include backdoors, Trojan horses, and viruses. The other class, independent malware, has no other function than to perform malicious actions. Worms, botnet programs, and rootkits are examples of independent malware. The following sections describe some of these types of malware and give some illustrative examples.

Parasitic malware. A backdoor is code placed in a program that circumvents normal security functionality. It is not uncommon for programmers to place a backdoor in a system they are developing for testing purposes, so that they do not have to go through the authentication routine every time they want to test a different part of the system. Of course, such backdoors should be removed after testing is complete and before the system becomes operational. However, sometimes backdoors are left in place either accidentally or on purpose. Eventually, either the author of the backdoor or someone else who discovers it will use the backdoor for malicious purposes. A famous example of a backdoor is code that a programmer attempted to include in the Linux kernel that would have elevated a normal user to superuser privileges under certain circumstances.

A Trojan horse is a program with hidden (usually malicious) side effects. The program could be a tool, utility, game, or any piece of software that entices users to install and use it by offering interesting and seemingly benign functionality. The malicious code within the program surreptitiously performs actions, such as placing the user's machine under the remote control of a hacker, sending out spam e-mail, or destroying data. Other common Trojan horse payloads include adware and spyware. Adware is software that displays unwanted advertisements (usually for questionable products). Spyware steals private information, such as passwords and credit card numbers, and transmits them to a criminal for exploitation.

Viruses are pieces of code that seek to infect files and spread. One class of virus targets program files by copying the virus's code into an executable file, so that when the program is run it searches for other executable files, infects them, and then performs the program's normal functions. Some viruses (typically called macro viruses) infect documents, since many word processing programs have a macro functionality that allows malicious macro code to be embedded in the document. When an infected document is opened, the macro code runs, typically attempting to infect other documents and thereby spread. Many macro viruses use e-mail to spread very quickly. In addition to infecting new documents, the virus e-mails copies of infected documents to contacts found on the local computer. If a recipient opens an infected document, documents on his or her machine are infected and e-mailed to his or her contacts, propagating the virus. One of the earliest e-mail macro viruses was Melissa, which appeared in 1999. Melissa caused a great deal of damage by spreading very widely and very quickly and overloading Internet e-mail servers with a tremendous volume of mail messages. Newer viruses seek to spread via removable media or over peer-to-peer networks, social networking sites, mobile phones, and other platforms.

Independent malware. A worm is a self-replicating program whose main purpose it to spread. This article has already mentioned two very famous and very destructive worms: the Morris worm and Stuxnet. Worms use many of the same propagation techniques that viruses use to spread from system to system, including system vulnerabilities, removable media, and e-mail. Similarly, in addition to spreading, worms often perform other actions (called the payload), such as sending spam e-mail or installing backdoors, Trojan horses, adware, or spyware on the infected system.

An increasingly common action that some worms take is to install remote-control malware on a compromised host that allows a hacker to take control of the compromised host and turn it into what is called a zombie or bot (short for robot). Often, legitimate users do not even realize that their system is being used by, and under the full control of, someone else. Typically, the hacker, called a bot herder, will control thousands or even millions of machines worldwide in what is called a botnet. The bot herder can then use his or her botnet in a variety of ways, such as to send out spam, collect private information, or launch DDoS attacks. If the bot herder does not wish to perform these actions personally, he or she can still profit by renting out the botnet (usually by the hour or day) to someone else. A recent example is the Storm botnet that was estimated to have generated profits of almost $10,000 per day.

Closely intertwined with viruses, worms, and bot malware are rootkits. A rootkit is software that is designed to hide, from a machine's owner, the fact that it is infected and/or controlled by a hacker. Rootkits do this by hiding suspicious files, processes, network connections, and other information that might alert the owner that his or her machine is compromised. Rootkit functionality is being used in many types of malware to help it spread undetected and to allow it to evade anti-malware countermeasures.

Defending against malware. The best defense against malware is to prevent it from getting onto a system in the first place. This is the job of intrusion prevention systems, patches for system and application vulnerabilities, mail servers that scan incoming messages for malicious attachments and remove them, and other security best practices. However, prevention will not succeed all the time. Therefore,

there is a thriving industry in anti-malware programs that attempts to detect malware and remove it from systems. There are also online communities, such as the Malware Domain List, for tracking malware trends and developments and publicizing information about new malware and how to combat it.

Outlook. Malware is a serious problem. The amount of malware is growing every day, and new techniques are constantly being developed to spread it, hide it, and make it hard to defend against. The amount of damage done by malware is rising as well. There are tools and techniques that can be used to combat malware, but malware writers keep abreast of what they are up against and adapt their malware rapidly to circumvent any countermeasures that are employed. For the foreseeable future, it seems likely that the race between malware developers and defenders will continue.

For background information *see* COMPUTER PROGRAMMING; COMPUTER SECURITY; DIGITAL EVIDENCE; ELECTRONIC MAIL; INTERNET; PROGRAMMING LANGUAGES; SOFTWARE; SOFTWARE ENGINEERING; WORLD WIDE WEB in the McGraw-Hill Encyclopedia of Science & Technology.   Brett Tjaden

Bibliography. J. D. Aycock, *Computer Viruses and Malware (Advances in Information Security)*, Springer, 2010; M. Davis, S. Bodmer, and A. LeMasters, *Hacking Exposed: Malware & Rootkit Secrets and Solutions*, McGraw-Hill Osborne Media, 2009; Online exposure, *Consumer Reports*, June, 2011; J. C. Rebane, *The Stuxnet Computer Worm and Industrial Control System Security*, Nova Science Publishing, 2011; E. Skoudis and L. Zeltser, *Malware: Fighting Malicious Code*, Prentice Hall, 2003; E. H. Spafford, *The Internet Worm Program: An Analysis, Purdue Technical Report CSD-TR-823*, Department of Computer Sciences, Purdue University, 1988.

Membrane reactors

Reaction and separation are the most important parts of chemical processes. As shown in **Fig. 1**, reaction and separation in conventional chemical processes are carried out in separate units, whereas a membrane reactor (MR) combines the catalytic reaction and the membrane separation in one unit. Membrane reactors were proposed after membranes were developed in the 1960s. At present, the membrane bioreactor (MBR), in which porous membranes such as microfiltration and ultrafiltration membranes are combined with activated sludge processes for wastewater treatment, has been widely commercialized. Recent progresses in inorganic membranes, such as zeolite and silica membranes, which enable high-temperature operation, make membrane reactors highly promising for applications in the chemical industries.

A membrane reactor, which enables simultaneous reaction and separation processes in one unit, is a simple and compact configuration of reactors and separation units. More importantly, additional func-

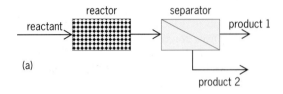

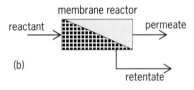

(a)

(b)

Fig. 1. Reaction and separation in (a) conventional reactor and (b) membrane reactor.

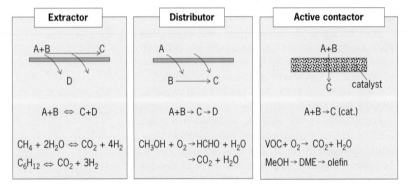

Fig. 2. Membrane reactor functionalities.

tionalities, including extraction, distribution, and active contact, can enhance reaction conversions and selectivity, compared with conventional reaction and separation processes.

Classification of membrane reactors. Membrane reactors can be categorized based on the function and configuration of catalysts and membranes. **Figure 2** summarizes the extractor, distributor, and active-contactor functions of membrane reactors.

Extractor. The extractor has been extensively investigated for membrane-reactor applications in chemical processes. Extraction of hydrogen in the steam reforming of methane is a typical application of a membrane reactor. The conversion can be increased by selective extraction of hydrogen (product) from the reaction system under thermodynamic equilibrium, according to Le Chatelier's principle. Another advantage is that purified hydrogen can be obtained in a permeate (product) stream through hydrogen semipermeable membranes, such as metal (palladium) and porous silica membranes. Typical applications include dehydrogenation of hydrocarbons such as cyclohexane to benzene ($C_6H_{14} \Leftrightarrow C_6H_6 + 3H_2$), steam reforming of methane (SRM) [$CH_4 + H_2O \Leftrightarrow CO + 3H_2$], and the water-gas shift reaction ($CO + H_2O \Leftrightarrow CO_2 + H_2$). Because most dehydrogenation reactions are endothermic, higher reaction temperatures are favored to obtain higher conversion. The reaction temperature of SRM, which is operated at approximately 700–900°C, can be lowered, typically to 500°C, by shifting the equilibrium of

reaction, which would enable the use of less expensive materials for reactor construction. Conventionally, Pd-membranes, which show infinite selectivities to hydrogen in principle, have been used for the membrane reactors. Recent progresses in the fabrication of porous silica membranes with high selectivity and permeability, makes available the application of silica membranes to membrane reactors. **Figure 3** shows the conversion of methane for steam reforming of methane using a porous SiO_2 membrane reactor, using a microporous silica membrane at 500°C. Conversion of methane, which was confirmed to show approximately an equilibrium conversion without membrane permeation, was increased by hydrogen extraction, and increased from 0.4 (equilibrium conversion) to 0.9 with a decrease in the CH_4 feed flow rate because of the larger effect of extraction of hydrogen. Calculated conversion (solid curve), based on the plug-flow model, shows good agreement with experimental results.

Another reaction type applied in extraction is esterification, in which acids and alcohols produce ether and water under thermodynamic equilibrium. The typical reaction is the production of ethyl acetate from acetic acid with ethanol ($C_2H_5COOH + C_2H_5OH \Leftrightarrow C_2H_5COC_2H_5 + H_2O$). Water semipermeable membranes, such as zeolite membranes, were used to shift the equilibrium conversion, because water, the smallest among the molecules involved in esterification, is relatively easy for selective extraction.

Distributor. The distributor, which distributes reactants gradually along the reactor, can be applied to consecutive reactions, including partial oxidation and oxidative dehydrogenation. Feeding one of the reactants, such as oxygen, to the other side of the reactor at a controlled rate through a membrane avoids the abrupt reaction caused by mixing reactants in high concentration, as well as flammability issues of conventional reactors. Ion transport membranes, such as perovskite membranes (typically $La_{1-x}Sr_xCo_{1-y}Fe_yO_3$), which have ion (oxygen) conductivity, as well as electron conductivity, show oxygen semipermeability at high temperatures (600–800°C). Promising applications include

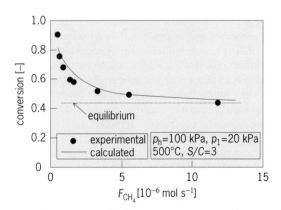

Fig. 3. Conversion of methane as a function of the feed flow rate of methane. (Reaction temperature = 500°C, feed flow rate of water/methane = 3; curve was calculated.)

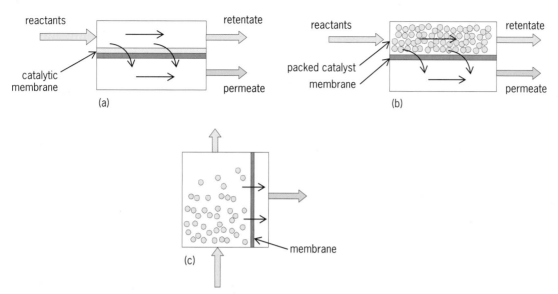

Fig. 4. Configurations of membrane reactor. (*a*) Catalytic membrane reactor (CMR). (*b*) Packed-bed membrane reactor (PBMR). (*c*) Fluidized-bed membrane reactor (FBMR).

the production of synthetic gas (syngas) through the partial oxidation of methane by the controlled addition of oxygen ($CH_4 + 0.5O_2 \rightarrow CO + 2H_2$), where complete oxidation would occur ($CH_4 + 2O_2 \rightarrow CO_2 + 2H_2O$) in a consecutive reaction with excess amount of oxygen. Recently, hydrogen permeating through Pd membranes was found to be active and effective in producing phenol in a one-step reaction from benzene.

Active contactor. In the active contactor mode, the membrane acts as a diffusion barrier and does not need to be permselective, but catalytically active. The catalytic reaction occurs inside membrane pores. The typical applications include the total oxidation of volatile organic compounds (VOCs), in which all the feed gas is fed through the membrane pores in a forced-flow mode. VOCs are decomposed just after contacting oxygen on the catalysts dispersed in the membrane pores. Because all gases are forced through the pores, the residence-time contact with the catalysts is more uniform than in the conventional packed-bed reactor, resulting in increased catalytic efficiency. Another feature of the active contactor is the possible control of the residence time of the reaction, which enables production of intermediates in consecutive reactions, because the residence time can be controlled by pressure difference in the viscous-flow mechanism.

Classification based on the configuration of membrane and catalysts. Membrane reactors can be categorized, based on the configuration of the catalysts, as packed-type membrane reactors, fluidized membrane reactors, and catalytic membrane reactors. It is also possible to categorize membranes based on their separation ability (permselective or nonpermselective) and catalytic activity (catalytically inert or active). **Figure 4** shows the typical configurations of these membrane reactors. Packed-bed membrane reactors (PBMR) consist of catalysts for reaction, as well as membranes for separation. Both catalyst parti-

cles and membranes are packed into one unit, but reaction and separation occur separately. For example, particulate catalysts impregnated with Ni-catalysts are used for the steam reforming of methane, and

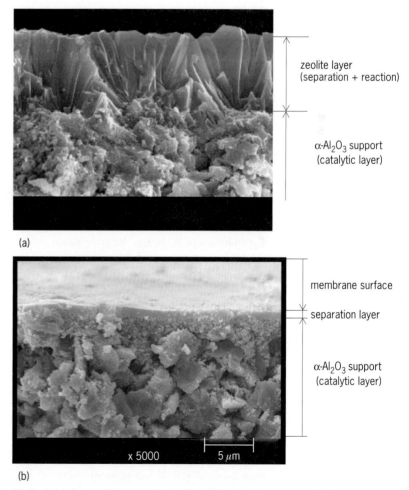

Fig. 5. Catalytic membranes. (*a*) Zeolite (inherently catalytic active) and (*b*) hydrogen separation layer (SiO$_2$ membrane) on catalytic supports.

palladium or silica membranes are used for hydrogen separation. The catalytic activity of the membrane is not necessarily required in this configuration. Catalysts could be dispersed in a fluidized bed. One of the typical applications commercialized on a large scale is the membrane bioreactor (MBR), in which activated sludge is fluidized using air bubbling. Porous membranes, such as microfiltration membranes, are installed outside the reactor or inside the reactor, which is called an immersed MBR. Both systems retain the activated sludge in the bioreactor and allow the permeation of water.

Other catalytic membranes have catalytic activity as well as separation ability in one membrane. Zeolite membranes (such as ZSM-5 and TS-1) are inherently catalytically active, whereas another type of catalytic membrane consists of a separation of the top layer and catalytically active supports. **Figure 5***a* shows ZSM-5 zeolite membranes (inherently catalytically active) in which reaction and separation occur simultaneously. In the methylation of toluene with methanol (CH_3-C_6H_5 + CH_3OH → *p*-CH_3-C_6H_4-CH_3 + H_2O), *p*-xylene can be selectively produced in the permeate stream through ZSM-5 membranes because of steric hindrance, in comparison with thermodynamic equilibrium. Another type of catalytic membrane is a silica-hydrogen separation membrane on a catalytically active support, which can be prepared using catalytically active powders, such as nickel oxides, or the impregnation of a Ni catalyst inside porous α-alumina supports.

For background information *see* CATALYST; CHEMICAL REACTOR; CHEMICAL SEPARATION (CHEMICAL ENGINEERING); DEHYDROGENATION; MEMBRANE SEPARATIONS; PEROVSKITE; REFORMING PROCESSES; ULTRAFILTRATION; WASTEWATER REUSE; ZEOLITE in the McGraw-Hill Encyclopedia of Science & Technology.   Toshinori Tsuru

Bibliography. A. Julbe, D. Farrusseng, and C. Guizard, Porous ceramic membranes for catalytic reactors—overview and new ideas, *J. Membr. Sci.*, 181(1):3–20, 2001; J. G. Sanches and T. T. Tsotsis, *Catalytic Membranes and Membrane Reactors*, Wiley-VCH, 2002; T. Tsuru, K. Yamaguchi, T. Yoshioka, and M. Asaeda, Methane steam reforming by microporous catalytic membrane reactors, *AIChE J.*, 50(11):2794–2805, 2004.

MicroRNA-33 (miR-33)

The metabolism of cholesterol, an essential component of many biochemical pathways, is tightly regulated at the cellular level. Insufficient or excess cholesterol levels can be detrimental to cells and are associated with disease states such as atherosclerosis (the hardening of arteries). Many regulatory mechanisms exist to ensure that cholesterol levels are balanced. In particular, recent findings have revealed that small noncoding RNAs (microRNAs) have a crucial role in the posttranscriptional control of cholesterol- and lipoprotein-related genes.

Of note is microRNA-33 (miR-33), which is an intronic microRNA (miRNA) located within a sterol regulatory element–binding protein (SREBP) gene, one of the master regulators of cholesterol and fatty acid metabolism. MicroRNA-33 regulates cholesterol efflux and high-density lipoprotein (HDL) formation in concert with the SREBP host genes, suggesting an important role for miRNAs in the epigenetic regulation of cholesterol metabolism and highlighting the clinical potential of miRNAs as novel therapeutic targets in treating cardiovascular-related diseases.

Classical regulation of cellular cholesterol homeostasis. Animal cells acquire cholesterol endogenously, from acetyl–coenzyme A (acetyl-CoA, the thioester of acetic acid with coenzyme A) through a highly regulated enzymatic pathway, or exogenously, from circulating low-density lipoproteins (LDLs) in a receptor-specific manner. Intracellular cholesterol levels are tightly controlled by feedback mechanisms that operate at both transcriptional and posttranscriptional levels (**Fig. 1**). When intracellular cholesterol levels are low, there is an increase in the cell surface expression of the LDL receptor (LDLr) and the activity of 3-hydroxy-3-methylglutaryl coenzyme A reductase (HMGCR, the rate-limiting enzyme of cholesterol biosynthesis). Conversely, when cells accumulate excess sterols, there is a decrease in the activity of HMGCR and the expression of the LDLr. This coordinated process is regulated by a group of membrane-bound transcriptional activators—the aforementioned SREBPs. Vertebrates have two SREBP genes. *Srebp-2* preferentially controls the synthesis and uptake of cholesterol, whereas *Srebp-1* preferentially activates the genes involved in fatty acid metabolism.

The other major transcription factor that regulates cholesterol homeostasis is the liver X receptor (LXR). LXRs belong to the nuclear receptor superfamily and are activated by endogenous oxidized metabolites of cholesterol (oxysterols). When cholesterol levels increase, LXRs activate the expression of genes involved in the removal of cholesterol from peripheral (nonhepatic) cells, such as the adenosine triphosphate (ATP)–binding cassette (ABC) transporters, ABCA1 and ABCG1. ABCA1 and ABCG1 primarily function in macrophages and hepatocytes. In macrophages, ABCA1 and ABCG1 promote cellular efflux of excess cholesterol and phospholipids. In the liver, they are essential players in the formation of HDL, also known as "good cholesterol." Cholesterol efflux to HDL and its associated apolipoprotein, apoA1, is considered a critical step in the initiation of reverse cholesterol transport (RCT), which is the process that delivers excess cholesterol to the liver for excretion into bile.

MicroRNAs and lipid metabolism. In addition to the classical transcriptional regulators of cholesterol metabolism (SREBPs and LXRs), miRNAs have been shown to regulate the expression of key genes involved in lipid homeostasis. MicroRNAs are small, noncoding, double-stranded RNAs (approximately

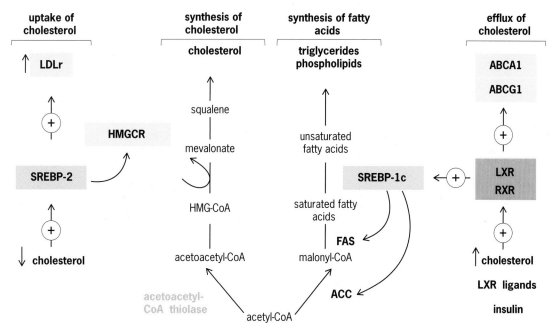

Fig. 1. Gene regulatory network of lipid homeostasis. When intracellular concentrations of sterols are low, sterol regulatory element–binding protein 2 (SREBP-2) is activated and enhances the expression of genes that control cholesterol synthesis (3-hydroxy-3-methylglutaryl coenzyme A reductase, HMGCR) and uptake (low-density lipoprotein receptor, LDLr). SREBP-1c is activated by liver X receptor (LXR) ligands (oxysterols) and insulin, and regulates the expression of genes required for fatty acid synthesis, including fatty acid synthase (FAS) and acetyl–coenzyme A carboxylase (ACC). In cholesterol-saturating conditions, the oxidized cholesterol derivatives activate LXR, leading to an increase in ABCA1 and ABCG1 and cellular cholesterol efflux. [Note that LXR binds as a heterodimer with another nuclear receptor, specifically retinoid X receptor (RXR).]

23 nucleotides in length) that control the expression of protein-coding genes and are involved in a variety of important biological processes. Primarily, miRNAs act as sequence-specific inhibitors of messenger RNA (mRNA) by binding to the 3′-untranslated regions (3′UTRs) of the mRNAs of their target genes, causing mRNA destabilization or protein translation inhibition.

Different aspects of lipid metabolism have been described as being under posttranscriptional regulation by several miRNAs: miR-122, miR-370, miR-378/378*, miR-335, miR-125a-5p, and miR-33. Among these, miR-122 was the first to be described in terms of its role in regulating total serum cholesterol and liver metabolism; overall, it has been the most widely studied miRNA in the liver. More recently, the miR-33 family has been identified as a key regulator of cholesterol homeostasis.

Regulation of cholesterol homeostasis by miR-33. MicroRNA-33 is expressed in a variety of cell types and tissues and consists of two intronic microRNAs, *miR-33a* and *miR-33b*, which are encoded within the introns of *Srebp-2* and *Srebp-1*, respectively. Whereas miR-33a and miR-33b differ in only 2 of their 19 nucleotides in the mature form and have the same targets, they differ in their pattern of evolutionary conservation. Thus, *miR-33a* is encoded within intron 16 of the human *Srebp-2* gene and is conserved in many animal species, including small and large mammals, chickens, frogs, and flies. In contrast, the conservation of *miR-33b*, which is found within intron 17 of the human *Srebp-1*

gene, is lost in many species. The *Srebp-1* genes of larger mammals contain *miR-33b*, but this miRNA is absent from the *Srebp-1* genes of rats, mice, and chickens.

Metabolic stimuli that activate *Srebp* gene expression also regulate *miR-33a* and *miR-33b* expression, indicating that miR-33 is coexpressed with its host genes. In mammalian cell lines, miR-33a is upregulated in response to cellular cholesterol depletion and downregulated in response to cholesterol loading. When intracellular cholesterol levels are low, the expression of *Srebp-2* and miR-33 increases. In this condition, miR-33 represses the genes involved in cholesterol trafficking and efflux, most prominently ABCA1, which has three highly conserved binding sites for miR-33 in its 3′UTR. In concordance with this, cultured mammalian cells that overexpress miR-33 show a decrease in the mRNA and protein levels of ABCA1, as well as a decrease in cholesterol efflux to apoA1, an essential step in the production of HDL. More interestingly, the opposite occurs when cells are transfected with "anti-miRs" or "antago-miRs," that is, RNA molecules that specifically reduce endogenous levels of miR-33. Antagonism of miR-33 in vitro and in vivo significantly increases ABCA1 expression, promotes cholesterol efflux to apoA1, and increases HDL plasma levels in mice. Furthermore, genetic deletion of miR-33 in mice causes an increase in hepatic ABCA1 expression and a 25% increase in circulating levels of HDL.

In addition to ABCA1, miR-33 also targets ABCG1, which is a sterol transporter that promotes

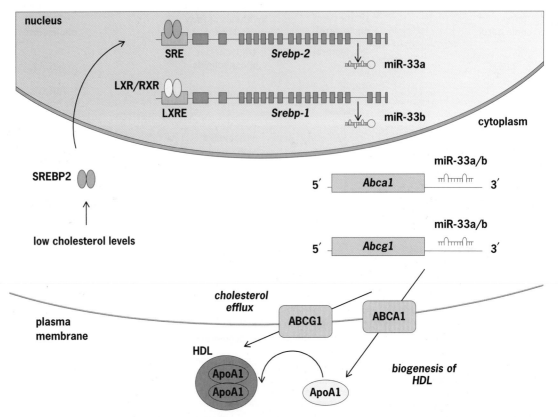

Fig. 2. MicroRNA-33 regulation of cholesterol efflux. Cellular stimuli (for example, LXR ligands or low-sterol conditions) activate the transcription of *Srebp-1* or *Srebp-2* by binding to the LXRE (liver X receptor element) or the SRE (sterol response element), respectively. This induces the cotranscription of miR-33b or miR-33a. These miRNAs simultaneously inhibit the expression of the genes involved in cholesterol transport (ABCA1 and ABCG1) by binding to the 3′-untranslated regions (3′ UTRs) of their mRNAs. The outcome of miR-33a/b activation is reduced cholesterol efflux and reduced lipidation of apolipoprotein A1 (apoA1) and high-density lipoprotein (HDL) particles. [Note that LXR binds as a heterodimer with another nuclear receptor, specifically retinoid X receptor (RXR).]

cholesterol efflux to prelipidated apoA1 (HDL). Unlike ABCA1, however, the targeting of ABCG1 by miR-33 appears to be significant only in mouse cells. In this context, miR-33 reduces ABCG1 protein and decreases cholesterol efflux to HDL in mouse macrophages and hepatocytes. Taken together, this confirms the physiological effects of miR-33 on cholesterol and lipoprotein homeostasis. Under low-sterol conditions, *miR-33* is coincidently generated with *Srebp-2* and coordinates cellular cholesterol homeostasis by simultaneously initiating transcription of cholesterol uptake and synthesis pathways and posttranscriptionally repressing genes involved in cellular cholesterol export (**Fig. 2**).

Concluding remarks. The role of miR-33 in regulating lipid metabolism highlights the ability of miRNAs to both fine-tune and dramatically alter cell signaling pathways. Two central pathways that control HDL cholesterol are regulated by miR-33: (1) the generation of HDL in the liver and (2) cellular cholesterol efflux to macrophages (which is the first step in reverse cholesterol transport). Because HDL levels correlate inversely with atherosclerosis susceptibility, there is an increasing interest in raising HDL levels (particularly ABCA1) to promote cholesterol efflux. The discovery that miR-33 regulates cholesterol homeostasis by targeting genes involved

in cholesterol trafficking opens new avenues to develop novel therapeutic strategies. In this context, using anti-miR-33 treatment to elevate ABCA1 levels and increase HDL levels would hold tremendous therapeutic potential for the medical care and prevention of cardiovascular disease.

For background information *see* ABC LIPID TRANSPORTERS; ARTERIOSCLEROSIS; CHOLESTEROL; GENETICS; LIPID; LIPID METABOLISM; LIPOPROTEIN; MICRORNA; MOLECULAR BIOLOGY; PROTEIN; RIBONUCLEIC ACID (RNA) in the McGraw-Hill Encyclopedia of Science & Technology.

Leigh Goedeke; Cristina M. Ramírez;
Carlos Fernández-Hernando

Bibliography. D. P. Bartel, MicroRNAs: Target recognition and regulatory functions, *Cell*, 136:215–233, 2010; M. S. Brown and J. L. Goldstein, The SREBP pathway: Regulation of cholesterol metabolism by proteolysis of a membrane-bound transcription factor, *Cell*, 89:331–340, 1997; S. H. Najafi-Shoushtari et al., MicroRNA-33 and the SREBP host genes cooperate to control cholesterol homeostasis, *Science*, 328:1566–1569, 2010; K. J. Rayner et al., MiR-33 contributes to the regulation of cholesterol homeostasis, *Science*, 328:1570–1573, 2010; A. R. Tall et al., HDL, ABC transporters, and cholesterol efflux: Implications for the treatment of atherosclerosis, *Cell Metabol.*, 7:365–375, 2008.

Microscopic calculations of heavy-ion fusion reactions

Nuclear fusion refers to the process in which two atomic nuclei combine to form a single larger nucleus. The two fundamental forces that determine the probability of nuclear fusion are the electrostatic Coulomb force and the strong nuclear force. The Coulomb force acts only between protons; it has a long range and is repulsive. On the other hand, the strong nuclear force acts between any combination of protons and neutrons; it has a short range (about 1.4×10^{-15} m) and is attractive. The fusion process is hindered by the Coulomb repulsion between the protons of the two colliding nuclei. To overcome this repulsive force, one must supply kinetic energy to bring the nuclei into close contact. Only when the nuclear surfaces are almost touching does the attractive strong nuclear force set in and cause the nuclei to "snap together," that is, to fuse. Thermonuclear fusion occurs naturally in the interior of stars. As explained by Hans Bethe in 1938, the Sun produces the energy it radiates by burning hydrogen into helium nuclei. In this case, the kinetic energy of the hydrogen nuclei is supplied by a conversion of gravitational stellar energy into thermal energy. The source of the energy released in the fusion of light nuclei is the difference between the nuclear binding energies of the reaction partners. Significant progress has been made in recent years to achieve controlled thermonuclear fusion in a fusion reactor. In the magnetic confinement method, a plasma confined in a magnetic "bottle" is heated to very high temperature, and in the inertial confinement approach the initial kinetic energy is supplied by powerful laser beams. Another frontier is the production of new superheavy elements in heavy-ion fusion reactions, in particular around the predicted "island of stability" with proton numbers $Z = 114$, 120, and 126, and neutron number $N = 184$. *See* DISCOVERY OF ELEMENT 117.

One major challenge that is common to many areas of theoretical physics and other natural sciences is the quantum many-particle problem. Theorists working in atomic, molecular, and condensed-matter physics, in nuclear physics, and in some areas of astrophysics face a similar challenge; how to describe the features of interacting quantum many-particle systems in terms of suitable constituent particles and the fundamental interactions between them. Such theories are referred to as "microscopic theories." The aim of nuclear theory is to study the quantum many-particle aspects of the strong interaction. At relatively low energy, an atomic nucleus may be viewed as a system of protons and neutrons that interact via Coulomb and strong nuclear potentials. Presently, a practical ab-initio many-body theory for fusion does not exist. Instead, the fusion of two nuclei is reduced to the problem of finding the interaction potential between the nuclei as a function of the distance between their centers, R. The fusion probability is then obtained by solving the quantum mechanical Schrödinger equation for this potential.

Approaches to calculating ion–ion potentials. Among the various approaches to calculating ion–ion potentials are the following.

1. *Phenomenological models such as the Bass model, the proximity potential, and potentials obtained via the double-folding method.* Here, one either assumes a prescribed mathematical form for the ion–ion potential or one uses empirical nuclear densities that contain a number of free parameters. Some of these potentials have been fitted to experimental fusion barrier heights and have been remarkably successful in describing scattering data. These calculations can be further improved by considering excitations of the nuclei as they approach each other and/or include the possibility of transferring a few neutrons, using the so-called coupled-channels formalism. In the double-folding method, the empirical parameterized nuclear densities may be replaced by densities calculated using a microscopic theory. One common physical assumption in such semimicroscopic calculations is the use of the frozen-density approximation. As the name suggests, in this approximation the nuclear densities are unchanged during the computation of the ion–ion potential as a function of the internuclear distance.

2. *Microscopic calculations such as the Hartree-Fock method with a constraint on various moments of the density or some suitable definition of the internuclear distance.* These microscopic calculations follow a minimum energy path and allow for the change of the nuclear densities as the new minimum energy configuration for the nuclear system is computed as a function of the chosen collective coordinates (a term used for the moments of the density characterizing its shape). Some of the disadvantages of these microscopic approaches are the confinement of the motion to a configuration space described by a few collective coordinates, while the actual densities can have an infinite number of nonzero moments, and the fact that the process is assumed to be static and adiabatic in character. The dynamics of the nuclear collision and the rearrangement of the nuclear densities during the collision process are not included. This is a disadvantage in particular for the description of fusion at collision energies below the ion–ion barrier peak, because it has been demonstrated that the inner part of the potential barrier (the barrier for small internuclear separations) plays a very important role while the outer part of the barrier is determined largely by the early stages of the collision process. The inner part of the potential barrier is strongly sensitive to dynamical effects such as particle transfer and neck formation during the collision.

TDHF method. All of the remarks made above suggest that a time-dependent and microscopic approach for calculating fusion barriers may lead to a better description of the fusion process. The most established microscopic theory for studying the low-energy collisions of nuclei is the time-dependent Hartree-Fock (TDHF) method. In TDHF, two static many-body states calculated in the Hartree-Fock approximation are boosted with a relative kinetic

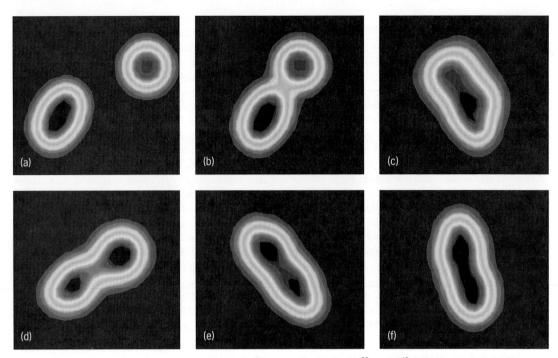

Fig. 1. Nuclear density contours obtained during the TDHF time evolution for the ^{22}Ne and ^{16}O collision at an impact parameter of 6 fm and initial neon orientation angle of 60°. Times T are in fm/c ($\sim 3 \times 10^{-24}$ s) [the time it takes light to travel 1 fm]. (a) $T = 0$ fm/c. (b) $T = 60$ fm/c. (c) $T = 140$ fm/c. (d) $T = 528$ fm/c. (e) $T = 936$ fm/c. (f) $T = 1170$ fm/c.

energy to initiate a nuclear collision in time via the time-dependent Schrödinger equation. This evolution results in a self-organizing system that selects its evolutionary path by itself following the microscopic dynamics. Some of the effects naturally included in the TDHF calculations are neck formation, mass exchange, internal excitations, deformation effects to all orders, as well as the effect of nuclear alignment for deformed (nonspherical) nuclei. **Figure 1** shows a few snapshots of the evolving nuclear densities for a noncentral collision of ^{22}Ne + ^{16}O. Note that ^{22}Ne is a strongly deformed nucleus. In the calculation shown in Fig. 1, the deformed density has been initialized at a 60° angle with the collision axis. To compute the total fusion cross section, fusion calculations are performed for a complete range of initial orientation angles.

DC-TDHF method. Recently, a new method has been developed to extract ion-ion interaction potentials directly from the TDHF time evolution of the nuclear system. This approach relies on a novel numerical method for minimizing the energy of a system subject to a constraint that the density function for the system takes on a prescribed distribution. The procedure for the so-called density-constrained TDHF (DC-TDHF) method is as follows. The TDHF time evolution takes place with no restrictions. At certain times during the evolution, the instantaneous density is used to perform a static Hartree-Fock minimization while holding the neutron and proton densities constrained to be the corresponding instantaneous TDHF densities. In essence, this provides the ion-ion interaction potential directly from the dynamical TDHF densities. The advantages of this method in comparison to other mean-field-based mi-

croscopic methods such as the constrained Hartree-Fock method are obvious. First, there is no need to introduce artificial constraining operators which assume that the collective motion can be described in terms of a few shape degrees of freedom; second, the static adiabatic approximation is replaced by the dynamical analog, where the most energetically favorable state is obtained by including sudden rearrangements, and the dynamical system does not have to move along the valley of the potential energy surface.

Applications. From the DC-TDHF calculations, one obtains heavy-ion interaction potentials, fusion/capture cross sections, and dynamic excitation energies during the collision. So far, the theory has been applied to the reactions ^{64}Ni + ^{132}Sn, ^{64}Ni + ^{64}Ni, ^{16}O + ^{208}Pb, ^{70}Zn + ^{208}Pb, ^{48}Ca + ^{238}U, and ^{132,124}Sn + ^{96}Zr. **Figure 2** shows an example of potential barriers calculated for the capture process during the collision of the ^{48}Ca + ^{238}U system. The figure shows the dependence of the potential barrier on the orientation of the deformed ^{238}U nucleus with respect to the collision axis, as well as the energy dependence of the potential barriers. Also shown for comparison is the curve for the Coulomb potential corresponding to two particles with the same charge. This reaction was one of the primary reactions leading to the new element with $Z = 112$.

Figure 3*a* shows similar potential barriers but for the ^{16}O + ^{208}Pb collision. Here, one can clearly see that as the bombarding energy gets closer to the barrier top, the inner part of the barrier thickens and develops a curved structure. This is due to the fact that at smaller energies the system has enough time to rearrange its density, whereas at higher energies

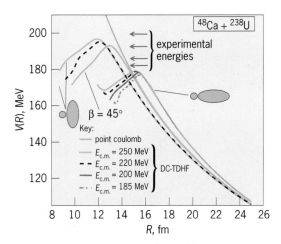

Fig. 2. Potential barriers for the ^{48}Ca + ^{238}U system obtained from DC-TDHF calculations as a function of energy and for selected orientation angles of the ^{238}U nucleus. Also shown are the experimental bombarding energies. β denotes the initial orientation of the ^{238}U nucleus with respect to the collision axis.

there is substantially less rearrangement. Figure 3*b* shows the comparison of calculated fusion cross sections to the experimental measurements. There is a very reasonable agreement with data.

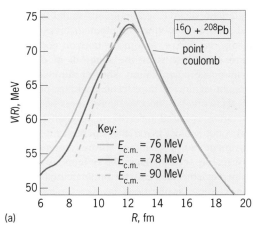

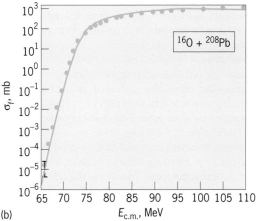

Fig. 3. Properties of the ^{16}O + ^{208}Pb collision. (*a*) Potential barriers obtained from DC-TDHF calculations at three different energies. (*b*) Calculated fusion cross sections using these potentials compared to experimentally measured ones.

In conclusion, the fully microscopic, dynamical, and parameter-free DC-TDHF method for the calculation of potential barriers yields excellent results when the underlying microscopic theory gives a good description of the static nuclear properties. These calculations demonstrate that for the proper description of fusion cross sections, especially at deep subbarrier energies, the dynamical effects such as neck formation and mass transfer must be included to modify the inner part of the potential barrier. Also interesting are the implications of the energy-dependent barriers and the dependence of barriers on the orientation of the colliding nuclei.

For background information *see* ATOMIC THEORY; NONRELATIVISTIC QUANTUM THEORY; NUCLEAR FUSION; QUANTUM MECHANICS; SCHRÖDINGER EQUATION in the McGraw-Hill Encyclopedia of Science & Technology.   A. Sait Umar; Volker E. Oberacker

Bibliography.   R. Bass, *Nuclear Reactions with Heavy Ions*, Springer-Verlag, Berlin, 1980; J. W. Negele, The mean-field theory of nuclear structure and dynamics, *Rev. Mod. Phys.*, 54(4):913–1015, 1982; A. S. Umar and V. E. Oberacker, Density-constrained time-dependent Hartree-Fock calculation of ^{16}O + ^{208}Pb fusion cross-sections, *Eur. Phys. J. A* 39:243–247, 2009; A. S. Umar and V. E. Oberacker, Heavy-ion interaction potential deduced from density-constrained time-dependent Hartree-Fock calculation, *Phys. Rev. C* 74:021601(R), 2006; S. Umar et al., Entrance channel dynamics of hot and cold fusion reactions leading to superheavy elements, *Phys. Rev. C* 81:064607, 2010.

Mining automation

The implementation of automation in a number of industries has brought significant benefits in terms of higher productivity (through improved performance and reduced downtime), enhanced safety (removing human personnel from hazardous and stressful working environments), and better cost effectiveness. In surface applications, automation is usually employed in well-defined manufacturing cycles of identical components in a well-defined, well-structured, and well-controlled environment. In underground mining, however, the environment is highly variable, unpredictable, and harsh. Currently, the deepest mines in the world, reaching 4000 m, are in South Africa. Geomechanical hazards, limited space, low light, poor air quality (presence of dust, humidity, exhaust gases from diesel engines, and so on), and often intense heat released by the surrounding rock mass in deep mines not only have a negative impact on the health and safety of workers, but also on productivity, equipment effectiveness, and production costs. In this context, enhancement of workers' safety, improvement of the working conditions, and the need to increase productivity and reduce production costs are the main drivers behind attempts to automate underground operations. However, underground mining consists of a series of discrete steps or unit operations (such as drilling, blasting, loading, haulage, conveying, and hoisting) by groups of different equipment types working usually in tandem. Because of the character of the working environment, successful execution of each operation or sequence of unit operations may be compromised. Autonomous equipment must be able to sense, reason, and adapt to this peculiar environment, but many well-proven technologies are not easily transferable to underground mines. As a result, it has been quite difficult for automation to make inroads into underground mines. However, some remarkable progress has been made, particularly in the last 10 years or so. Some interesting technologies that have been developed and deployed underground will be briefly discussed in the next sections.

Examples of underground equipment automation. Teleoperation is the ability to operate robotic machines from a remote location by putting people "virtually" inside the machine. Teleoperation and automation in underground mining were first implemented for some stationary equipment and processes, where duty cycles are repetitive and working conditions are relatively stable. The Kiruna iron-ore mine in Sweden was among the first mines to have successfully implemented automated underground train operations toward the end of the 1970s. Hoisting and conveying automation followed. Today, in practically every modern underground mine, hoist operators act only as supervisors. Subsequently, significant progress has taken place in continuous operations, such as those in longwall hard-coal mining. In addition, automation of mobile production equipment operating in metal mines has been attempted.

Remarkable progress has been achieved in the past 5–10 years mostly due to the ability to construct robust communications systems, capable of handling data, voice, and video signals, enabling the development of a new generation of mining equipment with on-board computers and sensors. For example, an autonomous LHD (load-haul-dump machine) may have around 150 sensors. Vision systems provide three-dimensional views of the mining area and the machine itself. In the following section, we will briefly review some of the recent achievements in underground mining automation.

Longwall face automation. Longwall mining is one of the most frequently used methods to mine underground coal seams. It is essentially a continuous method. Longwall faces are equipped with hydraulically powered supports (shields) that are progressively moved to support the newly extracted face as slices of coal are cut, allowing the section where the coal had previously been excavated and supported to collapse (**Fig. 1**). They also protect face crews and face equipment. Modern faces are several hundred meters long, and they may exceed 200 individual active shields. The drum shearer extracts coal along the full length of the face and loads it onto an armored face conveyor (AFC), which handles the coal excavated by the shearer from the face (Fig. 1). It also serves as tracks for the shearer.

The longwall face should be as straight as possible, and perpendicular to gate (or access) roads.

Fig. 1. General view of a modern longwall face, showing the hydraulic shields (seen at the top), drum shearer (center), and armored face conveyor (AFC). (*Courtesy of http://www.coaleducation.org*)

The process that keeps it that way is called alignment. Because it is critical to maximize coal extraction and minimize extraction of waste rock, the shearer should operate so that roof- and floor-cutting horizons are entirely within the seam (horizon control). Continuous automatic face alignment can virtually eliminate stoppages caused by current stringline-based manual alignment processes. Optimal alignment of AFC components minimizes wear and reduces consequent equipment breakdown or change-out delays. Automatic horizon control can more effectively steer the longwall in the seam to minimize coal dilution. Shearer automation has brought significant improvement in face productivity. Its work has become consistent, because its operation is based on a fixed program sequence and is not subject to variable operator inputs. Face alignment and programmed shearer automation are now mature technologies. The systems are commercially available and can be implemented in most mines. Some applications involve "on-face observation," in which the face equipment is fully automated, but local operators manage exceptional conditions. Longwall face automation allows mine operators to remove miners from hazardous areas and minimize exposure to coal dust, heat, noise, and danger from roof and face falls.

Automation of haulage and loading equipment. With regard to haulage and drilling equipment, teleoperation and automation offers the potential of a "zero-harm" working environment for miners (no rock falls, less noise, no noxious gases and dust exposure, no risk of high-pressure fluid injections, no struck by and crushing injuries, etc.) and higher productivity. In the last 20 years or so, the emphasis in research and development has been mostly on the teleoperation of LHDs, as they often operate in highly hazardous places. Up to now, the autonomous systems have reached the stage where driving (tramming) and dumping are fully automated, while bucket loading is performed using teleremote operation. (Although technically possible, autonomous bucket loading is not yet a proven technology that would be able to

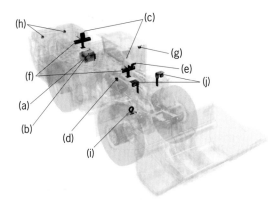

Fig. 2. Loader automation components. (*a*) Control box: collects and processes data from sensors. (*b*) Video communication gateway (VCG): transmits data and video between vehicle and operator's station. (*c*) Front and rear lasers: scan the mine environment. (*d*) Angle sensors: provide data for controlling the steering and the bucket position. (*e*) Inertia measurement unit (IMU) sensor: enables the loader to detect acceleration and deceleration forces. (*f*) Front and rear antennas: for vehicle communication redundancy. (*g*) Onboard cameras: provide the operator with live video. (*h*) Odometer: measures the distance travelled. (*i*) Xenon high-beam lights: for excellent video image. (*Courtesy of Atlas Copco*)

adapt to variations in rock fragmentation during loading.) Once the bucket is filled, the operator hands the control over to the automated system, and a loader travels along the programmed route, empties its bucket into an ore pass or truck, and then returns to a given loading point. The autonomous vehicles are equipped with scanning lasers (indispensable for navigation), on-board cameras, and antennas (**Fig. 2**).

An operator located in a control room receives video images from cameras installed on the machines and at workplaces. For most of the cycle (except bucket loading), the operator's role is limited to supervision, but, if necessary, the operator may take control of the system at any time, and execute all the actions remotely. One operator may control up to three machines.

As of June 2011, there were three commercially available autonomous haulage systems: AutoMine™ from Sandvik, MineGem™ (Caterpillar), and Scooptram Automation (Atlas Copco). AutoMine is the only system that also offers autonomous truck operation. In this case, the role of operator is limited to filling a truck box with ore from chutes or silos. At present, there are more than 20 mines using these systems, with about 60 LHDs and eight trucks working in an autonomous mode. In the Finsch diamond mine in South Africa, loaders and trucks work together within the same AutoMine system. According to mine regulations and safety rules, all autonomous vehicles must operate within zones isolated from personnel and other equipment, and must be protected by security gates and sensors (**Fig. 3**). Any breach of the restricted zone results in an immediate shutdown of all autonomous traffic.

Each commercially available system enables the removal of an operator from the machine to a comfortable remote-control station (**Fig. 4**), which is usually

located on the surface and is sometimes far away from the mine site.

Codelco's El Teniente copper mine in Chile will soon operate its loaders working in the Pipa Norte and Pilar Norte sections from a control station located about 50 km (31 mi) away in the city of Rancagua, where most of the mine's employees live.

Automation of drilling operations. There are two main types of underground drilling: face drilling and longhole production drilling. Face drilling is usually done with "jumbos" (machines) that drill an entire round of typically 40–50 holes (a few meters deep), depending on the face size. Autonomous jumbos equipped with on-board computers may drill an entire round using an electronic card with a drill pattern previously designed by a planner on the surface. The role of operator is limited to setting up the equipment, inserting the card, and supervising the drilling. In longhole drilling, the holes are much deeper (several tens of meters), and drill rods or tubes must be added as drilling progresses. This process has been successfully automated (by Sandvik and Atlas Copco), and autonomous rigs can drill an entire ring or fan of several holes (again using a preprogrammed card with a drill pattern) with an operator acting only as a supervisor. Just as with LHDs and trucks, operators may work at a remote control room on the surface.

Fig. 3. Autonomous mine truck and the security gate in the Finsch mine in South Africa. Please note the presence of a scanning laser (above the machine's nameplate), antennas (on both sides of the grille), and cameras (one in the cabin and another on the left, behind the antenna). (*Courtesy of Sandvik Mining and Construction*)

Fig. 4. LHD operator in a surface control station. (*Courtesy of Sandvik Mining and Construction*)

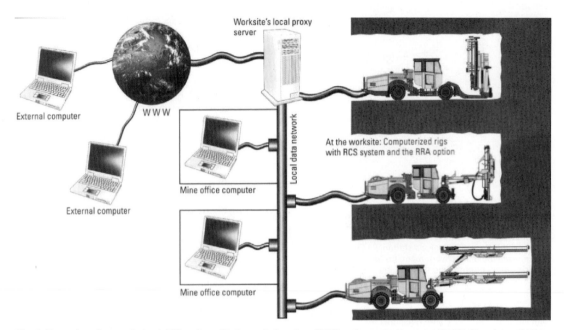

Fig. 5. Examples of computerized drilling rigs with rig control system (RCS) and remote rig access (RRA). (*Courtesy of Atlas Copco*)

However, an underground crew is needed to move the rig to a new position and set it up for the next ring or fan of holes. The results obtained by mining companies using this technology confirm high productivity, while maintaining high precision in terms of hole depth and deviation.

Modern drill rigs may be also equipped with production-grade computerized control and guidance systems, automatic drill-bit changers, tunnel-profiling systems, and "measurement while drilling" instrumentation. The latter logs rock strata characterization using a rock drill as a sensor that measures the penetration rate, rotation speed, thrust force, oil pressure, and so on. This is helpful not only in production planning, blast design, and ground control, but also with regard to equipment performance and maintenance. **Figure 5** shows Atlas Copco's rig control system (RCS) combined with remote rig access (RRA).

The system enables remote control of the operating machines and the sending of data and information about their working parameters to distant locations. This way original equipment manufacturers (OEMs) can assist mine personnel in optimizing their operating parameters and in remote troubleshooting to diagnose technical problems, even from another continent.

Outlook. Teleoperation and automation of individual operations and whole processes in underground mining may bring substantial benefits, including enhanced safety, improved access, reduced production cost, and higher operating precision. Removing mine workers from hazardous areas or situations enhances safety and improves the working environment. It is also possible to mine in the areas previously considered inaccessible (unfavorable or too dangerous in physical or geotechnical conditions). In addition, production costs are reduced through the higher productivity and effectiveness of the equipment. This is possible through higher reliability (less collisions, less operator induced wear, failures, etc.), availability, and utilization (increases in actual operational time, with less time lost due to shift changes, personnel travelling, regulatory breaks, evacuation of noxious gas after blasting, and so on). As a result, fewer machines and operators are needed to meet production targets, and maintenance costs are lower. Finally, there is higher precision in the operation and consistent productivity throughout a shift (as opposed to decreased productivity when operators get tired).

Given the harsh and hazardous working environment, the ultimate goal would be to have a mine operated entirely from surface, with very limited human presence underground. That was initially the target of complex automation programs, such as Mine Automation Program (MAP) of Inco Canada (now part of Vale) and Intelligent Mine in Finland. Achieving this will be very difficult and it will not happen in the near future. Furthermore, since partial automation achieves 90% of the benefits of full automation, it is highly unlikely that human intervention underground will not be required at all. Instead, equipment and methods likely will be automated to an observable extent, relying less on human intervention and operators, particularly at the individual equipment level. This will probably achieve the majority of the productivity and safety benefits gained from full automation.

For background information *see* AUTOMATION; COAL MINING; MINING; REMOTE-CONTROL SYSTEM; ROBOTICS; UNDERGROUND MINING in the McGraw-Hill Encyclopedia of Science & Technology.

Jacek Paraszczak

Bibliography. T. J. Horberry, R. Burgess-Limerick, and L. J. Steiner, *Human Factors for the Design,*

Operation, and Maintenance of Mining Equipment, CRC Press, 2011; P. Lever, Automation and robotics, in *SME Mining Engineering Handbook*, P. Darling (ed.), Society for Mining, Metallurgy, and Exploration, 2011; L. Lien, Mining's new future, *Mining Engineering*, 63(2):40–46, 2011; P. Moore, The future mine, *Mining Magazine*, 201(3):25–32, 2010.

Mirror neurons and social neuroscience

How do we understand the actions and intentions of others? The traditional view is that such understanding occurs through reasoning not much different from that used to solve logical problems. When witnessing the actions of others, we process them with our sensory systems; this information is then elaborated by some cognitive centers. At the end of this process, we know what others are doing and why.

It is likely that such operations occur in some situations (for example, when the behavior of the observed person is unusual or bizarre). Yet, the lack of effort with which we understand others in everyday life suggests the existence of another more basic, but less traditionally espoused, mechanism. The actions of others, after being processed in the observers' sensory systems, are directly mapped on their motor representations. This direct mapping allows immediate understanding of others' actions and motor intentions.

Mirror mechanism. Evidence in favor of the existence of a direct mapping mechanism for action understanding comes from the discovery of mirror neurons in the monkey premotor cortex (brain area F5). Mirror neurons are a distinct class of visuomotor neurons that discharge both when individuals perform a specific motor act and when they observe the same motor act done by another individual. Grasping, holding, and manipulating are among the motor acts that F5 mirror neurons encode. Some mirror neurons show congruence between their visual and motor responses in terms of the goal (for example, grasping) and in terms of the way in which the goal is achieved; others show a similarity, but not identity, between the observed and executed effective motor acts.

Mirror neurons have been demonstrated (by single neuron recordings and by noninvasive techniques such as transcranial magnetic stimulation, electroencephalography, and functional magnetic resonance imaging) in the premotor cortex and parietal lobe, as well as areas mediating emotions (for example, the insula and cingulate gyrus), in both monkeys and humans. The presence of mirror neurons in areas with radically different functions suggests that the mirror mechanism is a general mechanism that transforms the sensory aspects of actions done by others into a motor format. When this mechanism becomes active in emotional areas, this enables the observer to feel the emotions of others; when it becomes active in the motor circuits, it enables the observer to understand the actions and the intentions of others.

The peculiar role of the mirror mechanism is that of linking biological movements with our experiences. In virtue of this mechanism, we can understand others "from inside." How does this mechanism work? When mirror neurons become active, they cause the activation of large numbers of neurons located in different motor cortical areas and subcortical motor centers. These complex, externally activated, motor schemata are similar to those activated when the observer decides to perform the same actions. Thus, without any inferential reasoning, we know what the others are doing.

The same is true for emotional understanding. Recent work has indicated that the electrical stimulation of the insula in the monkey elicits, according to the stimulation site, different emotional behaviors, such as affiliative behaviors (that is, behaviors that promote social cohesion) or disgust. The injection of tracers in the stimulated regions has shown that each of these sites is connected with specific motor and vegetative centers. The insula and other emotional centers are activated by specific natural stimuli. Most interestingly, because of the mirror mechanism, when an individual observes an emotion expressed by another person, there is an activation of the same sites that are active during naturally elicited emotions. This matching allows an immediate empathic comprehension of the emotions of others.

Intention understanding. Let us imagine the following scene. You are in a café and your attention focuses on a customer grasping a cup of coffee. You immediately understand what this person is doing and, according to how he grasps the cup (by the handle or by the body, you also understand his intention: for example, if he wants to drink the coffee or if he wants to give it to a friend. As described here, the observed motor act (grasping) enters into our motor system and provides us with the observed motor act representation, enabling us in this way to understand its meaning. How though do we understand the intention of the observed agent? Some recent data provide a neurophysiological explanation of this ability.

Single neuron recordings from the monkey inferior parietal lobule (area PFG) have shown that the discharge of many PFG neurons (about 65%) encoding grasping is affected by the type of overarching action in which the grasping is embedded (for example, grasping for eating or grasping for placing). Panel *a* of the **illustration** shows examples of neurons exhibiting this property. These data suggest that the "action-constrained" PFG neurons are embedded in prewired chains, with each encoding a specific action (for example, eating). This organization is obviously very appropriate for providing fluidity to action execution because each neuron not only encodes a specific motor act but (being embedded into a specific action) is also linked with neurons coding the next motor acts, thereby facilitating them.

Further experiments showed that these "action-constrained" neurons have also mirror properties

Motor responses of action–constrained mirror neurons

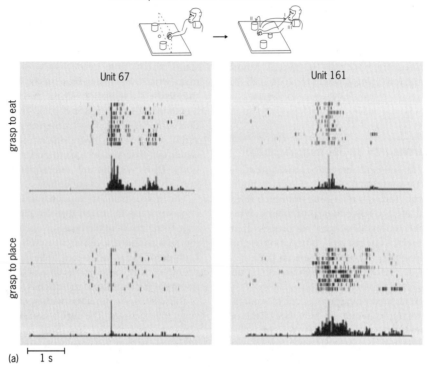

(a) ├─ 1 s ─┤

Visual responses of action–constrained mirror neurons

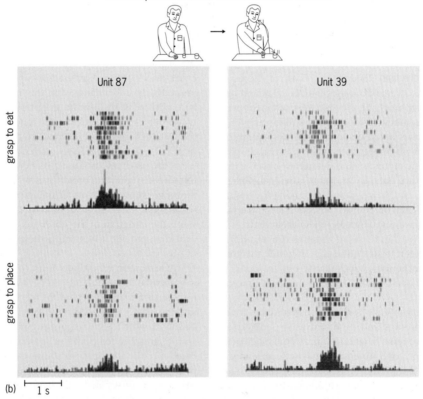

(b) ├─ 1 s ─┤

Action-constrained neurons recorded from the monkey inferior parietal lobule. (*a*) *Top*—Motor task: The monkey, starting from a fixed position, grasps a piece of food or an object when the barrier is lifted. Then, in one condition (I), the monkey brings the food to the mouth and eats it; in another condition (II or III), the monkey grasps the object and places it into a container. *Bottom*—Each panel shows, for each task condition, the neuron response during individual trials and the histogram representing the average response frequency. Unit 67 responds stronger when the monkey executes grasping for eating compared to when it executes grasping for placing. Unit 161, on the contrary, responds better during grasping for placing. (*b*) *Top*—Visual task: The monkey observes the experimenter performing the same two actions of the motor task. *Bottom*—Unit 87 responds stronger when the monkey observes a grasp-to-eat action compared to when observing a grasp-to-place action. Unit 39 presents the opposite behavior. Other conventions are the same as in panel *a*. (*Modified from G. Rizzolatti, M. Fabbri-Destro, and L. Cattaneo, Mirror neurons and their clinical relevance, Nat. Clin. Pract. Neurol., 5:24–34, 2009, with assistance from D. Mallamo in preparing the iconographic material*) and that their visual response during grasping observation is influenced by the overarching action in which the grasping is embedded (see illustration, panel *b*). Therefore, when an action-constrained grasping neuron is activated by the observation of a grasping motor act, inserted into a specific overarching action, its discharge likely triggers the whole motor chain underpinning that action. In this way, the observer activates the motor representation of the action that the observed agent intends to do. As a result of this mechanism, the observer understands the agent's motor intention.

Mirror mechanism and autism. A fundamental consequence of the discovery of the mirror mechanism is the demonstration of a dichotomy between the comprehension of others "from inside" and the comprehension of others based on inferential processes. A deficit in the capacity of comprehension from inside is one of the landmarks of autistic spectrum disorder (ASD) and appears to be related to a malfunctioning of the mirror mechanism. Evidence for such a malfunctioning comes from various studies, including electroencephalography, magnetoencephalography, transcranial magnetic stimulation, and brain imaging.

ASD is a neurodevelopmental disorder characterized by impairment in communication, deficits in social interactions, and restricted, stereotyped behavior. Recently, converging evidence from a variety of studies has suggested that an important feature associated with ASD consists of disturbances in motor behavior. These disturbances encompass a variety of deficits, including clumsiness, postural instability, and disturbances in motor coordination.

What then is the relationship between motor disturbances and the core symptoms of ASD? Some recent studies have shown that deficits in praxis (action or practice of a skill) in children with ASD correlate positively with social, communicative, and behavioral impairments. Until a few years ago, the possibility of a causal correlation between motor disturbances and cognitive deficits would have sounded implausible. However, the motor system, by virtue of the mirror mechanism, is involved in the understanding of actions, intentions, and emotions.

In this respect, it is important to mention a study on typically developed (TD) children and high-functioning autistic children. The children were required to execute and observe two different actions: grasping (with the hand) an item placed on a plate and bringing it to the mouth, or grasping an item and placing it in a box. During the execution and observation of both actions, the activity of the mouth-opening mylohyoid muscle (MH) of the participants was recorded using electromyography (EMG) surface electrodes. During the execution as well as the observation of the eating action, there was a sharp increase of MH activity in TD children, which started well before the food was grasped. No increase of MH activity was present during the observation of the placing action. Together, these findings indicate that the motor system plans the final goal of an action both when the action is executed and when it is observed. In contrast with TD children, children with ASD showed a much later activation of the MH muscle during action execution and no activation at all during eating action observation. These impairments in the capacity to organize a sequence of motor acts within an intentional overarching action most likely account for the difficulty that children with ASD have in directly understanding the intentions of others' actions.

In conclusion, it is well established that children with ASD have a deficit in understanding others from inside. Certainly, a disruption in the mirror neuron system is not the only cause of the core deficits characterizing the ASD syndrome. Our hypothesis is that the primary deficit in ASD is a genetically determined malfunctioning of the motor system and that this malfunctioning is responsible for most (albeit not all) of the core disturbances of this syndrome. The demonstration of this link as well as the isolation of cognitive motor-related deficits from those that may have another genetic origin (for example, repetitive, stereotyped behavior) is a challenge for investigators and will be a fundamental step for establishing specific, neurophysiologically based, rehabilitation protocols.

For background information *see* AUTISM; BRAIN; COGNITION; DEVELOPMENTAL PSYCHOLOGY; EMOTION; INFORMATION PROCESSING (PSYCHOLOGY); LEARNING MECHANISMS; MEDICAL IMAGING; MOTOR SYSTEMS; NERVOUS SYSTEM (VERTEBRATE); NEUROBIOLOGY; NEURON in the McGraw-Hill Encyclopedia of Science & Technology.

Giacomo Rizzolatti; Leonardo Fogassi

Bibliography.   L. Cattaneo et al., Impairment of actions chains in autism and its possible role in intention understanding, *Proc. Natl. Acad. Sci. USA*, 104:17825–17830, 2007; G. Rizzolatti and C. Sinigaglia, *Mirrors in the Brain: How Our Minds Share Actions and Emotions*, Oxford University Press, Oxford/New York, 2008; G. Rizzolatti, L. Fogassi, and V. Gallese, Mirrors in the mind, *Sci. Am.*, 295:54–61, 2006; G. Rizzolatti, L. Fogassi, and V. Gallese, The mirror neuron system: A motor-based mechanism for action and intention understanding, pp. 625–640, in M. Gazzaniga (ed.), *The Cognitive Neuroscience*, MIT Press, Cambridge, MA, 2009.

Molecular diffractograms

The fastest changes in the structure or shape of a molecule happen on time scales shorter than a picosecond (one picosecond is one trillionth of a second, $1 \text{ ps} = 10^{-12} \text{ s}$). For example, the first step in both the photosynthesis and the vision processes is a change in molecular structure triggered by the absorption of a photon, and this happens on a time scale of 100 femtoseconds ($1000 \text{ fs} = 1 \text{ ps}$). Investigating these types of processes in their natural spatial and temporal scale is essential for understanding and controlling dynamics at the molecular level. One of the goals of current research is to visualize the changes in the molecular structure as they happen, which requires an imaging system with a spatial resolution on the atomic scale and a temporal resolution of femtoseconds.

Atomic scale. The distance between atoms in a molecule is of the order of a few tenths of a nanometer (1 nm = 10^{-9} m, 1 billionth of a meter). Imaging systems based on visible light cannot reach sufficient spatial resolution to image on the atomic scale because the wavelength of the light ultimately limits the resolution to about 500 nm, which is 5000 times larger than the atomic scale of 0.1 nm. The required resolution can be reached only by using waves with very short wavelengths. Electromagnetic radiation of sufficiently short wavelength (and correspondingly high photon energy) falls in the x-ray region of the spectrum. In fact, high-energy x-rays generated at synchrotron accelerators can have sufficiently short wavelengths. Alternatively, an electron beam can be used instead of an x-ray beam. Electrons behave both as particles and as waves, and as such can have a wavelength, called the de Broglie wavelength. Electrons have de Broglie wavelengths of the order of picometers (1 nm = 1000 pm), even for moderate kinetic energies. For example, an electron accelerated by a potential of 30,000 V will acquire a kinetic energy of 30,000 eV and thus will have a wavelength of 7 pm. Conventional electron microscopes and diffractometers use electron beams with kinetic energies in the range of tens to a few hundred thousand electronvolts. In an electron microscope, electron optics are used to create an image of the object in the detector, while in electron diffraction, the scattering pattern is recorded directly on the detector without any intermediate optics, and the image is reconstructed in postprocessing.

Even though both x-rays and electrons can provide sufficient spatial resolution, they have significant differences. The scattering cross section (the probability that an incident x-ray photon or electron will be scattered by a target atom) is much higher for electrons. This means that for a given number of scattering atoms, the fraction of incident electrons that will be scattered is much higher than that for incident x-ray photons. For this reason, x-rays have a longer penetration depth and can be used to study thicker samples, while electrons are more effective for scattering from diffuse samples, such as molecules in the gas phase. X-ray crystallography (diffraction from crystals) is the most widely used method to determine the structure of molecules. In this technique, the molecules are crystallized to form a periodic structure from which x-rays are diffracted. The periodic ordering of the molecules is critical for recovering the structure from a set of diffraction patterns. Even though this technique has proven very useful, it has limitations. There are molecules that are very difficult (or currently impossible) to crystallize. Furthermore, crystallization hampers the study of structural changes in the molecule, as the molecules are tightly constrained within the crystal. In order to image isolated molecules, it is necessary to have the molecules in the gas phase, where the density is low, and in this case it is more efficient to use electron diffraction.

Diffraction from aligned molecules. Molecules in the gas phase are oriented randomly. Therefore, the resulting diffraction pattern is the sum of the diffraction of single molecules with all possible orientations. Such a diffraction pattern contains information only about the distances between atoms in a molecule, with no geometrical information; that is, there is no information about bond angles in the molecules. For small molecules, this information is sufficient to fully reconstruct the structure, but such a reconstruction is not possible for larger molecules. In order to retrieve the full molecular structure, the molecules need to be aligned, that is, pointing in the same direction. **Figure 1** shows a simulated electron diffraction pattern of ozone molecules (O_3) in the gas phase. The ozone molecule is composed of 3 oxygen atoms that form the vertices of a triangle with a bent angle of 116.8°. Figure 1a shows the diffraction pattern that would be generated by randomly oriented molecules in the gas phase, while Fig. 1b shows the pattern that would be obtained if the molecules were aligned. Both diffraction patterns were divided by the scattering pattern of isolated oxygen atoms in order to highlight the diffraction rings. In Fig. 1a, the diffraction contains only circularly symmetric rings with low contrast; there is no geometrical information. In Fig. 1b, the triangular geometry of the molecule is reflected in the diffraction pattern. In addition to containing more information, Fig. 1b displays greater contrast in the diffraction pattern, making detection easier. This example assumes perfect alignment of the molecules, but even for the case of partial alignment, the information content of the diffraction pattern increases significantly. It should be noted that diffraction from aligned molecules in the gas phase is fundamentally different from diffraction from a crystal. In the gas phase, the molecules are separated by a large enough distance that waves scattered from different molecules do not interfere; that is, the diffraction pattern is an incoherent sum of the diffraction patterns of individual molecules. In a diffraction pattern from a crystal, the periodicity is

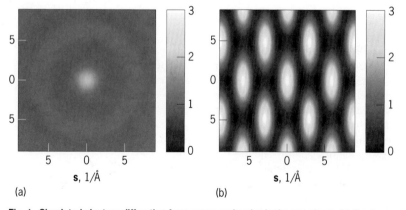

Fig. 1. Simulated electron diffraction from ozone molecules in the gas phase. (*a*) Random orientation. (*b*) Aligned molecules. The parameter *s* is approximately proportional to the scattering angle divided by the wavelength of the electrons, and is given in units of inverse length, where 1 Å = 0.1 nm. The gray scale corresponds to the intensity in the diffraction pattern, in arbitrary units.

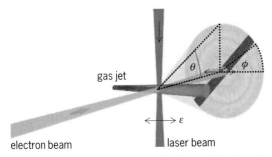

Fig. 2. Experimental setup. The molecules are introduced in a gas jet. Both the electron and laser pulses intersect the gas jet at right angles. The diffraction pattern is recorded on a two-dimensional detector. (*From P. Reckenthaeler et al., Time-resolved electron diffraction from selectively aligned molecules, Phys. Rev. Lett., 102:213001, 2009*)

imposed by the crystal, while in this case it reflects only the molecular structure.

The question, then, is how to have molecules that are isolated but aligned. A short, intense laser pulse can be used to align the molecules in a gas along the direction of the laser polarization. If the laser pulse is much shorter than the rotational period of the molecules, it imparts a "kick" to the molecules that causes the molecules to be aligned for a short time afterward. This method, called nonadiabatic laser alignment, has been shown to produce a high degree of alignment, particularly if the molecules are initially at low temperature. A short electron pulse can then be used to capture a diffraction pattern while the molecules are aligned. The alignment is expected to last only a few picoseconds, so a sufficiently short electron pulse must be used to capture the diffraction pattern. The ideal experiment would first align the molecules, then record the molecular dynamics (which happen on femtosecond time scales) while the molecules are aligned. The first step in this direction is to demonstrate that it is possible to capture a diffraction pattern of aligned molecules.

It has recently been shown in a proof-of-principle experiment that this is indeed possible.

Experimental results. A femtosecond laser pulse with a frequency in the ultraviolet (wavelength of 267 nm) was used to trigger a dissociation reaction in $C_2F_4I_2$ molecules in the gas phase, and a 3-picosecond electron pulse was used to capture the diffraction pattern. The experimental geometry is shown in **Fig. 2**. This photoreaction is not isotropic, as the molecules that happen to be oriented parallel to the direction of the polarization of the laser pulse dissociate with a higher probability. This creates a selective alignment of the molecules; that is, the laser dissociates preferentially the molecules that happen to be oriented in a given direction at the moment the laser arrives. This method was chosen for a first experiment, rather than the active alignment, because it creates an anisotropic distribution and a change in the structure that can be detected more easily. Although this method of selective alignment does not produce a high degree of alignment, the lifetime of the effect is comparable to that of the nonadiabatic alignment. One can safely assume that the molecules do not rotate while the laser pulse is present because the pulse duration is much shorter than the rotational period of the molecules. The dissociated molecules, that is, the product state, are generated with a preferred orientation. The molecules then rotate and the alignment disappears.

The electron pulse was used to capture the diffraction pattern while the molecules were aligned. The telltale sign of the alignment is that the circular symmetry of the diffraction is broken, and the pattern becomes anisotropic. This is shown in **Fig. 3**, where each panel shows the experimental diffraction pattern and the theoretical prediction for a given time delay between the laser and electron pulses. The diffraction patterns show a clear anisotropy that vanishes after a few picoseconds.

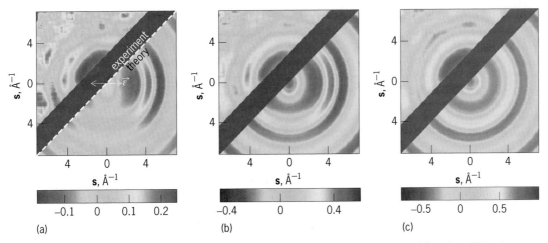

Fig. 3. Diffraction patterns. Each panel corresponds to a diffraction pattern at a different time *t*. (*a*) $t = 1$ ps. (*b*) $t = 4$ ps. (*c*) $t > 8$ ps. Each pattern is the difference in the diffraction pattern before and after the laser beam illuminates the molecules. The left part of each panel corresponds to the experiment, while the right part is a theoretical prediction. The parameter *s* is approximately proportional to the scattering angle divided by the wavelength of the electrons, and is given in units of inverse length, where 1 Å = 0.1 nm. The gray scale corresponds to the intensity in the diffraction pattern, in arbitrary units. (*From P. Reckenthaeler et al., Time-resolved electron diffraction from selectively aligned molecules, Phys. Rev. Lett., 102:213001, 2009*)

Two main conclusions can be drawn from these results. The first is that bond breaking in a molecule during a photoreaction is not isotropic, and a preferred orientation is defined by the direction of the laser polarization. This is the first direct visualization of this effect. Second, the experiments established the time duration of the effect to be a few picoseconds, and thus within the window in which they can be probed with electron pulses. In terms of technology, these results demonstrate that it is possible to capture the diffraction pattern of aligned molecules, but in order to retrieve the three-dimensional molecular structure, a higher degree of alignment is necessary.

Prospects. As mentioned before, with nonadiabatic laser alignment, it is possible to achieve a high degree of alignment. Coming experiments using this alignment method should result in a diffraction pattern with more features that reflect the molecular geometry. At the same time, algorithms are being developed to reconstruct the molecular structure for the case in which the alignment is not perfect. Recently, electron pulses from tabletop sources shorter than 100 fs have been demonstrated. Incorporating these shorter pulses would allow scientists to measure not only the molecular structure, but also the dynamics on very short time scales. The future of tabletop ultrafast electron diffraction experiments is bright; technological improvements will allow for very high resolution in space and time. These technological advances will allow scientists to probe molecules with unprecedented detail and provide new insights into the structure of molecules and their interaction with light.

For background information *see* ELECTRON DIFFRACTION; LASER; OPTICAL PULSES; ULTRAFAST MOLECULAR PROCESSES; X-RAY CRYSTALLOGRAPHY; X-RAY DIFFRACTION in the McGraw-Hill Encyclopedia of Science & Technology.   Martin Centurion

Bibliography. M. I. Davis, *Electron Diffraction in Gases*, Marcel Dekker, New York, 1971; P. Reckenthaeler et al., Time resolved electron diffraction from selectively aligned molecules, *Phys. Rev. Lett.*, 102:213001 (4 pp.), 2009; H. Stapelfeldt and T. Seideman, Colloquium: Aligning molecules with strong laser pulses, *Rev. Mod. Phys.*, 75:543–557, 2003.

Mushroom sclerotia

Mushrooms are known for their nutritional values and for biopharmacological activities that are beneficial to human health. Traditionally, mushroom fruiting bodies (fleshy or woody aerial caps) have been used as sources of bioactive ingredients for dietary supplements or nutraceuticals. Currently, mushroom sclerotia (aggregated mycelia) are being investigated intensively in functional food research because of the unique profile of the bioactive components in the sclerotia. In particular, the cell wall polysaccharides (β-glucans) that are found abundantly in the mushroom sclerotia are the most potent components.

Mushroom life cycle. The developmental stages in the mushroom life cycle may vary according to different genera of mushrooms and different environmental conditions (**Fig. 1**). Spores are first released and then can germinate to form filamentous structures called hyphae. The hyphae continue to extend; upon meeting, they fuse and reproduce mitotically so that the genetic materials can be combined. The resulting mated mycelium grows much faster than an unmated one from single spores. When the mycelium expands, it gradually forms a web of cells called the mycelial network. In addition, the thickening of the mycelial mat will occur when side branches of the mycelium arise. The mycelium can be triggered to form fruiting bodies from primodia in a stepwise manner under favorable conditions, such as appropriate humidity and light intensity. However, under unfavorable climatic conditions, including cold temperatures or drought, sclerotia are formed (**Fig. 2**). Sclerotia are made up of a fungal mass of hyphae and resemble a hardened tuber with a woodlike texture. When mushrooms form sclerotia, they enter a resting phase in their life cycle and remain dormant. Once the conditions become favorable again, the sclerotia will germinate, and mushroom fruiting bodies will emerge. Formation of sclerotia allows the mushroom to survive through harsh weather and is thus one of the important stages in its life cycle. In some fungal species (especially ascomycetes and basidiomycetes), production of mushroom sclerotia is an effective strategy for survival under undesirable conditions.

Structure of mushroom sclerotia. Mushroom sclerotia are compact masses of hardened mycelia that are spherical in shape; their sizes range from a few millimeters to several tens of centimeters in diameter. The outer woodlike wall of a sclerotium is called the rind; it is a continuous layer of tightly packed hyphal tips that become thick-walled and pigmented to form an impervious outer surface layer impregnated with melanins (oxidized phenolic pigments). The inner part of a sclerotium is called the medulla; it constitutes the main part of a sclerotium (consisting of interwoven hyphae) and is the main storage area for intracellular reserves. In some sclerotia, a narrow layer of close-fitting hyphae, namely the cortex, is discernible between the rind and the medulla. **Figure 3** shows a scanning electron micrograph of the hyphal-like structure of a typical sclerotium from *Pleurotus tuber-regium*.

Sclerotia from edible and medicinal mushroom species. Common sclerotia-forming edible and medicinal mushroom species include *Grifola umbellata*, *Omphalia lapidescens*, *Xylaria nigripes*, *Wolfiporia cocos*, *Polyporus rhinocerus*, and *Pl. tuber-regium*. Among them, *W. cocos*, *Pl. tuber-regium*, and *P. rhinocerus* (or *Lignosus rhinocerus*) are the most economically important because of their successful cultivation under controlled artificial conditions, especially in China.

Wolfiporia cocos is also known as fu ling or hoelen in China and is a brown rot fungus that grows in association with the roots of conifers,

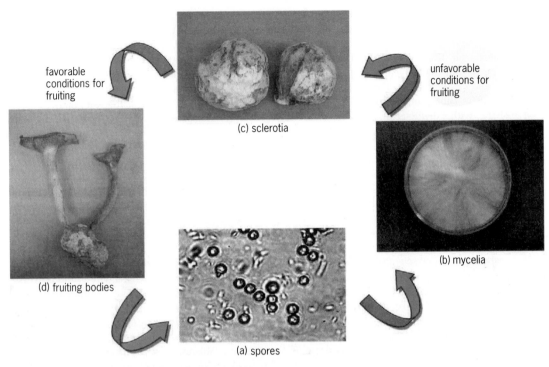

favorable
conditions for
fruiting

unfavorable
conditions for
fruiting

(c) sclerotia

(b) mycelia

(d) fruiting bodies

(a) spores

Fig. 1. Life cycle of a typical sclerotium-forming mushroom.

Wolfiporia cocos (Schw.) Ryv. et Gilbn.,
[*Poria cocos* (Schw.) Wolf] (茯苓)

Pleurotus tuber-regium (Fr.) Sing. (虎奶菇)

Polyporus rhinocerus (虎乳苓芝)

(a)   (b)   (c)

Fig. 2. Mushroom sclerotia: (*a*) *Wolfiporia cocos*, (*b*) *Pleurotus tuber-regium*, and (c) *Polyporus rhinocerus*.

especially pines and oaks. The sclerotium from *W. cocos* is subterranean and is spherical to oval in shape, with a diameter of 10–30 cm (4–12 in.) [Fig. 2*a*]. Its rind is brownish-yellow in color and is

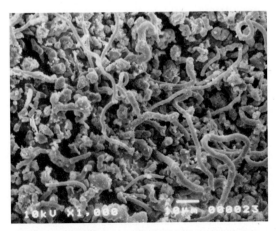

Fig. 3. Scanning electron micrograph of the sclerotia from *Pleurotus tuber-regium*.

widely used in Chinese medicine as a diuretic and a decoction for coughs. The whole sclerotium is often used to treat jaundice and induce menstruation. The polysaccharides known as pachyman that are extracted from the sclerotia have potent antitumor and immunomodulatory effects, whereas the low-molecular-weight tetracyclic triterpenes have antiviral and tumor-inhibitory effects.

Pleurotus tuber-regium, which is known as tiger milk mushroom in China, is a wood-rotting fungus that can utilize a wide range of broad-leafed and needle-leafed trees. The sclerotium of *Pl. tuber-regium* is also subterranean and is spherical to oval in shape, with a diameter of 10–25 cm (4–10 in.) [Fig. 2*b*]. The rind of the sclerotium is dark brown in color, whereas its medulla is white. The whole sclerotium of *Pl. tuber-regium* is traditionally used in Nigeria both as a food ingredient and as a medicinal material for treating headaches, stomach ailments, colds, constipation, fever, asthma, and high blood pressure. Polysaccharides extracted from its sclerotium have been shown to have antitumor and immunopotentiating effects.

Polyporus rhinocerus, which is commonly known as hurulingzhi in China, is a white-rot fungus that can be found in China, Malaysia, Sri Lanka, Australia, and East Africa. The sclerotium of *P. rhinocerus* is subterranean and is spherical to oval in shape, with a diameter of 5–10 cm (2–4 in.) [Fig. 2c]. The rind of its sclerotium is pale brown in color, whereas its medulla is white. The whole sclerotium of *P. rhinocerus* is used in Chinese medicine for treating liver cancer, chronic hepatitis, and gastric ulcers. Compared to those of *W. cocos* and *Pl. tuber-regium*, the polysaccharides extracted from the sclerotia of *P. rhinocerus* are the least studied. Preliminary investigations have shown that they have both antitumor and immunomodulatory effects.

Chemical composition of sclerotia. Mushroom sclerotia are primarily made up of compact mycelial hyphae, whose main components are fungal cell walls. The structural fibril and matrix components in fungal cell walls consist of large amounts of polysaccharides [up to 90% dry matter (DM)], including mainly β-glucan, chitin, α-glucan, and glycoproteins (mannoproteins), as well as small amounts of cell wall proteins. Melanins are found in the rind. Depending on the different stages of differentiation and maturity of the sclerotia, reserves consisting of glycogen, polyphosphates, protein bodies, and lipids are present in the cytoplasm of cortical and medullary hyphae.

The three aforementioned sclerotia (that is, *W. cocos, Pl. tuber-regium,* and *P. rhinocerus*) all share a similar chemical composition, with 90–98% DM of carbohydrates (predominantly polysaccharides), low levels of proteins (0.7–6.7% DM) and ash (1.1–2.8% DM), and trace amounts of lipids (0.02–0.14% DM). Because of their chemical profiles, these three sclerotia are a target of study as a rich source of bioactive polysaccharides, especially β-glucans.

Bioactive components in mushroom sclerotia. From the chemical composition of mushroom sclerotia, it can be seen that β-glucans are the predominant bioactive components as compared to other minor components. Mushroom β-glucans obtained from fruiting bodies, mycelia, culture medium, and spores have been demonstrated previously to be potent immunomodulating agents or biological response modifiers (BRMs), and they are used as therapeutic adjuvants or health-food supplements. BRMs can act on different human immune cells to initiate a cascade of signal transduction pathways that are responsible for the enhancement and potentiation of the immune responses for cancer treatment. As very large molecules of at least a few thousand sugar units, mushroom β-glucans are recognized by the surface receptors in the immune cells, and such binding triggers the signaling cascade for the immune responses. In addition to functioning as BRMs, β-glucans in mushroom sclerotia can stop the growth of cancer cells by blocking the progression of certain phases in their cell cycle and thus hindering their proliferation. Moreover, sclerotial β-glucans can kill cancer cells directly by inducing apoptosis or programmed cell death.

Mushroom sclerotial β-glucans are nondigestible carbohydrates that are resistant to human digestive enzymes. They have been shown to be ideal prebiotics or selective substrates for the beneficial bacteria, especially bifidobacteria, in the human colon and therefore can promote colonic health through the selective stimulation of the growth of these healthy bacteria. Bifidobacteria possess specific enzymes (β-glucanases) that can preferentially utilize sclerotial β-glucans, whereas other harmful bacteria do not have such enzymes. Moreover, mushroom sclerotial β-glucans can scavenge a number of different free radicals that can cause oxidative damage to human cells, possibly preventing degenerative diseases and aging in humans.

Sclerotia as functional food ingredients. Mature mushroom sclerotia, often in a white solid powdery form, are edible and bland in taste. They have physical properties (for example, water-holding capability, oil absorption, and emulsion formation) that are comparable to those of commercial dietary fiber ingredients. Mushroom sclerotia can be added directly to food products as a functional food ingredient without the need for further processing. Alternatively, sclerotia by themselves can be used as nutraceuticals.

Future perspectives. The biological activities of β-glucans in mushroom sclerotia, especially the anticancer and immunomodulatory activities, are influenced by a number of factors, including size (molecular weight), the linkages between sugar units, the branching and linearity of the molecules, and the conformation or three-dimensional orientation of the molecules. With recent advances in functional mushroom genomics, the complex picture of how the formation of sclerotia is governed by genetic control is starting to be revealed more clearly. Most important, the synthesis of "tailor-made" sclerotial polysaccharides (β-glucans) with the most desirable structural characteristics that can have the optimum biological activities may become a reality in the not too distant future.

For background information *see* CANCER (MEDICINE); FOOD MANUFACTURING; FOOD MICROBIOLOGY; FUNGAL BIOTECHNOLOGY; FUNGAL ECOLOGY; FUNGI; GENOMICS; MUSHROOM; MYCOLOGY; NUTRITION; POLYSACCHARIDE in the McGraw-Hill Encyclopedia of Science & Technology.

Peter C. K. Cheung

Bibliography. I. Chet and Y. Henis, Sclerotial morphogenesis in fungi, *Annu. Rev. Phytopathol.*, 13:169–192, 1975; S. P. Wasser and A. L. Weis, Therapeutic effects of substances occurring in higher Basidiomycetes mushrooms: A modern perspective, *Crit. Rev. Immunol.*, 19:65–96, 1999; H. J. Willetts and S. Bullock, Developmental biology of sclerotia, *Mycol. Res.*, 96:801–816, 1992; K. H. Wong and P. C. K. Cheung, Sclerotia: Emerging functional food derived from mushrooms, pp. 111–146, in P. C. K. Cheung (ed.), *Mushrooms as Functional Foods*, Wiley, Hoboken, NJ, 2008.

Nanoparticle risk informatics

The potential of risk from chemical ingredients in products is affected by the physical and chemical properties of the compound and the other substances in the product. For example, a chemical compound may be more or less hazardous, depending on its product formulation (that is, an impregnated solid suspension versus a chemical solution). Recently, there is been much interest in the role that the size of a particle plays in chemical hazard and exposure potential. In particular, do very small particles cause greater chemical risk than larger particles with the same chemical composition? For example, nanoparticles are less than 100 nm in at least one dimension. Do particles comprised of titanium dioxide (TiO_2 that are smaller than 100 nm exhibit different toxicity or exposure than larger TiO_2 particles?

Nanoparticles and derivative nanomaterials are pervasive in modern consumer products and are growing in number at over 243 new products per year, and currently estimated at 1317 products manufactured by 587 companies in over 30 countries. These consumer products span several major categories (health and fitness, home and garden, automotive, food and beverage, electronics and computers, appliances, goods for children), yet their full impact on product life-cycle and human health remains poorly understood, with a lack of harmonized standards on safety characterization and eco/human health assessments.

These emerging material additives have attracted considerable attention from scientists from the environmental exposure and health hazards communities. The source of these emerging materials in products, and their overall environmental fate and transport, environmental degradation, biodegradation, biological uptake and disposition, and effects remain poorly characterized. In fact, physiologically based pharmacokinetics and pharmacodynamics (PBPK/PD) research is ongoing in both the medical and environmental sciences. The manner and rates at which chemicals are absorbed, distributed, metabolized, and eliminated (ADME) are crucial in determining whether a molecule will reach the desired target, such as a chemotherapeutic compound reaching the tumor. Similarly, ADME informs the environmental toxicologist about the manner and rate that a chemical (xenobiotic) may reach an undesired target; for example, lead (Pb) reaching neurons in the brain and adversely affecting normal nerve activity.

Chemical manufacturers, pharmaceutical companies, and environmental researchers look for similar characteristics in chemical compounds that may allow them to "hide" from physiological processes as they find their way to the target cells. For example, nanoparticles for drug delivery consist of various biological substances such as albumin, gelatin, and phospholipids for liposomes and abiotic substances, including various polymeric and solid metal-containing nanoparticles. The potential interaction of these nanoparticles with tissues and cells is under investigation, but there appears to be both great promise and concern, depending on whether the target is drug delivery or potential toxicity. One aspect of more adequately addressing emerging concerns of nanomaterials is being able to properly characterize both their core and surface properties, variables that ultimately delineate many down-stream phenomena found in biological systems that have remained elusive in public health and environmental characterization.

There is a great need for reliable information as to whether a chemical ingredient may provide benefits, risks or, as is usually the case, both. The typical means of gaining such information is to compare a chemical to other chemicals with similar structures (known as quantitative structure activity relationships, QSAR). Unfortunately, only a small number of chemicals have been studied sufficiently from which to draw such comparisons. More importantly, QSAR cannot account for unique differences in even very similar compounds. For example, it is well known that chemicals with identical formulas, but with different configurations (for example, enantiomers) may either be efficacious or toxic as ingredients in medicines. Likewise, chiral compounds with the same composition but different handedness, will have very different environmental persistence (for example, the right-handed chiral may take years to biodegrade whereas the left-hand compound may degrade in a few weeks).

Surface-core characterization. A particularly challenging knowledge gap is the role nanoparticle composite constituencies (that is, surface versus core) play in determining efficacy and toxicity. The particle may be a composite, with two distinct compartments: surface and core. It may also change in time and space, so that the core constituencies become surfaces with time, as the original surface is degraded (for example, during metabolism and environmental degradation).

Comparison of the key factors that dominate the toxic responses of nanomaterials and the bulk materials of the identical chemical composition

Bulk material or nanomaterials	Nanomaterials
• Chemical composition*	• Nanostructure*
• Dose (mass concentration)*	• Particles concentration*
• Exposure route*	• Particle size/distribution*
• Reactivity	• Particle number
• Conductivity	• Aggregation/ agglomeration*
• Morphology (crystalline, amorphous, Shape, *etc.*)	• Surface adsorbability*
	• Surface areas*
• Physical form (solid, aerosol, suspension, *etc.*)	• Surface charge*
	• Self-assembly*
• Purity/Impurities	• Quantum effects
• Solubility	• ...

Reproduced by permission of the Royal Society of Chemistry: L. Yan et al., Low-toxic and safe nanomaterials by surface-chemical design, carbon nanotubes, fullerenes, metallofullerenes, and graphenes, *Nanoscale*, 3(2):362-382, 2011.

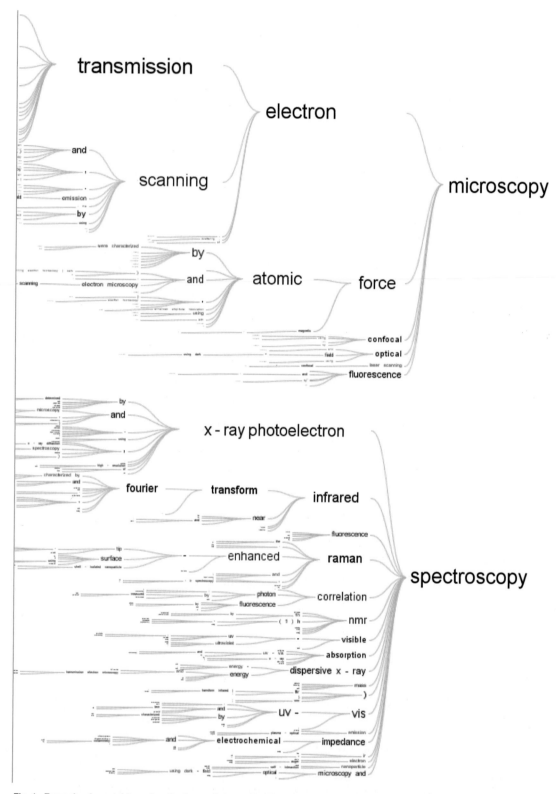

Fig. 1. Example of a word-tree visualization and abstraction from http://www.many-eyes.com on the domain specific corpus provides the "seeker" with information related to characterization techniques. For instance, nanomaterial core and surface characterization is enabled primarily by a variety of microscopy [transmission electron, scanning electron, atomic force, fluorescence, and both confocal and optical microscopy] and spectroscopy techniques [x-ray photoelectron, FT-IR (Fourier transform-infrared), NIR (near-infrared), SERS (surface enhanced Raman scattering), NMR (nuclear magnetic resonance), and UV-VIS (ultraviolet-visible) spectroscopy]

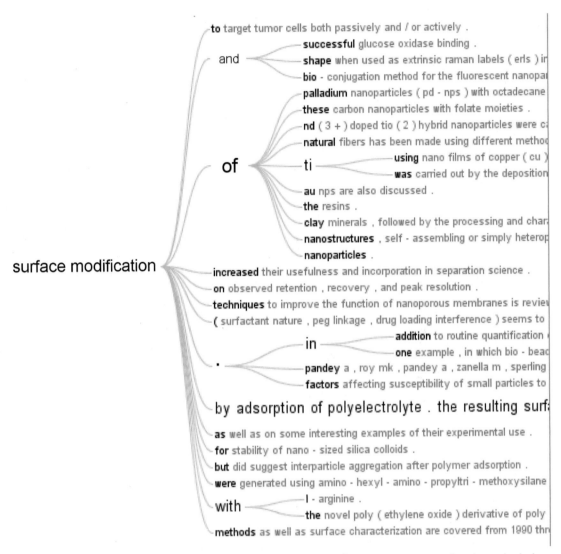

Fig. 2. Example of a word-tree visualization of the search corpus from http://www.many-eyes.com using phrases beginning with "surface modification." This visualization or abstraction suggests that surface modification of nanomaterials and nanoparticles for a variety of different applications can be used on clay, carbon, metal, or metal-oxide nanoparticles to modify their stability and biocompatibility by adding natural fibers, polymer, folate, and bioconjugation.

Usually, the toxicity and other hazards associated with chemical ingredients is a function of the physical composition of the substance. This means that the properties of both the surface and the substance core help to determine the chemical risk. For example, surface properties drive the mobility and toxicity of engineered nanomaterials/particles, but the makeup of the material construct that has been coated is also important. To this extent, it has become more apparent that there are significantly more than nine factors used for similar-composition bulk materials alone that affect nanomaterial ADME and toxicity, some of which are nanomaterial specific are listed in the **table**.

Scientists are searching for reliable means of simultaneously characterizing core-surface coupled characteristics. Surface modifiers and passifiers must be added to engineering nanomaterials to reduce interparticle or material aggregation properties and to tailor the nanomaterials for their desired applications. That is, very small particles have large relative surface areas, so that they tend to aggregate readily compared to larger particles comprised of the same substances. Thus, engineers must coat these materials changing the electromagnetic properties, so that they repel one another (for example, coatings of electronegative cores with neutral organic compounds). However, the newly formed passivated nanomaterial core has a completely unique morphology and surface chemistry that modifies not only molecular driving forces from a materials integrity perspective, but elicits novel biological kinetics and responses as aggregation and surface modification or core leaching occurs.

Biological macromolecules (such as liposomes and micelles) can help provide clues to engineered molecular aggregates. For example, nanoparticles take on different forms, morphologies, phases, and shapes depending on their environment. Surface-core characterization is used in biomedical applications. For example, the use of liposomes in drug delivery is based on their ability to evade immunoresponse, allowing selective bioavailability of the drug at the target site. Methods for characterizing nanoparticles exist, such as electron microscopy. However, these methods tend to be limited to characterizing total morphology and chemical composition. Thus, there is a need to holistically capture core and surface responses and to develop models for testing the variety of surface modifiers or core components (including morphology and symmetry dimensions) to characterize both the environmental and physiological transport mechanisms, that is, through core-leaching, trans-surface modification.

One form of holistic characterization critical to environmental toxicology is to predict a substance's ADME. The typical means of gaining information about ADME, aside from developed QSAR models and other tools for predicting whether a bulk material or compound will present risks, is to conduct in vitro, in vivo, and, more recently, in silico experimental methods. In vitro studies may look at cellular activity in a petri dish, whereas in vivo studies may expose rats to the chemical ingredient and observe effects. In silico studies use computational methods such as genomic and proteomic studies, data mining, molecular modeling, and quantitative structure–activity relationships. The PBPK/PD of a

chemical or its modified progeny must be understood in order to determine how the particle and its components become bioavailable. Similarly, using biomimetics methods (such as lung-on-a-chip technology) and core-surface models such as the chirally perturbed particle in a box model (CPPIB), a variety of accelerated degradation studies can be performed on varied core-surface combinations and morphology domain size optimization that will call on microscopy and chromatography specialists and methods. These studies fully integrate a variety of core activities, research streams, and facilities to take on a major emerging challenge for this novel class of materials, from both a modeling and characterization perspective.

Nanomaterial domain-specific scoping and knowledge audits. Both Web 2.0 text and visual analytics tools in conjunction with open-access databases, such as PubMed or ChemSpider, can be combined to scope and outline the state of the science in any domain-specific research stream. Used appropriately, these informatics approaches can be used for nanomaterials to help study and to predict exposures and hazards from chemical ingredients. Such approaches

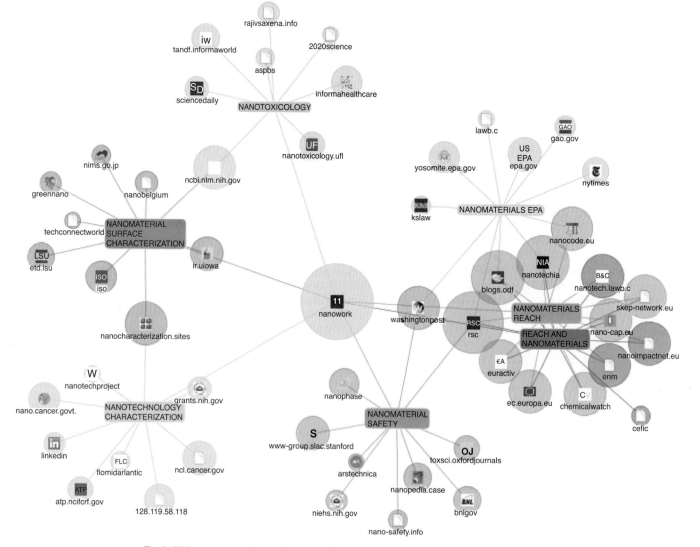

Fig. 3. Web resources related to [nanotoxicology OR (nanomaterials AND {characterization OR Reach OR safety OR EPA})] as rendered in http://www.touchgraph.com.

can target information and enhance knowledge prior to an ingredient reaching the marketplace. That is, domain-specific scoping, or knowledge auditing, can capture key information assets needed by researchers. As such, informatics tools can prevent duplication of effort and can help to consolidate scarce research expertise. The domain-specific corpus of interest is the greater body of literature (**Fig. 1**) that relates to nanomaterial surface characterization methods. This query on pubmed.org provided a total of 790 articles (removing review articles). Extracting the abstracts and titles of all 790 articles and pasting the domain-specific corpus (DSC) into IBM's Many-Eyes (http://www.many-eyes.com) allows one to visualize a large amount of data in a streamlined manner (Fig. 1), focusing on key points and themes in nanomaterials characterization research. One may observe that there are two predominant techniques used to characterize nanomaterials: spectroscopy and microscopy. Curiously enough, microscopy is generally surface-related [for example, atomic force microscopy (AFM) and scanning tunneling microscopy (STM)] and spectroscopy is more intrinsically core related (for example, x-ray diffraction and plasmon resonance) in addition to surface characterization methods [for example infrared (IR) and Raman) that all work in unison to probe different compositions of the core-surface composite. In a similar context, the variations of surface modifications accessible from a Many-Eyes text analysis of a nanotoxicology research article (for example, L. Yan and et al., 2011) can also be useful to provide immediate insight into key research areas outlined within an article. For instance, in **Fig. 2** the case of "surface modification" as applied to nanotoxicology becomes apparent from the context of both core-modifier details and the potential applications for which the materials were tailored. Briefly, both core and surface modifiers that are being actively explored for their intended purposes [(1) cores = titanium, palladium, gold, clay, silica colloids, and carbon; (2) surface modifiers = glucose, folate, octadecane, polyethylene glycol, resins, surfactants, amino-hexyl amino propyl methoxysilane, poly(ethylene oxide), and polyelectrolytes] can be rapidly outlined.

Another scoping technique is to compartmentalize the expertise and discipline domains of all web-accessible resources for a given subject area. For example, **Fig. 3** shows at least seven different areas of interest in the topic of nanotechnology. From this display, more intensive investigation or broader searches can be made, depending on the question at hand. In this instance, since the interest is in core and surface properties of nanoparticles, the left bottom domain would be further explored at a higher level of detail.

[*Disclaimer*: The U.S. Environmental Protection Agency through the Office of Research and Development funded and managed some of the research described here. The present article has been subjected to the Agency's administrative review and has been approved for publication.]

For background information *see* CHEMOMETRICS; ENVIRONMENTAL ENGINEERING; ENVIRONMENTAL TOXICOLOGY; MESOSCOPIC PHYSICS; NANOPARTICLES; NANOTECHNOLOGY; RISK ASSESSMENT AND MANAGEMENT; SPECTROSCOPY; STEREOCHEMISTRY; SURFACE PHYSICS; TOXICOLOGY in the McGraw-Hill Encyclopedia of Science & Technology.

Michael R. Goldsmith; Daniel Vallero

Bibliography. Analysis of the nanotechnology consumer products inventory http://www.nanotechproject.org/inventories/consumer/analysis_draft/; W. H. De Jong and P. J. A. Borm, Drug delivery and nanoparticles: Applications and hazards, *Int. J. Nanomedicine*, 3(2):133–149, 2008; M.-R. Goldsmith et al., The chiroptical signature of achiral metal clusters induced by dissymmetric adsorbates, *Phys. Chem. Chem. Phys.*, 8:63–67, 2006; Nanotechnology Consumer Products Inventory http://www.nanotechproject.org/inventories/consumer/; L. Yan et al., Low-toxic and safe nanomaterials by surface-chemical design, carbon nanotubes, fullerenes, metallofullerenes, and graphenes, *Nanoscale*, 3(2):362–382, 2011.

Neandertal genome

Since the discovery of the first Neandertal fossil bones in the mid-1800s, Neandertals have been an enigma. Morphologically, their bones are very similar to ours (that is, modern humans), yet distinct enough to demand an explanation. Following the earliest finds in Belgium, Gibraltar, and Germany, scientists were in disagreement as to whether they represented a separate group that had long since gone extinct or whether these were simply diseased human individuals whose bones were oddly shaped. Compounding the mystery was the lack of any reliable method to determine the age of these bones.

In the decades that followed, conceptual and technological breakthroughs have allowed us to understand the Neandertals in great detail. Dozens of other Neandertal sites have been discovered across Europe and the Middle East. Careful examination of the bones and stone tools recovered at these sites has revealed that Neandertals were in fact a culturally and morphologically distinct hominin group. They evolved this distinctness in Eurasia starting several hundred thousand years ago. Neandertals persisted until about 30,000 years ago and then mysteriously disappeared at roughly the same time that modern humans were migrating out of Africa and entering the Neandertal geographic range.

Ancient DNA. This understanding of the Neandertals, largely the work of archeologists and anthropologists who find and examine bone remains and stone tools, is now being augmented by genetic analysis. For several decades, it has been known that, in special circumstances, DNA can survive in bone and other biological tissues. In 1997, DNA was recovered for the first time from a Neandertal, specifically the Feldhofer 1 type specimen (the sample that morphologically typifies the group). Through painstaking amplification and sequencing of overlapping regions of the maternally inherited mitochondrial genome (mtDNA), a small segment of the Neandertal mtDNA

Fig. 1. Vindija Cave, Croatia. The three bones used to initially determine the Neandertal genome were excavated from this cave. (*Image copyright* © *Johannes Krause*)

genome was determined and compared to that of humans alive today. The analysis of this Neandertal and the dozen or so from whom mtDNA has since been recovered showed that Neandertals carried mtDNA sequences that are extinct; that is, no living human carries the Neandertal mtDNA sequence. This finding is consistent with no interbreeding between humans and Neandertals, although it could not be ruled out entirely.

Within the past several years, the field of ancient DNA has undergone a renaissance of sorts, driven by technological advances that now make it possible to sequence complete genomes of extinct species. Whereas previous versions of DNA sequencing machines, such as those used to sequence the human genome in 2001, could read about 100 short regions of DNA in one machine run, current machines can read about 100 million regions. Although the length of reads is shorter, DNA recovered from ancient bones is typically fragmented to such an extent that this does not matter; longer reads are only useful if the DNA to be read exists in long stretches. In 2010, using this new technology, the genome sequence and its analysis were reported for the Neandertal.

The Neandertal genome sequence was derived primarily from DNA recovered from three bones found in Vindija Cave in Croatia (**Fig. 1**). These bones were radiocarbon-dated to approximately 38,000 years old. Thus, they came from late Neandertals, who lived in the last few thousand years before Neandertals went extinct and after humans had migrated out of Africa and began their range expansion throughout Eurasia. Importantly, however, there is no evidence that humans had yet made their way to Croatia or any region in central Europe as of 38,000 years ago. More than one billion nucleotides of human DNA (the smallest units of the DNA polymer) were recovered and sequenced from each of these bones.

Challenges for Neandertal genomics. Several technical hindrances must be addressed in genome-scale sequencing from ancient samples. First, the majority of DNA recovered from most fossil bones does not derive from the individual who originally possessed the bone. Rather, most of the recovered DNA derives from soil-living microbes that have colonized the bone in the tens of thousands of years that it remained in the ground. These extraneous sequences, often comprising more than 99% of the data, ne-

cessitate special sequencing and analysis to identify the bona fide Neandertal sequences. Second, DNA becomes chemically damaged as time passes. This damage manifests as sequence errors, because the damaged bases are not read correctly. Third, DNA fragmentation occurs over time, resulting in DNA fragments that are typically about 50 nucleotides in length. These short fragments are often difficult to place correctly in the larger context of a genome that contains many repetitive sequences. Special computational methods are essential for addressing each of these challenges.

An added challenge for reconstructing the genome of extinct hominins is the ever-present danger of contaminating DNA from humans who may have had contact with the bones or lab reagents used in the sequencing process. Because Neandertals are so closely related to modern humans, this problem is particularly vexing. In most cases, it is impossible to determine whether a short sequence fragment derives from an error-riddled Neandertal sequence or a modern-human source. Clean room procedures can be employed to minimize the risk of human DNA contamination, but no methods are foolproof.

To assess the authenticity of the Neandertal data, three approaches have been taken. The first approach used the known sequence of the Neandertal mtDNA and the known variation in mtDNA in living humans to identify about 130 positions where Neandertals differ from living humans. These are contamination diagnostic positions: any sequence fragment that covers one of these positions has either the human- or Neandertal-specific nucleotide. Of 27,000 such fragments, human contaminants were found in less than 1%. A second contamination assay was devised that relies on the fact that each of the three Vindija bones derives from a female Neandertal. Thus, there should be no Y chromosome sequence, because this chromosome is present only in males. Of the roughly 80 million Neandertal sequence reads, four unambiguously Y chromosome sequences were found, again consistent with a level of contamination of less than 1%. Finally, the Neandertal sequences were compared to the genomes of several modern humans that were concurrently sequenced. Positions where all humans carried the same base but where this differed from two of the three Neandertals were used to then judge the third Neandertal. Rotating which two Neandertals were used to discover such positions and which was being evaluated for contamination also revealed a level of human contamination of less than 1%. From these analyses, the Neandertal genome sequence was deemed to contain less than 1% of modern-human contamination.

Genome results. The genomes of the Neandertal, human, and chimpanzee were compared to determine the overall level of divergence between humans and Neandertals. The chimpanzee was required, because a simple pair-wise comparison between the human and Neandertal genomes would be dominated by any sequencing errors present in chemically damaged Neandertal sequences. By using the chimpanzee genome as an out-group, it is possible to determine which evolutionary changes happened since our divergence from Neandertals. This methodology neatly sequesters a large fraction of the Neandertal sequence errors onto the Neandertal lineage. From this analysis, it is possible to calculate that the human–Neandertal genomic divergence is roughly 12% of the human–chimpanzee divergence. Although this number gives some impression of the closeness of Neandertals and humans, it obscures some important details.

Because Neandertals are so closely related to humans, simple genome-to-genome comparisons are not sufficient to probe the Neandertal–human genetic relationship. The reason for this is that there is, in fact, not one single human genome. The human genome sequence (more accurately called the reference human genome) is simply one possible sequence, pieced together from material donated by several anonymous people. Neandertals are so closely related to currently living humans that they often fall within the genealogies of living humans. Therefore, Neandertals must be considered a deeply diverging human population for the purposes of genetic analysis.

To understand the genetic relationship between Neandertals and humans, it is first helpful to consider how genetic information is passed from generation to generation (**Fig. 2**). Each individual receives a full complement of genetic information from both parents in the form of a unique version of each of the 22 chromosomes (and the sex chromosomes). These chromosomes are produced by recombining the two versions of each chromosome that existed in the parents. Thus, the chromosomes present in an individual are not those that existed in the parents; instead, they are a newly constructed version that melds genetic material from the parents' parents. Looking backward in time, the parents' chromosomes were constructed in the same way. Because the exact location of chromosomal recombination is random, the intact segments that are passed on are shorter as one goes back in time. Figure 2 illustrates this concept for two generations prior to the current generation. Of course, this process goes back in time many generations prior to our grandparents.

At any particular place in the genome, there is a common ancestor between any two individuals, who lived at some time in the past (**Fig. 3**). Because of the random nature of inheritance in each generation, the number of generations ago that this common ancestor lived varies across the genome. On average, for human populations, this common ancestor lived about 420,000 years ago. Thus, much of the variation among humans who live today was already in existence at the time of the common ancestor of humans and Neandertals, about 300,000 years ago. One consequence of this is that, at any particular region of the genome, a particular living human may be, by chance, more closely related (that is, have a more recent common ancestor) to a Neandertal than to another living human. If the genetic differences are compared between any two humans who are alive today, each should match the Neandertal equally often. Because the population that led to Neandertals diverged before the divergence

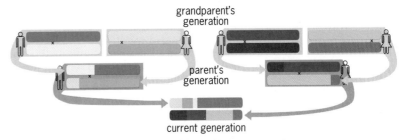

Fig. 2. Genetic inheritance in sexually reproducing species. Every individual carries two copies of the genome (one inherited from each parent). The version of the genome bequeathed by each parent is a recombined version of the two genomes present in each parent. Likewise, the parents' genomes were recombined versions present in each grandparent. Through the generations, contiguous segments intact from a single individual are shorter and shorter. At each place in the genome, every individual has a unique and random genealogy stretching back through the generations.

of any current human populations came into existence (and, in fact, before the onset of anatomically modern humans), each current human population should be equally related to Neandertals. Remarkably, the Neandertal genome revealed that this is not the case. Instead, the genomes of all non-Africans show a small, increased similarity to the Neandertal genome. Because this genetic similarity exists even in humans who live well outside the regions where Neandertals lived, for example, Oceania and the New World, the admixture between human ancestors and Neandertals must have occurred in the population ancestral to all non-Africans. The current model explains that humans who first migrated out of Africa, perhaps about 60,000 years ago, entering the Neandertal range in the Middle East, hybridized with Neandertals. Subsequently, this group spread out to colonize the rest of Eurasia and eventually the

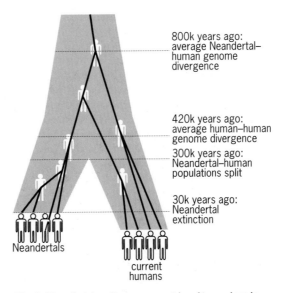

Fig. 3. Neandertal and human ancestries. At any place in the genome, there is a phylogenetic tree that relates all humans. In this tree, the common ancestor for two randomly selected humans lived about 420,000 years ago. This is more distant in the past than the population split that separated what would become Neandertals and what would become modern humans. Therefore, much of the genetic variation that exists between humans today was already present in the common ancestor of humans and Neandertals. For some genes, as shown here, some humans will be more closely related to Neandertals than to other humans.

New World. Thus, the Neandertal genetic legacy is present in roughly equal amounts within the genomes of all non-Africans.

Future outlook. In addition to addressing the historical question of what happened when humans and Neandertals came into contact with one another, the Neandertal genome also helps address important biological questions. For example, it is now possible to contrast the genetic variation present within living humans with the Neandertal genome to find regions that have experienced the last few genetic changes on our path to becoming fully modern humans. These regions, in contrast to much of the genome, are likely to be common to all humans. In other words, these will be genomic regions where all human genealogies have a common ancestor, that is, the ancestor who first had the beneficial mutation, who lived more recently than the common ancestor with Neandertals. Current efforts are focused on identifying these regions and interpreting the biological changes that they caused.

Since their first discovery, Neandertals, our enigmatic closest extinct relatives, have challenged and honed our ideas of what it means to be human. The secrets of human history written in our genome can now be probed more thoroughly using the Neandertal genome. We now know that our stories are not only similar, but intertwined in an unexpected way. Just as this remarkable and unexpected similarity provides a fantastic piece of the puzzle of human evolutionary history, finding and interpreting our genetic differences from Neandertals can now provide a new and important definition of human uniqueness.

For background information *see* ANTHROPOLOGY; BONE; CHROMOSOME; DEOXYRIBONUCLEIC ACID (DNA); EARLY MODERN HUMANS; EXTINCTION (BIOLOGY); FOSSIL; FOSSIL HUMANS; GENE; GENOMICS; HUMAN GENOME; MOLECULAR ANTHROPOLOGY; NEANDERTALS; PHYSICAL ANTHROPOLOGY in the McGraw-Hill Encyclopedia of Science & Technology.

Richard E. Green

Bibliography. A. W. Briggs et al., Targeted retrieval and analysis of five Neandertal mtDNA genomes, *Science*, 325:318–321, 2009; H. A. Burbano et al., Targeted investigation of the Neandertal genome by array-based sequence capture, *Science*, 328:723–725, 2010; R. E. Green et al., A complete Neandertal mitochondrial genome sequence determined by high-throughput sequencing, *Cell*, 134:416–426, 2008; R. E. Green et al., A draft sequence of the Neandertal genome, *Science*, 328:710–722, 2010; R. E. Green et al., The Neandertal genome and ancient DNA authenticity, *EMBO J.*, 28:2494–2502, 2009; D. Reich et al., Genetic history of an archaic hominin group from Denisova Cave in Siberia, *Nature*, 468:1053–1060, 2010.

Oceanic islands: evolutionary laboratories

Oceanic islands comprise only about 5% of the Earth's land surface area, yet they have long captivated the human imagination for their splendid isolation and their unusual, sometimes bizarre, plants and animals. Well-known examples include the finches and huge tortoises of the Galápagos and the magnificent silversword plants of Hawaii. Oceanic archipelagos were of keen interest to the two formulators of the theory of organic evolution, Charles Darwin and Alfred Russel Wallace, both of whom recognized the significance of oceanic islands for understanding organic evolution. Interest in islands as ideal settings for studying evolution has remained strong from the time of Darwin to the present.

Getting there. Although oceanic islands may be highly isolated from the nearest continent, "natural" dispersal must have occurred (discounting humans as dispersal agents in historical times). Despite being formed by volcanic activity and being initially barren of life, such islands are inhabited by animals and plants. Both the isolation and climate of islands act as filters so that only those organisms able to disperse to and establish on an island will occur there. These two filters render island biotas "disharmonic" relative to the animals and plants of continental source areas; that is, they are not representative samples of the continental organisms. For example, bats and land birds can disperse over water for greater distances than large mammals. The airborne spores of ferns and the seeds of many flowering plants can disperse great distances. DNA sequence data are constantly providing insights into unlikely source areas and improbable closest relatives of island plants; a noteworthy example is the origin of Hawaiian violets from Arctic ancestors.

Arrival and getting started. Once a colonizer arrives on an island, it faces both unique opportunities and challenges. It is often free of its competitors and predators from the original continent, and it finds new open niches. This "new freedom" has been termed ecological release. However, colonizers face formidable challenges. Because only one or a few individuals (seeds in the case of plants) are likely to be dispersed to an island, they will carry only a fraction of the genetic variation from the ancestral source population. Various molecular markers typically show reduced diversity in island endemics. The ancestors of island organisms thus face their new world with limited genetic resources. Finding a compatible mate may be a major challenge on a remote island. Animals may lose their former food resources, and plants will often leave behind the pollinators that facilitated sexual reproduction. The first evolutionary experiment in the island laboratory is how organisms "make do" with what is available on an island. Plants, for example, may turn to wind for transferring pollen, while still retaining some of the ancestral attributes of animal-pollinated flowers. Comparisons of island animals and plants with their closest continental relatives offer insights into how they have adapted, or are in the process of adapting, to their new insular environment.

Laboratory paradigm. Oceanic islands have several attributes that qualify them as natural laboratories for studying organic evolution. They are precisely delimited spatially, because they are completely surrounded by water and have never been in direct contact with continents. The distances of islands from landmasses effectively preclude genetic contact with their continental relatives. The ages of islands may be estimated from radiometric (radioactive) dating of the volcanic lava, thus setting the oldest possible age for their endemic biota. Indeed, much like a laboratory experiment carried out in a confined, closed system under controlled conditions, evolution can be studied on islands. Laboratory conditions may vary among islands, depending on factors, such as size (area), elevation, and climate. In addition, oceanic islands are often part of an island system or archipelago, with islands of different ages (for example, the Galápagos, Hawaiian, and Canary Islands).

Individual islands can act as evolutionary laboratories for a rather short time relative to most continental situations, because they may be above sea level for only several million years. Erosion and subsidence (sinking) continually decrease the heights and areas of islands, ultimately resulting in their disappearance below sea level. The short life of an island means that any evolutionary change in its biota must have taken place not only on a limited spatial scale, but also relatively rapidly. When an island recedes under water, the biota must go extinct or disperse (in the latter case, it will most likely disperse to another island in the same archipelago). Therefore, any observed evolutionary changes must have been telescoped in both time and space because of the small size and short life of islands, making the remarkably diverse products of the evolutionary process both noteworthy and amenable to study.

Adaptive radiation and speciation. Adaptive radiation is a widely used term, although it has been variously defined. Central to the concept is the evolution of ecological diversity within a lineage (relatives sharing a common ancestor) by occupying (radiating into) new niches. The evolution of novel characters within a lineage may facilitate the occupation of a new niche.

Spectacular examples of adaptive radiations are seen in oceanic archipelagos around the world, where morphological and ecological diversity has evolved from a common ancestor. A well-known radiation is Darwin's finches in the Galápagos; these birds are so named because Charles Darwin collected them on the voyage of the *Beagle*. After returning to England with the specimens, it became evident that the 15 species were different, yet had similar attributes, suggesting derivation from a single group of ancestors. All subsequent data, including DNA sequences and other molecular markers, have confirmed this hypothesis, and they indicate that diversification occurred over the past 2–3 million years. Peter and Rosemary Grant have studied the finches for decades, and their efforts provide some of the very best examples of using islands as laboratories for studying adaptive radiations. For example, variation in beak shape ostensibly evolved as an adaptation to food sources; species feeding on seeds have broad triangular beaks, whereas species utilizing plant nectar

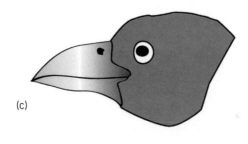

Fig. 1. Variation in beaks of Darwin's finches. (*a*) Large ground finch, crushes large seeds. (*b*) Cactus ground finch, feeds on pollen and pulp of prickly pear cactus. (*c*) Woodpecker finch, extracts insects from bark. (*d*) Warbler finch, gleans insects from plant surfaces and sips nectar from flowers. (*Drawings courtesy of Gilbert Ortiz*)

have long narrow beaks (**Fig. 1**). During their three decades of study, the Grants have observed episodic changes in beaks after fluctuations in precipitation altered available food sources.

The silversword alliance (within the sunflower family, Asteraceae) of Hawaii is arguably the most spectacular example of a plant adaptive radiation in an oceanic archipelago. Molecular data show that

(a)

(b)

Fig. 2. Life forms of silverswords. (*a*) Cushion plant *Dubautia waialealae* grows in bogs. (*b*) *Dubautia laxa* subsp. *hirsuta*, a shrub up to 15 ft (4.6 m) tall, occurs in wet forests and other habitats. (*Photographs courtesy of Ken Wood*)

on Earth [having more than 500 in. (1270 cm) annual rainfall]. Comparative studies of plant structure and function indicate that these features are related to water availability in their habitats and thus are adaptive.

Darwin's finches and the silverswords thus illustrate rapid adaptive radiations into diverse habitats within the small confines of oceanic archipelagos. Within an archipelago of two or more islands, the initial arrival of a colonizer on an island may be followed by adaptive radiation into diverse habitats on that island, with subsequent dispersals from the different habitats to similar habitats on another island in the archipelago (**Fig. 3***a*). Alternatively, initial dispersal to an island may be followed by dispersal from that island to one or more other islands in an archipelago, with parallel adaptive radiations into similar habitats on each of the islands (Fig. 3*b*). DNA sequences have been useful for distinguishing between the two scenarios. With the first scenario, organisms from similar habitats on different islands should be more similar in DNA sequences; in the second case, the organisms from each of the islands should have more similar sequences. Both processes have been detected in island lineages, further illustrating the different evolutionary experiments possible in island laboratories.

The origin of new species (speciation) is arguably the fundamental process in evolution. Oceanic islands offer unparalleled advantages for studying speciation, because, as mentioned previously, evolution is telescoped in time and space. As a result, it is often feasible to infer process from pattern with some confidence. For example, the occurrence of

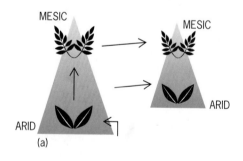

(a)

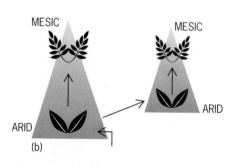

(b)

Fig. 3. Scenarios for adaptive radiation in an archipelago. (*a*) Radiation on one island, followed by dispersal to similar habitats [arid (extremely dry) or mesic (characterized by a moderate amount of water)] on another island. (*b*) Parallel radiations on two islands.

the silverswords originated from a common ancestor, probably about 5 million years ago, and diversified into many life forms, ranging from cushion plants to vines and trees (**Fig. 2**). They occur from near sea level to more than 5100 ft (1550 m) and inhabit locations ranging from very dry [having less than 17 in. (43.2 cm) annual rainfall] to some of the wettest areas

species in different habitats suggests that adaptive radiation into novel habitats reduces or precludes hybridization between different distinct forms recognized as species. In the case of Darwin's finches, the Grants have observed directly the dynamics and complexity of speciation in the Galápagos. They demonstrated that changes in food composition may either promote divergence or facilitate convergence by hybridization. These observations amount to watching evolutionary experiments in a natural laboratory under varying conditions over a very thin slice of time.

Future studies. The ever-increasing use of DNA markers has facilitated more sophisticated studies of the evolution of insular biota. DNA-based hypotheses of evolutionary relationships within island lineages provide insights into the patterns and timing of radiations and speciation in archipelagos, and they allow all observations to be placed within a robust evolutionary framework. DNA markers are also valuable for studying processes, such as hybridization, as well as dispersal within and between islands. The use of new methods for the generation and analysis of molecular data will continually enhance the value of islands as natural laboratories for the study of evolution.

For background information *see* ADAPTATION (BIOLOGY); ANIMAL EVOLUTION; BIODIVERSITY; BIOGEOGRAPHY; DEOXYRIBONUCLEIC ACID (DNA); ECOLOGICAL COMMUNITIES; GENOMICS; ISLAND BIOGEOGRAPHY; OCEANIC ISLANDS; ORGANIC EVOLUTION; PLANT EVOLUTION; POPULATION DISPERSAL; POPULATION DISPERSION; SPECIATION; SPECIES CONCEPT; ZOOGEOGRAPHY in the McGraw-Hill Encyclopedia of Science & Technology.   Daniel J. Crawford

Bibliography. S. Carlquist, *Hawaii: A Natural History—Geology, Climate, Native Flora, and Fauna above the Shoreline*, National Tropical Botanical Garden, Lawai, Hawaii, 1980; P. R. Grant and B. R. Grant, *How and Why Species Multiply: The Radiation of Darwin's Finches*, Princeton University Press, Princeton, NJ, 2008; L. R. Walker and P. Bellingham, *Island Environments in a Changing World*, Cambridge University Press, Cambridge, U.K., 2011; R. J. Whittaker and J. M. Fernández-Palacios, *Island Biogeography: Ecology, Evolution, and Conservation*, 2nd ed., Oxford University Press, New York, 2007.

Origin of the heavy elements

One of the most human of all endeavors is to wonder and to ask questions about our origins: the origin of life, the building blocks of nature, or the beginning of the universe. Exploration of space has shown us how vast the universe is around us; it is full of galaxies, stars, and lots of empty space. Science explorations through the ages have led us to our present-day understanding of the basic components of the universe.

We can verify the existence of a type of matter (dark matter) whose gravitational pull is felt and observed although we know nothing of its constituents.

The visible matter of the universe, conversely, is approximately 4% of the total. All the light, stars, planets, globular clusters, and cosmic dust that we see is the result of the interactions of atoms, or matter, that we are familiar with. At the center of each atom is a nucleus. Understanding nuclear properties and their various interactions (fusions, captures, decays, and so forth) in specific astrophysical scenarios allows us to explain the origin of the elements, as well as the evolution of all matter in the visible universe. Observations of the relative abundances of the elements in outer space, in young stars, or in more mature stars, where matter created in former cataclysmic events is also incorporated to form new elements, provide important information about the history of astrophysical events.

Elemental abundances. Figure 1 shows the elemental abundances as a function of cosmic evolution from the big bang to the present. Each part of the figure shows the abundances as a function of the mass number, A, which is the sum of Z protons and N neutrons. The big bang is thought to have created the very lightest elements: hydrogen, helium, and lithium. These very light elements were then combined in stars by fusion reactions, capturing protons, alpha particles (helium nuclei), and other heavier nuclei to form the elements up to iron. (Figure 1*b* shows the elemental abundances in early stars.) Nuclear physics basically determines this upper limit of iron synthesis. Binding energies of the protons and neutrons inside the nucleus and nuclear reactions show that it is not energetically favorable to make the elements beyond iron by fusion reactions. Observations of the elemental abundances in the very oldest stars confirm this.

A comparison of the elemental abundances of the human body against the abundances of the universe is proof of Carl Sagan's statement that "we are made of star stuff." The elemental abundances of the human body and universe are shown in **Tables 1** and **2**. Each of the heavy elements in our bodies had to be processed in several stellar explosions and supernovae from the very beginning of time to get the abundances that are the building blocks of our bodies. For example, iodine is present in our bodies in very small quantities, but it is present in every cell and essential for regulating metabolism and maintaining overall health.

The solar elemental abundances are shown in Fig. 1*d*. They show elements up to the mass region of uranium (approximately $A = 250$). These are the elemental abundances that resulted from the embedded astrophysical history of our solar system. Implicit in this history is nuclear physics that guides all the reactions and decays that have resulted in the elemental abundances that we see today in the solar system.

One way we can imagine that these elements above iron could be made is in very neutron-rich environments, because charged particle reactions are no longer energy sustainable. The challenge for astrophysical science is to understand what sort of astrophysical conditions would provide a major abundance of neutrons and what conditions would lead to

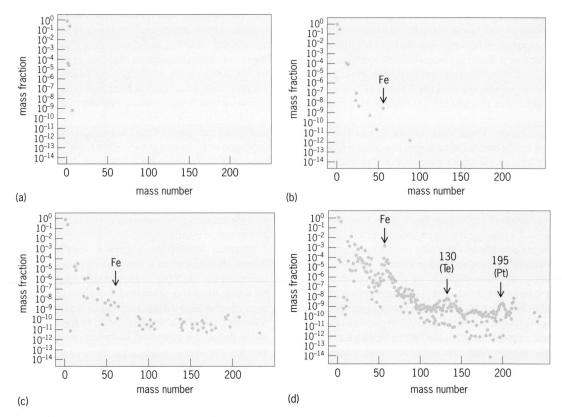

Fig. 1. The evolution of elemental synthesis from the big bang to early stars and what we find today for solar abundances. (*a*) Big bang. (*b*) Oldest stars. (*c*) Early *r*-process. (*d*) Today (Sun). (*From H. Schatz, Rare isotopes in the cosmos, Phys. Today, 61(11):40–45, November 2008*)

successive captures before the nucleus has a chance to decay, while on the nuclear side, the challenge is to determine the laws of nuclear physics that would apply to very neutron-rich nuclei far from stability, where the range and impact of the nuclear force is less well known.

Slow and rapid neutron capture processes. The question is entangled in complexity because the heavy elements are thought to be made through both slow- and rapid-neutron-capture processes (*s*- and *r*-processes). The *s*- or slow capture refers to capturing a neutron and giving it enough time to beta decay or change a neutron into a proton (increase *Z* by 1), whereas the *r*- or rapid captures refer to sequentially capturing neutrons before a nucleus beta decays. The end result of the *s*-process is a network of nuclei near

stability, while the *r*-process allows the production of nuclei with increasing neutron numbers before a beta decay occurs. The resulting path or network of nuclei and isotopes lies much further from stability and produces nuclei with far greater numbers of neutrons than protons in comparison to nuclei around stability. Near stability, much is known about nuclear physics, but there are still experimental challenges in measuring neutron capture cross sections, and far from stability, the nuclear physics of nuclei that are created and decay in fractions of seconds is also unknown.

Figure 2 is a schematic of the chart of nuclides. It shows the roughly 200 stable nuclei in black and the area of radioactive nuclei that have been measured, along with the terra incognita of the region of unknown nuclei. The chart shows schematically where we expect the paths of the *s*- and *r*-processes to lie. The creation in the laboratory of these nuclei far from stability and the study of the nuclear force when there is a large imbalance of neutrons and protons in the nucleus is one of the most prominent frontiers of nuclear science today. The scientific goal here is to learn enough about the nucleus to be able to put restrictions on the astrophysical models.

Astrophysical models. Astrophysical models describing scenarios that might provide the necessary conditions for the rapid neutron capture process to take place and to result in the incorporation of the product nuclei into the stellar medium necessarily include natural explosions. The most promising ones

TABLE 1. Elemental abundances of the universe	
Element	Abundance in the universe, %
Hydrogen	74
Helium	24
Oxygen	1.07
Carbon	0.46
Neon	0.13
Iron	0.11
Nitrogen	0.10
Silicon	0.06
Magnesium	0.06
Sulfur	0.04
All others	0.06

TABLE 2. Elemental abundances of the human body

Element	Abundance in the human body, %
Oxygen	65
Carbon	18
Hydrogen	10
Nitrogen	3
Calcium	1.5
Phosphorus	1.2
Sulfur	0.2
Chlorine	0.2
Sodium	0.1
Magnesium	0.05
Iron, cobalt, copper, zinc, iodine	0.05
Selenium, fluorine	0.01

are in core collapse supernovae or the merging of two neutron stars. Core-collapse supernovae are violent explosive events associated with the release of remarkable energy and light, as well as a tremendous neutrino wind where we expect neutrino-nucleus reactions to occur. There are many challenges to understanding what causes a supernova to explode, but the explosions are one of the most likely sites for the *r*-process to take place. Neutron star mergers are another possibility for the *r*-process to occur. Neutron stars are typically formed when a star runs out of fuel, forcing its constituents to collapse into neutrons.

New facilities. On the nuclear science side, there is a global competition to access the most exotic nuclei that existing facilities can reach, while simultaneously building new, more powerful accelerators to make the even more exotic nuclei that existed very briefly in nature. In Europe, SPIRAL II at the Grand Accélérateur National d'Ions Lourds (GANIL) laboratory in the Normandy region France will be starting construction in 2012; and Facility for Antiprotons and Ions Research (FAIR), a project at Gesellschaft für Schwerionenforschung (GSI) in Darmstadt, Germany, is expected to be operational in 2018. The Chinese have started construction of such a facility, and in the United States the Facility for Rare Isotope Beams (FRIB) is being planned, with the Department of Energy and Michigan State University aiming to be ready in 2017 (**Fig. 3**). The most advanced operating facility is the Japanese facility at RIKEN, which started taking data in 2009.

Measurements of *r*-process nuclei. While models are being developed and facilities are being built, the experimental challenge is to measure nuclei along the *r*-process path. These nuclei are at the very extremes of what is possible at the existing facilities today. A recent measurement of the doubly closed–shell nucleus nickel-78 (^{78}Ni) at the Coupled Cyclotron Facility at Michigan State University's National Superconducting Cyclotron Laboratory (NSCL) yielded 11 events and led to the first measurement of the half-life of ^{78}Ni, an important *r*-process nucleus.

Simulations of *r*-process nuclei. It is a particular challenge to identify nuclei that are most critical to the *r*-process. A group at the University of

Notre Dame has done this using an *r*-process simulation code. The *r*-process simulation starts with "seed" nuclei and an abundance of neutrons. The initial conditions for the simulation can be correlated with different astrophysical scenarios. The nuclear physics components of the simulation depend on the neutron-capture cross sections, the beta-decay rates, and the separation energies of the nuclei that lie along the *r*-process. Figure 2 schematically shows the path of the *r*-process. The rates for beta decay are certainly an important component of the simulation and have the greatest impact for nuclei with the very shortest half-lives (less than seconds). Where possible, measured decay half-lives are incorporated into the simulation. The neutron capture cross sections themselves depend on level densities and have shown some sensitivity near closed shells. The remaining aspect is the separation energies of the various nuclei along the *r*-process path. The separation energies or mass differences between nuclei along the *r*-process path impact both the beta decay (which increases the proton number *Z* by 1) and the cross sections for neutron capture (which increase the neutron number *N* by 1). In many cases, where the experimental numbers are known, they have been incorporated into the simulation. In other cases, it is necessary to rely on theoretical estimates to get the masses or mass differences (separation energies) of many nuclei that have yet to be measured in the laboratory. Typically, most mass models use the known mass measurements and predict masses for nuclei far from stability.

The goal was to isolate the sensitivity of the *r*-process to the masses of individual nuclei in terms of the resulting elemental abundances under various

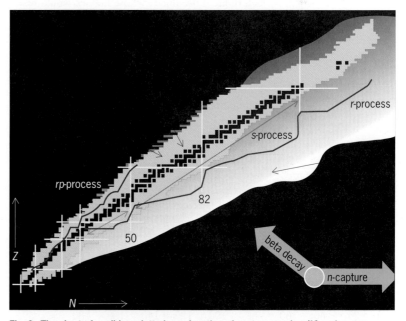

Fig. 2. The chart of nuclides, plotted as a function of neutron number (*N*) and proton number (*Z*). The black boxes represent the 200 or so stable nuclei. The colored regions of the chart represent unstable nuclei that have been measured in the laboratory; they can live from many years to fractions of seconds. The gray regions on the map are those regions of the chart of nuclides that have yet to be studied. How far the gray area stretches will be determined by the properties of the nuclear force.

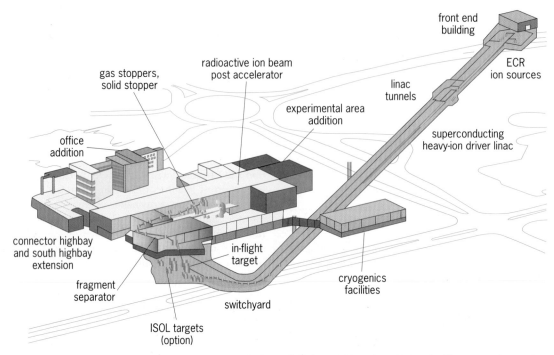

Fig. 3. Diagram of the proposed Facility for Rare Isotope Beams (FRIB) at Michigan State University. The expected date of operation for this facility is 2017.

astrophysical conditions and then to evaluate what the differences were between the existing mass models in an effort to identify the most robust model to use. The approach was to start with the finite-range droplet model (FRDM) to fill in for the thousands of nuclei along the *r*-process for which there are no measurements. In this case, the theoretical masses were taken as a starting point, and the nuclear masses for thousands of nuclei were varied individually to determine those nuclei whose masses would result in the biggest changes in the *r*-process abundances. The solar abundances that are observed today hold information about the nuclear physics and the history of the astrophysical scenarios, and show the resulting abundances of the elements. The Notre Dame group attempted to use the *r*-process abundances resulting from the simulation as a guide for identifying

the nuclei within a given mass model that would make the greatest impact. Initially, calculations were done for a fixed astrophysical scenario with varying mass models. In turn, the mass model was also fixed and the astrophysical scenarios were varied. For the fixed astrophysical scenarios, the same set of studies was carried out with different mass models to differentiate nuclear theoretical effects coming from mass models from those of the astrophysical scenario. The individual mass models were also fixed and the astrophysical conditions varied in order to better understand the resulting abundance differences that are dependent on the astrophysics conditions.

The simulations resulted in a list of nuclei that were essentially the same. Surprisingly, there was a very large overlap of most critical nuclei that emerged in all the hundreds of simulations. **Figure 4** shows the summary of one such set of calculations for a fixed astrophysical scenario and a fixed mass model, where each experimentally unknown nuclear mass was varied $\pm -25\%$. The equation below shows the absolute value of the

$$\Delta_{\pm 25\%}(N, Z) = \Sigma_A \frac{|Y(A) - Y(A, \pm 25\% S_n(NZ))|}{Y(A)}$$

fractional change in the abundances of the 3000 or more nuclei in the network due to a small change of the mass of an individual nucleus. The quantity (N,Z) is that fractional difference, where $Y(A)$ and $(YA \pm 25\% S_n(N,Z))$ are the abundances resulting from no change of separation energy, $S_n(N,Z)$, from the FRDM model and one that shows a 25% change in the theoretical FRDM separation energy for a given nucleus of (N,Z), respectively, normalized to

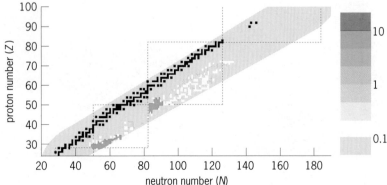

Fig. 4. This figure shows by color intensity the nuclei that show the greatest impact on the resulting *r*-process simulations.

$Y(A)$. The goal is to show, in intensity, the change in r-process abundances, summed over all A in the network, that results from changing one mass or separation energy by 25%. The intensity of color or shading in Fig. 4 shows the nuclei that caused the largest variations in the resulting abundances against a control case. That is, it shows the nuclei that had the most impact on the resulting abundances, and which are therefore the critical nuclei to measure. Not surprisingly, they cluster around the closed shells of nuclei. The nuclear force determines the magic numbers of nuclei, and many of the nuclear processes are dependent on their proximity or distance to these magic numbers.

The conclusion is that measurements of nuclei near the closed shells are the most critical to understanding the nuclear properties and therefore for setting restrictions on the proposed astrophysical models as potential sites for the r-process.

Simulation studies like this one will be very important to determining the priorities for measurements of nuclei along the r-process until the various facilities around the world are able to produce many of the nuclei along the r-process in abundance.

Scientific endeavors yield both expected and unexpected results. The community of nuclear scientists is looking forward to what lies on the neutron-rich frontiers of the chart of nuclides.

This work was supported by the NSF under contract numbers 07-58100 and 08-22648.

For background information *see* BIG BANG THEORY; DARK MATTER; ELEMENTS, COSMIC ABUNDANCE OF; EXOTIC NUCLEI; MAGIC NUMBERS; NEUTRON STAR; NUCLEAR REACTION; NUCLEAR STRUCTURE; NUCLEOSYNTHESIS; PARTICLE ACCELERATOR; SUPERNOVA in the McGraw-Hill Encyclopedia of Science & Technology.   Ani Aprahamian

Bibliography. J. Duflo and A. P. Zuker, Microscopic mass formulas, *Phys. Rev. C*, 52:R23–R27, 1995; P. Hosmer et al., Half-life of the doubly magic r-process nucleus ^{78}Ni, *Phys. Rev. Lett.*, 94:112501 (4 pp.), 2005; P. Moellerw, J. R. Nix, and K.-L. Kratz P. Moeller, Nuclear properties for astrophysical and radioactive-ion-beam applications, *At. Data Nucl. Data Tables*, 66:131–343, 1997; J. M. Pearson, R. C. Nayak, and S. Goriely, Nuclear mass formula with Bogolyubov-enhanced shell-quenching: Application to r-process, *Phys. Lett. B*, 387:455–459, 1996; H. Schatz, Rare isotopes in the cosmos, *Phys. Today*, 61(11):40–45, November 2008; R. Surman and J. Engle, Changes in r-process abundances at late times, *Phys. Rev. C*, 64:035801(8 pp.), 2001.

Origins of agriculture

One of humankind's most significant accomplishments was the transition from hunting and gathering to food production and the establishment of permanent settlements, commonly known as the Neolithic Revolution. Occurring independently in several parts of the Old and New Worlds, it is oldest and best documented in the Near East (**Fig. 1**), beginning about 10,000 years ago and ultimately spreading to Europe and elsewhere. The Near East region (also known as the Middle East in contemporary times) is vast and includes the modern countries of Israel, Iran, Iraq, Jordan, Lebanon, Syria, and Turkey. Although different classifications are applied to the Near Eastern Neolithic, one that has prevailed is a division into three Pre-Pottery Neolithic phases (PPNA, PPNB, and PPNC) as well as a later Pottery Neolithic. In addition, an earlier period, the Natufian (and variants), set the foundations for subsequent developments (see **table**).

Environment. The Near East is a remarkably diverse landscape; approximately 10,000 years ago, it was characterized by dramatic climatic fluctuations that resulted in the expansion and contraction of favorable environments suitable for farming. Based on current data, the first experiments with sedentary living, but not agriculture, started during the Early Natufian under optimal environmental conditions. Slightly later, a global cooling and drying episode (the Younger Dryas) may be linked to a more mobile pattern for many Late Natufian groups. During the actual Neolithic, climatic conditions improved and were optimal for farming (perhaps being better than today's climate). About 7000 years ago, conditions approximated those of the present.

Cultural periods. To understand the Neolithic, it is important to look first to the Natufian and similar groups. Their material culture was rich and included elaborate burials (some with grave goods) and portable art depicting animals. There is evidence of sedentism during the Early Natufian in form of small hamlets. More mobile adaptations occur during the Late Natufian. The exceptions to this pattern were large and sophisticated settlements such as Tel Abu Hureyra and Mureybat in the Euphrates Valley of Syria.

Natufians practiced an array of economic options, reflecting a broad-spectrum exploitation pattern. Specialized gazelle exploitation provided most protein, whereas many Natufian groups harvested wild cereals seasonally. Deteriorating ecological conditions during the Younger Dryas may have required intensification of cereal exploitation, setting the stage for subsequent Neolithic domestication. However, there is no evidence that Natufians actually domesticated plants or animals, with the apparent exceptions of dogs and, perhaps, rye.

The Neolithic itself begins with the PPNA, whose features included the establishment of actual villages, the intense use of key economic resources, and the elaboration of social organization and ritual behavior. One of the earliest PPNA settlements that has been investigated is Jericho, and its sophistication has both impressed and confused scholars for decades. As additional PPNA villages have been documented, more is now understood about Jericho, but it still ranks as a quantum leap in complexity from previous Natufian settlements, both in size [more than 10 acres (4 hectares)] and in material culture.

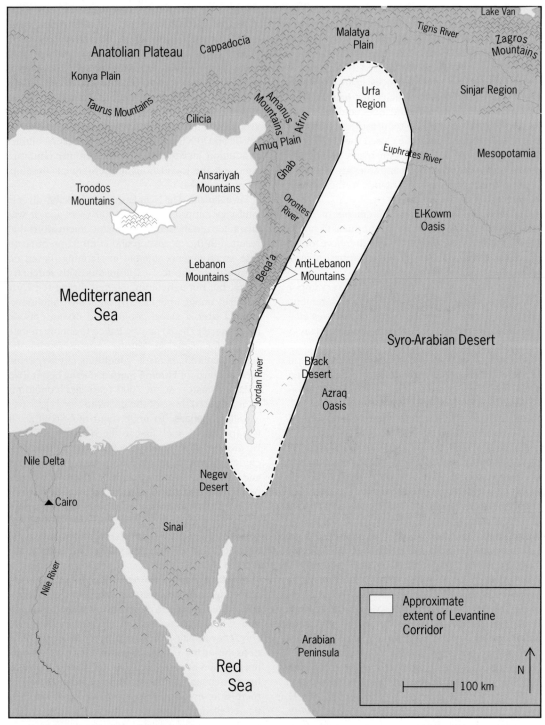

Fig. 1. Map of the Near East, showing the "Levantine Corridor," one of several likely "cores" for Neolithic development. (*Adapted from A. H. Simmons, The Neolithic Revolution in the Near East: Transforming the Human Landscape, Fig. 1.2, p. 8, University of Arizona Press, Tucson, 2007*)

The PPNA was a "point of no return" and comprised regionally identifiable groups and a wide range of site types. Intensive cultivation of plants, primarily cereals, and hunting supported much of the population. Again, however, there is no clear evidence for morphologically domestic plants or animals, although these resources clearly were being culturally manipulated.

Mortuary practices now included ritualized decapitation, a pattern that became more formalized during the PPNB. Domestic architecture varied, although the basic dwelling form was circular or oval. A few sites include public architecture, such as the walls and tower at Jericho and the massive engraved pillars at Göbekli Tepe in Turkey. The latter stands out particularly, as it appears to have

Approximate dates for major periods related to the Near Eastern Neolithic*		
Period	Conventional date (uncalibrated), B.P.	Calibrated date, B.P.
Natufian	12,800–10,500	15,000–11,700
PPNA	10,500–9500	11,700–10,500
PPNB	9500–7900	10,500–8700
PPNC	7900–7500	9250–8250
Pottery Neolithic	7500–6100	8250–6900

*Note that these periods vary by region. The three Pre-Pottery Neolithic phases are designated PPNA, PPNB, and PPNC. Dates of periods are given in years before present (B.P.).
SOURCE: Adapted from A. H. Simmons, *The Neolithic Revolution in the Near East: Transforming the Human Landscape*, pp. 50, 89, 124, and 201, University of Arizona Press, Tucson, 2007.

been a large ritual site rather than a domestic settlement. Both sites continued to be occupied into the PPNB.

Although evidence for PPNA social structure is elusive, major reorganizations were necessary to cope with expanding populations and the need for communal activities. There was a pronounced change in ritual behavior during the PPNA, reflected by burials and a change to human female figurines from the zoomorphic figurines common in the Natufian. In addition, both short- and long-distance trade occurred. Indeed, a PPNA site has been recently reported on the island of Cyprus, as has at least one equivalent Natufian site, suggesting considerable seafaring skills. This is important because earlier views postulated that the Mediterranean islands were not colonized prior to the late PPNB.

The full flowering of the Neolithic was reached in the PPNB. Society was radically transformed, and several dramatic achievements occurred. These included the definitive domestication of sheep, goats, pigs, cattle, barley, emmer wheat, and einkorn wheat; the elaboration of ritual behavior; widespread trade; and the emergence of large villages as well as "mega-sites," including 'Ain Ghazal and others in Jordan (**Fig. 2**). The latter sites often exceeded Jericho's size, even though, unlike Jericho, they were permanently abandoned after the Neolithic. Despite the domestication of key plants and animals, wild resources continued to be significant. Although the PPNB is characterized by village society, some PPNB peoples continued hunting and gathering; however, they ultimately adopted domesticates (especially animals), likely establishing the classic Near Eastern dichotomy of "the desert and the sown," or village dwellers and pastoral nomads.

Architectural complexity is a hallmark of the PPNB. Communities consisted of multiroom rectangular structures. Nondomestic structures with communal or ritual significance also appeared. There are a few examples of sophisticated nonresidential sites such as Göbekli Tepe or the elaborate burial site of Kfar Hahoresh in Israel. Mortuary patterns became much more standardized, with a widespread "skull cult." The PPNB is rich in symbolism, reflected by

figurines, masks, statues [including the remarkable 'Ain Ghazal statues (Fig. 2)], tokens, an apparent reverence of cattle, and the use of ritual sites.

PPNB society was complex and diverse, and older ideas of a few "core zones" (such as the Levant, depicted as the "Levantine Corridor" in Fig. 1) are giving way to a new understanding that there were multiple points of origin and development. The Neolithic world was widespread, with trade and other social interactions across essentially the entire Near East, as well as Cyprus, where recent research has documented some sites dating to the early PPNB.

The final Pre-Pottery Neolithic phase is the PPNC. Initially identified at 'Ain Ghazal, it is now documented at other sites. Its primary significance is that it showed an uninterrupted transition into the Pottery Neolithic. Earlier research proposed a "Great Neolithic Gap" between the Pre-Pottery and the Pottery Neolithic, suggesting that the later represented "new" populations. Although there may have been

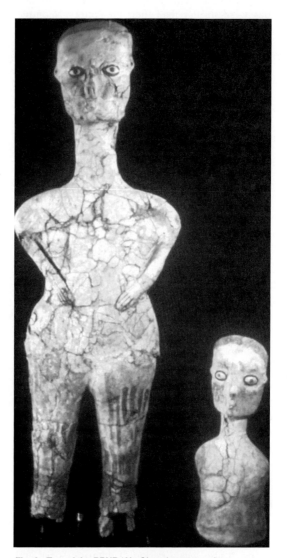

Fig. 2. Two of the PPNB 'Ain Ghazal statues, after conservation. (*Photo courtesy of Alan H. Simmons*)

some localized abandonments and population relocations, several sites show an unbroken linkage between the Pre-Pottery and the Pottery Neolithic.

The subsequent Pottery Neolithic (or Late Neolithic) is, obviously, defined by the incorporation of ceramic technology. Ceramics may have initially served ceremonial functions rather than utilitarian ones. Although the PPNB shared many similarities over a wide geographic range, a pan–Near East interaction sphere appears to have broken up during the Pottery Neolithic. Different parts of the Near East now followed very distinctive trajectories. Parts of the northern Near East and Turkey elaborated on earlier developments, leading ultimately to the classic urban societies in Mesopotamia. In the Levant, however, there was more of a rural adaptation that increasingly came under influence from both the north and the south.

Architectural and village variability during the Pottery Neolithic is heterogeneous. Huge, densely packed settlements occurred in Turkey, such as Çatalhöyük. With the notable exception of a few sites, though, many Pottery Neolithic communities were smaller than their Pre-Pottery Neolithic counterparts. Economically, the Pottery Neolithic reflects an agropastoral system with increased reliance on domesticated plants and animals. There also is evidence for wine production. Ritual or symbolic behavior is especially expressed through elaborate figurine imagery, particularly of corpulent females. Earlier claims of cultural deterioration during the Pottery Neolithic are unsubstantiated. Although parts of the region, such as the southern Levant, did not reach the elaboration of settlements such as Çatalhöyük, this may reflect an efficient readaptation to new conditions.

The southern mega-sites were permanently abandoned at the end of the Pottery Neolithic. Many explanations have been proposed for this. One widely held concept is that their populations grew rapidly and overexploited fragile landscapes. Others have posed social tension as a reason, whereas other explanations, such as organized violence, disease, and sanitation, may have been relevant. There is, however, little evidence of trauma on most Neolithic remains. Many scholars hint that the mega-site experiment was a failure or that this type of settlement experienced a collapse; however, sites such as 'Ain Ghazal were continuously occupied for at least 2000 years, suggesting a remarkable resiliency and adaptability.

The period marking the end of the Pottery Neolithic and the beginning of the subsequent Chalcolithic (Copper Age) is vague. Overall, the Pottery Neolithic may be characterized as an adaptation of local settlements with strong tribal ties, forming a pattern that has endured in the Near East over several millennia.

Conclusions. Although it is undeniable that the Neolithic was an economic and technological milestone, it also was a period of dramatic social and symbolic transformation. Without the security and surplus provided by food production during the Neolithic, subsequent cultural achievements that ultimately culminated in contemporary society would never have occurred. The Neolithic dictated changes in how humans interacted with each other and with the environment. The benefits, as well as the liabilities, of the Neolithic molded the world in which we now live.

For background information *see* AGRICULTURE; ANTHROPOLOGY; ARCHEOLOGICAL CHRONOLOGY; ARCHEOLOGY; CEREAL; DOMESTICATION (ANTHROPOLOGY); NEOLITHIC; POTTERY; PREHISTORIC TECHNOLOGY; SOCIOBIOLOGY in the McGraw-Hill Encyclopedia of Science & Technology.

Alan H. Simmons

Bibliography. G. Barker, *The Agricultural Revolution in Prehistory: Why Did Foragers Become Farmers?*, Oxford University Press, Oxford, U.K., 2006; P. Bellwood, *First Farmers: The Origins of Agricultural Societies*, Blackwell, Oxford, U.K., 2005; J. Cauvin, *The Birth of the Gods and the Origins of Agriculture* (translated by T. Watkins, originally published in 1994), Cambridge University Press, Cambridge, U.K., 2000; I. Kuijt (ed.), *Life in Neolithic Farming Communities: Social Organization, Identity, and Differentiation*, Kluwer Academic/Plenum, New York, 2000; E. Peltenburg and A. Wasse (eds.), *Neolithic Revolution: New Perspectives on Southwest Asia in Light of Recent Discoveries on Cyprus*, Levant Supplementary Series 1, Oxbow Books, Oxford, U.K., 2004; A. H. Simmons, *The Neolithic Revolution in the Near East: Transforming the Human Landscape*, University of Arizona Press, Tucson, 2007.

Oxytocin and autism

The capacity to interact with others in complex social situations constitutes one of the key elements of our daily life. Patients suffering from high-functioning autism spectrum disorders (HF-ASDs) are not capable of interacting spontaneously with others in social situations, despite their normal intellectual abilities. They avoid eye contact and have difficulties in understanding the intentions of others. Autism is classified as a pervasive developmental disorder because symptoms emerge early in infancy, or perhaps from birth, affecting many aspects of cognition and behavior. Many studies have investigated autism, covering several domains of research: neuroscience, genetics, neurobiology, and the social sciences. However, there are still no promising cures for autism. Recently, one uniquely mammalian hormone, oxytocin, which probably evolved along with parturition and lactation approximately 200 million years ago, has attracted considerable attention because of its potential use in the treatment of autism. There has been much research exploring the link between oxytocin and autism at the genetic and behavioral levels, especially because researchers reported that children with autism had reduced levels of oxytocin in their plasma as compared to age-matched controls. This has opened an interesting avenue of investigation for the treatment of autism. Thus, what is oxytocin?

Oxytocin as a key facilitator for life. The process of evolution has endowed oxytocin with a remarkable capacity to be released in the blood as well as in the brain. Although best known for its role in birth delivery and breast feeding, oxytocin is also a fundamental mediator for a large range of emotions and behavior. Researchers have found that oxytocin initiates maternal behaviors such as nest building even in virgin animal models. Recently, in humans, it has been shown that the simple presence of the mother's voice represents an impressive support for the child, releasing the child from stress and increasing the child's levels of oxytocin. As such, oxytocin is a mediator of maternal attachment through its action on the reduction of stress and anxiety.

Also best known for its role in sexual activities, oxytocin is related to pair attachment and love bonding. Elegant studies on prairie voles have shown that the distribution of oxytocin's receptors in the brain can predict the degree of social behavior of these rodents. (Unlike most mammalian species, prairie voles form long-term pair bonds and show more oxytocin release during the sex act compared to their relatives, the meadow voles. Thus, the prairie vole is an attractive subject of research on oxytocin.) Administration of oxytocin to social species (such as prairie voles) induced a partner preference formation, and blocking oxytocin prevented formation of pair bonds even after mating. Thus, oxytocin has been associated with sexual activity and approach behavior, and does not only mediate the birth-life process but also promotes the fundamental basis of attachments, affiliation, and social interactions. Overall, a substantial body of research has established that oxytocin increases trust, empathy, eye contact, memory for faces, and generosity in humans.

Oxytocin and social deficits. In the light of the aforementioned findings, scientists have started to explore the implications of oxytocin treatment in patients with autism. For example, researchers have looked at the effect of an intravenous administration of oxytocin on the modulation of repetitive behavior in 14 adults with HF-ASDs. Indeed, these patients also suffer from the development of rigid adherence to routines or rituals and repetitive motor mannerisms, such as stereotypes. It was found that oxytocin decreased the severity as well as the number of different types of these repetitive behaviors (need to know, repeating, ordering, need to tell/talk, self-injury, and touching). In another study, the administration of

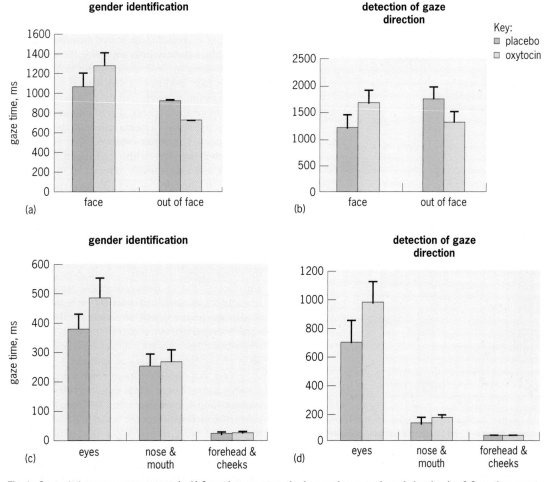

Fig. 1. Oxytocin increases eye contact. (*a, b*) Gaze time spent on the face under oxytocin and placebo. (*c, d*) Gaze time spent on different regions of interest under oxytocin and placebo. (*Reprinted with permission from E. Andari et al., Promoting social behavior with oxytocin in high-functioning autism spectrum disorders, Proc. Natl. Acad. Sci. USA, 107:4389–4394, 2010*)

oxytocin intravenously to 15 adults with HF-ASDs improved their comprehension of the emotional intonation of human voices. Finally, nasal administration of oxytocin in 16 youth patients with HF-ASDs led to improved capacity to recognize emotions expressed by the eyes.

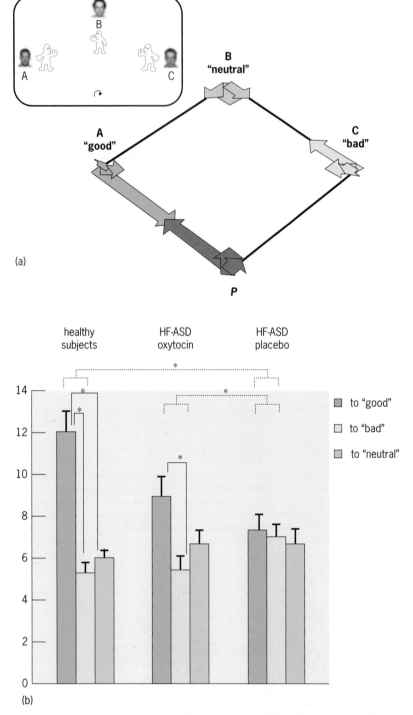

(a)

(b)

Fig. 2. Oxytocin improves social comprehension. (a) Computer ball game showing the amount of reciprocation of each partner toward the participant P (indicated by the tinted arrows). Gray arrows indicate a healthy subject's performance. (b) Number of balls sent toward each partner during the game for healthy subjects, patients under placebo, and patients under oxytocin. An asterisk indicates significant statistical differences. *(Reprinted with permission from E. Andari et al., Promoting social behavior with oxytocin in high-functioning autism spectrum disorders, Proc. Natl. Acad. Sci. USA, 107:4389–4394, 2010)*

However, because oxytocin plays a key role in social engagement and affiliation, and given that social deficits are significantly higher in patients with HF-ASDs, a key question will be whether oxytocin will improve social interaction in those patients. Will oxytocin improve one of the core deficits in autism, namely the lack of eye contact? Will oxytocin modulate social comprehension in these patients?

Oxytocin promotes social comprehension. Research has been carried out to see whether administration of oxytocin can improve the social capacities of patients with HF-ASDs, such as looking toward a face or interacting spontaneously with other people. In one test study, 13 adult patients with HF-ASDs received an intranasal administration of oxytocin (24 IU of synthetic oxytocin spray; three puffs per nostril) or placebo (no treatment). Administration of oxytocin or placebo was randomly assigned and separated by one week. In addition, another group of 13 healthy subjects, who only performed the behavioral experiments, were tested. The patients with HF-ASDs had significantly low plasma oxytocin concentrations, as compared to healthy subjects, before the nasal intake, which was followed by an increase in this hormone at 10 minutes after oxytocin inhalation. Fifty minutes after the nasal intake (the time needed for the molecule to reach the brain), the patients performed two experiments in front of a computer.

First, while patients were asked to detect the gender (male or female) of faces and the gaze direction (direct or averted), their eye movements were recorded by a camera inserted inside the computer. This was done in order to examine how patients looked at faces and to establish which parts of the face (for example, eyes, mouth, or nose) were fixated on more than others. Then, researchers tried to determine how oxytocin affected this perceptual process. The results showed that patients in the placebo condition (without treatment) spent significantly less time looking directly at the faces as compared to healthy subjects. In fact, they looked preferentially outside the face area. Strikingly, patients who were administered oxytocin looked significantly longer at the face region, particularly the eyes, as compared to the placebo condition (**Fig. 1**). No effects of oxytocin were observed with regard to the other regions of interest (mouth, nose, forehead, and cheeks) [Fig. 1]. Hence, oxytocin can promote a first level of prosocial approach by overturning what constitutes a core deficit of patients with HF-ASDs, namely the lack of eye contact. This is an important observation because eye contact between individuals is a basic element of social aptitude.

Second, how do patients interact with others in a social game where some implicit rules must be captured and learned over time? Will oxytocin play a role in the modulation of the social behavior of these patients? In order to investigate this, patients were asked to use a touch-screen computer to play an animated social ball-tossing game with three fictitious partners, who were depicted by cartoon characters and their corresponding photograph. Each time that the participant received the ball, he had to address

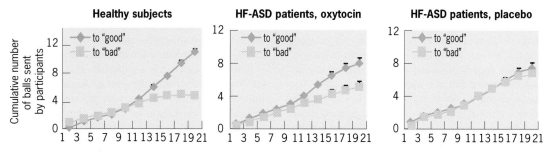

Fig. 3. Time course of ball tosses during the computer ball game. Cumulative number of balls sent by participants (vertical axis) after each trial played by the good partner (horizontal axis). *(Reprinted with permission from E. Andari et al., Promoting social behavior with oxytocin in high-functioning autism spectrum disorders, Proc. Natl. Acad. Sci. USA, 107:4389–4394, 2010)*

the ball to one of the three players by touching the corresponding photograph. Unbeknownst to the participant, the amount of reciprocation exhibited by the three players was predefined. At the start of the game, the participant had an equal probability of receiving the ball from any of the three players. As the game progressed, player profiles diverged such that player A (the good partner) sent 70%, player B (the neutral partner) sent 30%, and player C (the bad partner) sent 10% of its played balls to the participant. Accordingly, healthy subjects sent significantly more balls to the good player than to the bad or neutral player. However, patients under placebo responded in the same manner to all players through all the game. In striking contrast, oxytocin intake led patients to engage more often with the good player and to send significantly more balls to this player as compared to the bad one (**Figs. 2** and **3**).

Interestingly, after oxytocin treatment, patients modified their behavior and responded according to the partner profiles during a time window as healthy subjects did. Following the game, patients estimated their sentiments of trust and preference with respect to the three players using a subjective seven-point rating scale. Whereas feelings expressed toward the three partners did not differ in the placebo condition, patients who received oxytocin reported that they trusted more and showed stronger preference for the good than the bad player. Thus, oxytocin enhanced the ability of patients to process socially relevant cues and acquire their meaning in an interactive context by facilitating learning. These results are consistent with previous studies with animals and healthy subjects that have enlightened the role of oxytocin in social recognition and social memory. Hence, oxytocin improved social interaction in patients with HF-ASDs. Caution though should be exercised in oversimplifying the role of oxytocin in enhancing social engagement. Recent research also indicates that oxytocin may simply amplify affiliative responses, whether positive or negative. For example, a rise in oxytocin was found in women in distressed relationships that was comparable to the rise of the hormone in women in affectionate relationships; and among men who inhaled oxytocin, those who reported being insecure in their current intimate relationship recalled a worse childhood relationship with their mothers, compared to those who reported being more secure in their current relationship.

Nevertheless, the findings on oxytocin's role in enhancing social behavior in patients with autism highlight a therapeutic potential of oxytocin through its action on core deficits, such as social comprehension, emotion recognition, and repetitive behavior, of patients with HF-ASDs. However, these studies were limited to small samples of patients with high-functioning autism and Asperger disorder. Another interesting question concerns the neural mechanisms activated by oxytocin to facilitate a patient's prosocial behavior. It has been proposed that oxytocin enhances affiliation and attachment partly by reducing the levels of stress and anxiety. It is possible that patients with autism possess latent social skills and that oxytocin may thus favor social engagement behavior by suppressing fear and mistrust. More work will be needed to understand the relationship between changes in social behavior induced by oxytocin administration in individuals with autism and local changes in brain oxytocin metabolism. This could be accomplished using functional imaging techniques (among other methods). Finally, an important goal is to investigate whether a long-term intake of oxytocin may improve the functioning of patients in real social life.

For background information *see* AUTISM; COGNITION; DEVELOPMENTAL GENETICS; DEVELOPMENTAL PSYCHOLOGY; HORMONE; LACTATION; NERVOUS SYSTEM (VERTEBRATE); NEUROHYPOPHYSIS HORMONE; NEUROSECRETION; STRESS (PSYCHOLOGY) in the McGraw-Hill Encyclopedia of Science & Technology.

Elissar Andari; Angela Sirigu

Bibliography. E. Andari et al., Promoting social behavior with oxytocin in high-functioning autism spectrum disorders, *Proc. Natl. Acad. Sci. USA*, 107:4389–4394, 2010; E. Hollander et al., Oxytocin increases retention of social cognition in autism, *Biol. Psychiatry*, 61:498–503, 2007; T. R. Insel, The challenge of translation in social neuroscience: A review of oxytocin, vasopressin, and affiliative behavior, *Neuron*, 65:768–778, 2010; C. Modahl et al., Plasma oxytocin levels in autistic children, *Biol. Psychiatry*, 43:270–277, 1998; J. R. Williams et al., Oxytocin administered centrally facilitates formation of a partner preference in female prairie voles (*Microtus ochrogaster*), *J. Neuroendocrinol.*, 6:247–250, 1994.

Paranoia: mechanisms and treatment

Paranoia denotes the unfounded (or exaggerated) fear that others are deliberately trying to harm you. It takes a variety of forms, all the way from the occasional worry that our friends are deliberately trying to irritate us, or are spreading malicious rumors about us behind our backs, to the anguished belief that someone is out to kill us. There are two key elements: the individual believes that harm is going to occur and that the perpetrator has the intention to cause harm. It is the second element—the intention of others to cause harm—that distinguishes perse-

cutory from anxious thoughts. In social anxiety, we fear being foolish; in paranoia, we fear others trying to make us appear foolish. The difficulty can be establishing when the thoughts are unfounded.

Recent research has established that there is a spectrum of paranoia in the general population, with mild suspicions at one end and the sort of severe delusions characteristic of serious mental illnesses like schizophrenia at the other end. Persecutory delusions are defined as such when the thoughts are held with high levels of conviction, interfere in daily life, and are very distressing. Many people have a few paranoid thoughts, and a few people have many (**Fig. 1**). This is similar to common psychological problems, such as a fear of heights, with many people having mild experiences and a few meeting diagnostic criteria for a phobia. Epidemiological and experimental studies indicate that paranoid thinking may be a regular experience in 1 out of 3 people from the general population, and at least 1 person in 20 has a persecutory delusion during his or her lifetime. This high prevalence is unsurprising if paranoia arises from the normal everyday decision-making process about whether to trust or mistrust. Paranoia in the general population has been linked with wealth inequalities, poverty, urbanicity, youth, poor physical health, less perceived social support, stress at work, and low social cohesion. Levels of trust and mistrust may well be markers of the health of a society.

Causes of paranoia. Over the past decade, cognitive psychologists have transformed the understanding of paranoia. (There is notably less research specifically on paranoia from a genetic or biological perspective.) What is clear is that paranoia, like all psychiatric problems, is caused by the complex interaction of several factors (**Fig. 2**).

Importantly, paranoid thoughts are now recognized as interpretations or judgments. They are individuals' explanations for their experiences. The sorts of experiences considered as the proximal sources of evidence for persecutory ideas are as follows:

Internal feelings. Unusual or anomalous experiences are frequently key to paranoid thinking. These include being in a heightened state or aroused; having feelings of significance; perceptual anomalies (for example, things may seem vivid, bright, or piercing; sounds may feel very intrusive); having feelings as if one is not really there (depersonalization); and illusions and hallucinations (for example, hearing voices). These sorts of experiences can be caused by stress, adverse events, insomnia, and illicit drugs (for example, cannabis).

External events. Ambiguous social information is particularly important. This includes both nonverbal information (for example, facial expressions, people's eyes, hand gestures, laughter, or smiling) and verbal information (for example, snatches of conversation or shouting). Coincidences and negative or irritating events also feature in persecutory ideation.

Typically, individuals vulnerable to paranoid thinking try to make sense of internal unusual experiences by drawing in negative, discrepant, or ambiguous

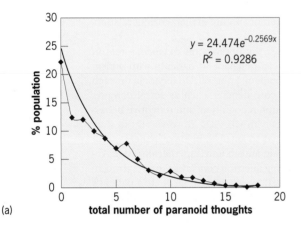

$$y = 24.474e^{-0.2569x}$$
$$R^2 = 0.9286$$

(a)

total number of paranoid thoughts

Self-assessment

How paranoid are you? Have a look at the statements below. Use the scale from 1 (not at all) to 5 (totally) to rate how strongly you agree with each of the statements in the light of your thoughts and feelings *during the last month*.

	Not at all		**Somewhat**		**Totally**
1. Certain individuals have had it in for me	1	2	3	4	5
2. I have definitely been persecuted	1	2	3	4	5
3. People have intended me harm	1	2	3	4	5
4. People wanted me to feel threatened, so they stared at me	1	2	3	4	5
5. I was sure certain people did things in order to annoy me	1	2	3	4	5
6. I was convinced there was a conspiracy against me	1	2	3	4	5
7. I was sure someone wanted to hurt me	1	2	3	4	5
8. I was distressed by people wanting to harm me in some way	1	2	3	4	5
9. I was preoccupied with thoughts of people trying to upset me deliberately	1	2	3	4	5
10. I could not stop thinking about people wanting to confuse me	1	2	3	4	5
11. I was distressed by being persecuted	1	2	3	4	5
12. I was annoyed because others wanted to deliberately upset me	1	2	3	4	5
13. The thought that people were persecuting me played on my mind	1	2	3	4	5
14. It was difficult to stop thinking about people wanting to make me feel bad	1	2	3	4	5
15. People have been hostile towards me on purpose	1	2	3	4	5
16. I was angry that someone wanted to hurt me	1	2	3	4	5

Once you have completed the questionnaire, add up your score. 50% of people score 16 or 17. 80% of people score between 16 and 25. 10% of people score above 34, putting them at the higher end of paranoia. People with severe clinical paranoia generally score between 40 and 70.

(b)

Fig. 1. (*Top*) Survey of paranoid thoughts in the general population. The total number of paranoid thoughts (as endorsed using a self-assessment checklist) is plotted against percentage of the population. The distribution fits a plotted exponential curve, showing that many people have a few paranoid thoughts, whereas a few people have many (*from D. Freeman et al., Psychological investigation of the structure of paranoia in a non-clinical population, Br. J. Psychiatry, 186:427–435, 2005*). (*Bottom*) Self-assessment questionnaire to assess paranoia (*copyright © 2008, reprinted with permission from Cambridge University Press*).

external information. For example, a person may go outside feeling in an unusual state; then, rather than label this experience as such (for example, "I'm feeling a little odd and anxious today, probably because I've not been sleeping well"), the feelings are instead used as a source of evidence, together with the facial expressions of strangers in the street, that there is a threat (for example, "People don't like me and may harm me"). Persecutory thoughts are explanations that contain ideas about physical, social, or psychological threat.

But why a suspicious interpretation of experiences? The internal and external events are interpreted in line with previous experiences, knowledge, emotional state, memories, personality, and decision-making processes, and therefore the origin of persecutory explanations lies in such psychological processes. Suspicious thoughts often occur in the context of emotional distress. They are frequently preceded by stressful events (for example, difficult interpersonal relationships, bullying, or isolation). Furthermore, the stresses may happen against a background of previous experiences that have led the person to have beliefs about the self (for example, as vulnerable), others (for example, as potentially dangerous), and the world (for example, as bad), thereby making suspicious thoughts more likely to occur. These sorts of negative beliefs about the self and others are associated with anxiety and depression, with anxiety being especially important in the generation of persecutory ideation. The theme of anxiety is the anticipation of danger, and it is the origin of the threat content in persecutory ideation. Anxious worry keeps the suspicions in mind and develops the content in a catastrophizing manner, leading to even more implausible ideas. Paranoid ideas are often an extension of anxious concerns about how the person is different, or vulnerable, and how the person could be rejected (**Fig. 3**).

The persecutory ideas are most likely to become of a delusional intensity when there are accompanying biases in reasoning. These include reduced data gathering ("jumping to conclusions"), a failure to generate or consider alternative explanations for experiences, and a strong confirmatory reasoning bias. Social isolation may also contribute to a failure to fully review paranoid thoughts. When reasoning biases are present, the suspicions are more likely to become near certainties; the threat beliefs become held with a conviction unwarranted by the evidence and may then be considered delusional. The fears persist as a result of processes, such as "safety behaviors," actions designed to reduce the threat, but which actually prevent disconfirmatory evidence from being received or fully processed. For example, individuals with severe paranoia may not travel on a bus because of fear of an attack and therefore fail to learn that they were safer than they had realized.

Treatment of paranoia. People with persecutory delusions, the most severe form of paranoia, are generally treated with antipsychotic drugs. However, because these drugs are strong tranquilizers, they can have unpleasant side effects, including

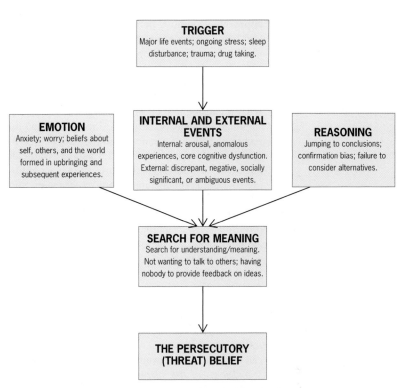

Fig. 2. Outline of factors involved in persecutory delusion development.

feeling disconnected from the world around you, impotence, weight gain, constipation, drowsiness, diabetes, heart problems, and (in rare cases) death. Antidepressants are often prescribed for people whose paranoia is less serious, but who still find that it interferes with their lives. The idea is that the antidepressant will help tackle the emotional factors contributing to the paranoid experiences. The side effects are usually much less troublesome than those of antipsychotics. Increasingly, a psychological

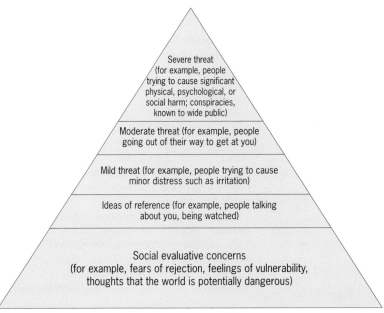

Fig. 3. The paranoia hierarchy. Paranoia may well build upon anxious social evaluative concerns.

approach—cognitive behavioral therapy (CBT)—is also being used and has been found to be effective in clinical trials. Some of the key strategies used in CBT for paranoia are as follows:

1. Understand what causes paranoia. Like all fears, paranoia loses much of its power if we know what is causing it. As part of that process of understanding, individuals are asked to focus in detail on recent episodes of paranoia, and they are helped to identify the role played by such key factors as difficult life events, emotions (particularly anxiety), confusing social encounters, anomalous experiences, and reasoning style.

2. Do not just accept the suspicious thoughts; question them. Paranoid thoughts often seem really compelling and plausible. Instead of taking them at face value, we need to challenge these thoughts, weighing the evidence for and against and seeking out alternative explanations for the way that we are feeling.

3. Put the paranoid thoughts to the test. Nothing disarms a paranoid thought so effectively as actually testing it out. It is not always easy to do, but people can test out and confront their fears in a real way.

4. Let go of the suspicious thoughts. It is unrealistic to think that we can put a complete stop to suspicious thoughts (after all, they have a purpose). However, we can improve the way that we deal with these thoughts when they do occur. The trick is to not focus on them, and instead to develop what is known as a mindful attitude. Individuals are encouraged not to fight the thoughts, but to watch the thought come to them, remind themselves that it does not matter, and let it go off into the distance.

5. Reduce the time spent worrying about paranoid thoughts. Worrying makes everything worse, including paranoia. The spotlight is put on catastrophizing (the "what if?" type of thinking that always foresees the worst possible outcome for any situation), and individuals are shown how to confine fretting to well-defined worry periods.

CBT draws upon the theoretical understanding of paranoia described previously. What unites the techniques is the underlying assumption that individuals' subjective experiences should be taken seriously and that they can be helped to make paranoid experiences less threatening, less interfering, and more controllable. This parallels the approaches taken with problems, such as anxiety and depression. However, CBT for paranoia is a relatively recent development, and clinical researchers are in a process of improving the intervention based upon the theoretical advances.

For background information *see* AFFECTIVE DISORDERS; ANXIETY DISORDERS; COGNITION; EMOTION; PHOBIA; PSYCHOPHARMACOLOGY; PSYCHOSIS; PSYCHOTHERAPY; SCHIZOPHRENIA; STRESS (PSYCHOLOGY) in the McGraw-Hill Encyclopedia of Science & Technology.　　　　　　　　Daniel Freeman

Bibliography. R. Bentall, *Madness Explained*, Penguin, London, 2004; D. Freeman and J. Freeman, *Paranoia: The 21st Century Fear*, Oxford University Press, Oxford, U.K., 2008; D. Freeman and P. A. Garety, Helping patients with paranoid and suspicious thoughts: The cognitive-behavioural approach, *Adv. Psychiatr. Treat.*, 12:404–415, 2006; D. Freeman et al., Concomitants of paranoia in the general population, *Psychol. Med.*, 41:923–936, 2011; D. Freeman, J. Freeman, and P. Garety, *Overcoming Paranoid and Suspicious Thoughts*, Basic Books, New York, 2008.

Piezotronics

Because of the polarization of ions in a crystal that has noncentral symmetry, a piezoelectric potential (piezopotential) is created in the crystal by applying a stress. For materials such as zinc oxide (ZnO), gallium nitride (GaN), and indium nitride (InN) in the wurtzite-structure family, the effect of piezopotential to the transport behavior of charge carriers is significant because of their multiple functionalities of piezoelectricity, semiconductor, and photon excitation. By using the advantages offered by these properties, a few new fields have been created. Electronics fabricated by using inner-crystal piezopotential as a "gate" voltage to tune (control) the charge transport behavior is named piezotronics, with applications in strain/force/pressure triggered (controlled) electronic devices, sensors, and logic units.

Piezoelectric potential. Piezotronics are based on an internal field created inside a crystal as a result of strain. We will use ZnO to elaborate the concept of piezopotential in the wurtzite family. Wurtzite has a hexagonal structure with a large anisotropy; that is, the *c*-axis direction properties are different from those perpendicular to the *c* axis. The lack of center symmetry, which is the core of piezoelectricity, is due to the intrinsic crystallographic structure. Simply, the Zn^{2+} cations and O^{2-} anions are tetrahedrally coordinated and the centers of the positive ions and negatives ions overlap with each other. Therefore, the crystal shows no polarization under strain-free conditions. If a stress is applied at an apex of the tetrahedron, the center of the cations and the center of the anions are relatively displaced, resulting in a dipole moment (**Fig. 1***a*). Summing of the dipole moments created by all of the units in the crystal results in a macroscopic potential drop along the straining direction in the crystal. This is the piezoelectric potential (piezopotential) [Fig. 1*b*]. The piezopotential, an inner potential in the crystal, is created by the nonmobile, nonannihilative ionic charges, and the piezopotential remains in the crystal as long as the stress remains. The magnitude of the piezopotential depends on the density of doping (impurities) and the strain applied.

The distribution of piezopotential in a ZnO nanowire has been calculated using the Lippman theory. For simplicity, we first ignore doping in ZnO, so that it is assumed to be an insulator. For a one-end-fixed free-standing nanowire that is transversely pushed by an external force, the stretched side and the compressed side surfaces exhibit positive and negative piezopotential (Fig. 1*b*), respectively, which can act as a transverse voltage for gating the charge transport along the nanowire. An alternative geometry is a simple two-end-bonded single wire with a length of 1200 nm and a hexagonal side length of 100 nm. When a stretching force of 85 nN is uniformly acting on the nanowire surfaces surrounded by electrodes in the direction parallel to the c axis, it creates a potential drop of approximately 0.4 V between the two end sides of the nanowire, with the +c-axis side of higher potential. When the applied force changes to a compressive, the piezoelectric potential reverses with the potential difference remaining 0.4 V but with the −c-axis side at a higher potential.

Piezotronic effect. The piezotronic effect is about the control (tuning) the charge carrier transport process in a semiconductor by the piezopotential created in the crystal. A better understanding of the piezotronic effect came from comparing it with the most fundamental structures in semiconductor devices: Schottky contact and pn junctions. When a metal and an n-type semiconductor form a contact, a Schottky barrier (SB) ($e\Phi_{SB}$) is created at the interface if the work function of the metal is appreciably larger than the electron affinity of the semiconductor (**Fig. 2a**). Current can only pass through this barrier if the applied external voltage is larger than a threshold value (Φ_i) and its polarity has the metal side positive (for an n-type semiconductor). If a photon excitation is introduced, the newly generated electron-hole pairs not only largely increase the local conductance, but also the effective height of the SB is reduced as a result of charge redistribution (Fig. 2b).

If a strain is created in the semiconductor that also has piezoelectric property, a negative piezopotential at the semiconductor side effectively increases the local SB height to $e\Phi'$ (Fig. 2c), while a positive piezopotential reduces the barrier height. The polarity of the piezopotential is dictated by the direction of the c axis for ZnO. The piezopotential effectively changes the local contact characteristics through an internal field; thus, the charge carrier transport process is tuned (gated) at the metal-semiconductor (M-S) contact. The change in piezopotential polarity can switch the strain from tensile to compressive, so the local contact characteristics can be tuned by the magnitude and the sign of strain. This is the core of piezotronics.

When a p-type and an n-type semiconductor form a junction, the holes in the p-type and the electrons in the n-type tend to redistribute to balance the local potential, and the interdiffusion and recombination of the electrons and holes in the junction region forms a charge depletion zone (Fig. 2d). With the creation of a piezopotential on one side of the semiconductor material under strain, the local band structure near the pn junction is modified. For easy understanding, we include the screening effect of the charge carriers to the piezopotential in the discussion, which means that the positive piezopotential side in the n-type material is largely screened by the electrons, while the negative piezopotential side is almost unaffected. By the same token, the negative piezopotential side in the p-type material is largely screened by the holes, but leaves the posi-

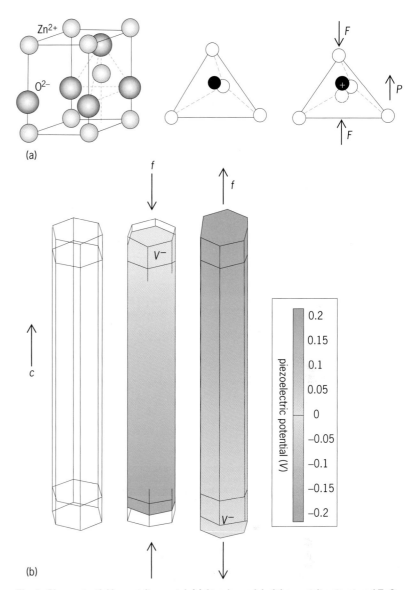

Fig. 1. Piezopotential in wurtzite crystal. (a) Atomic model of the wurtzite-structured ZnO. *F* is the force applied to the top and bottom of a single tetrahedral cation-anion unit, and *P* stands for the corresponding dipole moment created by the ion polarization in the tetrahedron unit. **(b)** Numerical calculation of the piezoelectric potential distribution in a ZnO nanowire under axial strain. The growth direction of the nanowire is along the c axis. *L* is the length of the nanowire, *a* is the side length of the hexagon at the nanowire cross section, and *f* is the force applied on normal to the cross section of the nanowire. *V* is the voltage. The dimensions of the nanowire are *L* = 600 nm and *a* = 25 nm; the external force is f_y = 80 nN.

tive piezopotential side almost unaffected. As shown in Fig. 2e for a case that the p-type side is piezoelectric and a strain is applied, the local band structure is largely changed, which significantly affects the characteristic of the charge carrier flow through the interface. This is the core of the piezotronic effect. The fundamental working principles of the pn junction and the Schottky contact provide an effective barrier that separates the charge carriers across the two sides. The height and width of the barrier are characteristic of the device. In piezotronics, the piezopotential effectively changes the width of pn junction or height of SB by piezoelectricity. In an n-type semiconductor, there is an abundance of electrons, and the density of the holes determines the light-emission intensity. The trapping of holes at the

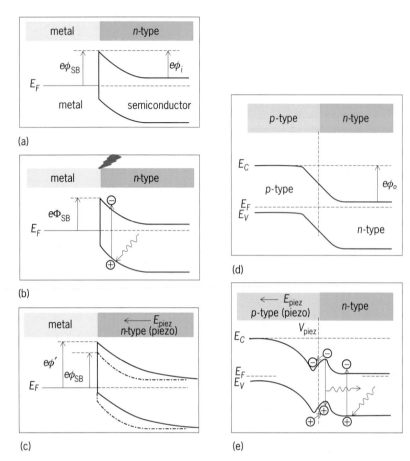

(a)

(b)

(c)

(d)

(e)

Fig. 2. Energy band diagrams to illustrate the effects of laser excitation and piezoelectricity on a Schottky contacted metal-semiconductor interface. (*a*) Band diagram at a Schottky contacted metal-semiconductor interface. (*b*) Band diagram at a Schottky contact after exciting by a laser that has a photon energy higher than the bandgap, which is equivalent to a reduction in the Schottky barrier height. (*c*) Band diagram at the Schottky contact after applying a strain in the semiconductor. The piezopotential created in the semiconductor has a polarity with the end in contact with the metal being low. (*d, e*) Energy band diagrams for illustrating the effect of piezoelectricity on a *pn* junction. (*a–c*) Band diagrams at a conventional *pn* junction made by two semiconductors have almost the same bandgaps. (*d, e*) Band diagrams of the *pn* junction with the presence of a piezopotential at the *p*-type side with a polarity of higher potential at the junction side.

channel near the interface is likely to result in a great enhancement of light-emitting intensity.

Piezotronic transistor. We have shown the piezopotential generated in the wurtzite structures and how to use it to serve as a "gate" voltage for fabricating new electronic devices. One of the most common electronic devices is a single-channel field effect transistor (FET) based on semiconductor nanowires, in which a source and drain are located at the two ends of the device and a gate voltage is applied to the channel and the substrate. For an *n*-channel metal-oxide-semiconductor FET (MOSFET) [**Fig. 3***a*], the two *n*-type doped regions are the drain and source; a thin insulator oxide layer is deposited on the *p*-type region to serve as the gate oxide, on which a metal contact is made as the gate. The current flows from the drain to the source under an applied external voltage V_{DS}, and is controlled by the gate voltage V_G, which controls the channel width for transporting the charge carriers.

The simplest FET is a two-ends-bonded semiconductor wire, in which the two electric contacts at

the ends are the source and drain, and the gate voltage can be applied either at the top of the wire through a gate electrode or at its bottom on the substrate. When a ZnO nanowire is strained axially along its length, two typical effects are observed. One effect is the piezoresistance effect, which is introduced because of the change in the band gap and possibly density of states in the conduction band. This effect has no polarity so that it has an identical effect on the source and drain of the FET. The other effect is that piezopotential is created along its length. For an axially strained nanowire, the piezoelectric potential continuously drops from one side of the nanowire to the other, which means that the electron energy continuously increases from the one side to the other. Meanwhile, the Fermi level (highest energy of the electrons in the material) will be flat over the nanowire when equilibrium is achieved, since there is no external electrical field. As a result, the effective barrier height or width of the electron energy barrier between ZnO and the metal electrode will be raised at one side and lowered at the other side; thus, it has a nonsymmetric effect on the source and drain. This is the piezotronic effect.

The gate voltage of a FET can be replaced by the piezopotential generated inside the crystal (inner potential) [Fig. 3*b*], so that the charge carrier transport process in the FET can be tuned (gated) by applying a stress to the device. This type of device is called a piezotronic device, as it is triggered by mechanical deformation. This structure is different from the complementary metal-oxide semiconductor (CMOS) design, as follows. First, the externally applied gate voltage is replaced by an inner crystal potential generated by the piezoelectric effect; thus, the "gate" electrode is eliminated. This means that the piezotronic transistor only has two leads: drain and source. Secondly, the control over channel width is replaced by a control at the interface. Since the current transported across a metal-semiconductor interface depends exponentially on the local barrier height in the reverse-biased case, the on and off ratio can be rather high because of the nonlinear effect. The voltage-controlled device is replaced by an external strain–stress-controlled device, which is likely to have complementary applications to CMOS devices.

A strain-gated transistor (SGT) or piezotronic transistor is made of a single ZnO nanowire with its two ends, the source and drain electrodes, fixed by silver (Ag) conductive adhesive on a polymer substrate (**Fig. 4***a*). Once the substrate is bent, a tensile (compressive) strain is created in the nanowire since the mechanical behavior of the entire structure is determined by the substrate. Utilizing the piezopotential created inside the nanowire, the gate input for a nanowire SGT is an externally applied strain rather than an electrical signal. I_{DS}-V_{DS} characteristic for each single ZnO-nanowire SGT is obtained as a function of the strain created in the SGT (Fig. 4*a*) before further assembly into logic devices. A nanowire SGT is defined as forward biased if the applied bias is connected to the drain electrode (Fig. 4*a*). For an

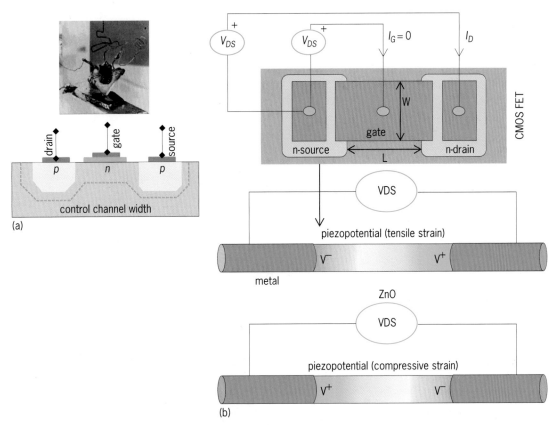

Fig. 3. Schematic of a (*a*) traditional CMOS device with three leads: source, drain, and gate. (*b*) A piezotronic transistor based on two-metal structure, in which the "gate" voltage is replaced by the piezopotential and its polarity depends on the sign of the applied strain. W = the channel width and L = the channel length.

SGT, the external mechanical perturbation induced strain (ε_g) acts as the gate input for controlling the on (off) state of the nanowire SGT. The positive (negative) strain is created when the nanowire is stretched (compressed). The SGT behaves in a similar way to an *n*-channel enhancement-mode MOSFET.

The working principle of an SGT is illustrated by the band structure of the device. A strain-free ZnO nanowire can have Schottky contacts at the two ends, with the source and drain electrodes but with different barrier heights of Φ_S and Φ_D, respectively (Fig. 4*b*). When the drain is forward biased, the quasi-Fermi levels at the source ($E_{F,S}$) and drain ($E_{F,D}$) are different by the value of eV_{bias}, where V_{bias} is the applied bias (Fig. 4*c*). An externally applied mechanical strain (ε_g) results in both band structure change and piezoelectric potential field in a ZnO nanowire. Since ZnO has a polar structure along the *c* axis, straining in the axial direction (*c* axis) creates a polarization of cations and anions in the nanowire growth direction, resulting in a piezopotential drop from V^+ to V^- along the nanowire, which produces an asymmetric change in the SB heights at the drain and source electrodes. Under tensile strain, the SB heights at the source side reduce from Φ_S to $\Phi_S' \cong \Phi_S - \Delta E_P$ (Fig. 4*d*), where ΔE_P denotes the effect from the locally created piezopotential and is a function of the strain, resulting in increased I_{DS}. For the compressively strained SGT, the sign of the piezopotential is reversed. Thus, the SB height at the source side is raised from Φ_S to $\Phi_S'' \cong \Phi_S + \Delta E_P'$ (Fig. 4*e*), where $\Delta E_P'$ denotes the piezopotential effect on the SB height at source side, resulting in a large decrease in I_{DS}. Therefore, as the strain ε_g is swept from compressive to tensile regions, the I_{DS} current can be effectively turned from on to off, while V_{DS} remains constant. This is the fundamental principle of the SGT. All of the logic operations have been achieved using the piezotronic transistor.

Piezotronic diode. A simple piezotronic device is a polarity switchable diode that is made of a ZnO nanowire in contact with metal at the two ends, on an insulator polymer substrate. From the initial current-voltage (*I-V*) curve measured from the device before applying a strain, as shown in **Fig. 5***a*, the symmetric shape of the curve indicates that the SBs present at the two contacts are about equal heights. Under tensile strain, the piezoelectric potential at the right-hand side of this nanowire was lower (Fig. 5*d*), which raised the local barrier height at the right-hand side of the device. Since the positive piezoelectric potential was partially screened by free electrons, the SB height at the left-hand side remained almost unchanged. As a result, under positive bias voltage with the left-hand side positive, the current transport was determined by the reverse biased SB at the right-hand side. However, under the reverse biased voltage with the right-hand side positive, the current transport depends on the reverse biased SB at the left-hand side, which had a much lower barrier

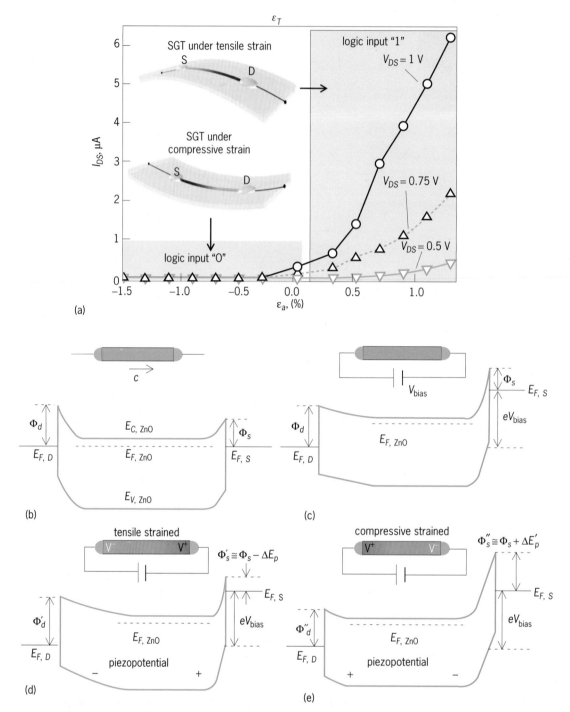

Fig. 4. Strain-gated transistor (SGT) or piezotronic transistor. (a) Current–strain (I_{DS} – ε_g) transfer characteristic for a ZnO SGT device with strain sweeping from $\varepsilon_g = -0.53\%$ to 1.31% at a step of 0.2%, where the V_{DS} bias values were 1, 0.75, and 0.5 V, respectively. (b–e) The band structures of the ZnO nanowire SGT under different conditions for illustrating the mechanism of SGT. The crystallographic c axis of the nanowire directs from drain to source. (b) The band structure of a strain-free ZnO nanowire SGT at equilibrium with different barrier heights of Φ_S and Φ_D at the source and drain electrodes, respectively. (c) The quasi-Fermi levels at the source ($E_{F,S}$) and drain ($E_{F,D}$) of the ZnO SGT are split by the applied bias voltage V_{bias}. (d) With tensile strain applied, the SBH at the source side is reduced from Φ_S to $\Phi'_S \cong \Phi_S - \Delta E_p$. (e) With compressive strain applied, the SBH at the source side is raised from Φ_S to $\Phi''_S \cong \Phi_S + \Delta E'_p$.

height than the right-hand one. Experimentally, the device exhibits rectifying behavior in the positive voltage region, and the I-V curve in the negative voltage region overlaps with that of the original curve without straining. Likewise, under compressive strain the device exhibits rectifying behavior in the negative voltage region (Fig. 5c), and the I-V curve in the positive voltage region overlaps with

that of the original curve without straining, as shown by the "compressed" line in Fig. 5a.

Outlook. Piezopotential is created in a piezoelectric material by applying a stress, and it is generated by the polarization of ions in the crystal. The introduction of this inner potential in semiconductor materials can significantly modify the band structure at a *pn* junction or metal-semiconductor Schottky

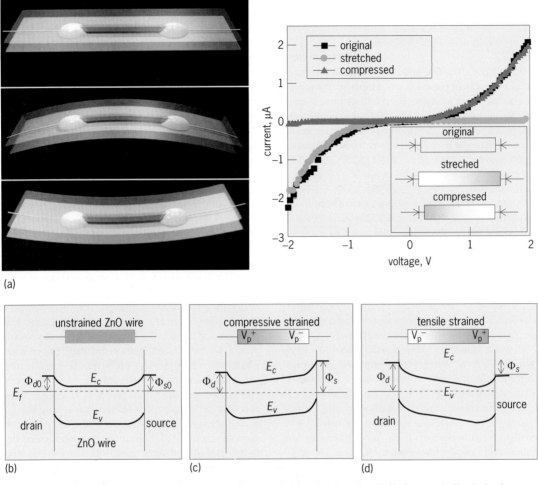

(a)

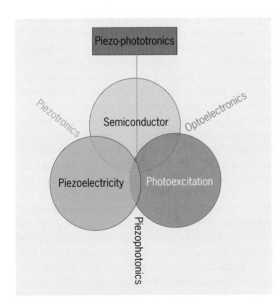

(b)   (c)   (d)

Fig. 5. Piezotronic strain sensor/switch. (*a*) Changes of transport characteristics of an Ag/ZnO-nanowire/Ag device from symmetric *I-V* characteristic (squares) to asymmetric rectifying behavior when stretching (circles) and compressing (triangles) the wire. (*b–d*) Band diagrams for interpreting the observed *I-V* characteristics.

Fig. 6. Schematic diagram showing the three-way coupling among piezoelectricity, photoexcitation, and semiconductor, which is the basis of piezotronics (piezoelectricity-semiconductor coupling), piezophotonics (piezoelectric-photon excitation coupling), optoelectronics, and piezo-phototronics piezoelectricity-semiconductor-photoexcitation). The core of these couplings relies on the piezopotential created by the piezoelectric materials.

barrier, resulting in significant changes in the charge transport properties.

For a material that simultaneously has semiconductor, photon excitation, and piezoelectric properties, besides the well-known coupling of a semiconductor with photon excitation process to form the field of optoelectronics, additional effects could be proposed by coupling a semiconductor with piezoelectrics to form the field of piezotronics, and piezoelectric with photon excitation to form the field of piezophotonics. Furthermore, coupling among semiconductor, photon excitation and piezoelectric provides piezo-phototronics, which can be the basis for fabricating piezo-photonic-electronic nanodevices. The piezo-phototronic effect concerns tuning and controlling of electro-optical processes by strain induced piezopotential (**Fig. 6**). Effective integration of piezotronic and piezo-phototronic devices with silicon-based CMOS technology offers unique applications in areas such as human-computer interfacing, sensing and actuating in nanorobotics, smart and personalized electronic signatures, and smart micro- and nano-electro-mechanical systems.

For background information *see* CRYSTAL STRUCTURE; ELECTRON AFFINITY; ELECTRON-HOLE RECOMBINATION; INTEGRATED CIRCUITS; IONIC CRYSTALS;

PIEZOELECTRICITY; SCHOTTKY BARRIER DIODE; SEMI-CONDUCTOR; SEMICONDUCTOR DIODE; TRANSISTOR; WORK FUNCTION (ELECTRONICS) in the McGraw-Hill Encyclopedia of Science & Technology.

Zhong Lin Wang

Bibliography.   Y. F. Gao and Z. L. Wang, Electrostatic potential in a bent piezoelectric nanowire: the fundamental theory of nanogenerator and nanopiezotronics, *Nano Lett.*, 7:2499–2505, 2007; Z. Y. Gao et al., Effects of piezoelectric potential on the transport characteristics of metal-ZnO nanowire-metal field effect transistor, *J. Appl. Phys.*, 105:113707, 2009; J. H. He et al., Piezoelectric gated diode of a single ZnO nanowire, *Adv. Mater.*, 19:781–784, 2007; Y. F. Hu et al., Optimizing the power output of a ZnO photocell by piezopotential, *ACS Nano*, 4:4220–4224, 2010; T. Rueckes et al., Carbon nanotube-based nonvolatile random access memory for molecular computing, *Science*, 289:94–97, 2000; S. J. Tans, A. R. M. Verschueren, C. Dekker, Room-temperature transistor based on a single carbon nanotube, *Nature*, 393:49–52, 1998; X. D. Wang et al., Piezoelectric field effect transistor and nanoforce sensor based on a single ZnO nanowire, *Nano Lett.*, 6:2768–2772, 2006; Z. L. Wang, Nanopiezotronics, *Adv. Mater.*, 19:889–992, 2007; Z. L. Wang, Piezotronic and piezophototronic effects, *J. Phys. Chem. Letts.*, 1:1388–1393, 2010; Z. L. Wang, Piezopotential gated nanowire devices: piezotronics and piezo-phototronics, *Nano Today*, 5:540, 2010; Z. L. Wang, Towards self-powered nanosystems: from nanogenerators to nanopiezotronics, *Adv. Funct. Mater.*, 18:3553–3567, 2008; Z. L. Wang, ZnO nanowire and nanobelt platform for nanotechnology, *Mater. Sci. Eng. R*, 64:33–71, 2009; Z. L. Wang and J. H. Song, Piezoelectric nanogenerators based on zinc oxide nanowire arrays, *Science*, 312:242–246, 2006; Z. L. Wang et al., Lateral nanowire/nanobelt based nanogenerators, piezotronics and piezo-phototronics, *Mater. Sci. Eng. R*, 70(3-6): 320-329, 2010; W. Z. Wu, Y. G. Wei, and Z. L. Wang, Strain-gated piezotronic logic nanodevices, *Adv. Mater.*, 22:4711–4715, 2010; J. Zhou et al., Flexible piezotronic strain sensor, *Nano Lett.*, 8:3035–3040, 2008; J. Zhou et al., Piezoelectric-potential-controlled polarity-reversible Schottky diodes and switches of ZnO wires, *Nano Lett.* 8:3973–3977, 2008.

Plant hybrid vigor

The continued growth in the world population constantly increases the demand for food, feed, fiber, and fuel, which requires increased levels of growth and productivity of crops and plants. To increase plant productivity, a common agricultural practice is the use of hybrids. Indeed, many crops, such as maize and sorghum, are grown as hybrids, and most crops, including wheat and cotton, are allopolyploids (hybrids formed between species followed by chromosome doubling). Hybrids also occur in animals. As examples, a mule is a hybrid between a horse and a donkey, and many farm and aquatic animals, including chicken, fish, and mussels, are raised as hybrids. The use of hybrid vigor or heterosis in maize (corn, *Zea mays*) has revolutionized plant breeding and production. Since the introduction of hybrid corn in the early 1920s, the yield of corn production in the United States has steadily increased sixfold. To date, about 99% of the corn in the United States is grown from hybrid seed. The same is true for many other crops and vegetables. Thus, what is hybrid vigor? Why do hybrids grow vigorously and produce larger and more fruits?

What is hybrid vigor? Hybrids are formed by cross-fertilizing between different individuals, strains, or species (see **illustration**). The opposite of cross-fertilization is inbreeding, namely, mating between close relatives (for example, siblings or cousins) in animals and within the same individual or genotype (genetic makeup) in plants. Hybrids often grow more vigorously and produce more seeds and offspring than the parents. This phenomenon, known as hybrid vigor or heterosis, is widespread in plants and animals and has intrigued scientists for centuries. After systematically documenting the growth patterns for hundreds of self-fertilized and crossed plants, Charles Darwin concluded that "the crossed plants when fully grown were plainly taller and more vigorous than the self-fertilised ones." In the early 1900s, two American corn breeders, George E. Shull and Edward M. East, discovered that selfing (or inbreeding) maize plants led to a reduction of overall growth vigor and yield, but the hybrids formed between the inbred lines grew more vigorously and produced larger and more numerous seeds.

In contrast, inbreeding or mating among related individuals leads to reduced fitness in a given population, which is a phenomenon known as inbreeding depression. The higher the genetic variation within a population, the less likely it is to suffer from inbreeding depression. It is generally assumed that hybrid vigor is opposite of inbreeding depression.

If the viable hybrids can be made, the degree of hybrid vigor is proportional to the genetic differences between parental strains. In other words, the levels of heterosis increase as the genetic distances between the parents increase. For example, rice hybrids between two subspecies show more heterosis than the hybrids between varieties within a subspecies. This notion may not be generalized across all hybrids. In maize, although inbred lines are genetically similar, the hybrids formed between different combinations of varieties show dramatic levels of heterosis. Heterosis is more predominant in outcrossing species (for example, maize) compared to inbreeding species, and the inbreeding populations do not have obvious heterosis of fitness. It is notable that hybrid vigor may not be a default state because many hybrids are incompatible. The incompatible hybrids are dwarfs and appear stunted, probably because genes and gene products that are highly diverged between the parents fail to compromise or function properly in the hybrids.

Although it may take many years to develop hybrid seeds, plant breeders have proven that hybrid

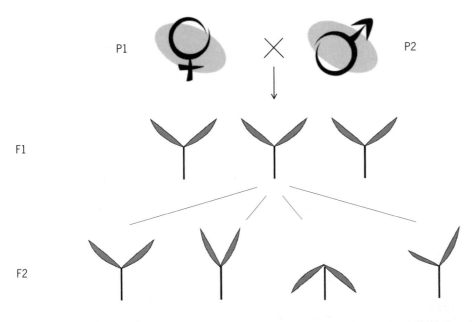

P1   ×   P2

F1

F2

A hybrid (middle) and its parents in *Arabidopsis thaliana*   An allotetraploid

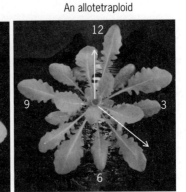

(*Top*) Diagrams of F1 hybrids and a segregating population in F2. An F1 individual is produced by crossing pollinating female parent 1 (P1) with pollen from male parent 2 (P2). Note that a reciprocal hybrid can be made by crossing pollinating P2 with pollen from P1. Seeds from the uniform F1 plants would generate a segregating population with variable phenotypes in the next generation (F2). (*Bottom*) A hybrid plant (middle) was produced between *Arabidopsis thaliana* Columbia (left) and C24 (right). An allotetraploid plant was produced between *A. thaliana* and its related species (*A. arenosa*). The numbers and arrows indicate the clock effect on growth vigor in hybrids and allopolyploids. *Arabidopsis* is a weedy plant in the mustard family, which is closely related to cabbage and oilseed rape (or canola).

benefits are worth the extra seed cost, whether used for small- or large-scale farming. Numerous commercial seed companies take advantage of hybrid vigor, and hybrid seeds are major business portfolios in many biotech companies. There are several reasons. First, hybrids have the vigor. Hybrids can result in more vigorous plants, better-quality vegetables, superior flavor, improved disease resistance, earlier maturity, and yield increases ranging from 30% to 100% compared to their open pollinated counterparts. Second, hybrids are uniform because their genetic composition is the same. Uniformity is suitable for large-scale farming, including machine operation and harvest, but it is not attractive to home gardening. Finally, it is profitable because of the high price for hybrid seeds. Hybrid genetics also favors commercial development because the uniformity and hybrid vigor in F1 (the first filial generation) will disappear in the next generation as a result of genetic segregation; thus, hybrid seeds cannot be saved for next-year crops (see illustration).

In addition to regular hybrids that are formed between strains or individuals, hybrids can be formed between species or genera. A synthetic species called triticale is a hybrid between wheat (*Triticum*) and rye (*Secale*). Estimates indicate that approximately 10% of animal species and approximately 25% of plant species hybridize with at least one other species. Some plant species such as wild sunflower exist as hybrids. In other plants, chromosomes in interspecific hybrids can be doubled to produce allopolyploids. Polyploids can be classified as autopolyploids (chromosome doubling) and allopolyploids (chromosome doubling coupled with interspecific hybridization). Many crops and fruit trees, including potato, sugarcane, and banana, are autopolyploids. Other important crops, including wheat, oats, cotton, and oilseed rape, are allopolyploids. Doubling of chromosomes through hybridization increases genetic potential for selection and adaptation. For example, tetraploid wheat is suitable for making spaghetti and pastry products because of the high glutenin content, but it is unsuitable for making bread. Therefore, introducing the genes for baking quality from another species into hexaploid wheat makes it into bread wheat. This process of interspecific hybridization and polyploidization occurs naturally and is ongoing, giving rise to the diversity of plant species through natural selection and crop domestication.

Why do hybrids and allopolyploids show vigor? The genetic basis for hybrid vigor or heterosis is largely unknown, but several hypotheses are available to explain hybrid vigor. Two genetic models, dominance and overdominance, have been debated for more than a century. An F1 hybrid inherits alleles from both parents. An allele is one member of a pair of genes that occupy a specific position (called the genetic locus) on a specific chromosome. A dominant allele has a corresponding phenotype (for example, flower color) in F1 hybrids, whereas a recessive allele does not produce a characteristic effect when present with a dominant allele. Both dominant and recessive alleles are transmitted to the offspring following the genetic principles discovered by Gregor Mendel. According to the dominance model, inbred parents contain some inferior or recessive alleles for genetic loci that inhibit overall superior performance; however, in the hybrids, these inferior alleles in one parent are complemented by the superior or dominant alleles from the other parent. As a result, the hybrids have an overall better performance than the parents. In theory, the parent that contains both superior alleles for all possible genetic loci would perform better than the hybrids, but this has not been shown in corn breeding. In spite of dramatic improvement of inbred parents by eliminating inferior alleles, the hybrids always perform better than the parents.

The overdominance model suggests that the superior performance of F1 hybrids is caused by the combination of different alleles in the hybrids, a condition called heterozygosity. Heterozygous allele combinations in the hybrids lead to superior function over the homozygous states in the inbred parents. This model is favored because hybrids always outperform the parents, despite the fact that the parents have been excessively inbred and selected and contain many superior genetic loci. A challenge for this model is to discriminate which heterozygous allele combination determines the levels of hybrid vigor. Using these genetic models, statistical methods can be applied to identify quantitative trait loci (QTLs) that are associated with a specific agronomic trait (for example, biomass or grain yield). Note that QTL refers to the location of a gene that affects a quantitative trait.

However, neither the dominance model nor the overdominance model explains allelic interactions among multiple loci, a genetic condition known as epistasis, which plays important roles in the determination of complex agronomic traits such as grain yield. Analysis of QTLs has shown that hybrid vigor in rice is associated with three different effects, namely, dominance, overdominance, and epistasis. Because of these complications, some scientists suggest the

abandonment of these terms that constrain data interpretation and instead desire to establish a quantitative genetic framework involving interactions in hierarchical biological networks.

At the molecular levels, both dominance and overdominance models can be explained by expression changes of the alleles in the hybrids relative to the parents. If gene expression were additive, the expression level of a gene in the hybrid would be 2 (1 + 1). If gene expression were nonadditive, the expression level of a gene in the hybrid would be larger or smaller than 2, suggesting allelic interactions (epistasis) and other changes in gene expression that cannot be explained by Mendelian genetics—hence, the term epigenetics. Both genetic and epigenetic changes in gene expression can lead to quantitative and nonadditive variation of phenotypic variation that is manifested in the hybrids. A molecular model of heterosis suggests that expression alterations of regulatory genes modulate expression of housekeeping (constitutive) genes and output traits in the hybrids. As a result, nonadditive expression of many genes collectively in various biological pathways gives rise to superior vegetative growth and higher yield. This model may also explain hybrid vigor in different stages of growth and development. For example, heterosis in biomass, such as vigorous growth in seedlings, roots, and other vegetative tissues, may not be directly translated into large fruits or seeds because different sets of genes in the biological pathways control vegetative growth and reproductive development, although some pathways are intricately related.

One such model involves circadian clock (daily rhythm) regulators. Most living organisms, including plants and humans, have adapted to 24-hour day-night cycles as a result of the Earth's rotation. Plants are sessile organisms and derive their energy from the Sun. In plants, synchronizing the internal clock period with the external light–dark cycle increases CO_2 fixation, chlorophyll and starch content, and survival rates, whereas disrupting circadian rhythms reduces growth and fitness. In the hybrids, the expression of clock genes is epigenetically modified and fine-tuned to allow high expression of downstream genes in biological pathways including photosynthesis and starch metabolism. As a result, the entire network is reset at a higher amplitude in the hybrids, increasing the flux of chlorophyll biosynthesis and starch metabolism. This provides a general mechanism for growth vigor and increased biomass in the hybrids that are produced within and between species.

Plant growth and development is also affected by flowering time. Late flowering often increases vegetative growth and biomass. In tomato, a gene related to flowering time affects flower and fruit numbers; in the heterozygous state, fruit yield is increased. A similar gene in sunflower is positively selected with an increase in seed size during sunflower domestication.

Gene expression changes in hybrids, especially in the interspecific hybrids, can be extended to include small RNA molecules that are present, absent, or at different levels in the respective hybridizing parents. The chaotic situation resulting from genomic hybridization is known as "genome shock." The "genome shock" may provoke a balance of small RNA molecules as well as metabolic pathways. Some small RNAs, known as small interfering RNAs (or siRNAs), suppress the mobile genes jumping from one location to another in the genome. Because of divergence between species or hybridizing parents, the siRNA pools are different and cannot suppress these unstable elements in the genome, which is predicted to cause changes in gene expression and genome stability in the hybrids. Other small RNA molecules such as microRNAs (miRNAs) can directly modulate transcript levels of the target genes that are important to growth and development in plants and animals. Many hybrids cannot survive probably because of genome instability and regulatory incompatibility that are associated with an imbalance of small RNAs or gene products (or both). In other hybrids, small RNAs, gene expression, and metabolic networks are reprogrammed to overcome deleterious effects, leading to growth vigor and increased fitness.

Can hybrid vigor be fixed? It is of practical importance to avoid making hybrid seeds every year, namely, to fix hybrid vigor. To prevent genetic segregation in F1 hybrids, one possibility is to produce seeds without fertilization; this process is known as apomixis. Apomixis occurs naturally in some plant species, and many of these species are polyploids. The genes responsible for apomixis have yet to be cloned and characterized. An alternative approach is to genetically engineer the plants that fail in meiosis and produce 2n gametes. Crossing this type of plant with another plant that induces haploid production would produce clones that carry the same genetic information as the female parent. Alternatively, F1 embryos can be produced as artificial seeds by tissue culture. Except for the production of embryos, this is similar to cutting plants through vegetative propagation. However, at present stage, producing these types of seeds would cost more than making hybrid seeds.

The hybrid vigor in interspecific hybrids can be permanently fixed through chromosome doubling to produce genetically stable allopolyploids; this is the natural process of polyploidization that has created many plants, including wheat, cotton, and canola. In the allopolyploids, because heterozygosity is maximized, hybrids between different varieties of allopolyploid wheat or cotton show little vigor. However, levels of hybrid vigor increase in the hybrids formed between distantly related allopolyploid species (for example, Upland and Pima cotton).

In conclusion, recent advances in the study of plant hybrid vigor and polyploids will offer a genetic solution to increased demand for productivity in plants and animals. The naturally occurring process of hybridization and chromosome doubling can be exploited to make new hybrid and polyploid species with increased levels of productivity. DNA and gene expression markers can be developed to select, predict, and make the best hybrids with superior performance. Finally, the genes and small RNAs that affect growth vigor in hybrids and polyploids can be used to improve production of many other crops without making hybrids.

For background information *see* AGRICULTURAL SCIENCE (PLANT); ALLELE; BIOTECHNOLOGY; BREEDING (PLANT); CHROMOSOME; DOMINANCE; GENE; GENOMICS; HETEROSIS; MENDELISM; POLYPLOIDY; TRITICALE in the McGraw-Hill Encyclopedia of Science & Technology.   Z. Jeffrey Chen

Bibliography. J. A. Birchler et al., Heterosis, *Plant Cell*, 22(7):2105–2112, 2010; Z. J. Chen, Molecular mechanisms of polyploidy and hybrid vigor, *Trends Plant Sci.*, 15:57–71, 2010; J. F. Crow, 90 years ago: The beginning of hybrid maize, *Genetics*, 148(3):923–928, 1998; C. R. Darwin, *The Effects of Cross and Self Fertilisation in the Vegetable Kingdom*, John Murray, London, 1876; M. P. Marimuthu et al., Synthetic clonal reproduction through seeds, *Science*, 331(6019):876, 2011; B. McClintock, The significance of responses of the genome to challenge, *Science*, 226:792–801, 1984.

Plant reproductive incompatibility

Most flowering plants have perfect (bisexual) flowers in which female (pistil) and male (anther) reproductive structures are found in close physical proximity, an arrangement that favors self-fertilization (**Fig. 1**). Self-fertilization, in which egg and sperm derived from the same plant fuse to produce viable offspring, may be advantageous in certain situations, such as when environmental conditions are stable or when mates or pollinators are scarce. However, self-fertilization can be disadvantageous under variable and unpredictable environments because the genetically identical offspring that are produced exhibit low genetic diversity and low capacity for adaptation. Consequently, flowering plants have evolved several mechanisms that promote outcrossing and allow them to avoid the potentially deleterious consequences of inbreeding. Among these, the most prevalent and best-understood genetic barriers to self-fertilization are the plant reproductive incompatibility systems known as self-incompatibility (SI). SI, which occurs in more than half of the approximately 250,000 species of flowering plants, is defined as the inability of plants having functional female and male gametes to self-fertilize and set seed in the absence of pollinators. SI acts as a prefertilization barrier that allows cells of the pistil to discriminate between "self" (incompatible) pollen (that is, pollen derived from the same flower, the same plant, or genetically related plants) and "cross" (compatible) pollen (that is, pollen derived from genetically unrelated plants of the same species). As a consequence of this discrimination, the germination of "self" pollen and elongation of "self" pollen tubes into the stigma, style, or ovary are inhibited. As a result, the sperm cells that

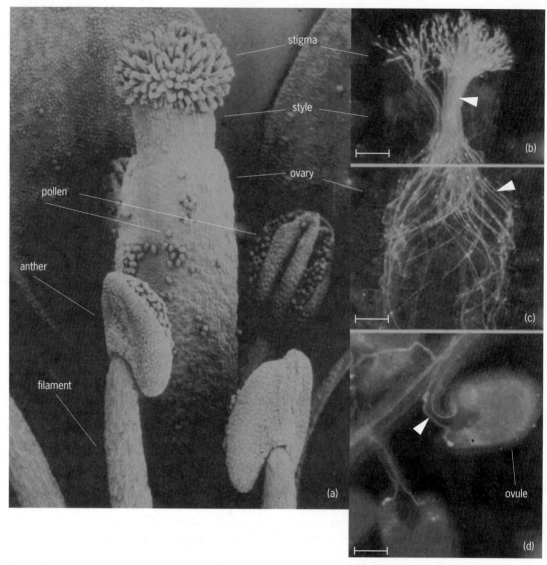

Fig. 1. Reproductive structures and the path of pollen tube growth in the crucifer flower. (*a*) A scanning electron micrograph of an *Arabidopsis thaliana* flower, showing the pistil and stamens. The pistil is subdivided into the apical stigma (which receives and screens pollen), the style (through which pollen tubes grow), and the ovary (which houses the ovules). The stamen consists of an apical anther, in which pollen grains develop, held up by a stalklike filament. Several mature pollen grains have been shed from the anthers. (*b, c, d*) The path of pollen tube growth in the *A. thaliana* pistil. The fluorescence of pollen tubes (arrowheads) is the result of binding of an aniline blue stain to polysaccharides in the tube wall. Pollen grains land on the stigma epidermal cells and, in a compatible pollination, they form pollen tubes that are guided down through subepidermal regions of the stigma and style (*b*) and into the ovary (*c*). In the ovary, a pollen tube is guided toward an ovule (*d*), into which it discharges its two sperm cells, which subsequently effect double fertilization. Scale bars: 100 μm (*b, c*); 20 μm (*d*).

are carried by the pollen tube never reach the ovule and fail to effect fertilization.

The occurrence of SI has been known for over two centuries and was most prominently highlighted by Charles Darwin. However, it was only in the first half of the twentieth century that scientists explained the genetic basis of SI. In many (but not all) plant families, it was noted that the specific recognition of "self" pollen is determined by a single highly polymorphic genetic locus, which was called the *S* (*sterility*) locus, and that "self" and "nonself" refer, respectively, to genetic identity and nonidentity at this locus. Subsequently, in the 1980s, molecular analyses of SI were initiated and culminated during the last decade in identification of the molecules that de-

termine how "self" pollen is specifically recognized by the pistil. It is now evident that, despite similar outcomes (that is, inhibition of self-pollination), the mechanism by which "self" pollen is recognized and inhibited can vary dramatically among plant families. Thus, the term "self-incompatibility" refers to molecularly diverse mechanisms for preventing self-fertilization that evolved independently during the diversification of flowering plants.

Different molecular mechanisms for inhibition of self-pollination. The pollen–pistil interactions that underlie SI are classical examples of cell–cell communication processes. To date, the mechanisms by which cells of the pistil discriminate between "self" and "nonself" pollen have been elucidated in the crucifer family [Brassicaceae, which includes cabbage, broccoli, and oilseed rape (canola)], the poppy family (Papaveraceae), the nightshade family (Solanaceae, which includes tobacco, tomato, potato, and petunia), the rose family (Rosaceae, which includes fruit trees), and the snapdragon family (Scrophulariaceae, which includes snapdragon). In all of these families, the *S* locus is highly polymorphic, with up to 100 different variants found in some species. These *S*-locus variants may be thought of as mating types that determine the outcome of pollination: inhibition, whenever pistil and pollen express the same variant (that is, in self-pollination); or success, whenever they express different variants (that is, in cross-pollination). In all cases, the *S* locus was found to contain two highly polymorphic "self-recognition" genes, one of which functions in the pistil and the other in pollen. These genes are inherited as a single genetic unit (that is, they do not typically segregate away from one another through the process of meiosis), and it is their allele-specific interaction that is responsible for the specific recognition of "self" pollen by cells of the pistil. This interaction triggers activation of a cellular response in the pistil or pollen (depending on the SI system), which culminates in inhibition of pollen tube development. Thus, a distinctive feature of SI systems is that, unlike other recognition phenomena, such as mammalian immunology, which are typically based on recognition and rejection of "nonself," they are based on the recognition and rejection of "self".

Receptor–ligand interactions at the stigma surface in crucifer SI. In crucifers, SI operates during the interaction between a pollen grain and a stigma epidermal cell, and a "self" pollen grain typically fails to hydrate or germinate and pollen tubes never grow into the pistil. Consistent with this rapid response, recognition of "self" pollen in this family is based on the activity of cell surface–localized receptors and ligands encoded by two *S*-locus genes: (1) the *S*-locus receptor kinase (SRK), a single-pass transmembrane protein kinase (an enzyme that exerts regulatory effects on cellular functions by addition of a phosphate group to proteins) localized in the plasma membrane of the stigma epidermal cell; and (2) the *S*-locus cysteine-rich protein (SCR), a small peptide that is located in the pollen coat and functions as the ligand for the SRK receptor (**Fig. 2**). When a pollen grain lands on a stigma epidermal cell, its SCR peptide is transferred to the stigma surface and transported across the cell wall, allowing for SRK–SCR interactions to take place. The *SRK* and *SCR* genes are highly polymorphic, and alleles of these genes can share as little as 40% and 20% overall amino-acid sequence similarity, respectively. As a result, an SRK variant is only bound and activated by the SCR encoded in the same *S*-locus variant. Thus, the SRK–SCR receptor–ligand system may be likened to the many odorant receptors found in animals: just as these receptors can "sniff out," and become activated by, specific odorant ligands, the SRK receptors can "sniff out," and become activated by, specific SCR ligands on pollen grains.

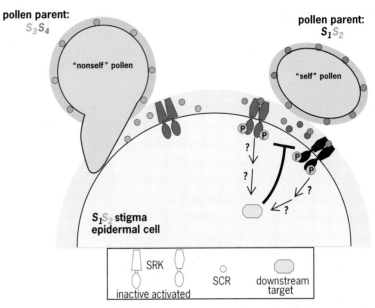

Fig. 2. Receptor-mediated recognition of "self" in crucifer SI. The diagram shows an S_1S_2 stigma epidermal cell interacting with a "self" pollen grain derived from an S_1S_2 plant and a "nonself" pollen grain derived from an S_3S_4 plant. The SI self-recognition molecules, that is, the stigma transmembrane SRK receptor and its pollen coat-localized SCR ligand, are shown. SRK and SCR molecules that are encoded in the same *S*-locus variant are shown in the same shade/color. Pollen grains are shown to display two SCR variants because self-incompatible plants are typically heterozygous at the *S* locus and SCR is produced by tapetal cells, which are derived from diploid cells of the anther; thus, assuming codominance of *S* haplotypes, the haploid pollen grains produced by an S_1S_2 plant will be phenotypically S_1S_2, despite being genotypically S_1 or S_2. SRK is shown as forming ligand-independent homodimers, which are maintained in an inactive state in the absence of "self" pollen. Allele-specific binding of the SRK extracellular domain to its cognate SCR ligand activates the receptor [shown by addition of a phosphate (P)] and triggers a poorly understood signaling cascade (arrows and question marks) within the stigma epidermal cell that leads to the inhibition of the "self" pollen grain. In contrast, the SCR derived from a "nonself" pollen grain neither binds nor activates SRK, and it can therefore germinate and produce tubes. For simplicity, several postulated components of the signaling cascade are not shown.

This allele-specific SRK–SCR interaction activates the receptor's kinase and initiates a signaling cascade within the stigma epidermal cell that culminates in pollen inhibition (Fig. 2). What biochemical events are triggered by the SRK–SCR interaction and how they lead to the inhibition of "self" pollen remain largely unknown. It is possible that these events use molecules that function in other plant signaling pathways. This possibility is suggested by the fact that the self-fertile model plant *Arabidopsis thaliana*, which evolved from a self-incompatible ancestor approximately 5 million years ago, can be reverted to a self-incompatible mode of mating by simply transforming it with an *SRK–SCR* gene pair from its self-incompatible close relative *A. lyrata*. This complementation experiment demonstrates that the SI signaling pathway has been maintained in *A. thaliana*, possibly because it is critical for survival.

Inhibition by pollen tube cell death. In contrast to the SI system of crucifers, the other molecularly studied SI systems are characterized by inhibition of pollen tubes after they have grown into subepidermal tissues of the pistil. By necessity, arrest of "self" pollen tube growth is associated with death of pollen tubes in these systems.

In the poppy (*Papaver rhoeas*), death of "self" pollen tubes occurs shortly after penetration into

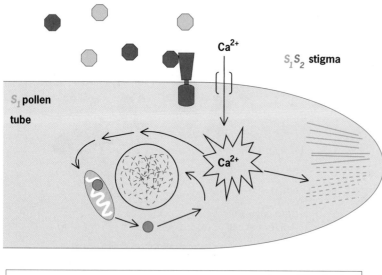

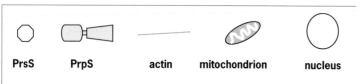

Fig. 3. Programmed cell death of "self" pollen tubes in poppy. An S_1 ("self") pollen tube is shown growing in an S_1S_2 stigma. The two PrsS glycoprotein variants secreted by diploid cells of the stigma accumulate in the extracellular matrix through which the pollen tubes grow. The haploid pollen tube expresses one PrpS variant (shown straddling the plasma membrane). PrsS interacts with the PrpS encoded in the same S-locus variant, and this interaction somehow initiates opening of calcium channels in the membrane of "self" pollen tubes and triggers a series of cellular responses that cause death of "self" pollen tubes. The simplified diagram omits several of the molecules known to contribute to programmed cell death in pollen tubes. Molecules that are encoded in the same S-locus variant are shown in the same shade/color.

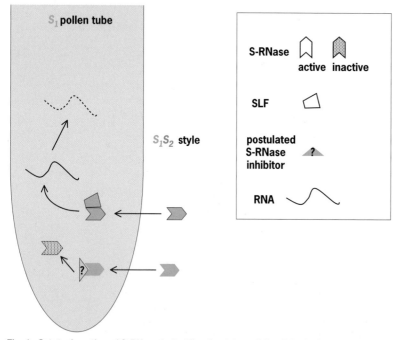

Fig. 4. Cytotoxic action of S-RNase in "self" pollen tubes of the nightshade, rose, and snapdragon families. The diagram shows a "self" S_1 pollen tube growing through the style of an S_1S_2 plant. The two S-RNase variants accumulate in the extracellular matrix of the transmitting tract. They are taken up nonspecifically by pollen tubes, where they interact with their cognate SLF. Each pollen tube expresses an SLF variant encoded in its own haploid genotype. The very simplified model shown here postulates the existence of an unknown general S-RNase inhibitor in pollen tubes, which would bind and inactivate S-RNases by either sequestering them or causing their degradation. Allele-specific interaction between S-RNase and its cognate SLF would prevent binding of this inhibitor, thus maintaining the S-RNase in an active form that degrades cellular RNA. Molecules that are encoded in the same S-locus variant are shown in the same shade/color.

the stigma. In this species, the two S-locus molecules that determine SI specificity are (1) a small secreted glycoprotein expressed in stigmas and called the PrsS protein (for *P. rhoeas* stigma *S* determinant) and (2) a plasma-membrane associated PrpS protein (for *P. rhoeas* pollen *S*) expressed in the pollen tube (**Fig. 3**). Allele-specific interaction between these two proteins triggers a rapid influx of calcium and a signaling cascade within the incompatible pollen tube that results in disruption of the actin cytoskeleton and culminates in loss of mitochondrial integrity, DNA fragmentation, and other hallmarks of programmed cell death (Fig. 3).

In the nightshade, rose, and snapdragon families, "self" pollen tubes stop elongating and their tips burst within the upper third of the style. In these plants, inhibition of pollen tube growth is effected by the S-RNase (S-locus ribonuclease), an abundant and highly polymorphic glycoprotein that is secreted by cells of the style into the extracellular matrix that lines the path of pollen tube growth. Despite being taken up by "self" and "nonself" pollen tubes alike, the S-RNase inhibits "self" pollen tubes by degrading their cellular RNAs, but does not affect "nonself" pollen tubes (**Fig. 4**). How S-RNases effect this specific inhibition of "self" pollen tubes is still a matter of debate. A clue derives from the nature of the pollen S locus–encoded partner of the S-RNase, *SLF* (S-locus F-box), which belongs to the F-box protein family, whose members are known to function in ubiquitin-mediated proteolysis. Thus, SLF is thought to target "self" S-RNase (that is, the S-RNase encoded by the same S-locus variant) for proteolysis. A current hypothesis invokes the involvement of an as yet unknown S-RNase inhibitor that inhibits all S-RNases by binding to their active site (Fig. 4). The interaction of S-RNase with its cognate SLF would somehow prevent binding of this inhibitor, allowing the protected S-RNase to degrade RNA.

Conclusions. Molecular analysis of SI in a few plant families has revealed the nature of the highly polymorphic pistil and pollen molecules encoded in the S locus and demonstrated that the specific recognition of "self" pollen by cells of the pistil is based on allele-specific interaction of these molecules. Current work is focused on genetic and biochemical studies aimed at explaining exactly how the recognition event determined by interaction of the S-locus molecules translates into inhibition of pollen germination or tube growth. A detailed mechanistic understanding of these processes is expected to facilitate the manipulation of pollination responses for crop improvement and the development of SI-based schemes for hybrid seed production on a commercial scale.

For background information *see* AGRICULTURAL SCIENCE (PLANT); ALLELE; FERTILIZATION (PLANT); FLOWER; GENE; GENETICS; PLANT ORGANS; PLANT PHYSIOLOGY; PLANT REPRODUCTION; POLLEN; POLLINATION in the McGraw-Hill Encyclopedia of Science & Technology.   June B. Nasrallah

Bibliography.   C. Darwin, *The Effects of Cross and Self Fertilisation in the Vegetable Kingdom*, John Murray, London, 1876; B. McClure, Darwin's foundation for investigating self-incompatibility and the progress toward a physiological model for S-RNAs-based SI, *J. Exp. Bot.*, 60:1069–1081, 2009; T. Tantikanjana, M. E. Nasrallah, and J. B. Nasrallah, Complex networks of self-incompatibility signaling in the Brassicaceae, *Curr. Opin. Plant Biol.*, 13:520–526, 2010; M. J. Wheeler, S. Vatovec, and V. E. Franklin-Tong, The pollen *S*-determinant in *Papaver*: Comparisons with known plant receptors and protein ligand partners, *J. Exp. Bot.*, 61:2015–2025, 2010.

Plant root hairs as a model cell system

Plant root hair cells [root epidermal cells characterized by polar growth (**Fig. 1***a*)] are emerging as a single cell experimental system by which to study plant cell biology at a systems level. This simple model is very attractive for investigating the response of a single plant cell type to various abiotic and biotic stresses. The integration of biological data collected using high-throughput technologies from isolated root hair cells and the use of innovative bioinformatic tools represent unique opportunities for understanding plant cell complexity.

Plant root hair cells: a multifunctional plant cell. The distinctive elongation of root hair cells (Fig. 1*a*) increases the surface exchange between the plant's root system and the soil environment (that is, rhizosphere). The main function of root hairs is to improve water and nutrient uptake. Consistent with this role, transcriptomic and proteomic analysis of root hairs has demonstrated the presence of a variety of water channels and ion transporters (for example, nitrogen, phosphate, calcium, potassium, and sulfate transporters). In legumes, root hairs are the site of infection by mutualistic symbiotic bacteria (that is, rhizobia). Infection leads to the development of nodules, which are the organs where bacteria fix and assimilate atmospheric nitrogen for the plant (Fig. 1*b*). Root hair cells, therefore, represent an excellent model for investigating fundamental plant–microbe interactions. Additionally, by virtue of their function and location, root hair cells are a good model for understanding how plants adapt to changes in the rhizosphere, including modifications to the chemical composition of the soil as a result of anthropogenic or microbial origins (for example, modification of pH, nutrient availability, soil salinity, and concentrations of heavy metals), fluctuations in temperature, and changes in water potential. Root hair cells can also be exploited to study plant cell determination, differentiation, elongation, polar growth, and biosynthesis, including the complex signaling cascades associated with these processes. Hence, root hair cells represent an extremely attractive model for probing fundamental questions of plant biology at the level of a single cell type.

Establishment of plant root hair cells as a model for systems biology. A fuller picture of plant root hair cell biology and responses to environmental changes will require the integration of multiple data sets into

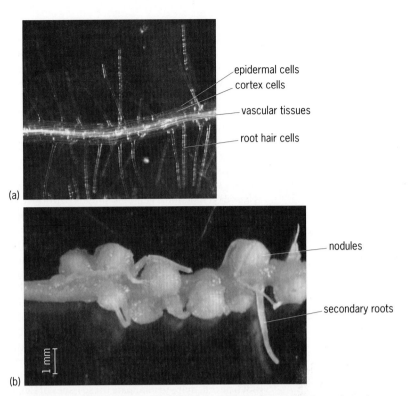

(a)

(b)

epidermal cells
cortex cells
vascular tissues
root hair cells

nodules

secondary roots

1 mm

Fig. 1. Polar growth elongation of root hair cells allows an increase of the surface of exchange between the plant root system and the rhizosphere to improve water and nutrient uptake (*a*). In legumes, root hair cells are also the first site of infection by mutualistic symbiotic bacteria. Several days after this infection, a new root organ named the nodule will emerge (*b*). The nodule is infected by differentiated bacteria that are beneficial to the plant as a result of their fixing of atmospheric dinitrogen.

a single model with the ultimate goal of being able to predict specific cellular responses (for example, in response to environmental changes). Systems biology can be defined as "a comprehensive, quantitative analysis of the manner in which all the components of a biological system interact functionally over time." A biological system can be understood at different levels of organization and complexity: ecosystem, organism, tissue, organ, and cell type. At the level of a single cell type, a systems biology approach is dependent on three conditions. First, it requires the ability to isolate large quantities of the purified single cell type in a way that is compatible with the application of modern, high-throughput technologies. Second, large, comprehensive data sets need to be developed using these tools derived from well-designed and controlled experiments. There are an increasing number of high-throughput tools that can be applied, resulting in an ever expanding and more detailed view of cellular function. These include not only transcriptomic, proteomic, and metabolomic studies, but also analyses of small RNA and various epigenetic features (for example, cytosine methylation, histone modification, and so on). Third, and perhaps most important, bioinformatic tools are essential in order to analyze, integrate, and visualize the multiple data sets generated. Indeed, of these three factors, the availability of bioinformatic approaches to truly integrate dissimilar data sets and infer new biological understanding is clearly the most challenging. The ultimate goal is to develop models that would allow accurate predictions of cellular function in response to a variety of treatments and environmental changes. The hope is that developing this detailed understanding of a single plant cell type will lead the way to ultimate understanding of plant tissues, organs, whole plants, and ecosystem interactions.

One of the greatest goals in plant systems biology is to draw a clear picture of the contribution of each plant cell type to the overall plant response to stress. Currently, our understanding of plant biology is based on functional genomic studies performed at the level of the whole plant, tissue, or organ. Such analyses, although essential and important, are inadequate for complete understanding of the contributions of different cell types that compose each organ, tissue, or plant to the biological response. Most functional genomic data derived from whole plants or tissues actually reflect the average response of all the cells within the specific plant or tissue. For example, it is impossible to distinguish between a gene that is expressed at a low level in all cells from one that is expressed at a high level but only in a few cells. Analysis of the response of single cells, such as the plant root hair cell, helps to eliminate these issues of signal dilution. A number of methods have been developed to conduct functional genomic studies of single cells (for example, laser microdissection). However, many of these techniques are limited to the application of only a few of the current high-throughput methods. In addition to root hair cells, other studies have been conducted on isolated trichomes (single cell protrusions from leaves) and germinating pollen. However, the largest data sets currently available have been derived from soybean root hair cells (**Fig. 2**).

Most of our knowledge of root hair cell biology is largely the result of recent developments in molecular, genetic, physiological, and microscopic tools. For example, the key regulatory genes involved in the determination and differentiation of root hair cells in the model plant *Arabidopsis thaliana* were characterized through the identification of mutants defective in root hair cell development. These and other forward-genetic studies have clearly established root hair cells as a reference for investigating cell–cell communication and positional effects on cellular fate and differentiation. Unlike most plant cells, root hair cells expand through polar elongation, making them particularly attractive to examine using different microscopic methods (for example, bright-field, fluorescent, and electron microscopy).

Soybean root hair cells are attractive for study because, given the size of the roots, root hair cells can be easily isolated in quantity using a very simple freezing method. Soybean roots are frozen in liquid nitrogen and then gently stirred. This mild shearing breaks the root hair cells and they can be subsequently isolated by filtration through a wire screen. The root hair cells remain frozen until analyses are conducted (Fig. 2). The recent publication of the complete soybean genome sequence has made possible the application of a variety of modern, high-throughput functional genomic approaches to study soybean root hair biology. For example, use of high-throughput complementary DNA (cDNA) sequencing methods was recently used to develop a gene expression atlas for soybeans involving multiple soybean organs and cells, including root hair cells. Similar approaches are now in progress to use high-throughput sequencing technologies to define key features of the soybean root hair epigenome (for example, chromosomal cytosine methylation), which plays a consequential role in the regulation of gene expression. Protein databases derived from the soybean genome sequence also support the application of proteomic methods to the study of soybean root hair cells. Likewise, the use of mass spectrometry is allowing metabolomic studies to be performed using this single cell model system. As discussed previously, the availability of bioinformatic tools to store data, as well as to analyze and visualize data, is essential. To serve this purpose, the Soybean Knowledge Base has been established. Although still under development, this database is already well stocked with data derived from studies of soybean root hair cells.

Perspectives and conclusions. Future challenges are awaiting plant biologists interested in systems biology. Given the complexity of living organisms, studies of well-defined single cell model systems provide a good starting point to develop large data sets and to overcome the issues of data integration to allow real advances in understanding. However, it is important to remember that a plant cell is composed of multiple well-defined compartments (for example, nucleus, mitochondria, amyloplast, and vacuole) that interact. Compartmentalization strongly increases the complexity of a eukaryotic cell. Thus, in addition to generating accurate quantitative and qualitative data sets, it is essential to also understand the distribution of transcripts, proteins, metabolites,

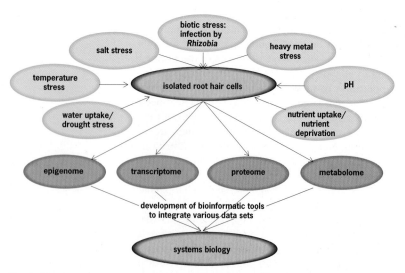

Fig. 2. Soybean root hair cells are a good model to investigate plant cell response to various abiotic and biotic stresses. Their isolation in gram quantities is compatible with the use of various "-omics" (for example, epigenomic, transcriptomic, proteomic, and metabolomic analyses). These different approaches must be integrated using bioinformatic tools to establish root hair cells as a system for plant biologists.

and ions within these compartments. This brings the important issue of structure, both molecular and cellular, into the discussion. It is only through the full integration of all aspects of cellular structure and function that a full understanding of cell biology is possible. These efforts will ultimately lead to the establishment of a systems biology model that can be used to understand and predict plant cell response. Clearly, these developments will require the continuing improvement and application of advanced bioinformatic tools. Given the effects of various biotic and abiotic stresses on plant productivity, it is important to attain this level of understanding. Any improvements in the adaptability of plants to varying biotic and abiotic stresses will help us to address some pressing societal challenges, including the feeding of a growing worldwide population while also meeting a growing demand for feed, fiber, and fuel in the midst of climate change.

For background information *see* CELL BIOLOGY; EPIDERMIS (PLANT); GENE; GENOMICS; NITROGEN FIXATION; PLANT MINERAL NUTRITION; PLANT PHYSIOLOGY; RHIZOSPHERE; ROOT (BOTANY); SOYBEAN in the McGraw-Hill Encyclopedia of Science & Technology.

Marc Libault; Gary Stacey

Bibliography. A. Aderem, Systems biology: Its practice and challenges, *Cell*, 121:511–513, 2005; L. Brechenmacher et al., Soybean metabolites regulated in root hairs in response to the symbiotic bacterium *Bradyrhizobium japonicum*, *Plant Physiol.*, 153:1808–1822, 2010; M. Libault et al., An integrated transcriptome atlas of the crop model *Glycine max*, and its use in comparative analyses in plants, *Plant J.*, 63:86–99, 2010; T. Nelson et al., Laser microdissection of plant tissue: What you see is what you get, *Annu. Rev. Plant Biol.*, 57:181–201, 2006; D. R. Page and U. Grossniklaus, The art and design of genetic screens: *Arabidopsis thaliana*, *Nat. Rev. Genet.*, 3:124–136, 2002; J. Schmutz et al., Genome sequence of the palaeopolyploid soybean, *Nature*, 463:178–183, 2010.

PML: promyelocytic leukemia protein

The promyelocytic leukemia protein (PML) is found in all mammalian cells that have been investigated, and the biomedical importance of PML arises from its role as a tumor suppressor. As such, PML plays roles in the control of cell proliferation, the suppression of oncogenic transformation, the promotion of apoptosis (programmed cell death), and cellular senescence. The PML protein is disrupted in 98% of patients with acute promyelocytic leukemia (APL) and is targeted by many different types of viruses, including human immunodeficiency virus (HIV) and lymphocytic choriomeningitis virus (LCMV), as a means to evade cellular antiviral responses. PML forms nuclear bodies (see **illustration**), and it is this ability to form these bodies that is linked to most of PML's biological functions. Many different biochemical activities have been proposed to underlie these processes.

PML protein. The PML protein is encoded by the *PML* gene on chromosome 15 (15q24.1) in humans. The protein is found in all mammalian cell types studied to date, but it has not been found in yeast, *Drosophila*, or plants, suggesting that PML, and its associated nuclear domains, may be specific to higher eukaryotes. There are several isoforms of PML that arise as a result of alternative splicing of the C-terminus. Isoforms can be nuclear- or cytoplasmic-specific (where some of these isoforms do not contain the conserved nuclear localization signal), although most isoforms are predominantly nuclear.

One of the defining features of the PML protein is a specific N-terminal domain known as the RBCC. The RBCC motif, which is found in all PML isoforms reported to date, contains a "really interesting new gene" (RING) type of zinc-binding domain, two related zinc-binding domains known as B-boxes, and a leucine coiled coil. The integrity of the RBCC motif in PML is required for the formation of PML nuclear bodies and is essential for most of the biological functions of PML.

The PML protein is subject to a wide variety of posttranslational modifications, including phosphorylation, acetylation, and sumoylation [addition of small ubiquitin-like modifier (SUMO) proteins]. Its role in sumoylation is postulated to be a key factor in the formation of PML nuclear bodies.

PML nuclear bodies. PML is distributed throughout the cell, but the vast majority of PML is located in spherical structures in the nucleus commonly referred to as PML nuclear bodies. Other names for these subnuclear structures include PML oncogenic domains (PODs), nuclear dot 10s (ND10s), and Kremer bodies. Although these structures contain many other proteins, PML is commonly used as their marker. On average, the interphase nucleus contains approximately 10–30 nuclear bodies ranging in size from 0.2 to 1 μm (0.000008 to 0.00004 in.)

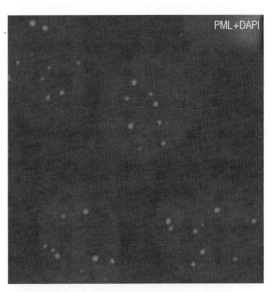

Immunofluorescence of the PML protein from human cancer cells (FaDu cells) as observed by confocal microscopy. DAPI (4′,6-diamidino-2-phenylindole) is used as a fluorescent nuclear stain. Magnification is 200×.

[see illustration]. These structures are attached to a nuclear substructure known as the nuclear matrix.

PML nuclear bodies are multiprotein complexes that are heterogeneous in composition relative to different cell types and cell cycle phases; they are even heterogeneous within the same interphase nucleus. This strongly suggests that the functions of PML nuclear bodies may fluctuate, depending on cellular and temporal contexts. Furthermore, there may be functional subtypes of PML nuclear bodies that coexist.

The position of PML nuclear bodies is distinct from other nuclear structures, but it is often found near Cajal bodies (small subnuclear membrane-less organelles) and splicing speckles (nuclear domains that contain components used for splicing), both of which are involved in RNA processing. Newly synthesized RNA is often found adjacent to PML nuclear bodies. However, RNA polymerase and U1 and U2 small nuclear RNAs (snRNAs) are absent from these bodies, suggesting that PML bodies are not the sites of active transcription or splicing; these events, though, could be occurring near or adjacent to the surface of PML nuclear bodies. Importantly, PML nuclear bodies are mobile and dynamic. For example, using real-time microscopy, PML nuclear bodies can be observed to divide or coalesce. These motions are dependent on adenosine triphosphate (ATP), but they appear independent of ongoing transcription.

How does PML form nuclear bodies? Electron microscopic studies have shown that the RING domain of PML is sufficient for the formation of bodies similar in size and shape to those observed in cells. These bodies can recruit other partner proteins, including eukaryotic translation initiation factor 4E (eIF4E, which is a protein encoded by the *eIF4E* oncogene), to the PML RING structures. Interestingly, these RING bodies appear to form catalytically active surfaces, enhancing the biochemical activity of PML with regard to its eIF4E target.

In vivo, the association of PML with other partner proteins is thought to be SUMO-dependent. Some groups consider PML as the key organizing component of the nuclear bodies and, certainly, some proteins do not form bodies in the absence of PML, as seen with PML knockout cells. However, SP100 (a PML body component) has been observed to form bodies in PML$^{-/-}$ cells, and eIF4E nuclear bodies are readily visible in PML$^{-/-}$ cells. In addition, cells lacking SP100 do not have PML nuclear bodies, suggesting that PML is not the most fundamental component for nuclear body formation.

PML in the nucleoplasm and cytoplasm. PML has been observed throughout the nucleoplasm in what is known as a diffuse pattern. Additionally, PML is found in cytoplasmic structures, and many different roles have been suggested for PML in the cytoplasm.

Biological effects of PML. The PML protein and PML nuclear bodies are often used interchangeably, and most studies cannot make a distinction between the function of the PML protein and the function of the PML nuclear body. Overexpression of the PML protein, and thus formation of larger and more numerous PML nuclear bodies, is associated with several biological phenomena, including suppression of tumor growth and promotion of apoptosis; hence, this protein is a tumor suppressor. Furthermore, PML and the integrity of PML nuclear bodies are important for the antiviral response of the cell. Therefore, any loss of the normal function of PML may in part contribute to a variety of disorders, which are discussed below.

Biochemical activities of PML. To date, PML has been directly implicated in two biochemical processes, as well as being indirectly implicated in many other pathways. First, PML directly binds the oncogene *eIF4E*, leading to a conformational change in the eIF4E protein. This conformation change inhibits the ability of eIF4E to bind and thereby export its RNA targets from the nucleus, suppressing the expression of these messenger RNAs (mRNAs). This is directly linked with the ability of PML to suppress eIF4E-mediated oncogenic transformation. The other biochemical activity of PML is observed when PML directly binds the ubiquitin-conjugating enzyme Ubc9, which is involved in SUMO conjugation. This results in increased sumoylation of the PML protein. It is postulated that PML promotes the sumoylation of other proteins, but this has yet to be established.

Aside from the aforementioned direct activities, PML has been implicated in numerous cellular processes. Functions are often deduced on the basis of association of specific protein partners for PML. To date, more than 70 proteins have been suggested to associate with PML nuclear bodies. Some of the processes derived from associated partners include DNA repair, transcription, and sites of sumoylation. In the cytoplasm, PML has been suggested to interact with many factors, including eIF4E, transforming growth factor (TGF), and p53 (a multifunctional tumor suppressor protein). As is evident, there is much to accomplish in the PML field prior to proving and understanding all relevant interactions.

Context-specific interactions: DNA repair, DNA replication, and mitosis. In replicating cells (middle to late S phase of the cell cycle), approximately 50% of PML is found adjacent to, but not overlapping with, DNA replication domains. In approximately 5% of telomerase-negative cells, PML is found associated with telomere-binding proteins and together with replication factor A, Rad52 (a protein important for repair of DNA breaks), and DNA repeats. These special nuclear bodies are referred to as alternative lengthening of telomeres (ALT)–associated PML bodies (APBs). PML also is associated with mitotic associations of PML proteins (MAPPs), which are structures distinct from PML nuclear bodies because of their content and dynamics. Disappearance of MAPPs coincides with the reappearance of PML bodies during the G$_1$ phase; thus, it is proposed that MAPPs serve as storage locations for PML protein during mitosis. These are all excellent examples of the importance of context with regard to PML nuclear body function; that is, PML function can vary depending on the functional state or condition of the cell.

PML knockout mice. Knockout of the PML gene in mouse models does not have a substantial effect on mouse development, and the knockout mice do not develop substantially more spontaneous tumors compared to littermate controls. These data suggest that there is significant regulatory redundancy in the pathways that are regulated by PML.

There are some phenotypes observed with the PML knockout mice. For example, PML$^{-/-}$ cells are more easily infected with some viruses compared to littermate controls, and PML$^{-/-}$ mice get stress-induced cancers more readily than littermate controls.

PML and APL. PML is translocated in approximately 98% of APL cases with the retinoic acid receptor alpha (RARA), leading to production of PML–RARA fusion proteins. It is this observation that has driven much of PML research. Treatment of APL patients with retinoic acid leads to remissions in most patients with the PML-RARA translocation. The PML-RARA fusion protein acts as a transcription repressor of retinoic acid–sensitive targets, leading to a block in differentiation. However, retinoic acid binds the PML-RARA fusion protein, leading to degradation of the PML-RARA fusion protein and return of normal transcriptional activity. These patients still contain a normal PML allele (only one allele is translocated), and thus normal cellular functioning can now occur. This treatment has turned APL from a devastating disease into a very treatable one.

PML in virus infections. PML nuclear bodies are targeted by a wide variety of DNA and RNA viruses. Different viruses use different strategies to target PML, but these are all likely means to evade host cell apoptosis. For example, the arenavirus LCMV moves PML nuclear bodies to the cytoplasm. In contrast, other viruses disrupt PML nuclear bodies; for example, herpes simplex viruses 1 and 2 (HSV-1 and HSV-2) induce proteasome-dependent degradation of PML protein and alternative splicing of PML pre-mRNA, resulting in enrichment of a cytoplasmic PML isoform. In addition, other viruses can reorganize PML nuclear bodies into different structures [such as elongated nuclear tracks (adenoviruses) or large dense aggregates (rabies virus)] or they can promote PML desumoylation (cytomegalovirus).

PML in neurological disorders. In a number of inherited neurodegenerative disorders, PML nuclear bodies are substantially enlarged, and the PML protein colocalizes with nuclear aggregates [nuclear inclusions (NIs)] formed by mutant polyglutamine-containing proteins. These features are characteristics found in Huntington's disease, dentatorubral-pallidoluysian atrophy (DRPLA) [a progressive brain disorder], and spinocerebellar ataxia (SCA) types 1, 2, 3, 7, and 17, which are caused by unstable expansion of CAG repeats in gene coding regions. However, it is not clear if PML is involved in formation or degradation of NIs.

For background information see APOPTOSIS; CANCER (MEDICINE); CELL (BIOLOGY); CELL CYCLE; DEOXYRIBONUCLEIC ACID (DNA); LEUKEMIA; ONCOGENES; ONCOLOGY; PROTEIN; RIBONUCLEIC ACID (RNA); TUMOR SUPPRESSOR GENES; TUMOR VIRUSES in the McGraw-Hill Encyclopedia of Science & Technology.   Katherine L. B. Borden; Biljana Culjkovic-Kraljacic

Bibliography. K. L. Borden, Pondering the promyelocytic leukemia protein (PML) puzzle: Possible functions for PML nuclear bodies, *Mol. Cell. Biol.*, 22:5259–5269, 2002; K. L. Borden, Pondering the puzzle of PML (promyelocytic leukemia) nuclear bodies: Can we fit the pieces together using an RNA regulon?, *Biochim. Biophys. Acta*, 1783:2145–2154, 2008; T. G. Hofmann and H. Will, Body language: The function of PML nuclear bodies in apoptosis regulation, *Cell Death Differ.*, 10:1290–1299, 2003; V. Lallemand-Breitenbach and H. de Thé, PML nuclear bodies, *Cold Spring Harbor Perspect. Biol.*, 2(5):a000661, 2010; G. G. Maul et al., Review: Properties and assembly mechanisms of ND10, PML bodies, or PODs, *J. Struct. Biol.*, 129:278–287, 2000.

Polymers from renewable resources

The biological activities associated with the animal and vegetable realms produce incessantly an immense variety of simple and complex and large and small molecular structures, which include macromolecular architectures that play fundamental roles in different aspects of life's requirements. These roles include, among others, basic genetic features such as DNA and RNA, essential and precise biological mechanisms such as the vast array of proteins, energy sources such as starch, and actual materials that ensure specific mechanical properties such as cellulose, lignin, and chitin or other functions, as with natural rubber. The importance of the multitude of small biological molecules is equally essential in all facets of the chemistry of life, including their role as precursors (monomers) to all the above polymers. In the present context, the term "polymers from renewable resources" refers to macromolecular materials derived, more or less directly, from natural compounds, constantly renewed, ultimately thanks to solar energy. It belongs therefore to a wider scientific and technological domain of polymers as materials, which were born around the mid-nineteenth century and boomed spectacularly a century later because of the rapid development of petroleum, gas, and coal chemistry (that is, an industrial activity based on nonrenewable fossil resources). A more colloquial but scientifically flawed definition for "polymers from renewable resources" is simply "making plastics using vegetable or animal sources," as an alternative to those based on fossil counterparts. It is particularly instructive to note that the very first macromolecular materials produced on an industrial scale were in fact derived from renewable resources, namely, cellulose esters, vulcanized natural rubber for tires and other commodities, and linoleum from vegetable oils, all still in widespread use today. The subsequent progress in the chemistry of both the synthesis of monomers from fossil resources and their polymerization processes, which culminated in the petrochemistry boom after World

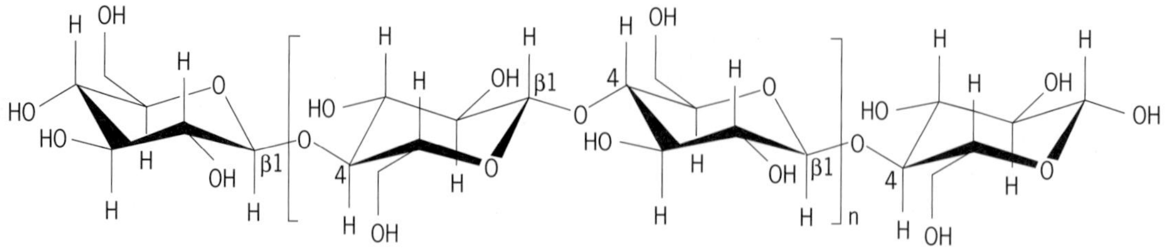

Fig. 1. Structure of cellulose, where *n* ranges from hundreds to thousands.

War II, spurred the polymer revolution that provided society with a startling range of new materials, from routine daily commodities to key components in the manufacturing of high-tech devices. At the beginning of the twenty-first century, the rising cost of petroleum and the realization that the availability of fossil resources was dwindling accelerated the interest in research on polymers from renewable resources. Within the first decade of the twenty-first century, this involvement increased exponentially in both the academic and industrial sectors. This article illustrates this burgeoning situation through a limited number of relevant examples, covering promising research and practical realizations.

Materials from polysaccharides. Of all the polysaccharides (polymers made up of sugar-type monomer units), cellulose is the most abundant natural polymer on Earth. Cellulose constitutes the basic structural element of woods and annual plants, where its fibers associate with lignin to generate a composite assembly that provides the essential mechanical properties of plants. Cellulose is a highly regular linear polymer (**Fig. 1**), whose macromolecules organize themselves in ordered bundles (microfibrils), which, in turn, join together to form the larger fibers (**Fig. 2**).

The interactions that provide the high cohesive energy within and among these filaments are the hydrogen bonds between the hydroxyl (OH) groups borne by each chain, resulting in two rather peculiar features: insolubility in most common solvents and the infusibility of cellulose. Papermaking exploits these features, whereby a thin disordered mat of cellulose fibers possesses the familiar strength of a sheet of paper. To turn these fibers into a thermoplastic ma-

terial, chemists in the 1850s converted a variable proportion of the hydroxyl groups into ester moieties, thus reducing, correspondingly, the tenacity of cellulose and creating the first plastic materials, which can be processed from a melt or a solution.

Research on cellulose esters has gained new momentum in recent years because of novel chemical approaches, providing more viable processes and a family of novel polymers. Other innovations have opened the way to original materials based on the exploitation of nanotechnology applied to cellulose fibers. The basic idea is a top-down dismembering of the fibers to isolate the highly crystalline fibrillar assemblies, which are a few nanometers thick and 100 nm in length (Fig. 2). These nanoscopic objects are exploited in various kinds of materials, including their incorporation in modest proportions into polymer matrices to prepare high-strength composites, the preparation of "nanopapers" with exceptional toughness, and transparent films with high gas-barrier properties. This field is presently in full expansion.

Starch is another widespread vegetable polysaccharide, which plays the role of an energy reservoir in the shape of granules. Its less regular macromolecular structure compared to cellulose (starch is composed of both linear and branched polymers) makes it a suitable source of thermoplastic materials after appropriate processing and plasticization. This has attracted considerable attention because of the renewable character of starch, its low price, its ubiquitous availability, albeit from different species, and its biocompatibility and biodegradability. Applications in packaging and short-life-cycle objects will be extended to other domains once the problem of reducing its moisture affinity is solved. Although starch from wheat, rice, potatoes, and other sources is a fundamental staple, its use as a precursor to polymer materials does not represent a potential interference with food availability because of the largely different quantities associated with the two uses, with the former representing only a very modest percentage of the latter. Moreover, starches from some tubers are not suitable for human consumption.

Chitin, the third important member of the polysaccharides, is the only one found in the animal realm, in the form of exoskeletons of crustaceans and insects, as a film or thicker coating. The key structural difference between chitin and the vegetable polysaccharide is the presence of an amide function replacing one of the three OH groups found in each monomer unit of cellulose and starch. This

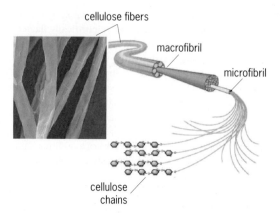

Fig. 2. Hierarchical morphology of a plant cellulose fiber.

cellulose fibers

macrofibril

microfibril

cellulose chains

substantial modification is associated with the higher cohesive energy of chitin, which is a rather intractable polymer in terms of processability because of its lack of solubility and the fact that, like cellulose, it decomposes before melting. It is, however, possible to convert chitin into chitosan by a simple hydrolytic process that converts the amide function into a primary amine group (NH_2), and this leads to a very interesting polymer in that it is readily solubility in mildly acidic aqueous media, making it easily processable and open to a multitude of interesting chemical modifications. Furthermore, the presence of the amino groups gives chitosan an additional positive feature, namely, its biocide property. In addition, its exceptional film-forming properties and biodegradability have generated enormous interest in the study and application of chitosan-based materials. Remarkable results have been achieved, and continue to flourish, in the biomedical, pharmaceutical, cosmetic, papermaking, and food-processing fields, as well as in wastewater treatment and more sophisticated technologies involving composites, nanoparticles, and nanocapsules, among others.

Materials from vegetable oils. Although vegetable oils have been used as components of paints and varnishes for centuries, new applications are steadily emerging because of a worldwide interest in novel chemical approaches aimed at converting them into a much wider family of materials. Plant oils bear the common structure of triglycerides, that is, glycerol esters of fatty acids with long aliphatic chains made up of 12 to 22 carbon atoms (**Fig. 3**, where R_1, R_2, and R_3 represent these chains). The large variety of fatty acids in nature results in a correspondingly rich choice of chemical peculiarities, which include mostly the number and position of alkenyl (C=C) unsaturations along their chains, but also other functions, such as hydroxyl groups; hence, the possibility of a number of specific chemical modifications that give rise to different properties of the ensuing polymers. The number of novel materials is thus increasing steadily, including polyurethanes, polyesters, photosensitive macromolecules for surface coating, and thermoreversible (that is, readily recyclable) polymers.

Materials from sugars. Whereas materials from polysaccharides require the modification of natural polymers and those from vegetable oils the chain extension of oligomers (low-molecular-weight polymers), sugars and their derivatives represent actual

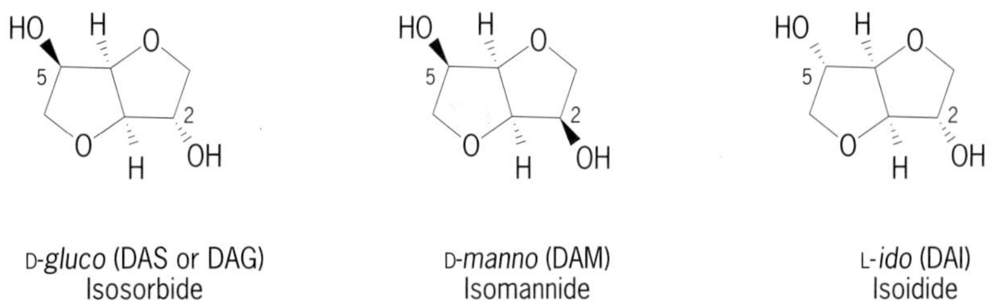

Fig. 4. Three dianhydroalditols, of which isosorbide is the most representative.

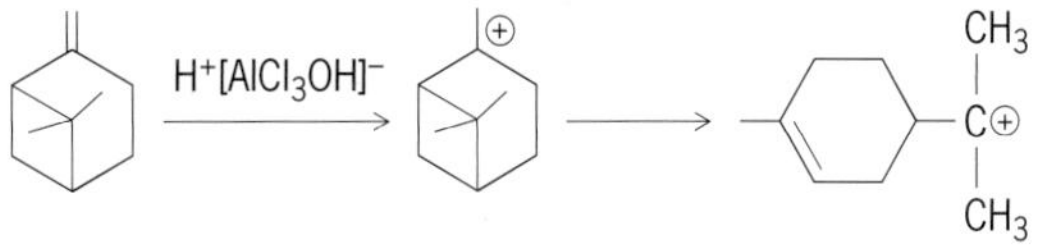

D-*gluco* (DAS or DAG)
Isosorbide

D-*manno* (DAM)
Isomannide

L-*ido* (DAI)
Isoidide

Fig. 5. Initiation step of the cationic polymerization of β-pinene.

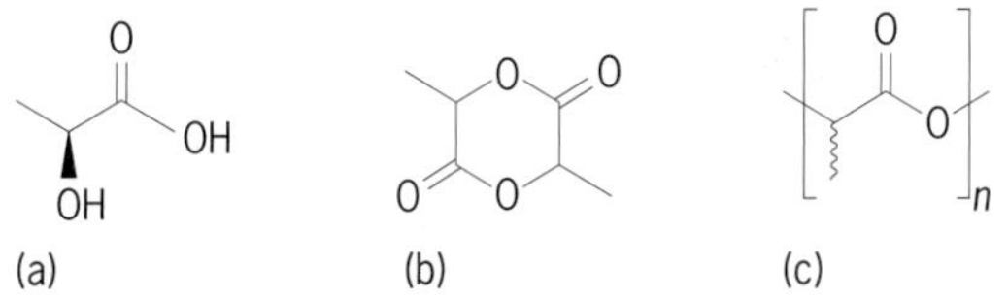

Fig. 6. Structure of (*a*) lactic acid, (*b*) lactide, and (*c*) poly(lactic acid).

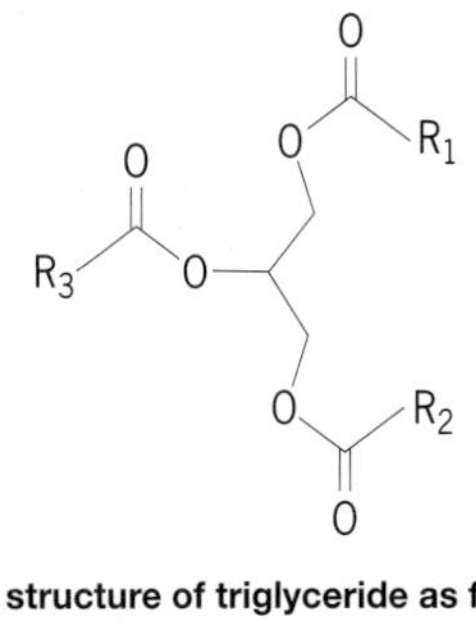

monomers that must be polymerized to generate macromolecular materials. Sugar molecules are rich in OH groups, and if linear polymers are desired from them, only two of these groups should be activated during chain growth. This can be accomplished by various chemical means, and polyurethanes, polyesters, and polyamides with similar or better properties have thus been prepared. Of particular interest in this context are the dianhydroalditols (**Fig. 4**), bifunctional sugar-based monomers obtained by the intramolecular dehydration of common sugars. Isosorbide (Fig. 4) is an industrial commodity that has received a great deal of attention as a co-monomer in the synthesis of polyesters, polyurethanes, and other materials. The stiffness of the isosorbide unit in the macromolecule imparts useful features, such as a high softening temperature. Depending on the complementary monomer used to build these novel materials, often another molecule derived from renewable resources, their properties can be tailored to satisfy a wide range of applications.

Materials from pine resin. Many conifer trees exude a resin that has been exploited for millennia for numerous applications, such as making wooden ships waterproof. The standard treatment of the resin calls for separation of the volatile fraction from the residue, with each fraction containing a complex mixture of homologous compounds, namely, the terpenes (unsaturated hydrocarbons made up of two isoprene units, with the general formula $C_{10}H_{16}$) and the resin acids (diterpenic monocarboxylic acids, $C_{19}H_{29}COOH$, bearing different molecular architectures). Terpenes are fractionated as individual compounds for various applications, particularly in

Fig. 3. Generic structure of triglyceride as found in vegetable oils.

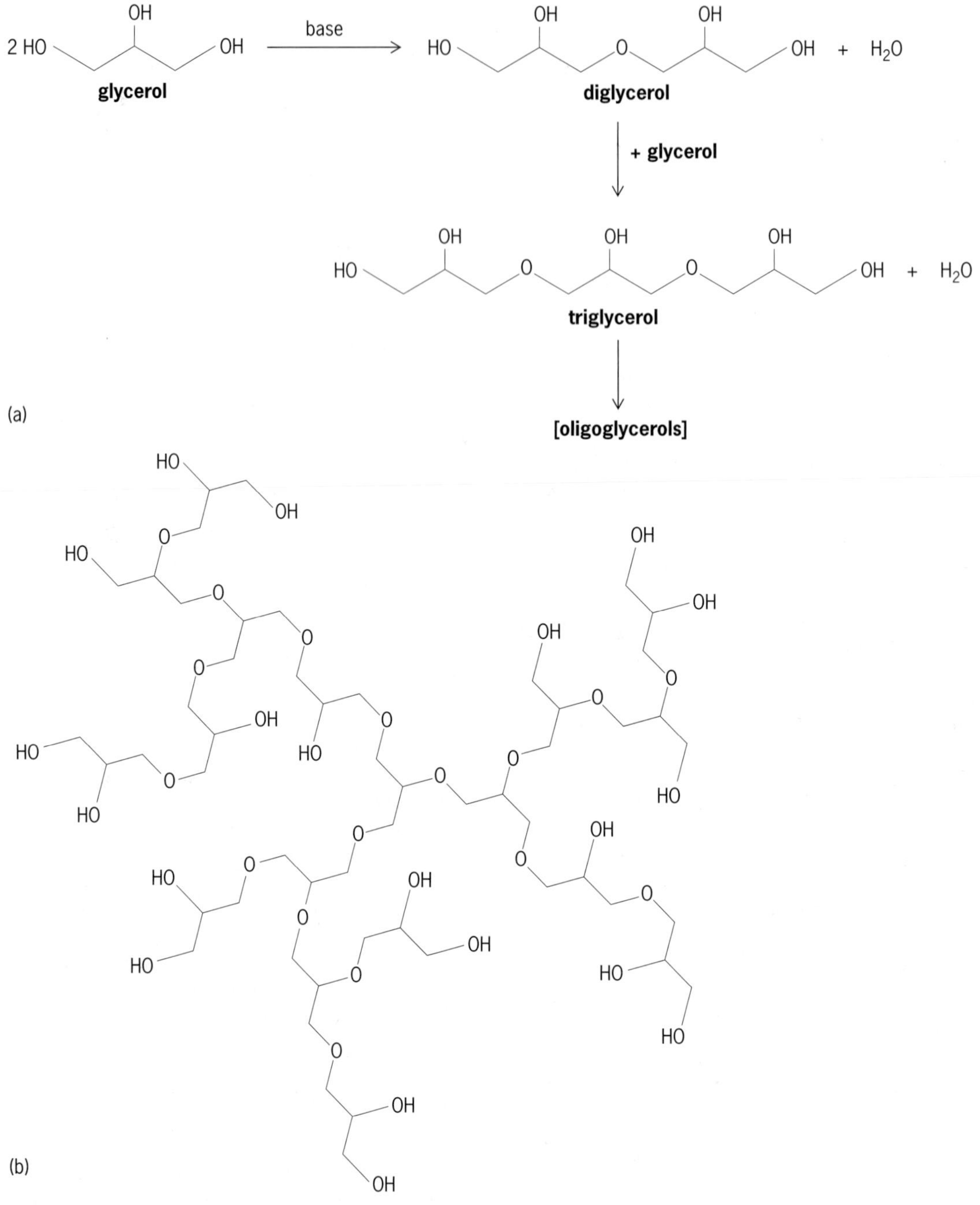

Fig. 7. Oligomers of glycerol: (*a*) linear and (*b*) hyperbranched.

perfumery. In polymer applications, only a few terpenes, notably β-pinene and limonene, are useful because of their pronounced reactivity as monomers. The cationic polymerization of the former proceeds through its initial protonation (**Fig. 5**) and the ensuing cation ensures chain growth to produce a material that is widely used as a very efficient adhesive.

Resin acids, also known as rosin or colophony, are usually chemically modified to prepare various monomers, whose polymers find applications in printing inks, paper additives, drug delivery, and epoxy adhesives.

Materials from poly(carboxylic acids), polyols, and hydroxyacids. The number of monomers from renew-able resources that can give rise to polyesters and polyethers is increasing steadily, and only some of them are highlighted here. Lactic acid (**Fig. 6**) and its cyclic form, lactide (Fig. 6), in both their L and D stereoforms, are by far the most relevant members of this family of compounds. They are prepared either by carbohydrate fermentation or by conventional chemical syntheses, because their polymerization has been optimized to the stage of an industrial process. Poly(lactic acid) [Fig. 6] represents one of the most successful examples of materials from renewable resources, with a yearly worldwide production capacity of about 250,000 tons. This semicrystalline thermoplastic and biodegradable polyester finds useful applications mostly as biomedical, textiles, and packaging materials that have good barrier properties.

Glycerol, a by-product of the biodiesel industry, has become a huge chemical commodity and hence there is growing interest in its use as a monomer for the synthesis of linear (**Fig. 7**) and hyperbranched (Fig. 7) polyether-polyols prepared by the intermolecular condensation of its OH groups. In another vein, chemists have been actively searching for new viable routes to convert glycerol into a variety of useful molecules, including monomers such as glycols, diacids, and acrylic derivatives. This field is therefore particularly promising in that it provides access to monomers that are normally synthesized from fossil sources.

Another monomer from renewable resources that has gained industrial relevance in recent years is succinic acid, which is readily prepared from the fermentation of glucose. Its structure, bearing two carboxylic (COOH) groups, is ideally suited for the synthesis of aliphatic polyesters, which are biodegradable, hydrophobic, and easy to process because of their low melting temperatures, among other notable properties.

Whereas all the above compounds are well-defined single molecules, mixtures of aliphatic diacids, diols, and hydroxyacids bearing long carbon chains, such as the plant-oil fatty acids, can be prepared by the simple chemical splicing of suberin, a natural polyester that is ubiquitous in the vegetable realm but is particularly abundant in two very different species, namely, cork and the outer bark of birch trees. In both instances, the very large amounts of residues generated either from cork manufacturing, or from papermaking using birch, give access to equally important quantities of suberin. This mixture of natural monomers can be polymerized to give biocompatible, biodegradable, and highly hydrophobic aliphatic polyesters with potential applications in packaging, coating, and biomedical materials.

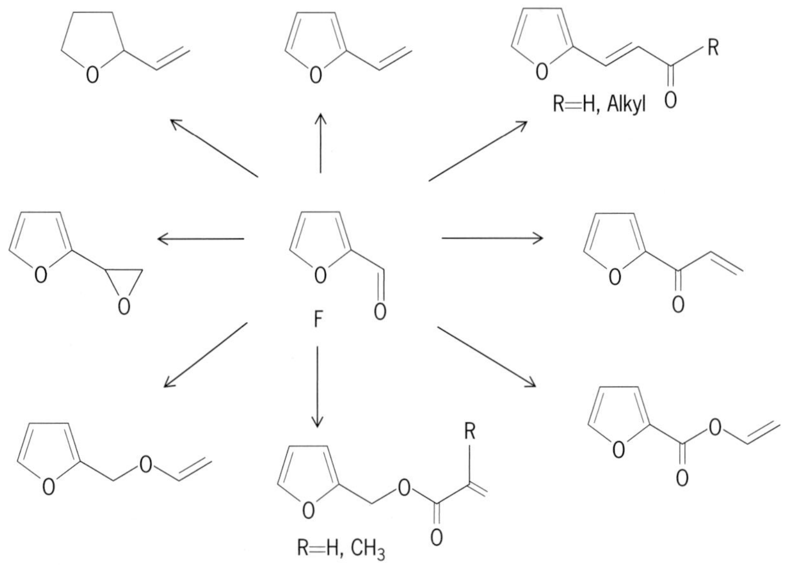

Fig. 8. Selection of monomers derived from furfural (F).

Materials derived from furan monomers. There is a qualitative difference between all the examples discussed so far and the field of furan polymers, because here the strategy is more comprehensive, with a large spectrum of different approaches, which makes it resemble the wide-ranging scope of polymers from fossil sources. The working hypothesis underlying this situation is based on the availability of two first-generation furan derivatives, furfural and hydroxymethylfurfural, which are easily prepared from sugars, oligosaccharides, or polysaccharides incorporating glycoside units made up of five and six carbon atoms, respectively (that is, abundant and cheap renewable resources). Each of these compounds is a progenitor to a family of monomers whose structures simulate those of common counterparts that are routinely prepared from fossil resources, except that they bear a furan heterocycle in their molecules,

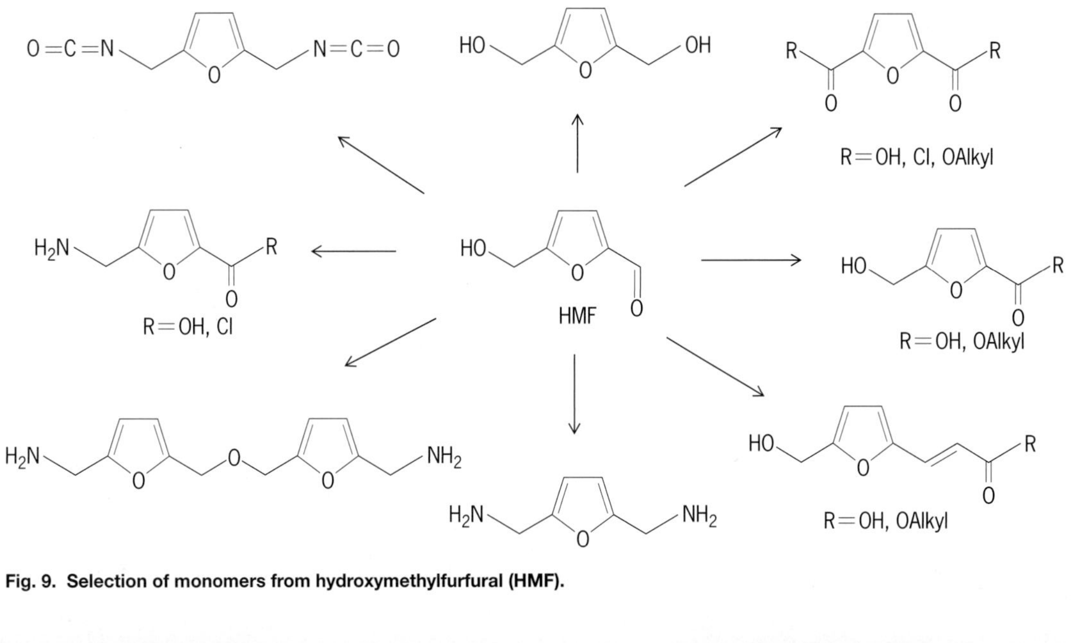

Fig. 9. Selection of monomers from hydroxymethylfurfural (HMF).

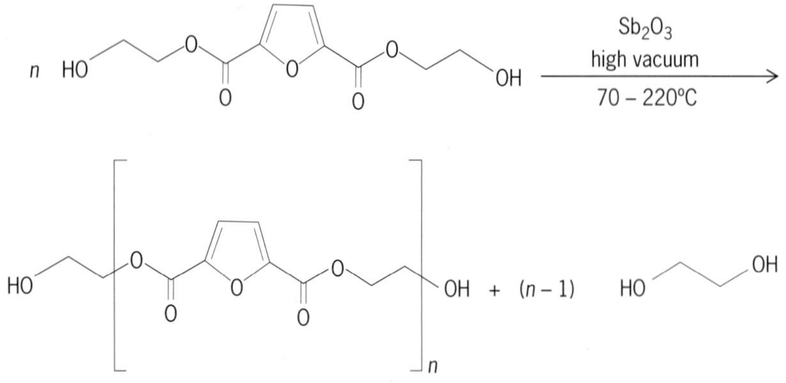

Fig. 10. Synthetic pathway leading to poly(2,5-ethylene furancarboxylate).

compared with the aliphatic or aromatic nature of the petroleum-derived monomers. With furfural, the ensuing monomers (**Fig. 8**) are suited for chain polymerization mechanisms and their structure mimics that of ethylene, styrene, and acrylic derivatives. Their polymers have been shown to be viable alternatives to those obtained from the latter counterparts, with the additional feature of exploiting the pendant heterocycles in their chains as a source of chemical modification.

Similarly, hydroxymethylfurfural is a precursor of numerous monomers (**Fig. 9**) that are suited for polycondensation reactions; that is, step-growth processes for the synthesis of polyesters, polyamides, and polyurethanes, whose structures again simulate those of the current aliphatic and aromatic monomers and polymers from petroleum and coal chemistry.

A single example illustrates this strategy, which has produced a vast array of novel polymers: the preparation of poly(2,5-ethylene furancarboxylate) [**Fig. 10**], the homolog of the most important commercial polyester from fossil resources, poly(ethylene terephthalate), by the replacement of the aromatic ring by the furan heterocycle.

A comparison of the properties of these two materials reveals only small quantitative differences, essentially associated with the higher stiffness of the aromatic moiety compared with that of the furan ring.

Furan chemistry has opened the way to a new class of materials built by successive couplings through the Diels-Alder reaction (**Fig. 11**), a type of click chemistry, which is thermally reversible. Linear, branched, and cross-linked polymers have been pre-

pared and shown to revert to their monomers by a simple heat treatment, which in the case of the furan–maleimide coupling (Fig. 11), requires temperatures of around 100°C.

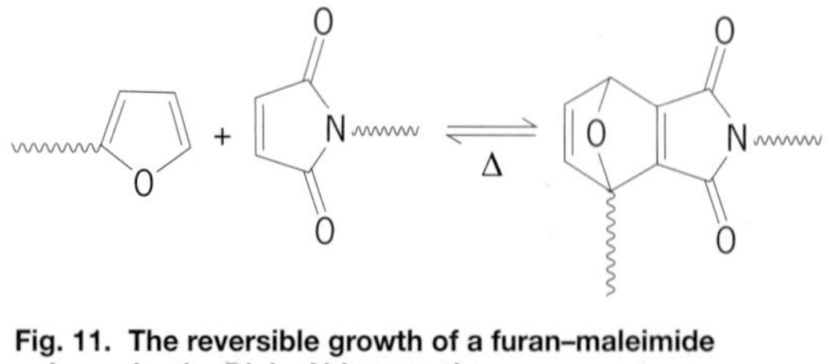

Outlook. The progressive development of polymers from renewable resources on the stage of macromolecular science and technology at a steadily growing pace confirms the need to prepare for the future by giving society a new source of materials that are sustainable and whose precursors are generally available in every corner of the planet.

For background information *see* CELLULOSE; CLICK CHEMISTRY; DIELS-ALDER REACTION; FURAN; GLYCEROL; LIGNIN; POLYAMIDE RESINS; POLYESTER RESINS; POLYETHER RESINS; POLYMER; POLYMERIZATION; POLYOL; POLYSACCHARIDE; POLYURETHANE RESINS; RING-OPENING POLYMERIZATION; ROSIN; STARCH; SUGAR; TERPENE; TRIGLYCERIDE (TRIACYLGLYCEROL) in the McGraw-Hill Encyclopedia of Science & Technology.   Alessandro Gandini

Bibliography. M. N. Belgacem and A. Gandini (eds.), *Monomers, Polymers and Composites from Renewable Resources*, Elsevier, Amsterdam, 2008; A. Gandini, Polymers from renewable resources: A challenge for the future of macromolecular materials, *Macromolecules*, 41:9491–9504, 2008; A. Gandini, The irruption of polymers from renewable resources on the scene of macromolecular science and technology, *Green Chem.*, 13:1061–1083, 2011; D. Plackett (ed.), *Biopolymers: New Materials for Sustainable Films and Coatings*, Wiley, Chichester, U.K., 2011.

Primary cilia

On nearly every cell type in the human body, a tiny hairlike organelle known as a primary cilium protrudes from the cell surface. These cellular appendages act as specialized antennae to survey the extracellular milieu and transmit signals into the cell that are essential for cellular homeostasis. Primary cilia were discovered in the late 1890s and were originally considered evolutionary remnants. However, interest in these organelles has increased dramatically over the past decade as a result of the recognized link between primary cilia and human disease. Improper formation or function of primary cilia can result in a myriad of human diseases and genetic disorders that are collectively called ciliopathies. Due to the ubiquity of cilia, ciliopathies can present with a wide range of clinical features, including obesity, cystic kidney disease, blindness, polydactyly (the condition of having supernumerary fingers or toes), cognitive deficits, behavioral disturbances, hypogonadism, and brain malformations (see **table**). The pathophysiological consequences of dysfunction of primary cilia highlight the important roles that cilia play during development and in the normal function of most tissues. Although great progress has been made in understanding the functions of primary cilia for some cell types, primary cilia function for most cell types is still not known.

Fig. 11. The reversible growth of a furan–maleimide polymer by the Diels-Alder reaction.

Clinical features of ciliopathies

	PKD	NPHP	BBS	JBTS	MKS	ALMS	SLSN	JATD	OFD	EVC
Cystic kidney	x	x	x	x	x	x	x	x	x	
Retinal degeneration			x	x		x	x	x		
Central nervous system malformations			x	x	x	x		x	x	x
Intellectual disabilities			x	x	x			x	x	x
Obesity			x			x				
Gonadal malformations			x		x					x
Polydactyly			x	x	x			x	x	x
Left-right asymmetry defects		x	x	x	x		x	x		
Diabetes			x			x				
Heart disease			x			x				x
Hepatic dysfunction	x	x	x	x	x	x	x	x	x	
Pulmonary dysfunction						x				
Skeletal defects								x	x	x

Abbreviations: PKD, polycystic kidney disease; NPHP, nephronophthisis; BBS, Bardet-Biedl syndrome; JBTS, Joubert syndrome; MKS, Meckel-Gruber syndrome; ALMS, Alström syndrome; SLSN, Senior-Løken syndrome; JATD, Jeune asphyxiating thoracic dystrophy; OFD, orofaciodigital syndrome; EVC, Ellis–van Creveld syndrome.

Structure of cilia. Cilia in the mammalian body are generally classified as either motile or primary. Although motile cilia possess sensory functions, their main responsibility is for movement, as in the case of a sperm flagellum, or to generate fluid flow, as in the case of cilia lining the airways and brain ventricles. Primary cilia, on the other hand, are generally nonmotile, solitary appendages that provide specialized sensory and signaling functions for the cell. All cilia contain a microtubule backbone known as an axoneme. The axoneme of motile cilia comprises nine outer microtubule doublets surrounding a pair of central microtubules (9 + 2 structure), whereas the axoneme of primary cilia lacks the central microtubules (9 + 0 structure). The axoneme is assembled from a modified centriole called a basal body. The centriole, which is perhaps more commonly known for its role in organizing the mitotic spindle during cell division, migrates to and attaches to the plasma membrane by transition fibers and initiates the assembly of the axoneme. Protein synthesis does not occur within cilia. Consequently, the structural proteins required for building and maintaining the cilium, as well as the proteins required for the cilium's sensory and signaling functions, are synthesized in the cell body and transported into and out of the cilium. This bidirectional transport of proteins into and out of cilia and along the axoneme is mediated by a highly conserved mechanism known as intraflagellar transport (IFT). IFT proteins, associated with molecular motors, form trains that move ciliary cargo from the basal body to the tip of the cilium (anterograde transport) and back to the cell (retrograde transport) [**Fig. 1**]. IFT is required for the formation and maintenance of all mammalian cilia.

Ciliary compartmentalization. As cilia grow, they are ensheathed by a membrane that is continuous with the plasma membrane. However, access to the cilium is restricted and there are selective barriers at the base of the cilium that allow only specific proteins to localize within the ciliary compartment.

Recent work has begun to reveal the molecular mechanisms that regulate ciliary localization. Proteins mutated in nephronophthisis (a genetic ciliopathy that affects the kidneys) appear to act as ciliary gatekeepers to regulate entry into the cilium. There are also mechanisms for retaining proteins within the cilium. Septin 2, a member of the septin family of guanosine triphosphatases (GTPases) that forms a

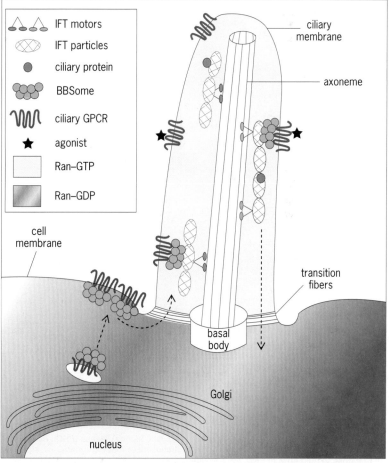

Fig. 1. Schematic of a primary cilium. The axoneme is a microtubule backbone that provides structure to the cilium and mediates trafficking of intraflagellar transport (IFT) particles via molecular motors (kinesin and dynein). Signaling proteins destined for the ciliary compartment are transported by trafficking machinery, such as the BBSome, in Golgi-derived vesicles and are exocytosed at the base of the cilium, where they associate with IFT particles. They are transported along the axoneme and presumably receive signals at the ciliary membrane. The basal body is attached to the membrane by transition fibers, which form a selective barrier to the ciliary compartment. Ciliary entry of some proteins utilizes a Ran-GTP/GDP gradient analogous to nuclear import, with higher proportions of the Ran-GTPase within the cilium containing GTP (compared to that with GDP) and with the opposite true in the cytosol.

diffusion barrier in budding yeast, creates a barrier at the base of the ciliary membrane to restrict the diffusion of ciliary membrane proteins between the ciliary and plasma membranes. Mutations in a protein that controls septin localization have recently been identified in patients with another genetic ciliopathy, Bardet-Biedl syndrome (BBS), which affects many parts of the body. Thus, the proper targeting and retention of proteins to the ciliary compartment are essential for establishing and maintaining receptor-signaling pathways in primary cilia, and disruption of this compartmentalization causes diseases.

Function of primary cilia. It is well established that primary cilia provide critical sensory and signaling functions in response to various stimuli. The outer

segments of photoreceptors in the retina are modified cilia that sense and respond to light. Disruption of the outer segment structure or function leads to retinal degeneration and blindness. Olfactory sensory neurons project olfactory cilia into the nasal cavity that allow for the recognition and response to odorants. Aberrant olfactory cilia formation or signal transduction leads to loss of smell. Primary cilia also can act as mechanosensors, sensing and responding to mechanical stress. In the kidney, epithelial cells lining the nephron project primary cilia that sense and respond to fluid flow. The bending of the cilium initiates various signaling pathways that are critical for cellular homeostasis. The common human disorder, polycystic kidney disease (PKD), is caused by dysfunction of renal primary cilia, which leads to uncontrolled cell proliferation and cyst formation. In each of these examples, the functions of primary cilia are defined by the complement of signaling proteins enriched within the ciliary compartment, and the precise signaling proteins enriched in cilia vary between different cell types.

Primary cilia also sense morphogens and growth factors and play crucial roles in coordinating several signaling pathways essential for growth and differentiation, including the Hedgehog, Wnt, and PDGFα (platelet-derived growth factor α) pathways. Indeed, cilia are absolutely required for proper development because mutations that disrupt formation of cilia are embryonic lethal. In the postnatal rodent brain, cilia are required for the expansion and establishment of neural precursor cells, which may explain the neurological phenotypes seen in several ciliopathies, including Joubert syndrome, Meckel-Gruber syndrome, and BBS. Given the obvious role of primary cilia in cellular differentiation and proliferation, it is perhaps not surprising that cilia have been recently implicated in cancer.

Ciliary protein localization. Although a great deal of information has been learned about primary cilia in a relatively short period of time, there are still many unanswered questions about the functions of these organelles. As the functions of primary cilia are defined by the proteins that localize to them, determining which proteins localize to cilia and the regulatory mechanisms that dictate this specificity will shed light on putative ciliary functions. Several recent discoveries lend significant insight into the molecular mechanisms that control ciliary import and export. Interestingly, there is no single mechanism for specifying ciliary localization of proteins, and a number of different ciliary localization signals (CLSs) have been identified within proteins destined for the cilium. Importantly, CLSs have been used to successfully predict novel ciliary signaling proteins, including melanin-concentrating hormone receptor 1 and dopamine receptor 1, thereby implicating primary cilia in new signaling pathways. Some CLSs contain sites for the covalent addition of fatty acids (myristic or palmitic acids), suggesting that targeting to lipids and perhaps lipid rafts (small regions of the plasma membrane enriched in sphingolipids and cholesterol) is required for ciliary localization of

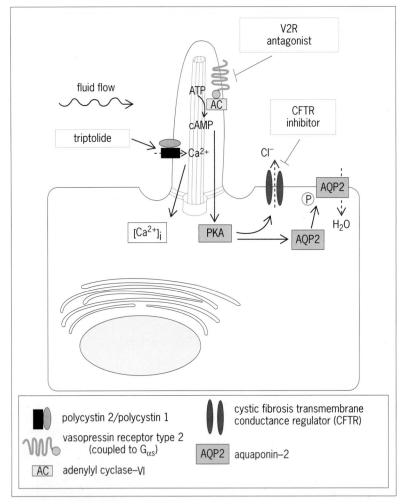

Fig. 2. Identification of therapeutic targets in polycystic kidney disease (PKD). Polycystin 1 (PC1) and polycystin 2 (PC2) form a ciliary mechanosensor that regulates Ca^{2+} signaling in response to fluid flow. In PKD, the function of this complex is altered, leading to a reduced level of Ca^{2+}. Triptolide (a diterpenoid triepoxide) acts on the calcium channel PC2 and restores Ca^{2+} levels in PKD cells. In animal models, triptolide reduces cyst formation and preserves renal function. Vasopressin receptor 2 (V2R) increases cAMP levels in kidney epithelial cells, leading to activation of protein kinase A (PKA) and further modulation of downstream targets, including the cystic fibrosis transmembrane conductance regulator (CFTR) and aquaporin-2 (AQP2). In PKD cells, increased cAMP/PKA signaling leads to alterations in fluid transport, which contributes to cyst formation. Tolvapton, a vasopressin receptor antagonist, is now in a phase III clinical trial for PKD. The CFTR is thought to be a primary route for chloride entry into cysts, leading to their expansion. CFTR inhibitors perturb cyst growth in PKD mouse models, providing a potential therapeutic avenue.

some proteins. Identification of the CLS in the kinesin motor protein KIF17 revealed similarity to a nuclear localization signal, which suggests that cilia possess an entry pathway for proteins that is analogous to nuclear entry of proteins. In agreement with this hypothesis, importin-β2 and GTPase Ran, which are trafficking molecules necessary for classical nuclear import, were found to play critical roles in ciliary import. A protein complex called the BBSome, which is mutated in BBS, is required for trafficking of G protein–coupled receptors (GPCRs) into and out of cilia (Fig. 1). The fact that proteins mutated in a ciliopathy mediate trafficking of proteins to cilia highlights the importance of protein ciliary localization for normal development and cellular homeostasis.

Identification of therapeutic targets. Determining the proteins that localize to cilia and the trafficking mechanisms involved will not only shed light on cell type–specific cilia functions but will also provide potential therapeutic targets. This strategy is illustrated in the development of drugs to treat PKD. Although there is currently no treatment or cure for PKD, the discoveries made over the past decade in understanding the disease pathophysiology and specifically the role of the primary cilium have led to many potential therapeutics (**Fig. 2**). The dominant form of PKD is caused by mutations in either the *PKD1* or *PKD2* gene. The gene products, polycystin 1 (PC1) and polycystin 2 (PC2), form a mechanosensor complex on renal epithelial cilia and regulate Ca^{2+} signaling. In addition, the GPCR vasopressin receptor 2 (V2R) localizes to cilia. V2R couples to $G_{\alpha s}$ (the alpha subunit of a stimulatory GPCR) and stimulates cyclic adenosine 3′,5′-monophosphate/protein kinase A (cAMP/PKA) signaling. PKA activity affects the fluid transport within the kidney epithelia by targeting various transporters, including aquaporin-2 (AQP2) and the cystic fibrosis transmembrane conductance regulator (CFTR). In concurrence with these findings, Ca^{2+} and cAMP levels are altered in PKD cells, potentially leading to cyst formation and PKD pathogenesis. By dissecting these signaling pathways mediated by primary cilia, researchers have been able to unveil many therapeutic targets and potential treatments for individuals affected by this ciliopathy.

In summary, primary cilia are specialized sensory and signaling organelles that perform important cellular functions throughout the human body. Studies in multiple tissues have demonstrated the critical role that primary cilia play in cellular homeostasis and the pathological consequences of primary cilia dysfunction. The study of primary cilia offers tremendous potential for determining basic biological processes as well as discovering novel therapeutic targets for human diseases.

For background information *see* CELL BIOLOGY; CELL MEMBRANES; CELL MOTILITY, CELL ORGANIZATION; CENTRIOLE; CILIA AND FLAGELLA; DEVELOPMENTAL BIOLOGY; HUMAN GENETICS; KIDNEY DISORDERS; PROTEIN; SIGNAL TRANSDUCTION in the McGraw-Hill Encyclopedia of Science & Technology.

Kirk Mykytyn; Jill A. Green

Bibliography. S. C. Goetz and K. V. Anderson, The primary cilium: A signalling centre during vertebrate development, *Nat. Rev. Genet.*, 11(5):331–344, 2010; J. A. Green and K. Mykytyn, Neuronal ciliary signaling in homeostasis and disease, *Cell. Mol. Life Sci.*, 67(19):3287–3297, 2011; P. C. Harris and V. E. Torres, Polycystic kidney disease, *Annu. Rev. Med.*, 60:321–337, 2009; F. Hildebrandt et al., Ciliopathies, *N. Engl. J. Med.*, 364(16):1533–1543, 2011; L. B. Pedersen and J. L. Rosenbaum, Intraflagellar transport (IFT) role in ciliary assembly, resorption, and signaling, *Curr. Top. Dev. Biol.*, 85:23–61, 2008.

Primate color vision

Color vision is a perceptual capacity that allows animals to discriminate reliably between objects or lights delivering different wavelength compositions to the eye, irrespective of their relative intensities. Color vision thus enhances significantly the visibility of objects in the environment and thus serves as a powerful aid in their detection, discrimination, identification, and evaluation. Undoubtedly, it is for these very reasons that many different species, including various primates, have evolved a capacity for color vision.

Basic physiology. Photopigments provide the critical first stage in the process leading to sight. Photopigment molecules consist of a transmembrane protein, an opsin, which is covalently bound to a chromophore, a vitamin A-based retinaldehyde. Conversion of the energy from a photon of light alters the configuration of the chromophore, initiating a signal that can subsequently be communicated to other cells that lie along the visual pathways of the brain. Photopigments are densely packed in photoreceptors (called cones in the case of color vision) located at the outermost portion of the retina, a tissue lining the back of the eye. Photopigments are most usefully characterized according to the relative efficiency with which they absorb lights of different wavelengths (an absorption spectrum).

To have color vision, an animal must possess multiple types of photopigments with different absorption spectra and also have neural mechanisms that compare the patterns of activation generated by these photopigments. An example of such an arrangement, in this case for human color vision, is illustrated in **Fig. 1**. The human retina has three classes of cones. Each contains one of the three types of photopigments, whose absorption spectra are designated as S, M, and L (Fig. 1*a*). In response to illumination, each type of photopigment generates a signal proportional to the number of photons absorbed. Signals derived from the three classes of cones are combined within the neural circuits of the retina (Fig. 1*b*). In one pathway, the signals derived from M and L cones are summed (L + M). This combination forms the luminance channel, which transmits information to mediate aspects of vision that do not involve color. The two other pathways comprise the chromatic channels; through their forward

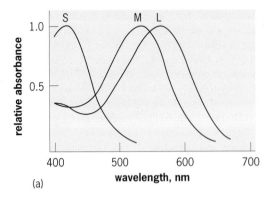

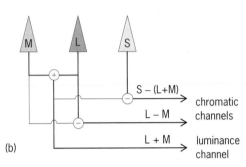

(a)

(b)

Fig. 1. (a) Absorption properties of the three classes of cone photopigments found in the human retina. The height of the curves at the various wavelengths indicates the probability that light of those wavelengths will be absorbed and thus contribute to vision. By convention, the three classes are abbreviated as S, M, and L, reflecting the fact that they are maximally sensitive in the short, middle, and long wavelengths of the visible spectrum. (b) Schematic diagram illustrating the way in which signals from the three cone classes are combined within the neural circuits of the retina. Signals derived from the M and L cones are summed to form the luminance channel (L + M); the two chromatic channels represent, respectively, the difference of the signals from L and M (L − M) and the difference of the signal from S and the sum of the signals derived from L and M [S − (L + M)].

projections and elaborations, these support all of the various manifestations of color vision. One circuit effectively signals the difference in activation between the L and M cones (L − M), whereas the second circuit computes the difference between the signal derived from the S cones and the sum of the signals coming from the L and M cones [S − (L + M)].

The number of types of cone photopigments as well as their relative spectral positioning can vary between species or, in some cases, between individuals within a species. All mammals have either one, two, or three classes of cone pigments and these alternatives condition the nature of their color vision. Three classes of pigments yield two chromatic signal channels, resulting in so-called trichromatic color vision, which is the arrangement characteristically found in people (Fig. 1); two types of cone pigments allow only a single chromatic channel and thus dichromatic color vision. If an animal has only a single type of cone pigment, no chromatic channels can be formed and hence there is no color vision (a condition called monochromatic). As a general rule, increasing the dimensionality of color vision greatly increases the number of items in the environment

that can be visually detected and evaluated and, accordingly, increases the importance of color vision to the animal.

Primate opsin genetics. Genes specifying the human photopigment opsins were first isolated and sequenced in the mid-1980s. The M and L cone opsin genes are positioned adjacently in a head-to-tail array on the q-arm of the X chromosome, whereas the S opsin gene maps to chromosome 7 (note that both localizations had been predicted previously from the pattern of inheritance of human color vision defects). M and L pigment genes have very high sequence identity (approximately 98%), indicating their recent common ancestry, whereas comparison of sequences of these two to the sequence for the S cone opsin gene reveals a more distant relationship. It turns out that only a small number of amino acid substitutions in these opsins are critical for determining the differences in spectral positioning of the cone photopigments; for example, of the 364 amino acids that comprise both the M and L primate cone opsins, substitutions at only 3 residues are principally responsible for the approximately 30-nm separation of the peaks of their absorption spectra. Alterations in the structure and the arrangements of opsin genes, particularly those that may occur during meiotic recombination, have been shown to largely explain the genetic underpinnings of human color vision defects.

Evolution and variation. All vertebrate cone opsins are products of four large gene families. Only two of these families are represented in eutherian (placental) mammals, with the other two having been lost during the early period of their evolution when mammals were consistently nocturnal. The surviving opsin gene families (termed SWS1 and LWS) specify cone pigments having absorption peaks in the respective ranges of approximately 360–440 nm and approximately 500–565 nm. Most mammals derive one cone pigment from each of these two and thus dichromatic color vision is the mammalian norm. Among the primates, though, there are striking departures from this general theme.

From direct measurements of color vision and cone photopigments and inferences drawn from opsin gene examinations, the nature of color vision has now been established for many primate species. Primates can be conveniently classed in three groups: catarrhines (Old World monkeys, apes, and humans), platyrrhines (New World monkeys), and the more primitive strepsirhines (lemurs and lorises, native to Africa, including Madagascar, and Asia). The four general patterns established for primate cone pigment complements are illustrated in the absorption spectra of **Fig. 2**. Some species have cone pigments similar to those described for normal humans, with two X-chromosome opsin genes and an autosomal gene that together specify three types of cone pigments and allow trichromatic color vision (Fig. 2a). All the catarrhines share this arrangement. In a second case, there is only a single X-chromosome opsin gene, but with alternative versions (alleles) in the population that specify differing cone pigments

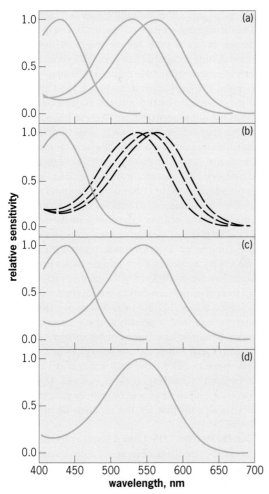

Fig. 2. The cone photopigment arrays found in all primates fall into one of four patterns. (*a*) All members of the species have three classes of cone photopigments, allowing trichromatic color vision. (*b*) The M and L pigments are polymorphic (dashed lines) such that individual animals (all of the males and those females that are homozygous at the X-chromosome opsin gene locus) get any one of the three classes and, in conjunction with the S cone pigment, have dichromatic color vision; or they derive any pair of the M/L pigments, which allows them (heterozygous females) to have trichromatic color vision. (*c*) All members of the species have an S pigment and a single M/L pigment and are dichromatic. (*d*) As a result of the genetic inactivation of the S cone opsin gene, some primates have only a single type of (M/L) cone pigment and therefore lack color vision (monochromatic).

these species are reduced to having only a single cone photopigment and, consequently, no color vision (Fig. 2*d*). This same fate has also befallen a number of nonprimate mammals, including nearly all the marine mammals as well as some rodents.

The details of the events that led to variations in primate color vision are unclear, but a number of suggestions have been offered. The earliest primates were most likely nocturnal; therefore, similar to other eutherian mammals, they probably had two cone pigments with the potential for dichromatic color vision. About 35 million years ago, a second X-chromosome opsin gene was added in the catarrhine lineage. This addition probably resulted from an unequal meiotic recombination event in animals that had previously acquired polymorphic L/M cone pigments. This change set the stage for all current members of this group to have trichromatic color vision. The X-chromosome opsin gene polymorphisms that are so characteristic of the platyrrhine monkeys seem to have been around since early in the history of this lineage. The timing of the loss of S opsin gene function in various nocturnal primates is not known and may likely be quite different for individual cases. How and when the various changes that led to color vision variations in the strepsirhines occurred is a topic still in its early stages of study.

Utility of primate color vision. To survive and prosper, members of a species need to find sustenance, avoid dangers, and acquire mates. Color vision is potentially a useful asset in the performance of all of these tasks; however, in considering the utility of primate color vision, almost all attention has historically been focused on the first of these (that is, finding sustenance). An early (and enduring) suggestion is that primate color vision may have evolved as a tool for detecting, evaluating, and harvesting the colorful fruits that make up a prominent portion of many primate diets. In recent years, researchers have tried to evaluate this possibility through a combination of modeling and empirical studies. In general, the results of these studies show that trichromatic color vision can in fact be a powerful asset in the business of visually detecting and evaluating red, orange, and yellow fruits from the green foliage backgrounds within which they are usually viewed. One problem with this idea is that such fruits comprise only a portion of the total diet in most primate species, and some species have diets that are almost completely fruit-free. These latter species often ingest leaves, which also often offer potential color cues whose detection might be facilitated by the presence of trichromatic color vision. Thus far, direct field studies of the platyrrhine monkeys whose polymorphic color vision makes them a natural resource for studies of the utility of color vision have yielded only inconsistent insights into this problem and therefore this important topic remains largely a matter for future study.

For background information *see* ANIMAL EVOLUTION; COLOR; COLOR VISION; EYE (VERTEBRATE); GENE; NERVOUS SYSTEM (VERTEBRATE); PHOTORECEPTION; PIGMENTATION; PRIMATES; VISION; VITAMIN

(Fig. 2*b*). Because these polymorphic genes are on the X chromosome, this means that heterozygous females get two different M/L pigments and have trichromatic color vision. Homozygous females and all males (with their single X chromosome) have only a single M/L pigment and dichromatic color vision. Most platyrrhine monkeys have polymorphic color vision. A few strepsirhines also have a somewhat similar polymorphic arrangement, whereas others from this group feature an opsin gene/photopigment arrangement like that of many nonprimate mammals, that is, with two types of cone pigments and dichromatic color vision (Fig. 2*c*). Finally, in a few nocturnal primate species, the S cone opsin gene has been inactivated as a result of mutational changes so that

A in the McGraw-Hill Encyclopedia of Science & Technology. Gerald H. Jacobs

Bibliography. D. M. Hunt, G. H. Jacobs, and J. K. Bowmaker, The genetics and evolution of primate visual pigments, pp. 73–97, in J. Kremers (ed.), *The Primate Visual System: A Comparative Approach*, Wiley, New York, 2005; G. H. Jacobs, New World monkeys and color, *Int. J. Primatol.*, 28:729–759, 2007; G. H. Jacobs, Primate color vision: A comparative perspective, *Visual Neurosci.*, 25:619–633, 2008; G. H. Jacobs, The comparative biology of photopigments and color vision in primates, pp. 79–86, in J. H. Kaas (ed.), *Evolution of Nervous Systems*, vol. 4: *The Evolution of Primate Nervous Systems*, Academic Press, Oxford, U.K., 2006; G. H. Jacobs and J. Nathans, The evolution of primate color vision, *Sci. Am.*, 300(4):56–63, 2009; R. D. Martin and C. F. Ross, The evolutionary and ecological context of primate vision, pp. 1–36, in J. Kremers (ed.), *The Primate Visual System: A Comparative Approach*, Wiley, New York, 2005; D. Osorio et al., Detection of fruit and the selection of primate visual pigments for color vision, *Am. Nat.*, 164:696–708, 2004; A. K. Surridge, D. Osorio, and N. I. Mundy, Evolution and selection of trichromatic vision in primates, *Trends Ecol. Evol.*, 18:198–206, 2003.

Priority emergency communications

Effective disaster response requires effective communications. Be it between first responders; between local, state, or federal government agencies; between equipment and service providers and their customers; between victims and emergency services; or between governments seeking to assist each other, communications is one of the key components, if not the key, in responding to an emergency. During a disaster, the systems that work so reliably during normal situations may be damaged and are subjected to heavier loads than normal, affecting their ability to support these vital communications. A priority communications system is needed to ensure that the most essential traffic is completed, while at the same time not denying service to nonpriority traffic, thus removing the barrier to implementing such a system in a time of crisis. Almost as important as implementing such a system on a national basis is ensuring that it works with the systems implemented by other countries of the world. The global economy and the nature of disasters demand it.

Need for communications in a disaster. The Cabinet Office of the United Kingdom clearly spells out the need for communications in a disaster in the first line on its Resilient Communications website: "Telecommunications are a fundamental enabler underpinning the effective response to any emergency." The need for communications in these situations begins with people within the disaster area having the ability to notify others of the situation and call for help, and continues with the need for emergency responders to coordinate their activities, contact those in need of rescue, and call for additional resources. It goes on to include communications between critical infrastructure sectors and secondary responders who are involved with recovering from the disaster and getting things back to normal. Unfortunately, when disasters strike, the communications systems are put at risk in two major fashions. First, they are apt to be damaged by the disaster, as components of communications networks are likely to be located in the area that suffered the attack, be it human-made (that is, terrorism) or a natural disaster (for example, a hurricane, flood, or earthquake). Second, the immediate need for communications, as described earlier, puts a stress on the communications systems that they are not engineered to handle. This overload can stymie communications, preventing necessary messages from getting through, or even cause systems to fail. An important aspect to understand is the notion of necessary messages. Certainly messages from first responders are considered necessary, as are calls for help to emergency answering points (that is, 911 in the United States, 112 in much of Europe). But what about the call from a person in the disaster area to apprise family or loved ones of his or her situation and well-being? Is that call any less necessary, at least to the person making the call? For this reason, the aim of priority communications systems should be to provide special treatment to priority calls without blocking all calls that are not officially identified as priority.

Current systems. The United States currently has a system for prioritizing emergency calls, called GETS (Government Emergency Telecommunications Service). With this system, priority users are assigned a GETS card with a personal identification number (PIN). To invoke GETS, the user dials an access code, then inputs his or her PIN and the called number. This call is then given priority treatment as it traverses the networks of the United States. Should an "all trunks busy" condition be encountered, which is common in a disaster situation, rather than failing to complete the call and directing it to an "all trunks busy" tone, the system places the call in a queue and assigns it to the next trunk that becomes available. In addition, GETS calls are not subject to any network management controls that may have been instituted in response to the overload situation. In this way, the call is given priority over other non-GETS calls, but doing so does not prevent other calls from being completed. The GETS system is designed to complete calls 90% of the time in overload situations, and it has consistently exceeded that performance in actual emergency situations. In a similar fashion, WPS (Wireless Priority Service) is a feature assigned to specific cell phones that gives them priority for access to available radio channels in a cell site. Again, it does not prevent other callers from gaining access to available channels, nor does it drop existing calls to give the WPS phone priority. Rather, it places the WPS phone in a queue and assigns it to the next radio channel that becomes available.

Priority communications schemes also exist in some European countries. For example, in 2009, the United Kingdom completed its deployment of a wireless priority scheme called Mobile Privileged Access Scheme (MTPAS), a system that is based on identifying certain handsets as privileged and giving them special treatment. The wireline priority scheme that is currently in use in the United Kingdom assures preferential treatment to registered users, but it does so by denying outgoing service to nonregistered users. A conscious decision must be made to invoke this type of priority feature, weighing the need of the priority calls against the general population's loss of access to the network. This type of decision cannot be made lightly, and the consequences of such a decision have caused this option to be used very rarely, thereby reducing its effectiveness.

There is a growing interest in establishing priority communications schemes in other parts of the world as well. In 2008, a workshop of the Institute of Electrical and Electronics Engineers (IEEE) held in Bratislava, Slovakia, explored the need for and impact of establishing a priority system on the existing networks. It was generally agreed that such a system was workable, and that it could be implemented by identifying specific wireless devices for priority treatment to speed the introduction of the service and reduce some of the burdens associated with administering PINs. Budget constraints prevented the initiation of the project, but it remains a consideration and a goal. Also in 2008, Bell Labs facilitated a workshop on priority communications in Australia. In 2010, an effort began in Spain, supported by the European Commission, to look at ways of addressing Priority Communications on Public Mobile Networks (PCPMN) in crisis situations. All of these examples illustrate the interest in and the obvious need for this type of service. However, as will be discussed later, there is a danger, or at least a missed opportunity, if nations develop and deploy priority communications systems internally without considering how they would interact with the systems of other nations.

Looking forward. Given the changing face of communications technology and its users, simply ensuring priority access to a voice or radio channel is no longer sufficient. Today's networks transport packets, generally treating each packet the same way, whether the contents of that packet support a voice call, an e-mail message, a text message, video, or even a game. The networks need the ability to prioritize packets, ensuring that packets supporting priority communications, in whatever form they take, receive priority handling. The first responders of tomorrow may be tweeting their counterparts at a disaster site or attempting to send video to remote experts for support. However they choose to use the network, a scheme must be established to identify the priority packets and ensure their timely delivery. Work has already begun in this area, but it is essential that it receive support from the entire communications community.

Need for international plans. As great as the need is for countries to establish priority schemes for emergency communications, that by itself is not sufficient. Commerce is no respecter of borders, with call centers for one area of the world often being located thousands of miles away on another continent, and suppliers of equipment and services being located anywhere in the world. Currently, however, priority schemes are bound by political borders. The GETS system previously mentioned helps priority calls traverse the networks in the United States, but that priority stops when calls go outside the U.S. borders. In our interconnected world, disaster situations demand international priority calling, whether it be communications to obtain assistance from other governments or from suppliers located in other countries, or calls to affected areas offering help. For this reason, an international effort is required to establish standards for an international priority communications scheme. People in each country may determine their own scheme for communications within their own country, but an agreement on how to handle priority communications from other countries, at international gateways, and within each country is needed. Without this type of agreement, nations that reap the benefit of common communications and international trade may be denied many of the benefits of international help in times of emergency, at exactly the time when it is needed the most.

In June 2011, the Worldwide Cybersecurity Summit was held in London. International priority communications was one of the topics addressed, and several breakthrough sessions on the topic were well attended and the issues hotly debated. While there was no agreement on the specifics of how such systems should be implemented, there was a general consensus on the need for them.

For background information *See* ENVIRONMENTAL GEOLOGY; MOBILE COMMUNICATIONS; TELECOMMUNICATIONS CIVIL DEFENSE SYSTEM; TELEPHONE SERVICE in the McGraw-Hill Encyclopedia of Science & Technology.   Richard E. Krock

Bibliography. Alcatel-Lucent Bell Labs, *Availability and Robustness of Electronic Communications Infrastructures—Final Report*, March 2007; IEEE Communications Society Technical Committee on Communications, Quality and Reliability (CQR), *Introduction to Priority Communications Workshop*, Bratislava, Slovakia, September 23, 2008; National Security Telecommunications Advisory Committee (NSTAC), *Next Generation Networks Task Force Report*, 2006; J. C. Oberg, A. G. Whitt, and R. M. Mills, Disasters will happen—are you ready? *IEEE Comm. Mag.*, 49(1):36–42, January 2011.

Quantitative assessment of MAV technologies

Micro Air Vehicles (MAVs) constitute a class of small flying systems intended to perform non-traditional tasks, such as indoor and/or urban surveillance and reconnaissance. A sense of vehicle size is given by the DARPA definition, which classifies MAVs to have a maximum span of about 15 cm (5 in.) and a gross take-off weight of less than 200 g (7 oz). At such small scales, new flight principles emerge, which are only now being explored from a vehicle standpoint. Consequently, the MAV arena is evolving rapidly as new technologies are conceived and investigated. The recent design and development of the Nano Hummingbird by AeroVironment challenges MAV researchers to understand the fundamental principles that govern MAV flight and to transition this understanding to new approaches for MAV design. This step is necessary to achieve levels of MAV performance on par with birds and insects.

MAV concepts. Different MAV concepts abound, and they may be roughly classified based on design of lift and propulsion systems. Variants include: (1) propeller driven fixed-wing MAVs, whose lift and propulsion systems are basically detached; (2) rotary-wing MAVs, whose lift and propulsion are integrated through the rotor assembly; (3) flapping wing MAVs, whose wing movements inspired by birds and insects also integrate lift and propulsion; (4) vertical-take-off-and-landing MAVs that combine characteristics of rotary-wing and fixed-wing MAVs, and (5) gliding (no-thrust) vehicles. Each concept offers advantages and disadvantages with regards to desired, but hard-to-achieve, capabilities, such as: fast, robust and enduring operation in very confined spaces; agile flight in the presence of large gusts, and quiet and inconspicuous flight.

Changes at small scales. One example of how governing physics change at small scales is a noteworthy drop in the lift-to-drag ratio exhibited by airfoils in steady flows at Reynolds numbers (Re) below about 70,000. Beyond steady flows, the description of the flight of insects and small birds requires not just an understanding of quasisteady aerodynamics, but also of unsteady aerodynamics. Other examples of physics or engineering processes that change with scale include: (1) behavior, efficiency, and design of mechanical systems at small scales (for example, very small rotorcraft transmissions or resonant transmissions for flapping); (2) manufacturing and fabricating mechanical systems and wings at small scales, (3) comparative sizes of inertial and aerodynamic forces and concomitant changes in the physics of aero/structure interactions, and (4) magnitude of gust velocity (which can become on the order of the flight speed). The highly effective flight of insects and birds certainly motivates the desire to design micro aircraft that flap. However, in this article the emphasis is not on the comparative assessment of different aircraft concepts, but on the need to develop models by which this comparative assessment can be carried out, particularly as a means to guide future technology investment at the system or subsystem level. The subsequent discussion is skewed toward flapping, but key concepts are also applicable to the other vehicle concepts.

Modeling. The ability to model flapping wing MAVs has improved with the development and integration of new computing methodologies and hardware. Further engineering and scientific studies are needed to adapt these techniques to MAV design. Modeling abilities are described in three broad categories: (1) aerodynamic and/or aeroelastic; (2) control, and (3) wing and mechanism. The ability to integrate the unsteady aerodynamic forces, including effects of wing shape and kinematics, with stability and control mechanisms is essential for ultimately producing efficient and reliable MAVs that flap. This is particularly important for achieving capabilities such as hover and maneuverability in the presence of gust and atmospheric turbulence. Work in low-Re aerodynamic modeling extends from quasisteady methods calibrated to computation or experiment, to analysis with lifting-line methods, to numerical analysis with computational fluid dynamics. Computational methods have been used to study unsteady effects such as wake capture and formation and detachment of vorticity, including the influence of wing flexibility, wing morphing, and variations in wing motion. Unsteady effects have also been injected into the quasisteady approach through inflow models.

Significant progress has also been made in conceiving and validating control strategies for MAVs, and developing physical models of mechanisms and wings used to produce wing flapping. The technique of "split-cycle constant-period frequency modulation" has been developed to achieve control in 5 degrees of freedom using only two controllers. The formulation of this control strategy is based on quasisteady aerodynamics, but use of the controller, integrated with an unsteady aerodynamic model derived from experiments, has been verified in simulation. In parallel with this development, work has been carried out to understand the influence of aerodynamic and inertial loads on the physical behavior of the mechanism that produces wing flapping, including

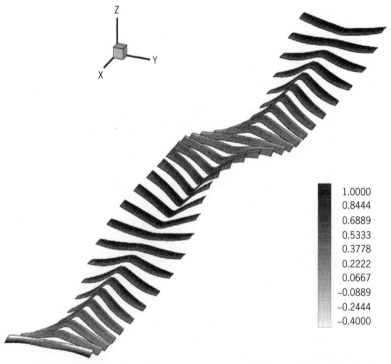

Fig. 1. Optimized pressure distribution of a flapping, morphing wing. The scale bar is for the (dimensionless) pressure coefficient. (*From M. Ghommem et al., Global optimization of flapping kinematics for micro air vehicles, 15th International Forum on Aeroelasticity and Structural Dynamics, Paris, France, IFASD-2011-150, June 2011*)

1.0000
0.8444
0.6889
0.5333
0.3778
0.2222
0.0667
−0.0889
−0.2444
−0.4000

modeling of a resonant thorax mechanism and a compliant mechanism to produce flapping motions. Finally, considerable attention has been given to the nonlinear modeling of flexible flapping wing structures.

Need for multidisciplinary assessment. Aerodynamic modeling, control, and mechanism and wing structural modeling need to be united during design in a manner commensurate with the highly integrated nature of biological systems that flap. Even with advances in computing power, simulations of detailed fluid physics remain very time consuming and are far from use in design or in the development of control strategies. A challenge is that MAVs can be sensitive to many design parameters in terms of both local and global response measures. These dependencies are perhaps best resolved long-term through multilevel, targeted simulations that can uncover physical interactions to exploit to improve MAV performance.

Steps toward a framework for assessment. Steps toward a design framework that can be used for quantitative technical assessment are now being taken at moderate levels of aerodynamic modeling fidelity. By assuming the flow to be inviscid, incompressible, and irrotational outside thin layers of vorticity bound to, and shed from, the wings, modeling can be greatly simplified while retaining key unsteady behavior. A well-established method based on these assumptions is the unsteady vortex-lattice method (UVLM), whose degrees of freedom are limited to the discretization of vorticity and thus small in number, thereby leading to a significant reduction in simulation time. M. Ghommem and colleagues combined UVLM with a global optimization to maximize the propulsive efficiency under lift and thrust constraints by dynamically changing the shape of the wing (morphing) during the flapping stroke (**Fig. 1**). They found that morphing could significantly improve thrust while keeping lift equal to the baseline (rigid) value.

Another important concept for achieving improved MAV performance is the passive tailoring (unlike the active morphing of the previous example) of the wing structure to respond favorably under inertial and aerodynamic loads. One recent approach involved the optimization of wing topology through a cellular division strategy. Aeroelastic interactions between the beam/membrane structure and the unsteady flowfield, including inertial effects, were modeled using an aerodynamic theory akin to UVLM. A genetic algorithm was used to reveal a wing layout that produced much higher levels of lift through adaptive cambering of the wing surface during the flapping stroke (**Fig. 2**). A similar study examined optimal energy transfers between the unsteady flowfield and a tailored, flexible, flapping structure in hover using a quasisteady aerodynamic formulation.

Prospects. In this short expose, the state-of-the-art for modeling flapping MAVs is reviewed from a multidisciplinary standpoint. Similar advances have taken place for the physics-based modeling of other MAV concepts. It is now time to fuse these pieces together in a validated design process that will enable the rel-

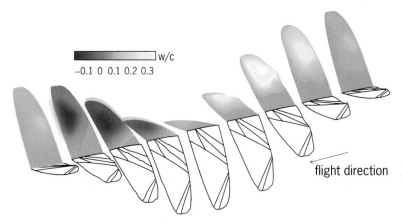

Fig. 2. Out-of-plane wing deformation, *w*, divided by wing chord, *c*, throughout the flapping cycle of a wing structure optimized for lift production.

ative strengths and weaknesses of these concepts, and subsystem technologies, to be assessed. Such a design capability will: more rapidly lead to closed designs, drive advancements in technologies fundamentally benefiting from coupling (for example, control facilitated by flexibility), and increase MAV performance and reliability.

For background information *see* ADAPTIVE WINGS; AERODYNAMICS; AEROELASTICITY; AIRCRAFT DESIGN; COMPUTATIONAL FLUID DYNAMICS; FLIGHT; GENETIC ALGORITHMS; REYNOLDS NUMBER; VERTICAL TAKE-OFF AND LANDING (VTOL) in the McGraw-Hill Encyclopedia of Science & Technology.

Muhammad Hajj; Philip S. Beran

Bibliography. B. H. Carmichael, Low Reynolds number airfoil survey, vol. I, NASA CR 165803, 1981; M. Ghommem et al., Global optimization of flapping kinematics for micro air vehicles, *15th Int. Forum on Aeroelasticity and Structural Dynamics*, Paris, France, IFASD-2011-150, June 2011; D. Pines and F. Bohorquez, Challenges facing future micro-air-vehicle development, *J. Aircraft*, 43:290–305, April 2006; B. Stanford, P. Beran, and M. Kobayashi, *Aeroelastic Optimization of Flapping Wing Venation: A Cellular Division Approach*, AIAA 2011-2094, April 2011; B. Stanford et al., *Shape, Structure, and Kinematic Parameterization of a Power-Optimal Hovering Wing*, AIAA 2010-2963, April 2010.

Reactor neutrino experiments

Nuclear reactors are prolific sources of electron antineutrinos. From the first detection of the neutrino in the 1950s to today's cutting edge experiments on the phenomenon of neutrino oscillations, reactor neutrinos have been a vital tool for the particle physicist.

Nuclear reactor cores harness heat generated when isotopes (mainly uranium and plutonium) contained in vast collections of fuel rods undergo nuclear fission. Radioactive daughter elements produced in these fissions, in turn, transform themselves through beta decay, emitting electron antineutrinos in the process. Roughly six electron antineutrinos are produced from the decay chain following each fission. Less than a kilometer from a typical reactor core, this human-made neutrino flux is comparable to the natural neutrino flux from the Sun. The energy spectrum of these reactor neutrinos peaks between 2 and 3 MeV and extends up to 8 MeV.

At these energies, reactor neutrinos are detected by taking advantage of the inverse beta decay reaction. An incoming antineutrino strikes a free proton in the neutrino detector. Typically, the proton is part of an organic molecule in a scintillating liquid. This reaction produces a positron and a neutron. By detecting these two particles in close-time coincidence, experimental backgrounds can be minimized. At a distance of 1 km (0.6 mi), roughly one neutrino per day will be detected through inverse beta decay per ton of scintillating liquid per gigawatt (GW) of thermal power released by the reactor core. Therefore, the figure of merit for a typical reactor neutrino experiment is the GW-ton-year.

Although the existence of the neutrino was first postulated in 1930, it remained only an idea until the 1950s, when neutrinos were first detected at a nuclear reactor at Savannah River in South Carolina. Detectors were placed near the reactor core, and signals consistent with inverse beta decay were seen to be significantly larger when the reactor was on compared to when the reactor was off.

Neutrino oscillations. Today, reactor neutrinos are crucial for experiments investigating the phenomenon of neutrino oscillations. Neutrinos come in three different types or flavors: electron neutrinos, muon neutrinos, and tau neutrinos. Over the last 20 years, physicists have learned that neutrinos can undergo the quantum mechanical phenomenon of mixing or oscillation. That is, a neutrino produced in one flavor state can be observed in another flavor state after it has traveled a given distance. This requires that at least two of the three neutrinos have nonzero mass. These oscillations are described by a mixing matrix that depends on three "angles" (θ_{12}, θ_{23}, and θ_{13}) and one "phase" (δ) related to CP violation. Between 1998 and 2008, two of the angles were measured in experiments using neutrinos of different energies from a variety of sources. These experiments used detectors situated at different distances from the neutrino sources, as oscillations are a function of the distance travelled (L) divided by the energy of the neutrino (E), thus L/E. It was a surprise that these two measured mixing angles were very large, because similar mixing angles that crop up in the theory of quarks are all quite small. An upper limit was placed on the third mixing angle.

Two different kinds of experiments seek to measure the third mixing angle, θ_{13}. In reactor neutrino experiments, electron antineutrinos transform to other flavors as they travel over short baselines. This manifests itself in a falloff of the number of detected electron antineutrinos that is faster than one over the square of the distance from the reactor. In long-baseline accelerator experiments, electron neutrinos can appear in a beam originally dominated by muon neutrinos. Each kind of experiment has its ad-

vantages. The interpretation of the results of a reactor experiment only depends on the difference of the squared masses of two of the neutrino states (Δm^2) and θ_{13}. The accelerator experiments also depend on the precise value of one other mixing angle, the sign of Δm^2, and the value of the CP phase. Thus, reactor experiments can give a cleaner measurement of θ_{13}, but accelerator experiments can also measure other important quantities if θ_{13} is large enough. Currently, a worldwide program is underway using both kinds of experiments.

To probe oscillations and to explore greater values of L/E, newer experiments need to be further away from the reactor core, where the neutrino flux is lower. Therefore, they measure fewer neutrinos and require additional shielding or placement underground to reduce backgrounds from cosmic rays.

In the 1990s, several experiments observed a deficit in the number of detected atmospheric neutrinos. While it is now known that this deficit was driven by neutrino oscillations involving the mixing angle θ_{23}, it was also theoretically possible that oscillations involving θ_{12} were the cause. Two reactor neutrino experiments, CHOOZ in France and Palo Verde in the United States, tested this idea and showed that it was not true. In the process, they set what remained the best limit on θ_{13} for many years.

Current experiments. Today, a new generation of reactor neutrino experiments is seeking to measure or to stringently limit the value of θ_{13}. They all have a similar design, employing multiple detectors. Near detectors situated very close to the cores precisely measure the neutrino flux before it has been altered by oscillations. Far detectors seek to measure a deficit in the number of reactor neutrinos. By making the detectors as identical as possible, many sources of systematic error can be made to cancel.

Each detector consists of three concentric volumes of organic liquids surrounded by an array of photomultiplier tubes to measure scintillation light. The innermost "target" volume is filled with scintillator that has been doped with a small amount of the rare-earth element gadolinium (Gd). The second volume is filled with undoped scintillator, and the outer volume is made of nonscintillating mineral oil. When an electron antineutrino undergoes inverse beta decay in the target, it produces both a positron and a neutron. The positron deposits its energy and promptly annihilates. The neutron thermalizes and captures on the gadolinium, which is useful for its enormous neutron capture cross section. The deexcitation of the gadolinium produces gamma rays with a unique energy signature. The middle scintillator volume ensures that the full energy of the shower can be recorded for showers induced by positrons or neutrons near the edge of the target. The final volume serves as a buffer for radioactivity in the photomultiplier tubes themselves. Outside these three volumes are other detectors to monitor and tag cosmic rays, cosmic-ray–induced particles, and radioactive backgrounds that might mimic the neutrino signature.

The Double Chooz experiment is located at the Chooz reactor complex in Northeastern France.

Using the same pit that was built for the original CHOOZ experiment, a new three-volume detector was installed with an 8-ton target volume. This far detector is located 1100 m (3600 ft) from two 4.3-GW reactor cores and started taking data in early 2011. The near detector is scheduled to be completed by late 2012. The experiment will be sensitive to 80% of the currently allowed range of θ_{13}.

Two larger experiments are being constructed in Asia, the RENO (reactor experiment for neutrino oscillation) experiment at the six-reactor Yonggwang site in South Korea and the Daya Bay experiment at a six-reactor site near Hong Kong in China. RENO has a simpler geometry. The six reactor cores are equally spaced on a line. The 16-ton near detector is located 290 m (950 ft) from the center line of the reactors under a 70 m (230 m) hill, and the similar far detector is 1380 m (4530 ft) away under a 200 m (650 ft) hill. RENO will be sensitive to 87% of the currently allowed range of θ_{13}.

The Daya Bay reactor site is a bit more complicated, with three pairs of reactors located up to 1600 m (5250 ft) apart. This complication is dealt with by building four near detectors in two locations, 20 tons each, and four more far detectors approximately 1800 m (5900 ft) from each reactor. Daya Bay will be the most sensitive reactor experiment, covering 94% of the currently allowed range of θ_{13}.

Future of reactor neutrinos. After the current generation of reactor neutrino experiments has run for several years, larger experiments could be built with further sensitivity to small values of θ_{13} if it has not been found, or to better measure the precise value of θ_{13} if it is found. Whether such experiments are desirable will depend on both the results from accelerator and reactor experiments and the rate of measured backgrounds in reactor experiments. The most important background is the rate of cosmic-ray–induced production of lithium-9, which has a decay process that exactly mimics the neutrino signal. Large reactor experiments at larger distances may also be desirable to better measure the mixing angle θ_{12}.

In addition, reactor neutrinos can be used to search for coherent neutrino-nucleus elastic scattering, a long-predicted phenomenon where the neutrino interacts coherently with the entire nucleus rather than one of its constituent nucleons. Other experiments seek to use reactor neutrinos as a way to monitor the nuclear fuel cycle with an eye toward nuclear nonproliferation activities.

For background information *see* GADOLINIUM; NEUTRINO; NUCLEAR REACTOR; PHOTOMULTIPLIER; RADIOACTIVITY; SCINTILLATION COUNTER; WEAK NUCLEAR INTERACTIONS in the McGraw-Hill Encyclopedia of Science & Technology.

Maury Goodman; Michelangelo D'Agostino

Bibliography. J. K. Ahn et al. for the RENO collaboration, *RENO: An Experiment for Neutrino Oscillation Parameter θ_{13} Using Reactor Neutrinos at Yonggwang*, March 2010; K. Anderson et al. for the International Working Group on Reactor Neutrinos, *White Paper Report on Using Nuclear Reactors to Search for a Value of θ_{13}*, January 2004; F. Ardellier et al. for the Double Chooz Collaboration, *Double Chooz, A Search for the Neutrino Mixing Angle θ_{13}*, June 2006; X. Guo et al. for the Daya Bay Collaboration, *A Precision Measurement of the Neutrino Mixing Angle θ_{13} Using Reactor Antineutrinos at Daya Bay*, December 2006.

Ribosome biogenesis and disease

Ribosomes, the cellular factories that make proteins, are essential for life in organisms ranging from bacteria to humans. Much of a cell's resources and energy are devoted to making ribosomes (ribosome biogenesis). In a rapidly growing human cell, more than 7000 ribosomes are made every minute. Ribosomes are so important that many of the antibiotics used to cure bacterial infections work to kill bacteria by interfering with ribosome function. Similarly, disruption of ribosome biogenesis in mammals, including humans, is likely to be lethal. However, recent studies have shown that mutations that partially perturb ribosome biogenesis lead to a variety of diseases called "ribosomopathies". Furthermore, dysregulation of ribosome biogenesis has been linked to cancer.

Ribosome biogenesis. Ribosomes are subcellular macromolecules composed of both RNA and proteins (**Fig. 1**). Ribosomes are made in two pieces: a small subunit (SSU, also called 40S) and a large subunit (LSU, also called 60S), which come together in the cytoplasm to form a mature ribosome (called 80S). The 80S ribosome is the functional unit that synthesizes proteins using the information encoded by messenger RNA (mRNA) in a process called translation. Ribosome biogenesis begins with the synthesis of the pre-ribosomal RNA (pre-rRNA) in a substructure of the cell nucleus called the nucleolus. The pre-rRNA is then cleaved and chemically modified to give rise to the mature rRNAs that are incorporated into the ribosome. In humans, the mature SSU contains the 18S rRNA and 33 ribosomal proteins, whereas the LSU contains the 28S, 5.8S, and 5S rRNAs and 47 ribosomal proteins (**Fig. 2**). More than 150 proteins and RNAs are involved in the process of making a ribosome.

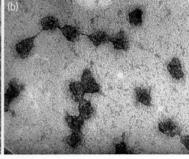

Fig. 1. Electron micrographs of mature 80S ribosomes translating mRNA into protein. The mRNA can be seen as the line connecting the ribosomes (in panel *b*). (*Reproduced, with permission of Elsevier, from J. R. Warner and P. M. Knopf, The discovery of polyribosomes, Trends Biochem. Sci., 27:376–380, 2002*)

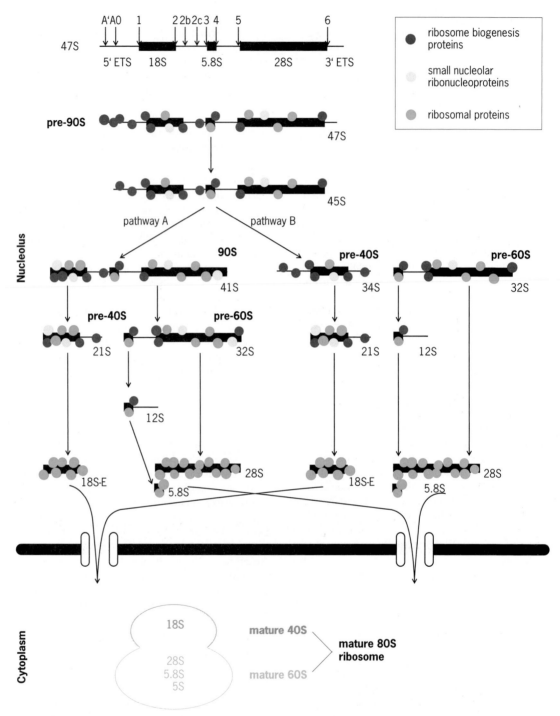

Fig. 2. Model of pre-rRNA processing in human cells. Processing can occur by either one of two pathways (pathway A or pathway B). (*Modified, with permission of the Royal Society of Chemistry, from E. F. Freed et al., When ribosomes go bad: Diseases of ribosome biogenesis, Mol. BioSyst., 6:481–493, 2010*)

Ribosomopathies. Disease-causing mutations have been identified in all stages of ribosome biogenesis, including changes in the synthesis of the pre-rRNA and structural changes in the ribosomal proteins themselves. The stage of ribosome biogenesis at which a particular protein acts can be determined by Northern blot, which is a technique that allows the detection of specific RNA molecules, including pre-rRNA processing intermediates and the mature rRNAs (**Fig. 3**). Although each riboso-

mopathy has a distinct set of clinical features, many of these diseases share symptoms, including bone marrow failure, anemia, predisposition to certain types of cancer, malformation of the head and face, limb abnormalities, short stature, mental retardation, immune deficiency, and skin pigmentation abnormalities (see **table**). Most of the ribosomopathies are clinically heterogeneous, that is, some patients are much more severely affected than others. Whereas multiple mutations in the same gene, or in related

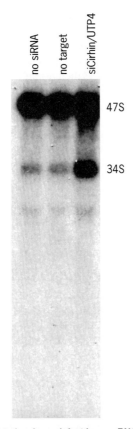

Fig. 3. **A Northern blot showing a defect in pre-rRNA processing when the ribosome biogenesis protein Cirhin/UTP4 (encoded by the *CIRH1A/UTP4* gene) is knocked down in human cells. A mutation in Cirhin/UTP4 is implicated in North American Indian childhood cirrhosis. Note the accumulation of the 34S pre-rRNA precursor in Cirhin/UTP4 knocked-down cells (siCirhin/UTP4) when compared to control cells that are expressing Cirhin/UTP4 [either no siRNA (small interfering RNA) or no target]. Refer to Fig. 2 for a diagram showing the 47S and 34S pre-rRNAs.**

genes, can cause the same disease, there is little correlation between the mutation and the symptoms observed. Therefore, the reasons for this heterogeneity in clinical signs and symptoms remain unknown.

One important genetic feature of the ribosomopathies is that the mutations causing these diseases do not completely abolish ribosome function. Most of the ribosomopathies are the result of haploinsufficiency, that is, the loss of only one of the two copies of each gene that are present in the cell. Other ribosomopathies result from mutations in both copies of a gene, but with only a partial loss of function of the encoded protein. This is expected because work in both yeast and mouse models suggests that complete loss of ribosome biogenesis factors or ribosomal proteins is lethal.

In all ribosomopathies, only particular organs or tissues are affected. This is unexpected because ribosome biogenesis factors and ribosomal proteins are expressed in all cells and because ribosomes are required in all tissue types. Although it is not known why only certain parts of the body are affected by disruption of ribosome biogenesis, several hypotheses exist. One possibility is that rapidly proliferating tissues, such as the bone marrow, require large quantities of ribosomes and are therefore particularly sensitive to perturbations in ribosome biogenesis. Another possibility is that ribosome biogenesis proteins are expressed at different levels in different tissues or that different versions of a protein are expressed in particular tissue types. Yet another possibility is that ribosome biogenesis factors have additional roles outside of ribosome biogenesis and it is the disruption of these functions that leads to disease. The determination of the molecular mechanisms of tissue specificity remains one of the largest challenges in the field of ribosomopathies.

Diseases of ribosome biogenesis

Disease	Mutated gene(s)	Stage of ribosome biogenesis affected	Symptoms
Treacher Collins syndrome (TCS)	TCOF1	Synthesis of rRNA, chemical modification of rRNA	Malformation of head and face
Male infertility	UTP14c	18S rRNA maturation	Reduced or absent sperm production
North American Indian childhood cirrhosis (NAIC)	CIRH1A/UTP4	18S rRNA maturation	Liver cirrhosis; lethal in adolescence
Bowen-Conradi syndrome (BCS)	EMG1	40S subunit maturation	Growth retardation, motor function delay, skeletal abnormalities; lethal in early childhood
ANE syndrome	RBM28	60S subunit maturation	Growth retardation, mental retardation, loss of motor function, skeletal and skin pigmentation abnormalities, hair loss, lack of steroid hormone production
Shwachman-Diamond syndrome (SDS)	SBDS	60S subunit maturation and export	Short stature, skeletal abnormalities, anemia, lack of digestive enzyme production, cancer predisposition
Skeletal dysplasias (CHH, AD, and MDWH)	RMRP	5.8S rRNA maturation	Extreme short stature, skeletal abnormalities, mild mental retardation, anemia, immunodeficiency, sparse hair, cancer predisposition
Dyskeratosis congenita (DKC)	DKC1, NOP10, NHP2	Chemical modification of rRNA	Bone marrow failure, skin pigmentation abnormalities, immunodeficiency, cancer predisposition
Diamond-Blackfan anemia (DBA)	RPS19, RPS7, RPS17, RPS24, RPL5, RPL11, RPL35A	Maturation of either the 40S (RPSs) or 60S (RPLs) subunit	Bone marrow failure, malformation of head and face, limb abnormalities, heart and urogenital abnormalities, cancer predisposition
5q- syndrome	RPS14	18S rRNA maturation	Bone marrow failure, cancer predisposition

Ribosomes and cancer. Dysregulation of ribosome biogenesis plays a role in cancer in at least two ways. First, many of the ribosomopathies lead to an increased risk of cancer. The most common cancers seen in the ribosomopathies are acute myeloid leukemia (AML) and non-Hodgkin's lymphoma.

Second, cancer cells, which grow very rapidly, have a high demand for ribosomes. The size and number of nucleoli, the subcellular compartments where ribosome biogenesis occurs, have long been used to grade the progression of cancer, with larger nucleoli or more nucleoli indicating a later stage tumor. Therefore, many ribosome biogenesis factors and ribosomal proteins are observed to be upregulated in cancers. Although it is unclear whether an increase in ribosome biogenesis is a prerequisite for tumorigenesis or a side effect of tumorigenesis, some evidence suggests that an increase in ribosome production may play an active role in cancer. For example, overexpression of a particular ribosomal protein, RPS3, causes tumors in mice. The drug, rapamycin, which downregulates ribosome biogenesis, is currently in clinical trials as a cancer treatment.

The mechanism (or mechanisms) connecting dysregulation of ribosome biogenesis to cancer is unknown; however, several ribosomal proteins have been shown to interact with tumor suppressor proteins or "cancer-promoting" proteins (called oncoproteins), providing a possible link. The oncoprotein, c-MYC, is able to regulate ribosome biogenesis and, in turn, c-MYC function is downregulated by the ribosomal protein, RPL11. Therefore, it is plausible that disruption of ribosome biogenesis could upregulate c-MYC activity, leading to cancer. In addition, ribosomal proteins can regulate the activity of the tumor suppressor protein, p53, which is discussed in more detail below. Alternately, dysregulation of ribosome biogenesis could sensitize cells to mutations in other genes and it is these additional mutations that lead to cancer.

p53 pathway. Mounting evidence suggests that activation of the p53 pathway may underlie some or all of the symptoms seen in the ribosomopathies. The p53 protein is a tumor suppressor and, when activated, causes cells to undergo apoptosis (programmed cell death). Studies in mouse models have shown that inactivation of p53 can prevent the anemia and malformations of the head and face seen in some ribosomopathies. It is easy to imagine that an increase in apoptosis in the bones, brain, and skin could also be responsible for the growth retardation, mental retardation, and skin pigmentation defects of the ribosomopathies. Indeed, mutations in some ribosomal proteins in mice lead to changes in skin pigmentation, and this is mediated by activation of p53.

Activation of the p53 pathway upon downregulation of ribosome biogenesis occurs, at least in some cases, through the interaction of ribosomal proteins with a protein called MDM2 (or HDM2 in humans). MDM2 normally causes the p53 protein to be degraded. However, when ribosome biogenesis is disrupted, excess ribosomal proteins accumulate and bind to MDM2. When MDM2 is bound to a ribosomal protein, it is no longer able to cause the degradation of p53, allowing p53 to become activated. The activation of p53 then results in increased cell death.

It is somewhat surprising that mutations that cause activation of a tumor suppressor and increased apoptosis also lead to a predisposition to cancer (which requires uncontrolled cell proliferation). One possible reason for this is that any type of nucleolar stress could lead to increased tumorigenicity, although the mechanism for this is unknown. As mentioned previously, the increased cancer risk could also result from activation of the c-MYC oncoprotein, or from additional mutations in other proteins. The reason why mutations in ribosome biogenesis factors predispose patients to cancer is another piece in the ribosomopathy puzzle that needs to be solved.

For background information *see* APOPTOSIS; CANCER (MEDICINE); CELL BIOLOGY; CELL ORGANIZATION; DISEASE; GENE; MUTATION; ONCOLOGY; PROTEIN; RIBONUCLEIC ACID (RNA); RIBOSOMES; TUMOR SUPPRESSOR GENES in the McGraw-Hill Encyclopedia of Science & Technology.   Emily F. Freed; Susan J. Baserga

Bibliography.   B. Ebert and J. M. Lipton (eds.), *Diamond Blackfan Anemia and Ribosome Biogenesis, Semin. Hematol.*, 48(2):73–144, 2011; E. F. Freed et al., When ribosomes go bad: Diseases of ribosome biogenesis, *Mol. BioSyst.*, 6:481–493, 2010; A. K. Henras et al., The post-transcriptional steps of eukaryotic ribosome biogenesis, *Cell. Mol. Life Sci.*, 65:2334–2359, 2008; H. Lodish et al., *Molecular Cell Biology*, 6th ed., W. H. Freeman, New York, 2008; A. Narla and B. L. Ebert, Ribosomopathies: Human disorders of ribosome dysfunction, *Blood*, 115:3196–3205, 2010; T. D. Pollard and W. C. Earnshaw, *Cell Biology*, 2d ed., Saunders/Elsevier, Philadelphia, 2008; D. Ruggero and P. P. Pandolfi, Does the ribosome translate cancer?, *Nat. Rev. Cancer*, 3:179–192, 2003.

Rinderpest eradication

Rinderpest, also known as *peste bovine* (French) and *peste bovina* (Spanish), was an economically serious, highly contagious disease of cloven-hoofed animals, both domesticated and wild, caused by a ribonucleic acid (RNA) virus of the genus *Morbillivirus* within the family Paramyxoviridae. The genus includes the viruses that cause peste des petits ruminants (PPR, which affects sheep and goats), measles, canine distemper, phocine distemper (which affects pinnipeds), and disease in cetaceans. Rinderpest virus strains that existed in the field have been grouped by molecular characterization into three lineages. These are known as African lineages 1 and 2 and Asian lineage 3.

The rinderpest virus is easily inactivated by heat, desiccation, and sunlight. Therefore, transmission was by direct contact between animals that inhaled aerosols from the mouth and respiratory tract, feces, and urine, or by ingestion of contaminated feed or water. Aerosol spread rarely occurred up to distances of 100 m (330 ft).

The disease is believed to have originated in Asia more recently than was originally proposed and is now thought to have occurred approximately 1500 years ago. It subsequently spread through trade in cattle and military campaigns. The ancestral virus for both rinderpest and measles seems to have been a virus related to the PPR virus. It spread throughout Africa after its introduction into present-day Eritrea at the end of the nineteenth century. Within five years, it was widespread throughout sub-Saharan Africa, leaving famine and massively denuded cattle and wildlife stocks in its wake. Once prevalent throughout Europe, Asia, and Africa, the virus is no longer present in its natural hosts anywhere on Earth and now exists only in a small number of laboratory virus archives. The World Organisation for Animal Health (OIE) and the Food and Agriculture Organization of the United Nations (FAO) issued joint declarations in mid-2011 stating that global eradication of rinderpest had been achieved. Rinderpest and smallpox now share the distinction as the first viral diseases to be eradicated worldwide. Preliminary data indicate that rinderpest eradication has yielded benefits surpassing virtually every other development program in agriculture, although the analysis so far is conservative and takes into account only those benefits that accrue directly from herd growth and excludes secondary effects on the economy, which will continue into the future. This magnificent achievement helps to protect the livelihoods of many millions of livestock-dependent farmers and removes a serious constraint to livestock trade.

The most recent Asian foci of infection existed in Yemen until 1997 and the Indus River buffalo tract of Pakistan until late 2000. The virus circulated in extensive pastoral cattle herds in southern Sudan and the Somali pastoral ecosystem until 1999 and 2001, respectively, and the virus was last detected in African buffaloes in northern Kenya in 2001.

Epidemiology. The host range of rinderpest was broad, with cattle, Asian water buffaloes, yaks, African buffaloes, lesser kudus, elands, warthogs, and giraffes being particularly susceptible. Clinical signs of disease appeared in as few as 3 days after exposure to strains that caused acute, severe disease, but the incubation period extended even to 15 days for milder strains. In fatal cases, most animals collapsed and died within 6 to 12 days after onset of fever; otherwise, a protracted convalescence ensued. In addition to high fever, the disease was characterized by serous ocular and nasal discharges, which rapidly became mucopurulent (containing mucus and pus). Gravid cows aborted and shed live virus for a brief period. Erosions throughout the intestinal tract caused profuse diarrhea and dysentery, and lymphoid tissue destruction caused immunosuppression, resulting in recrudescence of latent infections. The accompanying rapid fall in condition was a result of inappetance (lack of appetite) and protein loss in the feces; death resulted from severe dehydration (again, within 6 to 12 days from the onset of fever). Single cases were difficult to differentiate from bovine virus diarrhea, mucosal disease, malignant catarrhal fever, and occasionally even foot-and-mouth disease (when lesions were severe and limited to the mouth). Lesser kudus were rendered blind by severe keratoconjunctivitis, and African buffalo calves developed cataracts.

In its most severe form, when it was known as cattle plague, rinderpest was an acute disease that spread rapidly through herds and areas, causing high morbidity and mortality that could exceed 95%. Exceptionally, peracute (severe and brief) cases were reported in which death occurred before clinical signs developed. More commonly, mortality rates approximated 40–50%, with losses being concentrated in 2- to 3-year-old animals in endemic areas.

Infected sheep and goats rarely exhibited clinical signs. Earlier confusing reports of rinderpest in Indian sheep are now mainly considered to have been the result of the related PPR virus. Swine suffered severe disease and were important in disease transmission in East Asia. Wildlife populations could be seriously affected, and protracted epidemics were observed in African wildlife herds and Southeast Asian antelopes. However, wildlife populations did not act as long-term reservoirs of infection independently of cattle and did not represent an obstacle to rinderpest eradication as was once feared. Strains of virus causing only mild clinical reactions were reported.

Viruses of reduced virulence have been found in all three of the virus lineages. However, they have been most commonly associated with African lineage 2 in eastern Africa, where a virus so mild as to escape attention in cattle retained its virulence for wild animals, which thus acted as sentinels for occult virus circulation in cattle. An occult infection is one not accompanied by readily detectable signs of disease.

Immunity and prevention. Although existing as distinct virus clades, rinderpest viruses belong to a single serotype. Many vaccine formulations have been tried. A virulent virus/immune serum-simultaneous inoculation process helped to eliminate rinderpest virus from Russia and southern Africa. However, only modified (attenuated), live rinderpest vaccines were found to generate a strong, protective, lifelong immune response. Goat-adapted vaccines were widely used in Asia and Africa, but it was not until vaccines could be produced by growth in cell culture that sufficient quantities were manufactured to undertake mass vaccination campaigns. One such vaccine developed in East Africa in the 1960s by Walter Plowright was widely used in Africa and Asia and proved safe and efficacious in large-scale vaccination campaigns and outbreak control. A production protocol resulting in very low residual moisture produced a thermostable vaccine that could be used outside a cold chain (proper refrigeration). Its use in remote areas greatly facilitated the eradication of rinderpest from its last reservoirs.

Another vaccine used for decades in Russia may have reverted to virulence on several occasions. Recombinant vaccinia and capripox vaccines have been described but have not been licensed for general use. By virtue of the eradication resolutions passed by member countries of the FAO and OIE, vaccination against rinderpest is now outlawed.

Eradication. Rinderpest eradication was achieved through national control programs supplemented by a series of international programs that ran from 1945 to 2010. The FAO was very instrumental in leading initiatives in China, Southeast Asia, the Middle East, and Africa. Beginning in the 1960s, the African Union (formerly the Organization of African Unity) took on leadership of continental control programs through Joint Project 15, followed by the Pan-African Rinderpest Campaign and the Pan-African Control of Epizootics (PACE) program. The final stages of rinderpest eradication from 1994 to 2011 were overseen by the Global Rinderpest Eradication Programme (GREP), which was hosted by the FAO. Its function was to promote and foster global eradication, monitor progress, provide technical guidance, support regional coordination, and provide direct assistance to countries outside normal funding arrangements. Working in close collaboration with the FAO, the OIE oversaw the process of rinderpest-freedom accreditation. However, although rinderpest ceased to circulate in 2001, completion of the stringent, surveillance-based procedures required for countries to apply for recognition of rinderpest freedom was not attained until 2011.

Early programs relied excessively on annual, pulsed mass vaccination campaigns and did not succeed in eliminating infection, although they did result in a progressive reduction in outbreaks. However, once it was realized that eradication programs had to be based on sound epidemiological understanding, success came rapidly. The key to eradication was the recognition of discrete reservoirs of infection linked to outbreaks through trade and migration. The Intensified GREP, launched in 1999, focused intensive vaccination on the reservoirs of infection, achieving total eradication in 2001. Elimination of the last strongholds of infection in large herds of cattle that support pastoral farming in Africa was a result of development of community-based animal-health programs that achieved high levels of herd immunity in insecure areas.

For background information *see* ANIMAL VIRUS; BUFFALO; DISEASE ECOLOGY; EPIDEMIOLOGY; MEASLES; PARAMYXOVIRUS; RIBONUCLEIC ACID (RNA); RINDERPEST; VACCINATION in the McGraw-Hill Encyclopedia of Science & Technology.

Peter L. Roeder

Bibliography. P. L. Roeder and K. Rich, The global effort to eradicate rinderpest, IFPRI Discussion Paper 00923, International Food Policy Research Institute (IFPRI), Washington, D.C., 2009; P. L. Roeder, W. P. Taylor, and M. M. Rweyemamu, Rinderpest in the twentieth and twenty-first centuries, in T. Barrett, P-P. Pastoret, and W. Taylor (eds.), *Rinderpest and Peste des Petits Ruminants: Virus Plagues of Large and Small Ruminants*, pp. 105–142, Academic Press/Elsevier, London, 2006; P. B. Rossiter, Rinderpest, in J. A. W. Coetzer and R. C. Tustin (eds.), *Infectious Diseases of Livestock with Special Reference to Southern Africa*, 2d ed., vol. 2, pp. 629–659, Oxford University Press, Oxford, U.K., 2004.

Ring-opening metathesis polymerization of cycloolefins

Although a relatively new tool in the field of polymer chemistry, ring-opening metathesis polymerization (ROMP) has emerged as a powerful and broadly applicable polymerization method for synthesizing macromolecular materials with tunable sizes, shapes, and functions. The technique has found tremendous utility in preparing materials with interesting biological, electronic, and mechanical properties. ROMP is a chain-growth polymerization process in which a cyclic olefin is converted to a polymeric material. The mechanism of the polymerization, proposed by Y. Chauvin, is based on olefin metathesis and involves a unique metal-mediated carbon–carbon double-bond exchange process, leading to an unsaturated polymer. Initiation begins with the coordination of a cyclic olefin to a transition-metal alkylidene complex, forming a four-membered metallacyclobutane intermediate that breaks up productively to generate a new metal alkylidene (propagating alkylidene). Analogous steps are repeated during the propagation stage until polymerization ceases (that is, all the monomer is consumed or the reaction is terminated). The use of well-defined initiators combines rapid initiation with high propagation rates and often enables living polymerizations (that is, reactions for which there is no termination step to stop chain growth). Thus, precise adjustment of the molecular weight can be accomplished by simply varying the ratio of monomer to initiator. Polydispersity indices (PDI; molecular-weight distributions) of the obtained polymers are, in general, low (PDI < 1.5) and block-copolymer synthesis is feasible. The living character of ROMP enables the synthesis of well-defined, end-functionalized polymers as well as block and graft copolymers with complex architectures and useful functions. Living ROMP reactions are commonly quenched deliberately through the addition of specific reagents, depending on the type of the initiators used.

The most widely used well-defined initiators for ROMP are based on Schrock's tungsten or molybdenum (**Fig. 1a**) complex and Grubbs' ruthenium alkylidene complexes (Fig. 1b and c), which are cocatalyst-free systems. The most common monomers used in ROMP are monocyclic and bicyclic olefins, which possess a considerable degree of ring strain (>5 kcal/mol) such as cyclobutenes, cyclooctenes, cyclooctadienes or cyclooctatetraenes, norbornenes, norbornadienes, and 7-oxonorbornenes. Norbornene derivatives are the preferred monomers, as they can be easily prepared via the Diels-Alder reaction.

The temperature, solvent, and concentration of the monomer have strong influences over the

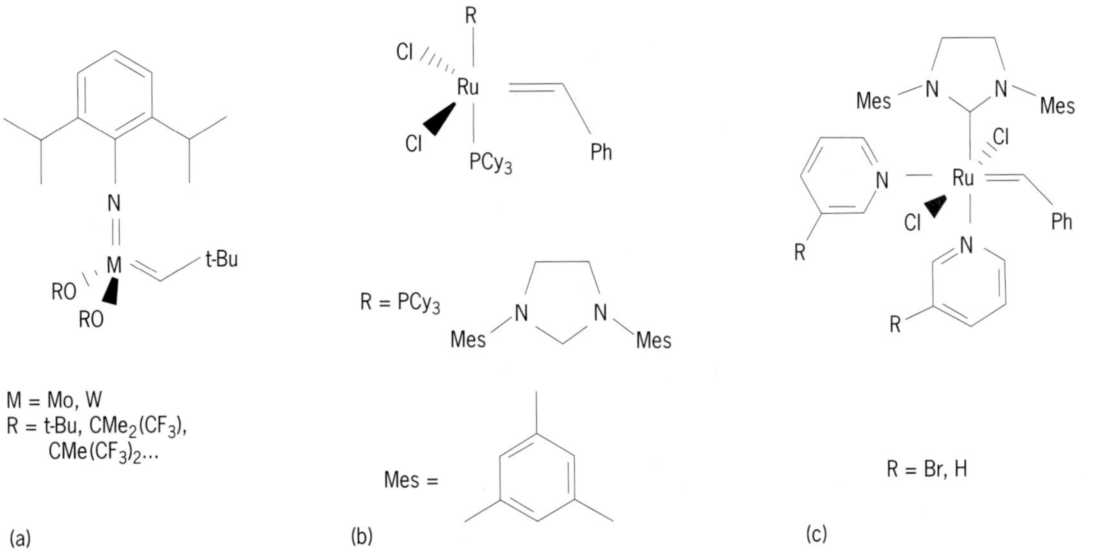

Fig. 1. Most common well-defined initiators for ROMP. (*a*) Schrock's complex and (*b, c*) Grubbs' ruthenium alkylidene complexes.

outcome of the reaction. The highest monomer concentration at the lowest possible temperature is generally the most favorable condition for monocyclic olefins and low ring strains (for example, cyclopentene). ROMP can be carried out in bulk, in solution, or in a heterogeneous system (emulsion or suspension). Various solvents have been used, such as benzene, toluene, dichloromethane, alcohols, and water.

Homopolymerization. The simplicity of the synthesis of different norbornene derivatives, the tolerance of initiators towards substituents, and mild reaction conditions allow the creation of advanced materials for a very broad spectrum of applications.

ROMP-derived monolithic capillary columns have recently been reported as excellent devices for the quantification of insulin and insulin analogues from interstitial (tissue) fluid. It has been reported that the introduction of bulky $SiMe_3$—side groups into the norbornene polymer chain significantly increased their gas permeability, making them attractive for gas membrane-separation processes.

N-hydroxysuccinimidyl ester-based ROMP polymers bearing cell recognition and signaling elements (R), because of their multiple binding modes, have been reported to be able to bind tightly to oligomeric proteins or even cell surfaces and thereby acted as potent inhibitors. Alternatively, they can also act as effectors by organizing proteins and promoting specific cell signaling processes (**Fig. 2***a*).

ROMP polymers containing a pendant galactose glyco group have been used as crosslinking agents for biologically derived extracellular collagen. The obtained hydrogel has physical properties suitable for corneal tissue engineering, including optical transparency, appropriate elasticity, and biological stability (Fig. 2*b*).

Copolymerization. There are multiple examples of using ROMP for the development of the different types of copolymers. Copolymerization was used for the synthesis of new phthalocyanine- and fullerene-

bearing polynorbornenes as new donor-acceptor materials for solar-cell application.

Organic light-emitting diodes (OLEDs) produced by ROMP of imide and dicarboxylic anhydride functionalized norbornenes exhibited good flexibility and optical transparency with a transmittance of around 70% at 400 nm as well as good thermal stabilities with a T_g at 276–300°C. A flexible OLED device was then fabricated from the indium-tin oxide (ITO)-coated polynorbornene-based copolymer film, which exhibited comparable performance to the corresponding glass-based OLED devices. Such a flexible OLED material could be a promising candidate as a substrate for flexible displays (**Fig. 3***a*).

Side-chain liquid-crystalline random terpolymers synthesized by ROMP (Fig. 3*b*) have been used to produce a crosslinked network with shape-memory properties.

The combination of random and block copolymerization resulted in sensor copolymers with covalently incorporated optical-sensor molecules which showed stimuli-responsive behavior and increased photochemical quantum yields and stability (Fig. 3*c*).

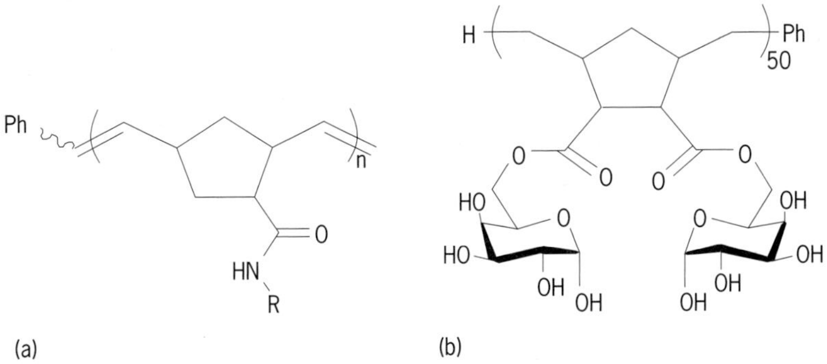

Fig. 2. ROMP polymers bearing (*a*) cell recognition and ROMP polymers containing (*b*) a pendant glyco group.

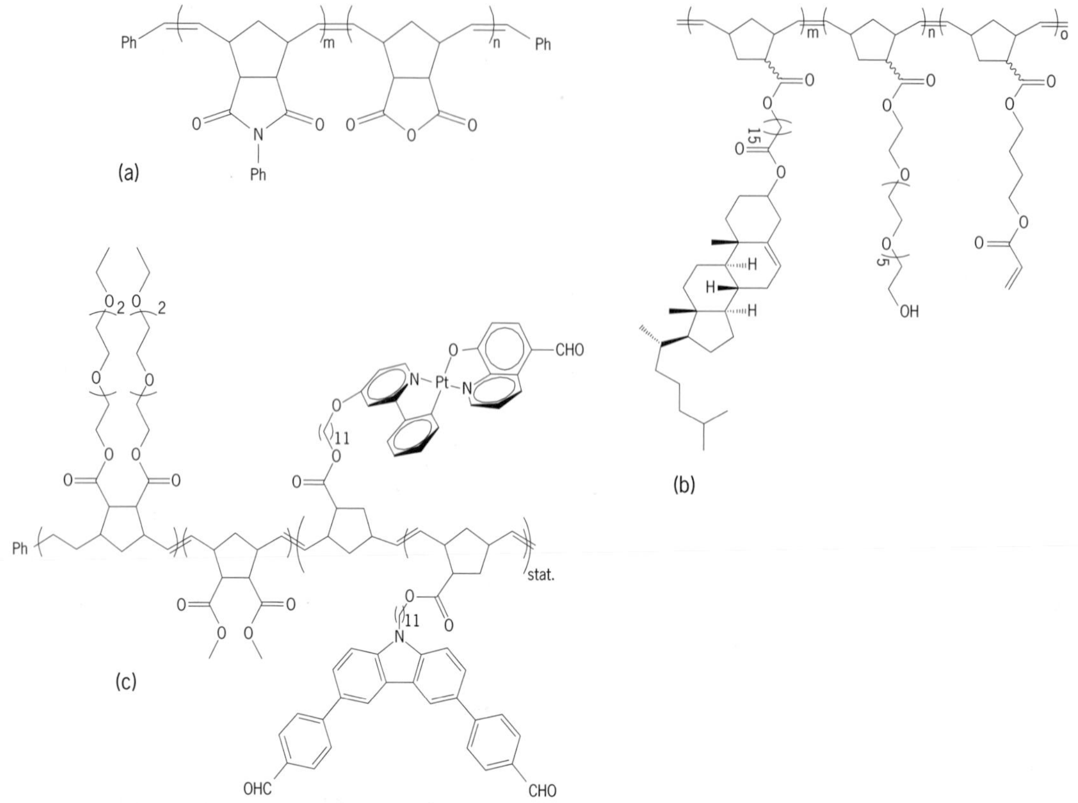

Fig. 3. Copolymers prepared via ROMP. (*a*) OLED material. (*b*) Crosslinkable side-chain random terpolymer containing liquid crystalline moieties. (c) Random and block copolymer containing optical sensor molecules.

Graft-through ROMP using ruthenium N-heterocyclic carbene catalysts and norbornene-containing macromonomers has enabled the synthesis of bottle-brush polymers that have shown promise as drug delivery systems. The first bivalent brush polymers were prepared by graft-through ROMP of drug-loaded, poly(ethylene glycol) [PEG]-based macromonomers (**Fig. 4***a*). Anticancer drugs doxorubicin (DOX) and camptothecin (CT) were attached to a norbornene-alkyne-PEG macromonomer via a photo-cleavable linker. The release of free DOX and CT from these materials was initiated by exposure to 365-nm light. The CT and DOX polymers were at least 10-fold more toxic to human cancer cells after photoinitiated drug release, while a copolymer carrying both CT and DOX displayed a 30-fold increase in toxicity upon irradiation.

ROMP of a cyclooctene-g-PEG macromonomer produced an amphiphilic graft copolymer with a polycyclooctene hydrophobic backbone and a PEG hydrophilic side chain. Protein absorption studies of the obtained poly(cyclooctene)-g-PEG (Fig. 4*b*), demonstrated a significant reduction in the absorption of bovine serum albumin onto polymer surfaces coated in the graft copolymer. Such coatings could be used in human implants, such as hip replacements, to reduce the absorption of human serum onto the surface of the implant which can affect the performance and lifetime of the replacement joint.

Polynorbornene containing dendrons are of great interest in biomaterials and drug delivery, as these types of linkages have been shown to have low toxicity and to be non-immunogenic (Fig. 4*c*).

While block, comb, and graft copolymers of well-defined architecture and narrow molecular-weight distribution can be achieved via ROMP, atom transfer radical polymerization (ATRP), click, and anionic polymerization, each mechanism is effective for only a limited range of monomers, greatly restricting the diversity of accessible block copolymers. Polymeric materials have been prepared by a combination of different polymerization mechanisms through end functionalization with complementary groups. The direct end-capping methodology for the synthesis of copolymers promises to find the use in a variety of areas, including bioconjugation, surface attachment, and in the synthesis of mechanistically incompatible block copolymers. End group functionalization has been used to place (among others) optical sensor molecules at the termini of the polymer chains. For example, europium complexes exhibit distinctly different luminescent lifetimes when attached at the hydrophilic or hydrophobic terminus of an amphiphilic block copolymer (Fig. 4*d*).

Norbornene macromonomers bearing polyacrylate, polystyrene, and polylactide chains have been synthesized. The random and sequential polymerization of the macromonomers via ROMP using ruthenium initiators resulted in well-defined brush block copolymers and brush random copolymers that self-assembled into nanostructures (**Fig. 5**).

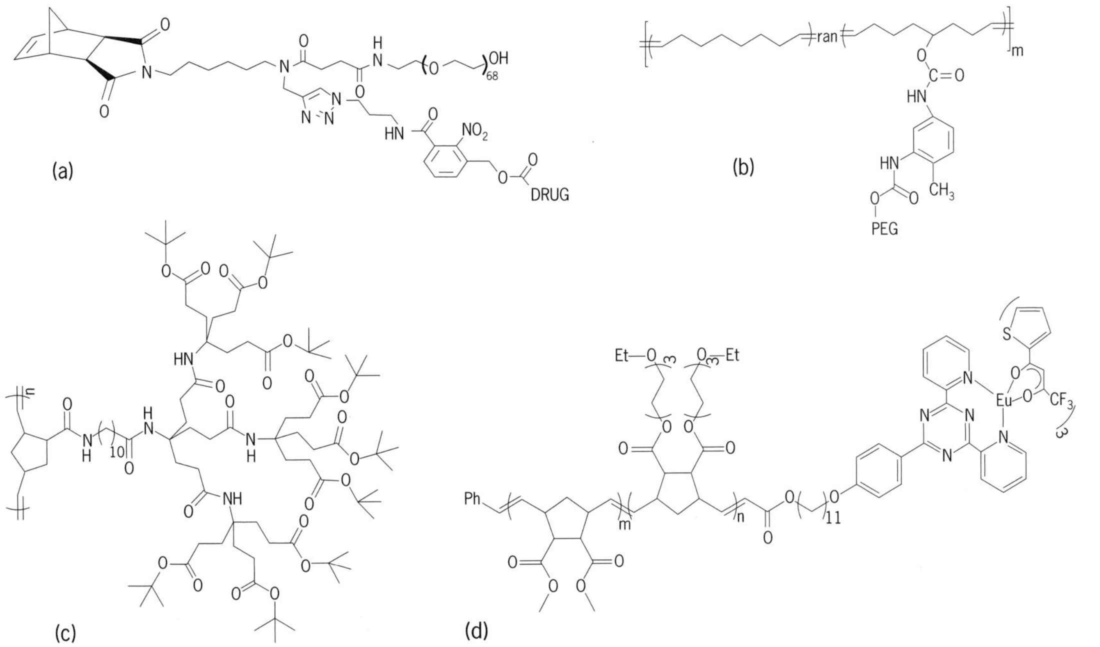

(a)

(b)

(c)

(d)

Fig. 4. Example (*a*) macromonomer for drug delivery. (*b*) Poly(cyclooctene)-g-PEG. (*c*) ROMP polymer containing dendrons for drug delivery. (*d*) ROMP polymer end-functionalized with optical-sensor containing molecules.

Crosslinking in ROMP. Thermosetting polymers exhibit excellent thermal and mechanical properties and have been used in a wide range of applications. However, the poor tractability, recyclability, and biodegradability of thermosetting polymers limit their use in applications where reworkability, recycling, and biodegradation are important. Incorporation of chemically cleavable linkages into monomer systems allows the synthesis of thermally breakable thermosetting materials. For example, ROMP of the difunctional norbornene dicarboximide monomers containing acetal ester groups has produced crosslinked materials that are stable up to 150°C but rapidly break down at about 200°C, as the result of the thermal decomposition of the acetal ester linkage (**Fig. 6a**).

There are some examples of the use of crosslinking ROMP in composite materials that can heal (repair) themselves (self-healing materials) when microcracks have been formed in the matrix. Epoxy resin containing urea-formaldehyde microcapsules containing dicyclopentadiene (DCPD) and ruthenium initiators have been shown to exhibit high self-healing efficiency. Furthermore, epoxy resin containing urea-formaldehyde microcapsules containing mixtures of monofunctional monomer (M) and difunctional monomer (D) [Fig. 6b], and ruthenium initiators have also been used in self-healing process with promising self-healing efficiencies.

Industrial applications. Norsorex® is commercially produced from the ROMP of norbornene and is used as a material for vibration, sound damping, and also as a soaking material for oil spills (absorbs up to 400% by weight oil). Hydrogenated polymers and copolymers of norbornenes, Zeonex® and Zeonor®, respectively, are amorphous, colorless, transparent

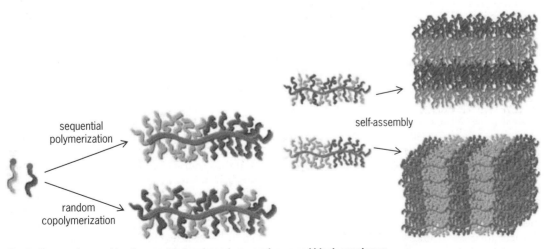

sequential polymerization

random copolymerization

self-assembly

Fig. 5. Proposed assembly of symmetric brush random copolymer and block copolymer.

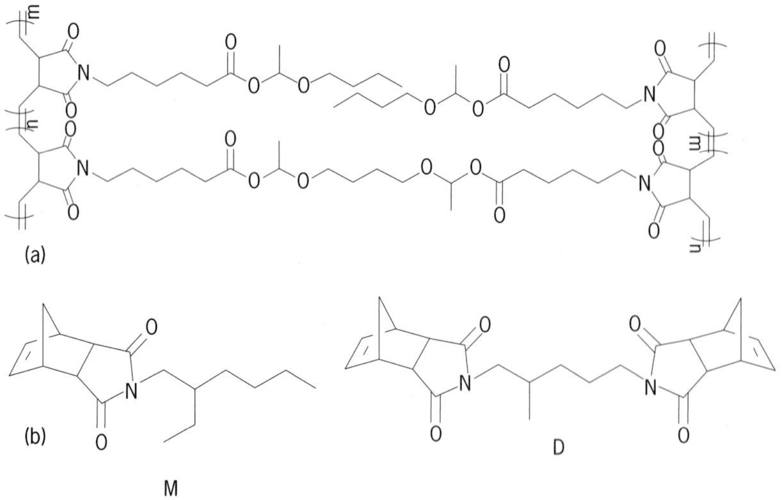

Fig. 6. Schematic representation of (a) crosslinked ROMP materials containing acetal ester groups and (b) structures of mono-functional monomer (M) and di-functional monomer (D) used in the self-healing process.

polymers with high T_g (140–160°C) and low moisture absorption, making them very suitable for optical applications such as disks, lenses, and displays. ROMP of dicyclopentadiene (DCPD) which leads to rapid formation of crosslinked materials, has been the most successful commercial process initially developed by Goodrich and Hercules as Telene® and Metton® using the earlier ill-defined ROMP initiators. The materials show superior mechanical and physical properties and have been used in the automotive, agriculture, construction, and leisure equipment. The application of Telene has been diversified by RIMTEC Corporation. The formation PDCPD resins from ROMP of DCPD using ruthenium well-defined initiators has also been exploited by Materia, Inc. for the production of body panels for truck, agricultural and earth-moving equipment, industrial applications, cell covers for chlor-alkali plants, and domestic wastewater treatment units.

For background information *See* CLICK CHEMISTRY; CONTROLLED/LIVING RADICAL POLYMERIZATION; COPOLYMER; DIELS-ALDER REACTION; MACROMOLECULAR ENGINEERING; POLYMER; POLYMERIZATION; RING-OPENING POLYMERIZATION in the McGraw-Hill Encyclopedia of Science & Technology.

Ezat Khosravi; Yulia Rogan; Barry Dean

Bibliography. Ch.W. Bielawski and R. H. Grubbs, Living ring-opening metathesis polymerization, *Prog. Polym. Sci.*, 32:1–29, 2007; M. R. Buchmeiser, Homogeneous metathesis polymerisation by well-defined group VI and group VIII transition-metal alkylidenes, *Chem. Rev.*, 100:1565–1604, 2000; R. H. Grubbs, *Handbook of Metathesis*, vol. 3, Wiley-VCH, 2003; A. Leitgeb, J. Wappel, and Ch. Slugovc, The ROMP toolbox upgraded, *Polymer*, 51:2927–2946, 2010; K. Nomura, M. M. Abdellatif, Precise synthesis of polymers containing functional end groups by living ring opening metathesis polymerisation (ROMP): Efficient tools for synthesis of block/graft copolymers, *Polymer*, 51:1861–1881, 2010.

RNA-Seq transcriptomics

Transcriptomics is the study of the entire complement of RNA that has been transcribed from DNA in a cell or, more likely, a population of cells. RNA-Seq, in turn, refers to the study of this RNA by use of sequencing technologies.

RNA-Seq is, at present, the most accurate way to describe the transcriptional landscape in a cell. This includes quantitating the amount of RNA in the cell, as well as determining its exact sequence. Determination of the exact sequence data, in turn, allows identification of any posttranscriptional modifications or rearrangements of the RNA, as well as identification of some differences between the organism's genomic DNA and the reference genome of the organism.

Historical development. The central dogma of biology holds that a cell's hereditary material, that is, its DNA, is transcribed into RNA and then translated into protein. Any scientist wishing to study this process at the RNA stage had several tools at his or her disposal up to the mid-2000s, including microarrays (introduced in the 1990s), quantitative polymerase chain reaction (PCR), Northern blots, and reverse transcription (converting the RNA to DNA) followed by Sanger sequencing. All of these methods have the major drawback of not directly characterizing the RNA itself. Microarrays, for example, measure how much RNA hybridizes to synthetically produced DNA molecules fixed to a glass slide, requiring, among other things, prior knowledge of the RNA being studied.

In the 2000s, however, several platforms for "next-generation" sequencing were developed, and the technology advanced to the point where it became possible to sequence reverse-transcribed RNA on a massive scale. The process of RNA sequencing still involves converting RNA to DNA (although new technologies may obviate this need in the near future), but it is more informative than the aforementioned methods. RNA-Seq is rapidly gaining adoption; from 2007 to 2010, the annual growth rate of papers referencing "rna seq" was 283%, whereas those papers referencing "microarray" was 0.76%.

Advantages of RNA-Seq. Compared to previous RNA analysis methods, RNA-Seq has several advantages. It has a higher dynamic range than microarrays, allowing quantitation of very highly and very lowly expressed transcripts simultaneously. In addition, by providing exact sequence data, it allows identification of any posttranscriptional modifications or rearrangements of the RNA. This includes splice junctions and introns; transcription start and stop sites, including polyadenylated end tags; and single-nucleotide variants, polymorphisms, and edits. If the organism's genomic DNA differs from the canonical reference genome, this can sometimes show up as well. For example, the Philadelphia chromosome in humans creates a cancer-causing gene (named BCR-ABL), and this transcript would be visible in an RNA-Seq experiment. An additional advantage of RNA-Seq is the ability to improve the resolution of the data by simply sequencing more of the sample because the reproducibility among technical replicates is very good.

RNA-Seq steps. Although several protocols exist, the process of RNA-Seq broadly involves the following five steps: (1) purifying RNA from cells; (2) removing ribosomal RNA; (3) preparing the RNA for sequencing, including (a) reverse transcription to complementary DNA, (b) library preparation, and (*c*) possible amplification; (4) sequencing the library; and (5) bioinformatic analysis.

Purifying RNA from cells can be done in a number of ways, but essentially involves breaking the cells and then extracting the RNA. At this point, most RNA is ribosomal RNA (rRNA), which is used by the cells in very large quantities to carry out translation of RNA to protein; it is not generally of interest and needs to be removed. The two most common ways to accomplish this are to select for polyadenylated RNA or to directly remove rRNA. Messenger RNA (mRNA) is the RNA that codes for protein and is the RNA that is most often of experimental interest. It is distinguished by having a polyadenylated tail, or a sequence of adenine bases added at its end [it should be noted that certain other types of transcripts may have this feature, including large intervening noncoding RNAs (lincRNAs)]. The sequence of rRNA for most organisms is also known. With knowledge of these sequences (either a long string of adenines or the rRNA sequence), it is possible to experimentally select for the polyadenylated RNA or against the rRNA in a solution. Selecting against rRNA leaves mRNA and all other types of RNA in the solution, which may or may not be desirable.

Preparing the RNA for sequencing with currently available platforms requires reverse-transcribing the DNA (henceforth called complementary DNA or cDNA) and then preparing a so-called library for sequencing. Different sequencing platforms have different requirements for how this library is produced, but typically they involve adding sequences to the beginning and end of the cDNA fragments to interface with the sequencing machines. At present, most sequencing platforms also require amplification of the library, or making many copies of everything that is found there. It should be noted that this step can introduce artifacts as a result of imperfect copying, and quantification of RNA levels may be hampered because of biased copying of certain sequences of RNA in comparison to others. Less amplification is therefore better.

Importantly, analysis of the resulting sequence data with computers is a major part of RNA-Seq research.

Choice of sequencing technology. Sequencing platforms are designed to sequence DNA; however, in principle and in practice, any sequencing platform will also sequence RNA that has been converted to cDNA. Future advances are likely to enable direct sequencing of RNA, but this is not widely available today.

A direct comparison of sequencing technologies is beyond the scope of this article, but it is worth noting that all sequencing platforms have advantages and disadvantages. In addition, no sequencing platform is perfect, and errors may be introduced. An error rate of 1–2% of the reported base pairs is typical. Different platforms will also have different biases as to where they introduce errors, complicating direct comparisons.

Analysis requirements. Bioinformatic analysis of RNA-Seq data requires a large amount of computing power in terms of storage and computational capacity; a single experiment may yield several gigabytes of data, and this number is growing exponentially. Generally, sequencing platforms are sold with computers and software that convert experimental results (often flashes of light) into sequence data. Then, the raw sequence data must be analyzed further, which is typically done by aligning the sequence to the known reference genome of the organism being studied.

In order to carry out this type of alignment, a known reference genome must be available. High-quality reference genomes exist for a variety of species commonly studied in laboratories, including humans, fruit flies, nematode worms, yeast, and *Escherichia coli*. Performing RNA-Seq on organisms for which no reference genome exists is difficult, but reference genomes can be generated by sequencing the genomic DNA of these organisms. Several software suites, both free and open source as well as proprietary, exist to do this in a computationally efficient manner.

In general, there are three types of reads: those that uniquely map back to the reference genome, those that redundantly map back (that is, they map back to multiple locations in the reference genome), and nonmapping reads.

Uniquely mapping reads are what allow us to quantify how much RNA is expressed from a given region in the genome. Comparison of known genes in the reference genome to how many reads mapped back to those genes allows quantitation of gene expression levels. The quantitation is relative, that is, how much of one gene is present relative to another gene. However, if a series of synthetically produced RNA sequences of known concentration has been added to the experiment, it is possible to quantify the absolute amount of RNA that was present; this is known as spike-in controls. The uniquely mapping reads with one or more nonmatching base pairs will either be sequencing errors or signals of a single-nucleotide variant (also known as a single-nucleotide polymorphism, or SNP). Because not all individual organisms have the exact same sequence, this is to be expected and can help characterize the organism being studied.

Redundantly mapping reads are typically rRNA that has not been fully removed or that comes from other repetitive regions of genomic DNA. If there are small areas in these repetitive regions that are not repetitive (down to single base pairs), quantification of expression in this region may still be possible. In this way, the specific alleles (alternate forms of a gene) that differ in as little as one base pair from each other can be quantified. However, reads from entirely repetitive regions are not greatly informative.

Nonmapping reads may result from sequencing errors or they may consist of primer sequences from poorly amplified cDNAs; still, they may be of interest. Any reads coming from an RNA transcript that has been posttranscriptionally modified or that comes from DNA that differs greatly from the reference genome will fall into this category. At present, the two categories of posttranscriptional modifications that are most studied are junction reads and end tags. Junction reads are reads that span over a section of the reference genome and do not match it exactly. They are good evidence of introns (intervening sequences), marking exactly where a splice event occurred. End tags are reads where there is an extra series of adenines that are not present in the reference genome; hence, they prevent alignment to the reference. However, they are good evidence of a polyadenylation event, marking the precise 3'-end of an mRNA transcript. Other analyses can be performed on these reads as well, including detection of fusion genes such as BCR-ABL.

Conclusions. RNA-Seq currently offers the best method to survey a transcriptome. It is highly technically reproducible and highly quantitative, and the availability of the exact RNA sequences allows a number of additional analyses to be performed. The technique is undergoing continuous refinement, and the experimental protocols, the sequencing platforms, and the software analysis routines are improving.

For background information *see* DEOXYRIBONU-CLEIC ACID (DNA); DNA MICROARRAY; GENE; GENE AMPLIFICATION; GENETIC MAPPING; GENETICS; GENOMICS; POLYMERASE CHAIN REACTION (PCR); POLYMORPHISM (GENETICS); PROTEIN; RIBONUCLEIC ACID (RNA) in the McGraw-Hill Encyclopedia of Science & Technology.   Karl Waern; Michael Snyder

Bibliography.   R. D. Hawkins, G. C. Hon, and B. Ren, Next-generation genomics: An integrative approach, *Nat. Rev. Genet.*, 11:476–486, 2010; S. Marguerat and J. Bähler, RNA-seq: From technology to biology, *Cell. Mol. Life Sci.*, 67:569–579, 2010; Z. Wang, M. Gerstein, and M. Snyder, RNA-Seq: A revolutionary tool for transcriptomics, *Nat. Rev. Genet.*, 10:57–63, 2009.

Robotic surgery

Medical robots autonomously removing an appendix of a child or pouring a glass of water and bringing it to grandma: some people will say this is science fiction, others will say it is the future. The reality is that in many parts of the world, there is an aging population, with relatively fewer people to take care of the elderly. A solution can be provided by autonomous robots, called care robots. Autonomous care robots are a research topic of different groups, and, in the case of a vacuum-cleaner robot, can already be found in shops. In addition, surgeons use "cure" robots to enhance their possibilities and working conditions for some types of surgery, such as the removal of

the prostate gland. Such robots are teleoperated or master-slave devices. For diagnosis, presurgical planning, and surgery these robots will be essential. This article focuses on cure robots, more specifically on the development of cure robots for keyhole surgery, also called minimally invasive surgery, in the abdominal and thoracic (chest) cavity. Sofie (Surgeon's operating force-feedback interface Eindhoven) is a next-generation system for this type of surgery and is taken as an example.

Need. The benefits of using a cure robot should take into account the often different needs of different stakeholders. Here the patient and surgeon are considered. The patient benefits from minimally invasive surgery (MIS), which is generally performed through at least three small incisions (approximately 10 mm) instead of one large incision. Surgeons use one endoscope to obtain visual information on the surgical field and two long and slender instruments to actually perform a MIS procedure in the conventional manner. MIS increases the patient value: the actual problem is solved (efficacy) while the secondary harm to the body (trauma) by keeping, for example, the surgical incision as small as possible. An increase of patient value typically means an earlier discharge from the hospital, reduced risk of infection, and less pain. However, increasing this patient value by performing conventional MIS does provide surgeons with some challenges. For surgeons, conventional MIS typically reduces the natural hand-eye coordination, the visual information to two dimensions, the comfort of the body posture, the dexterity (freedom to maneuver and to approach target organs), and the sense of touch. In addition, the ability to perform conventional MIS generally requires an extensive learning curve. Surgeons can overcome some of the above challenges and reduce the learning curve by using robots to perform MIS, a user's need. Note that here the system should be called a master-slave system, also called teleoperated system, although proximity still is the wish of the medical professional. The surgeon operates the master, which directly controls the slave performing surgery at the operating table. However, the term robot is adopted since it seems to be a generally accepted term in the surgical community for the system described above.

These robots fulfill a need of surgeons determined from performing conventional MIS and the implication is that the robots fulfill patients' needs as well since more difficult MIS procedures are possible. However, proof for improved efficacy has not been provided yet.

Table 1 shows cure robots intended for MIS. It is not intended to be complete, but merely provides an idea of the number of systems currently used and/or being developed. The table indicates the name of each system, the accompanying company or institute, and its status.

Currently, the only commercially available system is the da Vinci surgical system from Intuitive Surgical. To date, more than 1750 systems have been installed worldwide. In general, it is used for MIS in the

TABLE 1. Overview of cure robotic systems for minimally invasive surgery in the abdominal and thoracic cavity

Name of the system	Name of the company or institute	Status
Da Vinci surgical system	Intuitive Surgical (US)	Commercially available
Amadeus	Titan Medical (CA)	Human clinical studies mid-2013
Telelap Alf-X	SOFAR S.p.A. (I)	Preclinical studies, CE marking expected 2011
RobinHeart	Biocybernetics Laboratory of the Institute of Heart Prostheses (PL)	Unknown
MIROSURGE	Deutsches Zentrum für Luft und Raumfahrt (D)	Unknown
Sofie	Eindhoven University of Technology (NL)	Technology demonstrator

TABLE 2. The requirements of Sofie

Patient's requirements

Limit post-operative pain
Improve patient value for more difficult procedures

Surgeon's requirements

Small set-up and adjustment time of the robot
Adaptability to the procedures being performed and to the patient
Freedom to approach the patient and field of surgery and perform the (surgical) tasks
Possibility to work around a corner and behind organs
Ability to perform precise tasks
Haptic feedback to feel forces executed with the instruments
Safety for patient, people working with the system, and the system itself

abdominal and thoracic cavity, more specifically for prostatectomies (removal of the prostate) and hysterectomies (removal of the uterus).

The system Sofie will be discussed in more detail below (**Fig. 1**).

Medical robotic system: Sofie. Sofie is a typical example of a master-slave system for minimally invasive robotic surgery. The system has the status of a technology demonstrator and has not entered clinical trials yet. Sofie is currently used to perform simple pick and place tasks. The requirements used to develop Sofie will be discussed first, followed by a short description of the master and the slave. This description focuses on the design and realization.

Requirements. The requirements of Sofie are based on the needs of the different stakeholders (**Table 2**). These were gathered from literature, from talking to surgeons, and from observing more than twenty minimally invasive procedures either performed in the conventional manner or by means of the da Vinci surgical robot. The characteristics of this study are a mixture of characteristics of performing MIS and of the human operator. The slave should be able to at least match, but preferably exceed, the possibilities of the surgeon, since scaling (large hand movements resulting in small instrument movements) can be applied. The master should match the surgeon's possibilities.

Master. The surgeon controls the slave robot using the master. Two pen-shaped handles of the master form the interface to the surgeon (**Fig. 2**). In combination with visual feedback, the interface creates an intuitive working environment by virtually placing the hands of the surgeon on the grippers of the

distal part of the instrument inside the patient. The surgeon's hand movements are measured using high-resolution encoders, enabling accurate control of the slave robot movements. A motor for each degree of freedom in a handle of the interface provides haptic feedback of the instrument's forces to the surgeon. Where possible, a direct-drive minimizes friction.

Slave. Developments in robotic MIS often focus on compliant mechanisms for the slave robot, complicating the dynamic force measurements required for haptic feedback. In contrast, the basis of the slave of Sofie is stiff. The slave has a modular setup consisting of a presurgical set-up with three manipulators intended to manipulate one endoscope and two exchangeable instruments (**Fig. 3**).

The presurgical set-up consists of a single table-connected platform adjustment with one platform and three manipulator adjustments. The platform will be positioned near the field of surgery. The manipulator adjustments enable positioning of the manipulators inside the incision (position depends on both procedure and patient) and orienting them toward the target organ before surgery. Additionally, during surgery it provides a rigid frame connected

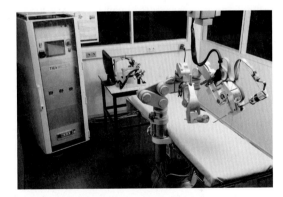

Fig. 1. Sofie, a technology demonstrator for robotic minimally invasive surgery in the abdominal and thoracic cavity. (*Photo by Bart van Overbeeke, Eindhoven, Netherlands*)

Fig. 2. Pen-shaped handles of the master device.

to the table by fixing all joints. This layout leads to a short and stiff force path between the instruments as well as between the instruments and the operating table. The short and stiff force path is advantageous for dynamic system behavior, facilitating accurate force measurements used for haptic feedback and high accuracy (50 μm at the instrument tip) handling up to 10 Hz.

Furthermore, the layout results in a compact, lightweight, and robust design. The compactness of the design improves access to the patient during surgery and reduces the system mass considerably. It improves the conditions of executing surgical and/or assistive tasks around the operating table. Moreover, the system is easy to handle and the presurgery set-up of the system can be done quickly. In addition, the table orientation can be changed during surgery to put, for example, the intestines in a different corner of the abdomen. The slave is connected to the table and will therefore move along with the table adjustment. This saves time since the instruments and slave robot of a system, standing beside the operating table, should be removed from the patient before the table adjustment and placed again after the table adjustment is finished.

The manipulators themselves have a kinematically fixed point of rotation, also called the remote center of motion. This point either coincides with the muscular layer of the abdominal wall or is positioned between the ribs of the thoracic cavity. This results in full support of the instrument, reducing the forces executed on the tissue surrounding the incision and reducing post-operative pain. The forces executed with the instrument are measured by force-sensors incorporated in the manipulator. These sensors are placed in the manipulator outside the patient to prevent electrical signals from being introduced into the patient and to reduce instrument costs. The force-information obtained is fed back to the surgeon to provide haptic feedback. The transmission ratios and encoders of the manipulator realize high-resolution position measurements.

Finally, an elbow-wrist configuration at the tip of an instrument, with an outer diameter of 8.5 mm, provides high dexterity in the surgical area (**Fig. 4**). This configuration increases possible approaches to organs, with the promise of allowing more difficult procedures. This should improve the patient value again.

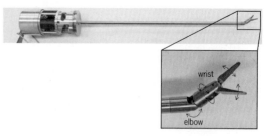

Fig. 4. Instrument with elbow-wrist configuration.

Future of medical robotics. Some trends and questions for the future of medical cure robots have been observed.

A trend toward even less invasive surgery (performed in the abdominal and thoracic cavity) exists. Techniques being researched and developed are (1) NOTES (natural orifice transluminal endoscopic surgery), which is surgery through natural orifices such as the anus, the mouth, or the vagina; (2) single-port surgery, that is, entering the body cavity through only one relatively small incision preferably (close to) the belly button; and (3) non-invasive treatments, for example, in which tumors are treated with a focused energy source from outside the body. The advantages mentioned will produce aesthetically superior results and faster recovery after surgery. Some disadvantages are increased challenges in performing the procedure, the need of applying an incision into a healthy organ (for instance, the stomach if performed through the mouth), and the relatively large dimensions of the single incision compared to the total length of incisions of conventional MIS. In any case, a robotic system providing surgeons with natural hand-eye coordination and improving the possibilities of performing surgery is desirable.

Outlook. Some initiatives in the development of cure robots for MIS in the abdominal and thoracic cavity are shown. The robots provide an answer to needs of both patients and surgeons. Since surgical techniques evolve, new robots will be developed as well to benefit the stakeholders involved.

For background information *see* REMOTE MANIPULATORS; ROBOTICS; SURGERY in the McGraw-Hill Encyclopedia of Science & Technology.

L. J. M. van den Bedem; G. J. L. Naus; M. Steinbuch

Bibliography.   L. v. d. Bedem, *Realization of a Demonstrator Slave for Robotic Minimally Invasive Surgery*, Eindhoven University of Technology, 2010; L. v. d. Bedem et al, *Design of a minimally invasive surgical teleoperated master-slave system with haptic feedback*, International Conference on Mechatronics and Automation, Changchun, China, pp. 60–65, 2009; R. Hendrix, *Robotically Assisted Eye Surgery: A Haptic Master Console*, Eindhoven University of Technology, 2011; P. Pott, H.-P. Scharf, and M. Schwarz, Today's state of the art in surgical robotics, *Computer Aided Surgery*, 10(2):101–132, 2005; R. Taylor and D. Stoianovici, Medical robotics in computer-integrated surgery, *IEEE Transactions on Robotics and Automation*, 19(5):765–781, 2003.

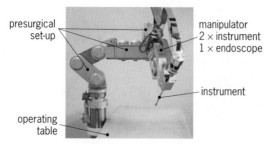

presurgical set-up

manipulator
2 × instrument
1 × endoscope

instrument

operating table

Fig. 3. Table-mounted slave system with one of three manipulator arms.

Rogue waves

"Rogue wave" is a popular term for an unusually large ocean wave that can cause damage to ships or endanger people. There are three issues to be discussed. First, some media accounts of rogue waves really just refer to very rough seas. Second, for the scientist, the term rogue wave refers to waves that are larger than the average in a particular sea state by a specified multiplier. The third issue is the most interesting: In a given sea state, do these large, rare, rogue waves occur more frequently than expected from standard dynamical or statistical models? An alternative version of this question is: Do waves that are larger than the average by a particular multiple occur more frequently than predicted? Each of these issues will be examined, along with further questions that arise.

Wave generation. Ocean waves are generated by the wind, first by random pressure perturbations disturbing the sea surface, and then in a feedback process in which the airflow over the small features amplifies them. Longer waves, which travel faster, then grow by nonlinear interactions that transfer energy from directly generated short waves.

A rough sea is not a regular train of sinusoidal waves, but rather an irregular jumble, albeit with a dominant wavelength and period. The "significant wave height" H_s was originally defined as the height of the highest one-third of the waves; the modern definition, which is close in practice, is that H_s is four times the root mean square vertical displacement of the sea surface. The ultimate size of the waves depends on the wind speed and on the length of time and the distance (termed the "fetch") over which the wind blows. With enough time and fetch (a day or more and several hundred kilometers, respectively, in strong winds), the dominant waves can have a speed as great as, or even slightly greater than, the wind speed U at 10 m (33 ft) above the surface. For such a "fully-developed" sea, H_s is given by $0.24U^2/g$, where g is the gravitational acceleration of 9.8 m/s^2 (32 ft/s^2). So for $U = 20$ m/s (66 ft/s or 45 mi/h), $H_s = 9.8$ m (32 ft). Many incidents experienced by ships and attributed to "rogue" waves have occurred where routine forecasts predicted very large waves in general as a consequence of strong winds and a large fetch.

Amplification by shoaling and currents. When waves generated in the open ocean encounter shallow water near the coast, they slow down and grow in size. They can also become larger as they encounter strong currents. One example is off southern Africa where waves generated by strong winds in the Southern Ocean meet the opposing Agulhas Current. Many reported incidents of very large and damaging waves occur in such situations.

Presentation of data or model results. The second issue concerns the frequency of occurrence of waves, in data sets or theoretical models, that are larger than the average by some multiple. This can be shown in a graph of the exceedance probability $P(H/H_s > M)$ or $P(\eta/H_s > M)$ as a function of the multiplier M, for both the trough-to-crest wave height H

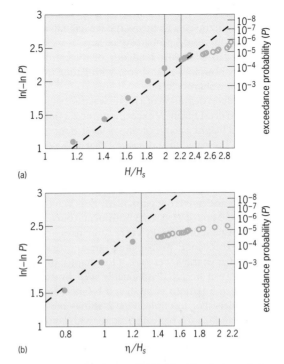

Fig. 1. The exceedance probability P, with a linear scale for $\ln(-\ln P)$ on the left-hand y axis but with the value of P itself shown on the right-hand y axis, for scaled (a) wave heights and (b) crest heights above the value shown with a logarithmic scale on the x axis. The broken lines are from the theoretical Rayleigh distribution for wave heights. The vertical lines indicate common definitions of rogue waves. The data are from the Gorm oilfield in the North Sea, with filled circles for averages over many data and open circles for single values. (*Data courtesy of Harald Krogstad*)

and the crest height η above mean sea level. In order to bring out the small tail of the exceedance probability, it is convenient to plot $\ln(-\ln P)$ against $\ln M$, where ln represents the natural logarithm. In such a plot, the smaller the value of P, the larger the value of the function. The graph is actually a straight line for several theoretically predicted forms of P. In particular, the simplest model of all predicts that P for both H/H_s and η/H_s is Gaussian, coming from the integral of a so-called Rayleigh distribution for the probability density of H/H_s and η/H_s. This is shown as the broken lines in **Fig. 1**.

The data were obtained from a vertically directed laser beam on an oil rig; the travel time from the instrument to the water surface and back gives the instantaneous elevation of the sea surface. The data may not be entirely reliable due to the influence of the oil rig itself in distorting the sea surface and the risk that the laser measurement is of spray rather than the true sea surface. There are also extensive data sets from offshore wave buoys, which measure wave height from the vertical acceleration of the buoy, but these are also open to question as the buoys can move around on their mooring lines and even dodge some wave crests.

For wave heights (Fig. 1a), the data tend to lie above the Rayleigh line, indicating a reduced probability of large waves. This is because a large crest is likely to be preceded, or followed, by a slightly smaller trough. For crest heights (Fig. 1*b*), on the other hand, the data lie below the Rayleigh line indicating an increased probability of large waves. This is likely caused by the tendency of waves to have a slightly non-sinusoidal shape, with flatter troughs and higher, sharper, crests.

A common definition of a rogue wave is that it has a trough-to-crest height *H* exceeding 2 or 2.2 times H_s, or a crest height η above mean sea level of more than 1.25 H_s. The data of Fig. 1 suggest that a rogue wave with $H/H_s > 2$ occurs once in every 10,000 waves. If the average wave period is a typical 10 seconds, this means that a rogue wave is likely to occur once every day or so, and thus is not particularly rare. Waves with $H/H_s > 2.2$ and $\eta/H_s > 1.25$ both occur once every 30,000 waves, or typically every three days or so.

For exceedance probabilities greater than the above (that is, the lower part of the graph), the data in Fig. 1 can be fitted reasonably well by straight lines. Such a pattern is predicted by most theoretical models. For example, **Fig. 2** shows the result of simulations for a typical sea state, with automatic allowance for a finite frequency bandwidth and also for the distortion away from a purely sinusoidal shape of individual waves. The predictions are above and below the Rayleigh line for wave height and crest height respectively, though by more than is apparent in the data. With the simplest and commonly used correction, termed "second order," the exceedance probability plots are fairly straight, not showing the downward curve for larger waves. Curvature is produced by an approximate implementation of a higher-order correction, termed fourth order. For crest height (Fig. 2*b*), the simulations show some curving over of the plots for exceedance probabilities less than one in 10^5 or so. This is in qualitative agreement with the data of Fig. 1, though the curvature is less in the simulation. The effect is much less marked for the trough-to-crest height (Fig. 2*a*).

This downward curve in the data and simulations is important in the context of our third issue, as it suggests that there is a tendency for extra large waves to occur significantly more frequently, or be considerably larger with respect to H_s, than implied by simple models, and it suggests that this can be explained by higher-order corrections to the wave shape. However, both the data sets and the fourth-order simulations are still open to question.

Other possible reasons for an exceedance probability for large waves being bigger than expected from simple theories include: (1) a phenomenon known as the Benjamin-Feir instability, whereby a group of long-crested large waves can become uneven, with one or more waves in the group growing at the expense of the others; and (2) the analysis, as if uniform, of a sea state that is actually patchy as a consequence of refraction by currents; such an analysis gives the impression of more large waves than expected.

Both of these hypotheses, however, still lead to rather straight plots in the suggested presentation, albeit with higher exceedance probabilities.

Directionality. A typical sea state is made up of waves propagating in a range of different directions. Similarly, large, or rogue, waves can come from a range of directions. This is important, as a large wave coming from an unusual direction may be more dangerous than one coming from the main direction of the seas, especially if it hits a ship on its beam. **Figure 3** shows the results of a recent analysis. The chances of a rogue wave approaching from more than 30° away from the main wave direction are only 0.8%, but not zero. These data were for waves from single storms; crossing seas from more than one storm increase the angular spread of rogue wave directions.

Persistence. The lifetime of an extreme wave is an unresolved issue. If the wave is the result of random superposition of waves coming from different directions, its lifetime may not be more than a few seconds. On the other hand, if the waves are long-crested, moving more or less in the same direction, and have a narrow range of periods so that the wave groups are made up of many large waves, one particular wave may persist, and be large, for several wave periods as it moves through the group. (In deep water, wave groups move at only half the speed of individual waves.)

Breaking. Ocean waves tend to break when their steepness, or ratio of height to wavelength, exceeds a particular value. Rogue waves tend to be steep and thus prone to breaking. This may make them more dangerous but may also limit their height. On the other hand, the brief random superposition of different waves to produce an extreme may not provide

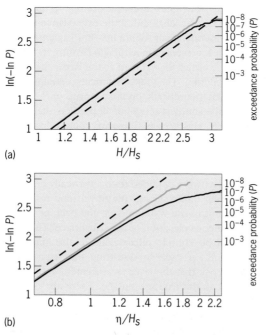

Fig. 2. Simulated exceedance probabilities for (*a*) wave height and (*b*) crest height, for waves from a typical model "JONSWAP" sea state at large fetch. As in Fig. 1, the broken lines are from the theoretical Rayleigh distribution for wave heights. The colored lines allow for so-called "second-order" corrections to the wave shape to give sharper crests and flatter troughs; the solid black lines use more complex "fourth-order" corrections.

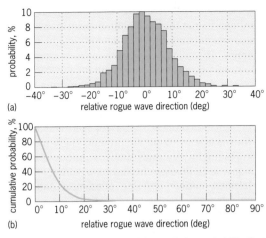

Fig. 3. The (*a*) probability and (*b*) cumulative probability that a rogue wave approaches from a particular angle relative to the direction of the predominant waves. (*Courtesy of B. Baschek and J. Imai, personal communication, using data from 16 wave buoys off the west coast of the United States*)

enough time for breaking to occur. As with the question of persistence, more observations and more theoretical studies are required.

"Unexpected" waves. Much research has emphasized the "rogues" that are large with respect to the average in that sea state. However, there are some situations where the unexpectedness of a large wave, as shown in **Fig. 4**, may be just as important. In observations as well as simulations, it has been found that a wave twice as high as any of the preceding 30 tends to occur approximately once a day, slightly more frequently as one approaches the shore.

Conclusions. In summary, rough seas with large waves occur when strong winds blow for a long time over large distances, particularly when the waves move into shallow water or are refracted and amplified in certain currents. In a given sea state, "rogue" waves with a size greater than the significant wave

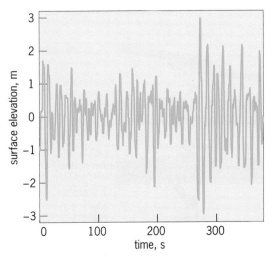

Fig. 4. Data from a buoy off the west coast of Canada on January 9, 1986. An "unexpected" wave at 270 s is much larger than any wave in the preceding 4 minutes. (*Courtesy of Integrated Science Data Management, Canada*)

height by a factor of 2 or 2.2 for wave height or 1.25 for crest height seem to occur with a frequency that is reasonably well understood. The major uncertainty, in observations and in theoretical understanding, is for the rare waves that are even larger. These may occur more frequently than predicted by simple theories, though some studies suggest that the cause may be associated with a change of shape of large waves, in particular the tendency of large waves to develop sharp crests. Further observations are required.

For background information *see* DISTRIBUTION (PROBABILITY); LOGARITHM; OCEAN WAVES; PROBABILITY; STATISTICS; SURFACE WAVES in the McGraw-Hill Encyclopedia of Science & Technology.

Chris Garrett; Johannes Gemmrich; Burkard Baschek

Bibliography. B. Baschek and J. Imai, Rogue wave observations off the US west coast, *Oceanography*, 24(2):158–165, 2011; K. Dysthe, H. E. Krogstad, and P. Müller, Oceanic rogue waves, *Annu. Rev. Fluid Mech.*, 40:287–310, 2008; J. Gemmrich and C. Garrett, Dynamical and statistical explanations of observed occurrence rates of rogue waves, *Nat. Hazards Earth Syst. Sci.*, 11:1437–1446, 2011; J. Gemmrich and C. Garrett, Unexpected waves: Intermediate depth simulations and comparison with observations, *Ocean Eng.*, 37:262–267, 2010; C. Kharif, E. Pelinovsky, and A. Slunyaev, *Rogue Waves in the Ocean*, Springer, Berlin, 2009.

Role of telomeres and centromeres in meiotic chromosome pairing

Meiosis is a process in which haploid cells are produced from diploid parent cells after two successive rounds of nuclear division. During the first division of meiosis, homologous chromosomes segregate to opposite poles of the dividing cell. Chromosome mis-segregation results in the eventual formation of aneuploid gametes, which may cause fertility problems and chromosomal abnormalities in the offspring, such as Down syndrome in humans.

Homologous chromosome pairing. To ensure correct chromosome segregation, homologous chromosomes must pair and synapse during an early stage of meiosis I, known as meiotic prophase. During chromosome pairing, two homologous chromosomes (one paternal and one maternal) recognize each other and align, forming a close physical association. The mechanisms of chromosome pairing remains poorly understood. Pairing is closely followed by synapsis, which is a process of installation of a proteinaceous structure, the synaptonemal complex, between the two paired chromosomes. Synapsis stabilizes the pairing interaction. However, by itself, synapsis does not require proper pairing. In pairing-defective mutants, the synaptonemal complex can be installed between nonhomologous chromosomes.

In most organisms, proper homologous chromosome pairing requires formation of DNA

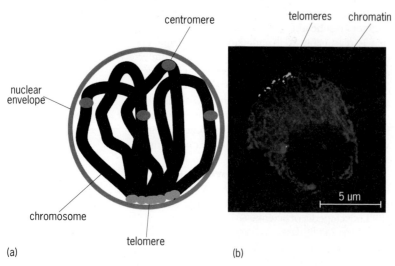

(a)

(b)

Fig. 1. (*a*) A diagram of the telomere bouquet. The telomeres of all chromosomes are clustered on the nuclear envelope. Centromeres are also shown. (*b*) The telomere bouquet in a wild-type maize nucleus during meiotic prophase. The telomeres and chromatin are indicated. (*Image courtesy of M. Sheehan; adapted from A. Ronceret and W. P. Pawlowski, Chromosome dynamics in meiotic prophase I in plants, Cytogenet. Genome Res., 129:173–183, 2010*)

double-strand breaks (DSBs) in chromosomal DNA catalyzed by a topoisomerase-like protein, SPO11, which is essential for the initiation of meiotic recombination. However, the DSB dependence of pairing and synapsis is not universal. In *Caenorhabditis elegans* (a small nematode worm) and *Drosophila melanogaster* (the fruit fly), chromosomes pair and synapse normally in the absence of DSB formation, suggesting that chromosome pairing in these species is controlled by a different mechanism. However, in all species surveyed, there is good evidence that chromosome dynamics play a critical role in homologous pairing. In most species, chromosome ends (telomeres) exhibit particular behavior, known as telomere bouquet formation, which facilitates chromosome alignment and pairing. In a few species, it also has been shown that chromosome centromeres (sites on chromosomes by which they attach to the spindle during cell division) are actively involved in chromosome pairing.

Telomere bouquet. At the beginning of meiosis, the telomeres of all chromosomes in most species of plants, animals, and fungi attach to the nuclear envelope and cluster, which leads to formation of the telomere bouquet (**Fig. 1**). Mutants defective in bouquet formation also show defects in homologous chromosome pairing. Although chromosomes in these mutants do pair in many cases, pairing is slow and inefficient. Consequently, it has been generally believed that telomere clustering facilitates homologous pairing by bringing chromosome ends together.

Bouquet formation has not been observed in some organisms, including *Arabidopsis thaliana* (thale cress) and *C. elegans*. In *C. elegans*, telomeres attach to the nuclear envelope during early prophase I, but do not cluster. In *Arabidopsis*, telomeres cluster during interphase and at the beginning of meiotic prophase on the nucleolus rather than on the

nuclear envelope. Telomeres of homologous chromosomes pair at the beginning of meiosis and dissociate from the nucleolus without forming the typical bouquet. However, during early meiotic prophase, telomeres become associated with the nuclear envelope and occasionally exhibit loose clustering. Overall, although a bona-fide bouquet is not formed in *Arabidopsis* or *C. elegans*, some features of the bouquet formation process, such as telomere attachment to the nuclear envelope, are present. These remnant features might play roles similar to the role of the bouquet in facilitating chromosome pairing.

Telomere–nuclear envelope attachment and telomere dynamics. Studies in several species, particularly fission yeast, have led to identification of a number of proteins involved in telomere attachment to the nuclear envelope (**Fig. 2**) and bouquet formation. In fission yeast, the Taz1 protein, which acts to maintain the proper copy number of telomeric repeats, has been shown to be required for bouquet formation. This protein interacts with three other telomere-associated proteins: Rap1, Bqt1, and Bqt2. The Rap1/Bqt1/Bqt2 protein complex forms a link between the telomeres and Sad1, a transmembrane protein located in the inner membrane of the nuclear envelope. Sad1 homologs have been identified in a number of diverse species, including budding yeast, mice, *C. elegans*, *Drosophila*, *Arabidopsis*, and maize. They are distinguished by the presence of an evolutionarily conserved SUN domain. Sad1 binds another transmembrane protein, Kms1, which is located in the outer membrane of the nuclear envelope and which interacts on its cytoplasmic end with the cytoskeleton and cytoskeletal motor proteins. Kms1 homologs have been also identified in many species, such as *Drosophila*, *C. elegans*, and mammals, and are collectively known as KASH proteins. The SUN/KASH protein pairs function in meiosis by linking chromosome ends with the cytoplasmic cytoskeleton (Fig. 2). The two proteins also function outside of meiosis, and their exact roles are still not well understood.

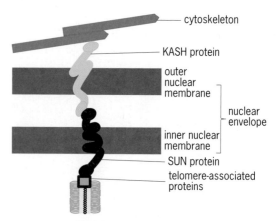

Fig. 2. A diagram showing the telomere–nuclear envelope attachment. Telomeres connect to the nuclear envelope through a telomere-associated and a SUN domain proteins. The SUN domain protein in the lumen of the nuclear envelope interacts with a KASH protein, which also interacts with the cytoplasmic cytoskeleton.

In a number of species, including budding and fission yeasts, rats, mice, *C. elegans*, and maize, it has been observed that chromosomes exhibit dynamic and complex movements during early stages of meiotic prophase. These movements are particularly dramatic in fission yeast, where they are known as "horse-tail" movements, in which the entire nucleus moves violently back and forth. In maize, several types of meiotic prophase chromosome movements have been observed. Rapid short-distance movements of fairly small chromosome segments adjacent to the telomeres coincide with chromosome pairing. At the same time, interstitial chromosome segments exhibit more restrained motility. After pairing is completed, these movements are supplanted by movements of much longer chromosome segments (sometimes entire chromosome arms) that exhibit slow, sweeping motions across large extents of the nucleus. In addition to the movements of individual chromosome segments, the entire chromatin exhibits oscillating rotations, often by as much as 90°. The rotational movements are present both during chromosome pairing and also after pairing is completed. The earlier, pairing-associated movements may facilitate chromosome homology recognition by allowing many pairing combinations to be tried until a proper homologous interaction is found. The later, slower movements may help to resolve chromosome entanglements that remain after the conclusion of chromosome pairing.

Meiotic prophase chromosome movements rely on the attachment of chromosome ends to the nuclear envelope, and forces generated in the cytoskeleton are necessary for the movements to occur. In budding yeast, cytoplasmic actin cables have been implicated in facilitating chromosome motility. In fission yeast, *C. elegans*, and rats, microtubules are involved. In maize, chromosome movements require both actin and tubulin.

Observations of the vigorous chromosome motility in early meiotic prophase and examinations of the role that these movements may play in chromosome pairing in several species, including *C. elegans*, which lacks the bouquet, shed a new light on the function of the bouquet. It is likely that the bouquet's main role is providing chromosome attachment to the nuclear envelope and transmitting forces that generate chromosome movements, rather than just bringing chromosome ends together.

Role of centromeres in chromosome pairing. In contrast to telomeres, much less is known about the role of centromeres in chromosome pairing. In some organisms with relatively large genomes, centromeres cluster opposite telomeres in interphase cells and at the onset of meiosis. This arrangement is called Rabl orientation (**Fig. 3**). The presence of Rabl may provide a role for centromeres in chromosome pairing. In tetraploid and hexaploid wheat, the centromere clustering is regulated by the *Ph1* locus. The functions of *Ph1* are to reduce association of nonhomologous centromeres and promote association of centromeres of homologous chromosomes. However, the *Ph1* locus does not affect telomere clus-

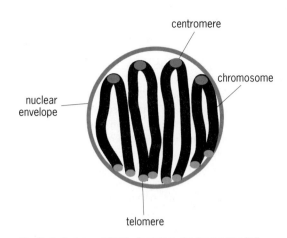

Fig. 3. A diagram of Rabl orientation. Centromeres of all chromosomes are located on the opposite side of the nucleus away from the telomeres. The nuclear envelope is also shown. (*Adapted from W. P. Pawlowski, Nuclear organization and dynamics in plants, Curr. Opin. Plant Biol., 13:640–645, 2010*)

tering or pairing of chromosome ends, suggesting that centromere pairing is independent of pairing of other chromosome regions.

Centromere associations that precede pairing of chromosome arms have also been described in budding yeast, a species with a relatively small genome that does not exhibit Rabl orientation. These associations are initially formed between nonhomologous centromeres but are converted to associations of centromeres of homologous chromosomes at the onset of chromosome pairing by a mechanism that depends on SPO11. Both nonhomologous and homologous centromere associations require installation of synaptonemal complex proteins in the centromere region. As observed in wheat, centromere associations in yeast are independent of the telomere bouquet formation. However, in contrast to wheat, chromosome arm pairing in yeast depends on the initial pairing at the centromere.

Conclusions. There are three main conclusions: (1) During early meiotic prophase, telomeres of all chromosomes attach on the nuclear envelope. In most species, the nuclear envelope attachment is followed by telomere clustering, known as telomere bouquet formation, which brings chromosome ends into close proximity. This configuration may aid chromosome interactions. (2) A SUN/KASH domain protein complex connects chromosome ends to the cytoplasmic cytoskeleton, which is essential for promoting chromosome movement. Telomere-led chromosome motility during early prophase I facilitates chromosome pairing. (3) In some species, such as budding yeast and polyploid wheat, chromosome pairing is promoted by centromere associations that precede pairing along chromosome arms. Centromere pairing is independent of both telomere bouquet formation and pairing of chromosome ends.

For background information *see* CELL BIOLOGY; CELL CYCLE; CELL DIVISION; CELL MOTILITY; CELL NUCLEUS; CHROMOSOME; CHROMOSOME ABERRA-

TION; CYTOSKELETON; GENETICS; MEIOSIS; PROTEIN in the McGraw-Hill Encyclopedia of Science & Technology.   Choon-Lin Tiang; Wojciech P. Pawlowski

Bibliography. N. Bhalla and A. F. Dernburg, Prelude to a division, *Annu. Rev. Cell Dev. Biol.*, 24:397–424, 2008; L. Harper, I. Golubovskaya, and W. Z. Cande, A bouquet of chromosomes, *J. Cell Sci.*, 117:4025–4032, 2004; R. Koszul and N. Kleckner, Dynamic chromosome movements during meiosis: A way to eliminate unwanted connections?, *Trends Cell Biol.*, 19:716–724, 2009; M. J. Sheehan and W. P. Pawlowski, Live imaging of rapid chromosome movements in meiotic prophase I in maize, *Proc. Natl. Acad. Sci. USA*, 106:20989–20994, 2009; T. Tsubouchi, A. J. MacQueen, and G. S. Roeder, Initiation of meiotic chromosome synapsis at centromeres in budding yeast, *Genes Dev.*, 22:3217–3226, 2008.

Sestrins

Sestrins are a family of proteins. Originally discovered in 1999, they are observed in most animals. There are three Sestrins—Sestrin1, Sestrin2, and Sestrin3—in mammals, whereas only a single Sestrin can be found in most invertebrates. Studies have shown that Sestrins accumulate in cells and tissues that have been exposed to environmental stresses, such as irradiation, oxidative damage, hypoxia (oxygen deficiency), and chemical stresses (**Fig. 1**). Sestrins can exert antioxidant and antigrowth effects that can protect cells from stress-induced damages and their detrimental consequences. Recent studies have also demonstrated that Sestrins have important physiological functions that protect animals from diverse age-associated diseases that are promoted by stresses.

Sestrin as an antioxidant molecule. It has been long anticipated that accumulation of oxidative damage can cause premature senescence of cells and organisms. Reactive oxygen species (ROS) damage many cellular macromolecules that are critical for physiological processes, including proteins, lipids, and nucleic acids. If unrepaired, these damages can cause physiological malfunction that can lead to cellular catastrophes such as apoptotic or necrotic cell death. Dietary supplementation of antioxidants or increased expression of ROS-scavenging proteins has been previously shown to increase life span and health in many model organisms.

The first biochemical activity of Sestrins to be described in the literature was the reduction of ROS by reinforcing the activity of peroxiredoxin, which is a critical antioxidant molecule that scavenges ROS. As an oxidoreductase, Sestrin can regenerate inactive peroxiredoxin to become a functional molecule, through its catalytic cysteine located in the N-terminal domain. Mutation of the cysteine residue abolished the role of Sestrin in peroxiredoxin regeneration, nullifying its antioxidant activity. In addition to acting as a direct oxidoreductase, Sestrin can strengthen mitochondrial quality-control mechanisms, removing dysfunctional mitochondria that produce pathogenic levels of ROS.

Moreover, production of Sestrin proteins is increased in cells challenged with oxidative stress. Thus, Sestrin induction after oxidative stress is very critical for cells to survive under the stressful condition. Indeed, cells that are not expressing Sestrins are very vulnerable to oxidative damage–induced cell death. Conversely, cells with high levels of Sestrin expression are protected from the oxidative insults. Sestrin's antioxidant role may also contribute to the antiaging activities of Sestrins.

Sestrin as a cell growth suppressor. Another important function of Sestrins is to suppress cell growth. By slowing down cell growth, Sestrins allow cells to gain sufficient time and resources for precise repair of damaged molecules. This growth-suppressive action of Sestrins is especially important for reducing DNA mutation as a result of genotoxic stresses, such as irradiation and chemical mutagens. Sestrin-mediated cell growth inhibition can also prevent the proliferation of genetically altered cells. Thus, Sestrin can suppress occurrence of cancer cells by reducing DNA damage and suppressing the proliferation of damaged cells.

Indeed, expression of Sestrins prevented tumorigenic processes in diverse experimental systems. In cultured cells, Sestrin expression suppressed the oncogenic transformation caused by constitutively active Ras protein. In *Drosophila* (fruit fly) tissues, Sestrin expression suppressed hyperplastic cell growth caused by chronic activation of target of rapamycin (TOR); in contrast, loss of Sestrin promoted TOR-stimulated cell growth. TOR is a cell growth-promoting kinase, and S6 kinase (S6K) and translation initiation factor 4E–binding protein (4E-BP) are targets of the TOR mediating cell growth control. Sestrin inhibits TOR activity through activation of adenosine monophosphate–activated protein kinase (AMPK) and tuberous sclerosis complex 2 (TSC2), which together antagonize TOR signaling. Interestingly, Sestrin itself can be induced by hyperactive TOR, through a signaling module containing c-Jun N-terminal kinase (JNK) and forkhead box O (FoxO) transcription factor. Thus, Sestrin can be viewed as a feedback regulator of the cell growth–stimulatory signaling pathway. Considering that TOR signaling

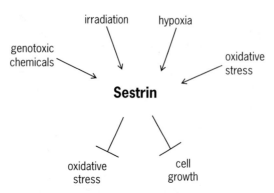

Fig. 1. Function of Sestrin at the cellular level. After being induced by a variety of stresses, Sestrins suppress cell growth and accumulation of reactive oxygen species (ROS).

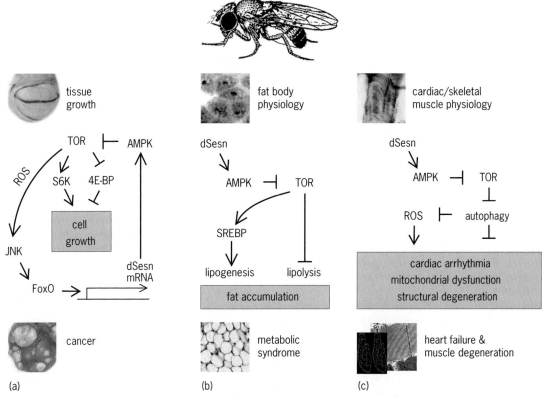

Fig. 2. Function of Sestrin at the organism level. In *Drosophila* (fruit fly), Sestrin-deficient mutants show (*a*) cell growth enhancement, (*b*) fat accumulation, and (*c*) cardiac and skeletal muscle degeneration. These phenotypes are very similar to human age-associated pathologies such as cancer, metabolic syndrome, cardiac dysfunction, and skeletal muscle degeneration. The diagrams between the *Drosophila* phenotypes and related human pathologies depict the role of *Drosophila* Sestrin (dSesn) in the biological signaling network. In the diagram, arrows indicate stimulation or activation between the two components or processes, whereas blunt-ended lines indicate inhibition or suppression. (*a*) TOR is a growth-stimulating protein kinase. TOR activates S6K, which stimulates cell growth, and inhibits 4E-BP, which suppresses cell growth. dSesn is transcriptionally activated by chronic TOR activity through ROS, JNK, and FoxO. Induction of dSesn in turn inhibits TOR activity through augmentation of AMPK, which suppresses TOR. Thus, the loss of dSesn liberates TOR from the feedback inhibition, resulting in enhanced cell growth. (*b*) TOR induces fat accumulation by stimulating lipogenesis through SREBP and suppressing lipolysis. Because dSesn inhibits TOR activity through AMPK, the loss of dSesn induces chronic TOR activation and fat accumulation. (*c*) Chronic TOR activation can inhibit autophagy, which is required for suppression of ROS accumulation and age-associated degeneration of cardiac and skeletal muscles. Loss of dSesn induces chronic TOR activation and the age-associated pathologies. Abbreviations: dSesn, *Drosophila* Sestrin; AMPK, adenosine monophosphate–activated protein kinase; FoxO, forkhead box O; JNK, c-Jun N-terminal kinase; ROS, reactive oxygen species; SREBP, sterol response element binding protein; S6K, S6 kinase; TOR, target of rapamycin; 4E-BP, translation initiation factor 4E–binding protein.

is chronically activated in many cancers, it is very likely that Sestrins may be a general attenuator of tumor cell growth (**Fig. 2***a*).

AMPK–TOR signaling: link between nutrients and metabolic homeostasis. AMPK–TOR signaling is important not only for control of cell growth but also for maintenance of cellular energy level (metabolic homeostasis). To perform this function, the activity of the AMPK–TOR signaling itself is regulated by cellular energy status. Adenosine triphosphate (ATP), the main cellular energy compound, is synthesized by the energy obtained from burning calories (in the form of fats, sugars, and proteins). ATP emits chemical energy when it is converted to adenosine diphosphate (ADP), and this energy fuels diverse life phenomena, including muscle contraction, replication of genomic information, and neuronal signal transmission. When cellular energy is very low, cells accumulate adenosine monophosphate (AMP), which is a derivative of ADP. The AMP ac-

cumulation activates AMPK, thereby inhibiting TOR. Therefore, AMPK is activated under nutrient-limiting conditions, whereas TOR is activated under nutrient-rich conditions.

As protein kinases, AMPK and TOR phosphorylate and regulate the activity of many metabolic enzymes and transcriptional factors that control the expression of enzymes and their cofactors. In nutrient-rich conditions, low AMPK and high TOR boost cellular synthesis of biomolecules, such as proteins and lipids, facilitating cell growth. On the other hand, during nutrient deprivation, high AMPK and low TOR stimulate the process known as autophagy (self-eating), which digests unnecessary cellular compartments. Importantly, autophagy is a critical process that maintains the cleanness of intracellular compartments by removing protein aggregates, excessive lipid droplets, and damaged or malfunctioning organelles. Mitochondria, which are the organelles that produce ATP by burning calories, are among the most important organelles subjected to an autophagy-mediated quality control. The autophagy-dependent mitochondrial quality control is very significant because malfunctioning mitochondria can produce numerous ROS that damage many biomolecules.

Recently, autophagy was suggested to mediate the beneficial effects of caloric restriction, which has been shown to extend life span and health in virtually every model organism (including primates). Caloric restriction results in a low energy level that activates AMPK and inhibits TOR. This in turn induces autophagy, which contributes to elimination of unnecessary and harmful deposits inside the cells, including toxic protein aggregates that can induce neurodegeneration, intracellular lipid that accumulates in metabolic syndrome (a cluster of metabolic risk factors for cardiovascular disease), and malfunctioning mitochondria that induce oxidative stress and facilitate aging. In contrast, overnutrition and lack of exercise, which result in obesity, can induce chronic AMPK inhibition and TOR activation, thereby suppressing autophagy. Chronic suppression of autophagy facilitates aging because it causes accumulation of dysfunctional mitochondria, producing pathogenic levels of ROS. Therefore, obesity is linked with various age- and ROS-associated degenerative diseases such as heart attack, metabolic syndrome, cancer, and muscle degeneration.

In many animal models, including worms, flies, and mice, genetic induction of AMPK or suppression of TOR was shown to extend both life span and health. Pharmacological administration of the TOR inhibitor, rapamycin, also extended life span in flies and mice. Elimination of autophagy in different mouse tissues resulted in a diverse range of age-associated pathological phenomena, including hepatic fat accumulation, neurodegeneration, and cardiac and skeletal muscle degeneration.

Sestrin: an AMPK–TOR controller that prevents age-associated pathologies. Sestrin's biochemical activities in inhibiting TOR and reducing ROS suggest that the molecule may have antiaging functions. Indeed, in the *Drosophila* model, Sestrin has been shown to suppress diverse age-associated pathological symptoms such as fat accumulation, cardiac arrhythmia, and muscle degeneration, which are very similar to the respective human diseases.

Drosophila has a metabolic organ named the fat body, which is homologous to the mammalian liver. Upon diet-induced or genetic obesity, *Drosophila* accumulates high levels of triglycerides in their fat body; this is similar to the obesity-induced liver fat accumulation in mammals such as mice and humans. Inhibition of AMPK or activation of TOR can also induce fat accumulation through the augmentation of the activity of sterol response element binding protein (SREBP), which is a lipogenic transcription factor. Interestingly, even without any dietary modulation, *Drosophila* Sestrin-null mutants spontaneously accumulate triglycerides in their fat body. This fat accumulation has been reduced by pharmacological treatments that activate AMPK or inhibit

TOR, meaning that Sestrin is needed to keep the signaling pathway under control for metabolic homeostasis (Fig. 2*b*).

As observed in humans and other animals, heart functionality in *Drosophila* gradually declines in accordance with aging, and this age-dependent decline can be facilitated by high TOR activity caused by genetic mutation or high caloric diet. The decline in heart functionality can be associated with cardiac arrhythmia, which is predictive of age-associated heart failure. Sestrin-null mutant flies exhibit strong cardiac arrhythmia at two weeks of age, which is not an age that shows extensive arrhythmicity. This is associated with structural degeneration of cardiac muscle. Interestingly, degeneration also has been observed in skeletal muscles, and this has been accompanied by accumulation of dysfunctional mitochondria and disorganization of myofiber microstructure. The degeneration of cardiac function and skeletal muscle structure was suppressed by AMPK activation, TOR inhibition, and ROS reduction, indicating that Sestrin-dependent regulation of the AMPK–TOR pathway and oxidative stress signaling are important for maintenance of muscle homeostasis (Fig. 2*c*).

Finally, recent studies in mouse models indicate that oxidative defense, autophagic machinery, and AMPK–TOR regulation are critical to attenuate aging and age-associated disorders. As discussed previously, caloric restriction, which activates AMPK and autophagy and reduces TOR and ROS, attenuates aging and prevents age-associated pathologies (**Fig. 3*a***). As a feedback regulator of AMPK–TOR

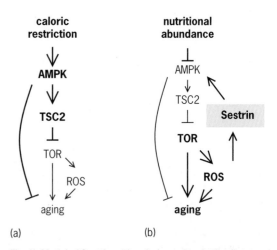

(a)　　　　　　　　(b)

Fig. 3. Model of Sestrin action during aging. In the diagram, arrows indicate stimulation or activation between the two components or processes, whereas blunt-ended lines indicate inhibition or suppression. Thick lines and large fonts mean that the interactions and components are active (operational) in the condition. Thin lines and small fonts mean that the interactions are inactive and signaling components are suppressed. (*a*) During caloric restriction, adenosine monophosphate–activated protein kinase (AMPK) and tuberous sclerosis complex 2 (TSC2) suppress target of rapamycin (TOR) and reduce reactive oxygen species (ROS), resulting in aging attenuation. (*b*) During nutritional abundance, which suppresses AMPK and TSC2, TOR is chronically activated and causes ROS accumulation and facilitation of aging. Sestrin, induced by chronic TOR activity, is critical to maintain the activity of AMPK and attenuate chronic TOR activation and aging during obesity.

signaling, Sestrins may have a very important role in attenuating aging and age-associated diseases, especially under conditions of nutritional abundance and obesity (Fig. 3b). Therefore, if the role of Sestrins can be clarified in mammalian pathogenesis models, it might be possible to devise a novel method to attenuate the age- and obesity-associated diseases that are prevalent in modern society.

For background information *see* AGING; AMP-ACTIVATED PROTEIN KINASE (AMPK); ANTIOXIDANT; AUTOPHAGY; BIOLOGICAL OXIDATION; CANCER (MEDICINE); CELL (BIOLOGY); HYPOXIA; METABOLISM; OBESITY; PROTEIN in the McGraw-Hill Encyclopedia of Science & Technology.   Jun Hee Lee

Bibliography. A. V. Budanov et al., Stressin' Sestrins take an aging fight, *EMBO Mol. Med.*, 2:388–400, 2010; M. Laplante and D. M. Sabatini, mTOR signaling at a glance, *J. Cell Sci.*, 122:3589–3594, 2009; J. H. Lee et al., Sestrins at the crossroad between stress and aging, *Aging*, 2:370–375, 2010; I. Topisirovic and N. Sonenberg, Cell biology: Burn out or fade away?, *Science*, 327:1210–1211, 2010.

Soft body armor

Human history is marked by the constant struggle to survive; that is, to overcome or defeat threats, such as natural environmental events, hostile animals, and human conflicts. Depending on the threat, protection has been sought in various ways, including shelter to avoid proximal threats, mobility to flee confrontation, and protective clothing and devices to directly engage the enemy. This investigation explores only protective clothing and focuses on the materials, manufacturing, and testing of fabric-based, protective clothing, namely soft body armors, used for ballistic protection.

Coping and survival instincts led to early developments in body armors often taking the forms of protective clothing and primitive shielding devices. Body armors—defined as any defensive coverings worn to protect the body from physical attacks—have evolved from readily available materials such as animal skins or natural fibers made from thatch, cotton, and silk that were (or are) often woven in textile forms to metals such as copper, steel, and iron that were used in plate and chainmail forms to the technologically complex armors that are used by today's armed services and law enforcement agencies. In short, sophisticated weaponry increases threat effectiveness levels, which, in turn, drives the search for enhanced body armors.

Recent innovations in materials and manufacturing technology during the twentieth century led to the discovery of advanced synthetic textile materials (such as nylon, fiberglass, Kevlar™, and many other synthetic fibers) that have provided body armor with extraordinarily improved ballistic protection levels at significantly reduced weight. This is a potent combination for enhancing the effectiveness and mobility of military troops, law enforcement officers, and security personnel. While those same demands

(increased protection at decreased weight) continue today, it is recognized that future improvements will be increasingly difficult to achieve because the financial costs associated with developing new fibers are becoming prohibitive, and the time-to-market for their commercialization remains long term.

Design criteria. Body armors must be worn to be effective. Weight, mobility, and comfort therefore are vital to ensuring their use. The armors must conform to a user's body, properly distribute their weight over the body to minimize user fatigue, provide sufficient breathability for extended use, especially during high temperatures, and must not interfere with or restrict a user's mobility. The significant challenge is to balance the level of protection required for specific threats against weight, comfort and flexibility, cost, environmental exposure (heat, ultraviolet light, moisture), and service life.

The principal factor that dictates the design of body armors is the types of threats for which protection is required, such as ballistic, fragment, blast, stab, slash, chemical, fire, and so on. Armors optimized for protection against one threat type may not be suitable for other threat types. For example, textiles designed for ballistic protection require sufficient yarn mobility within the weave to avoid premature failures and will not perform well for stab protection. Textiles designed for stab resistance require dense weaves to prevent yarns from being pushed aside from the tip of sharp-pointed objects such as knives, needles, awls, and ice picks. Dense weaves that prevent punctures can lead to premature or punch-through failures in ballistic impacts. Design parameters for optimizing ballistic versus stab defense often work against each other, as shown by **Fig. 1**. Multi-threat armors are commonly designed by integrating separate armoring solutions, a process that achieves only minimal synergistic efficiencies at best. Armors that combine multiple defeat elements are often categorized as "in-conjunction" armors, in which each component provides an enhanced level of protection for a given threat or multiple threat types.

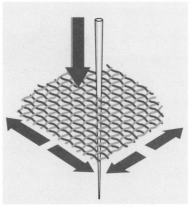 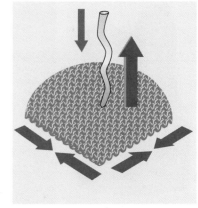

(a)   (b)

Fig. 1. Puncture behaviors of ballistic versus stab-resistant woven fabrics. (*a*) Ballistic fabrics enable yarn motions and do not prevent punctures. (*b*) Dense, tightly woven fabrics restrict yarn motions and prevent defeat punctures.

woven fabric

UD cross-ply laminate
(spectra shield™)

film

fibers and resin

film

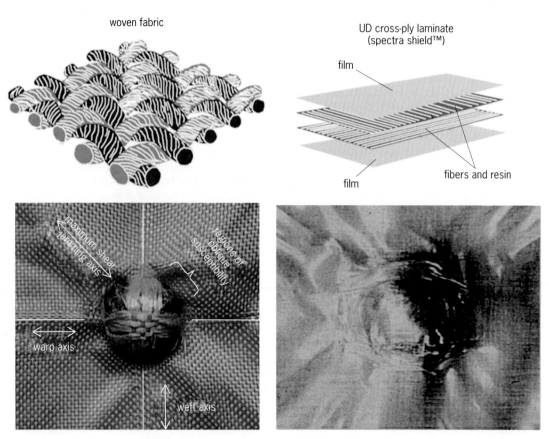

Fig. 2. Examples of woven and UD fabric laminate constructions with ballistic impact deformations shown. (*Courtesy of Honeywell Advanced Fibers and Composites, Inc.*)

Traditionally, soft body armors for ballistic protection were manufactured using layers of woven fabrics stitched together. Now, they include laminates stacked with nonwoven, unidirectional (UD) layers and combinations of woven and nonwoven laminates. Considering the UD laminates, fibers within each UD layer are aligned in a parallel arrangement and are reinforced with a compliant polymer resin or matrix, such as Kraton™ (styrene block copolymers; elastomers), that binds the fibers together. The UD layers are produced in very thin sheet forms

and are stacked in an alternating 0°/90° cross-ply fashion as shown in **Fig. 2**. Polyethylene films are added to protect the layers, and the final laminated shape is attained by applying heat and pressure. Commercial UD laminates used for ballistic protection include Honeywell's Spectra Shield™ [ultrahigh molecular weight polyethylene (UHMWPE) fibers], Gold Shield™ (Kevlar fibers), and DSM's Dyneema™ (UHMWPE fibers). In contrast, hard-textile or composite armors, such as helmets, are not flexible and are defined as those using a rigid resin material to bind the fibers together. Today's textile-based armors, such as bullet-resistant vests and helmets, integrate many sophisticated polymer materials and textile processing technologies that are optimized across multiple dimensional scales.

Standards for military and law-enforcement personnel. Current soft body armors used for ballistic protection are worn to protect the torso and extremity regions. They are developed in conjunction with rigorous standards and specifications to ensure proper performance and reliability levels against ballistic and fragment threats. For example, the National Institute of Justice (NIJ) prepared the *Ballistic Resistance of Body Armor NIJ Standard-0101.06* to categorize ballistic threats, including projectile types, sizes, and velocities; establish deformation limits; develop sample conditioning protocols; and specify acceptance testing procedures for nonmilitary body armors as shown in **Fig. 3**. **Table 1** lists the NIJ Standard-0101.06-specified projectile types

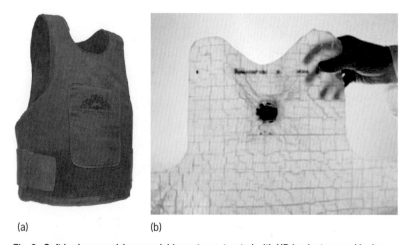

(a)   (b)

Fig. 3. Soft body armor (*a*) concealable vest constructed with UD laminates used by law enforcement officers for ballistic protection. (*b*) Ballistic test showing arrested projectile. (*Photograph courtesy of TurtleSkin*®)

TABLE 1. National Institute of Justice body armor standards

Armor type	Test round	Test bullet*	Bullet mass	Armor test velocity	Hits per panel at 0° angle	Maximum back face signature	Hits per panel at 30° or 45°
IIA	1	9 mm, FMJ RN	8.0 g (124 gr)	373 m/s (1224 ft/s)	4	44 mm (1.73 in)	2
	2	.40, S&W FMJ	11.7 g (180 gr)	352 m/s (1155 ft/s)	4	44 mm (1.73 in)	2
II	1	9 mm, FMJ RN	8.0 g (124 gr)	398 m/s (1306 ft/s)	4	44 mm (1.73 in)	2
	2	.357 Magnum, JSP	10.2 g (158 gr)	436 m/s (1430 ft/s)	4	44 mm (1.73 in)	2
IIIA	1	.357, SIG FMJ FN	8.1 g (125 gr)	448 m/s (1470 ft/s)	4	44 mm (1.73 in)	2
	2	.44, Magnum SJHP	15.6 g (240 gr)	436 m/s (1430 ft/s)	4	44 mm (1.73 in)	2
III	1	7.62 mm NATO FMJ	9.6 g (148 gr)	847 m/s (2,780 ft/s)	6	44 mm (1.73 in)	0
IV	1	.30 Caliber M2 AP	10.8 g (166 gr)	878 m/s (2,880 ft/s)	1 to 6	44 mm (1.73 in)	0

*AP = armor piercing, FMJ = full metal jacket, FN = flat nose, JSP = jacketed soft point, RN = round nose, SIG = sig Sauer, SJHP = semi jacketed hollow point, S&W = Smith & Wesson 1.0 gram (g) = 15.4324 grains (gr).

(deformable, steel-jacketed, high-hardness core, armor-piercing), velocities, and maximum allowable backface signature (BFS) depths.

Acceptance testing of soft armors determines their ballistic-limit velocities for prescribed projectiles, projectile velocities, and angles of incidence. A variety of ballistic-limit velocities are defined, with each having a statistical significance. These include the V_0, V_{50} and V_{100} ballistic-limit velocities and are designated as the maximum velocity at which no complete penetration will occur, the velocity at which a 50% probability of complete penetration will occur, and the minimum velocity at which 100% probability of complete penetration will occur, respectively. Ballistic tests are performed on both dry and wet body armors by firing a number of projectiles at prescribed locations apart from each other, at angles of incidence of 0° (normal) and 30° (oblique), at seams, and at specific distances from the edges. Testing has shown that ballistic-limit velocities are proportional to the areal weight density of the woven fabrics.

Ballistic testing of military armors for personnel, vehicles, and other systems subject to small-arms munitions is governed by *Army Military Standard MIL-STD-662F*. Fragment testing of military personnel armors resulting from fragmenting munitions, such as grenades and mortar rounds, is performed in accordance with the *North Atlantic Treaty Organization (NATO) Standard Agreement STANAG 2920*. Fragment-simulating projectiles (FSPs) are often used as test projectiles, with 2-, 4-, 16-, and 64-grain sizes. The FSPs are shaped as right circular cylinders (RCCs), as shown in **Fig. 4,** with a fixed length-to-diameter ratio equal to 1.0 and are made of hardened steel to resist deformations upon impact. A study by the U.S. Army's Ballistic Research Laboratory concluded that 95% of all bomb fragments under four grains (0.26 g) have a limit velocity of 3000 ft/s (914 m/s) or less. The study also determined that a textile system with a minimum areal weight

density of 1.1 lb/ft² (5.4 kg/m²) was required to defeat fragment threats of the complete grain series at the limit velocities. *NIJ Standard-0101.06, Army MIL-STD-662F, and STANAG 2920* use the V_{50} designated ballistic-limit velocity.

Results of ballistic-impact tests are often reported by plotting the energy absorbed by the fabric versus the initial projectile velocity, V_i, as shown in **Fig. 5**. The ballistic limit graphically corresponds to the highest initial projectile velocity that does not produce through-penetration failures in the fabric.

Although soft body armors are used to prevent penetration by specified small-arms projectiles, deformations in the form of indentations can occur to the extent that further life-threatening injuries remain possible. Impact deformation limits are often specified to help minimize indentation depths, or BFSs. The NIJ standard specifies a maximum BFS of 44.0 mm (1.73 in.). BFSs as shown in **Fig. 6** may lead to blunt trauma injury, which is also known as behind-armor-blunt-trauma (BABT). Serious injury to tissues, skeletal structures, and organs can occur and

Fig. 4. The 2-, 4-, 16-, and 64-grain size fragment simulating projectiles (FSPs). (*Photograph courtesy of Warwick Mills, Inc.*)

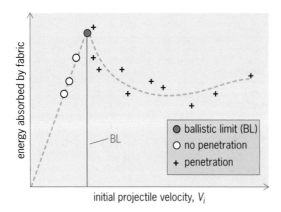

Fig. 5. Example of a ballistic-limit plot.

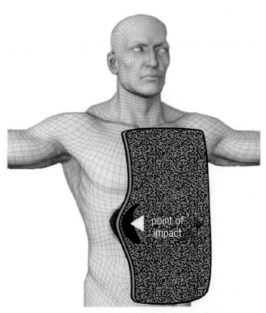

Fig. 6. Blunt trauma resulting from excessive impact deformation.

may be fatal. Blunt trauma may not be immediately detected and may manifest itself at a later time and can be damaging to organs remote from the impact site depending on the propagation of stress waves into the body.

BFS measurements are made during ballistic testing of vests backed with an oil-based modeling clay known as Roma Plastilina No. 1. This clay has mass properties similar to those of a human body and does not spring back after ballistic impact, a key feature for locking in the indentation depth for measurement purposes as shown in **Fig. 7**. The calibration process is used to qualify the clay for ballistics testing includes the following steps.

1. The clay must be maintained at a controlled temperature.

2. Five steel spheres, 63.5 mm ± 0.05 mm in diameter and 1043 g ± 5 g, are dropped from 2.0 m above the clay at defined spacing from the clay edges and each other as shown in **Fig. 8**.

3. The depth of clay deformation from each sphere is then measured.

The clay is qualified if the average depth measures at 19 mm ± 2 mm, with no depths measured less than 16 mm or greater than 22 mm.

Bullet-resistant soft body armor vests. Today's body armor vests are often constructed with lightweight breathable nylon or cotton outer shells that include ballistic packs or panels contained within carriers (pockets). The ballistic packs are assembled from woven, nonwoven, or combined woven and nonwoven fabrics and can prevent penetrations by NIJ threat categories IIA, II, and IIIA with a sufficient number of layers. For example, 20–30 layers of fabric may be used to arrest deformable projectiles fired from handguns, as shown in **Fig. 9**.

Further protection from blunt trauma is achieved through the addition of rigid trauma plates or inserts mounted in carriers of vests to further distribute the impact force in the plane of the armor. Threat levels III and IV, however, are designated for much higher velocity and hardness rounds fired from rifles that can easily penetrate fabric armors. Military body armors, such as the interceptor body armor (IBA) vest for Army personnel and the releasable body armor vest for the U.S. Special Operations Command (USSOCOM) Special Operations Forces, protect against these higher-velocity threats by integrating rigid plates known as small-arms protective inserts (SAPI) or enhanced SAPI (ESAPI) plates (inserts made of boron-carbide ceramics, for example), as shown in **Figs. 10** and **11**. These plates are positioned within the carriers in front of the various strike faces of the armor vests to force the

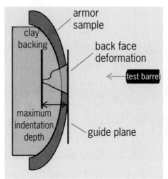

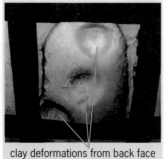

clay deformations from back face signatures of 3-shot placement test

Fig. 7. Clay-backed ballistic testing and back-face signatures (BFS).

projectile to erode (fracture) upon impact prior to any penetration in the fabric, thereby serving to spread the load throughout the armor plane, as shown in **Fig. 12**.

Maximizing energy-absorption levels. The design of woven fabrics for armor applications is complex because it requires an understanding of the related dynamics and the capability needed to optimize a system of systems. Numerous hierarchies are present; the smallest considered here is at the molecular level from which a single fiber is ultimately produced. Multiple fibers (or filaments) are bundled to form a yarn, yarns are woven to form a fabric layer, and fabric layers are stacked and joined to form body armors. Mechanical properties, however, do not efficiently translate across these hierarchies; that is, fiber properties do not directly translate to yarn properties, yarn properties do not directly translate to fabric properties and, likewise, single-ply behavior does not directly translate to multi-ply behavior for stacked layers. Quality-control testing, therefore, is typically done at each level.

To maximize energy-absorption levels, one must understand the materials and mechanics of (1) the fiber bundles within the yarns, (2) the type of woven architecture that forms the layers, and (3) the stacking arrangement and stitching patterns of the layers that form the ballistic packs. Engineers and scientists must consider the woven armor as a system of subsystems that span multiple dimensional scales in order to maximize protection levels, as shown in **Fig. 13**.

Polymer molecules. For convenience, the smallest scale considered in this review is the molecular level, having dimensional units described in nanometers (nm; 10^{-9} m). The molecular structures of polymer materials detail their building blocks, which are responsible for the desired fiber performance attributes, such as strength, stiffness, toughness, environmental and chemical resistance, and melting temperature. These attributes directly depend on the spatial arrangement and integrity of the chemical bonds formed during the polymerization process by which single monomer molecules are joined to form polymeric molecular chains. The types of organic polymers used in today's soft body armors include the aramids, polyesters, and polyethylenes.

Fibers. Next is the fiber (also referred to as filament) scale, in which fiber diameters are often measured in units of micrometers (μm; 10^{-6} m). Fibers weights are classified by denier, which is the linear density defined as the weight in grams of a 9000-m-long fiber (or yarn). Fiber tensile strength is defined as tenacity, having units of gram-force per denier (gpd). Tenacity generally increases with decreasing fiber diameter. The stiffness of a fiber is designated by its elastic modulus E. The elastic modulus is obtained from tensile tests of a fiber (or yarn) and has units of g-force per denier; it is computed as the initial slope of the tensile stress–strain curve. Many polymer fibers exhibit viscoelastic behavior (combined elastic and viscous traits) to the extent that tenacity and elastic modulus are sensitive to rates of loading;

Fig. 8. Calibration testing and depth measurements for Roma Plastilina No. 1 clay. (*Photograph courtesy of Warwick Mills, Inc.*)

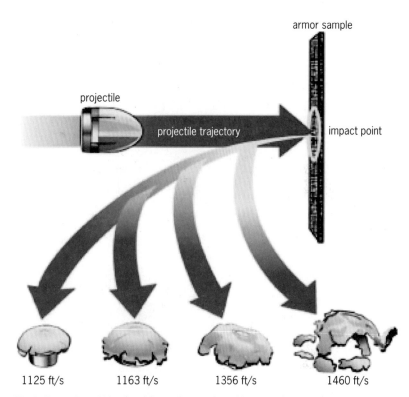

armor sample

projectile

projectile trajectory

impact point

1125 ft/s   1163 ft/s   1356 ft/s   1460 ft/s

Fig. 9. Formation of blunting deformations and petal fractures in bullet. (*Illustration courtesy of Warwick Mills, Inc. & Naval Undersea Warfare Center*)

that is, these properties can increase with increasing strain rates. Elongation at break is the amount of stretch that a fiber (or yarn) experiences during a tensile test at failure. Elongation is computed as a percent of the initial tested length. Additional properties helpful for weight-sensitive applications are specific strength and specific modulus—the strength and modulus values divided by the fiber density, respectively; both have units of length alone. Specific strength is also referred to as the breaking length, which is equivalent to the length of fiber required to break under its own weight when hanging vertically. Specific gravity is the ratio of the material density to the density of water. Fibers are buoyant if their specific gravities are less than one. The properties of various high-performance fibers are listed in **Table 2** in conjunction with those reported by H. H. Yang, with steel shown for comparative purposes.

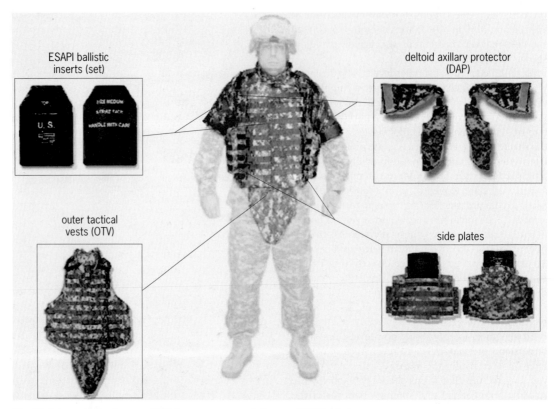

Fig. 10. Interceptor body armor (IBA) for army personnel. (*Source: PEP Soldier*)

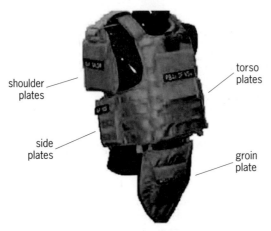

Fig. 11. Releasable body armor vest for USSOCOM special operations forces.

Fiber production. Polymer fibers are produced using a variety of spinning methods, including dry, wet, gel, and melt spinning, as depicted in **Fig. 14**. Spinning refers to the process of extruding fibers through a series of small holes in devices known as spinnerets. Spinnerets used to produce synthetic fibers are dies that closely resemble showerheads. The polymers (and solvents if present) are forced through holes in the spinnerets. As the polymer exits the spinneret, the polymer solidifies, forming fibers having controlled and consistent diameters and cross-sectional shapes with nearly unlimited lengths. The fibers are then stretched and drawn onto take-up rollers. Stretching further enhances the fibers' tensile strength and toughness properties by aligning the molecular chains along the fiber axis.

Fiber-spinning methods are selected based on polymer compatibility; for example, thermoplastic polymers require melt spinning, and thermoset

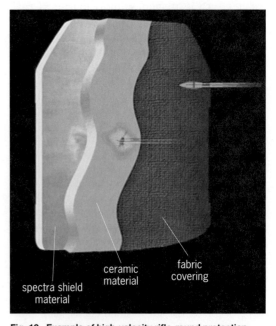

Fig. 12. Example of high-velocity rifle-round protection using a ceramic strike face backed with Spectra Shield™. (*Illustration courtesy of Honeywell Advanced Fibers and Composites, Inc.*)

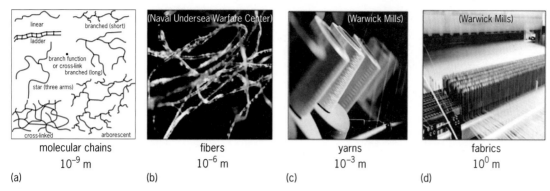

Fig. 13. Multiple dimensional scales for polymer material forms used in soft body armor. (*a*) Molecular chains (*courtesy McGraw-Hill*). (*b*) Fibers. (*c*) Yarns. (*d*) Fabrics.

polymers require dissolution in a solvent. Aramid (Kevlar™) and UHMWPE (Dyneema™ and Spectra™) fibers are produced by gel spinning; nylon (polyamide), Vectran™ (liquid crystal polyester), and poly(ethylene terephthalate) [PET] fibers are formed by melt spinning; and polybenzimidazole (PBI™) fibers are made by dry spinning. Additional postprocessing methods are also applied to fibers, including sizings. Sizings are surface-treatment agents applied to the fibers or yarns to (1) improve performance in the final product, (2) reduce abrasion for improved handling and weaving processes, (3) control moisture absorption, (4) protect the fibers from environmental effects, and (5) increase compatibility for bonding with matrix materials in fiber-reinforced composites.

Today's fiber research areas include (1) the development of next-generation ultrahigh performance fibers through advanced polymer chemistry, (2) carbon-nanotube reinforcement technologies to produce unprecedented fiber strengths, and (3) polymer-spinning processes capable of reducing fiber diameters from the micro- to the nanoscale.

Yarns. Production of today's woven body armors requires the bundling of tens to hundreds of continuous fibers to create a single yarn resulting in cross-sectional dimensions measured at the millimeter scale (mm; 10^{-3} m). The number of fibers within a yarn is referred to as the filament count. Yarns constructed of continuous filaments often align the fibers in a straight configuration or in a slightly twisted helix and are processed onto pirns (rods) or bobbins. The helical fiber arrangement results from the addition of twist. Twist is a mechanism that can significantly increase the tensile strength of staple (discontinuous) yarns; twist, however, is used only minimally for continuous fiber yarns to improve handling during weaving operations by restricting the lateral motions of individual fibers. Twist is measured by the number of turns per unit length of yarn. Yarns are often categorized by denier rather than filament count, with many woven body armors constructed of deniers from 500 down to as low as 70.

Examples of ongoing research of continuous filament yarns include (1) hybridization of yarns spun

TABLE 2. Properties of several high-performance fibers*

Fiber	Manufacturer	Grade	Polymer type	Spinning method	Density, g/cc	Strength, gpd	Modulus, gpd	Elongation at break, %	Specific strength, 10^6 in	Specific modulus, 10^8 in	Fiber diameter, $\times 10^{-6}$ in	Maximum temperature, °C
Dyneema	DSM	SK75, SK78	Polyethylene	Gel	0.97	38-45	1267-1552	3-4				
		SK60, SK62, SK65	Polyethylene	Gel	0.97	28-38	759-1158	3-4			12-21	
		SK25	Polyethylene	Gel	0.97	25	608	3-4	25.8			
Spectra	Allied Signal	1000	Polyethylene	Gel	0.97	35	2000	2.7	13.4	7.6	28	100
		900	Polyethylene	Gel	0.97	30	1400	3.5	11.5	5.3	38	100
		149	Aramid	Gel	1.47	18	1100	1.5	6.9	4.2	12	250
		129	Aramid	Gel	1.45	26.5	750	3.3	10.1	3	12	250
Kevlar	DuPont	119	Aramid	Gel	1.44	24	470	4.4	9.2	1.6	12	250
		49	Aramid	Gel	1.45	23	950	2.8	8.8	3.6	12	250
		29	Aramid	Gel	1.43	23	580	3.6	8.8	2.1	12	250
Nomex	DuPont		Aramid	Wet	1.38	5	140	22	1.9	0.5	121	250
Nylon		6,6	Polyamide	Melt	1.14	9	50	19	3.4	0.2	25	150
Technora	Tejin Aramid		Aramid	Dry	1.39	27	570	4.3	10.3	2.2	12	250
Twaron	Tejin Aramid		Aramid									
Vectran	Kuraray America	HT	Polyester	Melt	1.43	25.9	600	3.8	9.89	2.3	N/A (Chars >400)	
		UM	Polyester	Melt								276
PB	PBI Performance Products		Polybenzimidazole	Dry	1.43	3.1	45	30	1.2	0.2		250
PBO	Toyobo	HM	Polybenzobisoxazole		1.56	42	2034					
PBT			Polybenzobisthiazole		1.57	25	2690	1.3	9.6	10.3		350
PET					1.39	9.5	100					
E-Glass	Owens Corning		Glass	Melt	2.55	11.6	320	3	4.4	1.2	5-25	350
5-Glass	Owens Corning		Glass	Melt	2.48	21.9	390	5.3	8.4	1.5	5-15	300
Steel					7.8	11	220	4.8	4.2	0.8		500

*CONVERSIONS:
GPa = gpd × density (g/cc)/11.33.
Specific strength (in) = tenacity (gpd) × 3.82 × 10^5.
Specific modulus (in) = modulus (gpd) × 3.82 × 10^5.

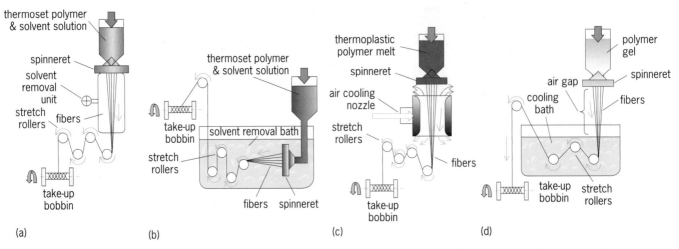

Fig. 14. Spinning methods for producing polymer fibers. (*a*) Dry spinning. (*b*) Wet spinning. (*c*) Melt spinning. (*d*) Gel spinning.

from comingling or coextruding multiple polymers, (2) nanotechnology reinforcements, and (3) strain-rate effects on tenacity and modulus properties.

Woven fabrics. A single layer of woven fabric has characteristic length and width, with dimensions on the meter scale (10^0 m). A plain weave is formed by interlacing yarns of two principal families, designated warp and weft, at right angles to each other, as depicted in **Fig. 15**.

The yarns of each family pass over and under yarns of the crossing family in a periodic fashion. Woven fabrics are referred to as crimped fabrics because yarns of one direction are bent around their cross-ing neighbor yarns. Warp yarns run parallel to the selvage (fabric edges) and are virtually unlimited in their length. The weft (or fill) yarns run across the fabric width. The undulations, which are referred to as crimp, are shown in Pierce's geometric fabric model (**Fig. 16**) for a plain weave. Pierce's geomet-ric model relates the fabric parameters as they are coupled among yarn families. The crimp height, h, is related to the crimp angle, α, and yarn length, L, as measured between yarns and the sum of yarn diam-eters at the crossover regions by the equations de-scribed by J. W. S. Hearle, P. Grosberg, and S. Backer in Fig. 16. Crimp, denoted as C, is the amount of waviness produced in a yarn when woven in fabric form, as shown in **Fig. 17**; it is a geometric prop-erty of the weave because of the woven architecture used. Crimp is obtained by measuring the length of a yarn in the woven state, L_{fabric}, and the length of that same yarn after being extracted from the fabric and straightened, L_{yarn}, and computed according to Eq. (1) as a percentage.

$$C = \frac{L_{yarn} - L_{fabric}}{L_{fabric}} \qquad (1)$$

Often, crimp contents are greater for warp yarns than for weft yarns, because of the differences in yarn-weaving tensions. **Figure 18** shows an ex-tracted warp yarn removed from the woven fabric of Fig. 17 to demonstrate the permanent crimping de-formation. The category of woven fabrics includes a variety of weaving architectures, such as plain, bas-ket, twill, satin, braid, leno, and triaxial weaves.

The architecture of the woven fabric is further described by the yarn's cross-sectional dimensions, number of warp yarns per unit fabric width, number of weft yarns per unit fabric length, and cover fac-tor, all of which affect the energy-absorption levels. Additionally, the weight of the fabric is defined by its areal weight density and often expressed in ounces per square yard.

Several woven architectures are used in soft body armors, including the plain and basket weaves

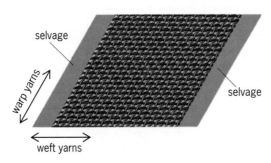

Fig. 15. Warp and weft yarn directions for a plain-woven fabric.

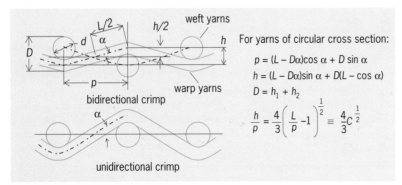

For yarns of circular cross section:

$$p = (L - D\alpha)\cos \alpha + D \sin \alpha$$
$$h = (L - D\alpha)\sin \alpha + D(L - \cos \alpha)$$
$$D = h_1 + h_2$$
$$\frac{h}{p} = \frac{4}{3}\left(\frac{L}{p} - 1\right)^{\frac{1}{2}} \equiv \frac{4}{3}C^{\frac{1}{2}}$$

Fig. 16. Examples of Pierce's geometric model for plain-woven fabrics with bidirectional and unidirectional crimp.

Fig. 17. Enlarged view of a unidirectionally crimped, plain-woven fabric with continuous multifilament yarns.

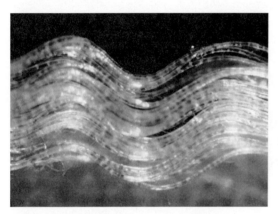

Fig. 18. Enlarged view of a continuous multifilament, crimped yarn extracted from a plain-woven fabric.

shown in **Fig. 19**, which differ only in the number of times or the frequency the yarns of one family cross those of the other family before the next undulation. The selected woven architecture influences the resulting protection levels.

Examples of ongoing research in woven fabrics for soft-body armors include (1) fabrics impregnated with shear-thickening fluids, (2) electrorheological fabrics, (3) bi-modulus fabric weaves, (4) improved experimental techniques using embedded sensors, and (5) advanced numeric modeling methods.

Woven-fabric kinematics. The energy absorbability of woven fabrics subjected to dynamic loading events, such as ballistic impact, stab penetration, and blast pressure, is significantly dependent on the ability of the fabric to enable or restrict yarn motions within the weave. Yarn motions, which are necessary for ballistic-energy absorption, occur because of the yarn-to-yarn interactions, such as crimp interchange, shearing (trellising), and friction.

Consider a plain-woven fabric subject to a tension along the warp yarns. The warp yarns attempt to straighten, decrease their crimp heights, and elongate their effective lengths. The weft yarns, however, are forced to increase their crimp heights, resulting in contractions of their effective lengths. This effect is referred to as crimp interchange and is analogous to Poisson's phenomenon exhibited in metals. Crimp

interchange is a coupling effect exhibited between warp and weft yarns and depends on the ratio of initial crimp contents between yarn families and the ratio of tensions between yarn families; it is a source of nonlinear load-extension behavior for woven fabrics.

Hearle, Grosberg, and Backer describe a limiting phenomenon to crimp interchange. Consider the case of biaxial tension. As the biaxial tensions continually increase for a given warp-tension-to-weft-tension ratio, yarn slip at the crossover regions initially increases and then ceases as the spacing between yarns reach their lowest limit. This configuration is referred to as the extensional jamming point, which can prevent a family of yarns from straightening and thereby not achieving its full strength.

Now, consider the plain-woven fabric subjected to pure shear, as shown in **Fig. 20**. The yarn families rotate at the crossover regions with respect to each other and become increasingly skewed with increasing shear load. The change in angle is referred to as the shear angle. At larger shear angles, the available space between yarn families decreases, and rotational jamming (locking) of the yarn families occur. This phenomenon is known as shear-jamming, and the angle at which the yarn families become jammed is referred to as the shear-jamming angle. The shear-jamming angle decreases with increasing yarn counts per unit length and can be estimated from Pierce's geometric fabric model or obtained experimentally with various trellising or biaxial test fixtures. Continued loading beyond the onset of shear-jamming produces shear wrinkling, a form of localized out-of-plane deformations.

It is important to determine the extension- and shear-jamming points for ensuring the proper amount of yarn mobility for optimized energy-absorption levels. In general, jamming is related to the maximum number of weft yarns per unit length that can be woven into a fabric for a given warp-yarn size and spacing.

Friction between yarns at the crossover regions can be used to minimize yarn migrations away from the impact site and to provide a dissipative energy-transfer mechanism.

Wave propagation in fibers. Consider a single fiber having restrained ends that is subjected to transverse ballistic impact at its midspan. The impact force produces a stress wave that travels along the longitudinal axis of the fiber away from the impact site. On

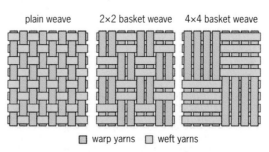

Fig. 19. Plain and basket weave architectures.

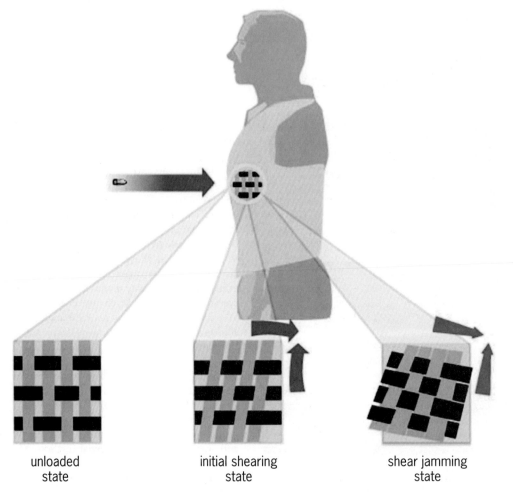

Fig. 20. Effect of shearing deformations on plain-woven fabrics.

unloaded
state

initial shearing
state

shear jamming
state

reaching the restrained ends, the wave reflects back toward the impact site because of the restoring forces resulting from the elasticity of the fiber. Stress waves travel at speeds equivalent to the speed of sound in the fiber material, which is computed by

Eq. (2), where E is the elastic modulus, ν is Poisson's

$$C_L = \sqrt{\frac{E(1-\nu)}{\rho(1+\nu)(1-2\nu)}} \qquad (2)$$

ratio, and ρ is the mass density.

A second type of wave develops and is known as the transverse displacement wave. This wave generates the observable deflections that travel at the same speed and direction as the projectile. The longitudinal stress and transverse displacement waves are depicted in **Fig. 21** by D. Roylance and S. Wang.

Impulse-momentum and energy-balance equations. The fundamental laws of motion as stated by Sir Isaac Newton relate to the masses, velocities, accelerations, and forces associated with interacting bodies. In addition, the laws of conservation of energy, mass, and momentum provide the governing equations used to fully describe these interactions. First, consider the impulse-momentum equation given by Eq. (3), which is particularly useful for characterizing

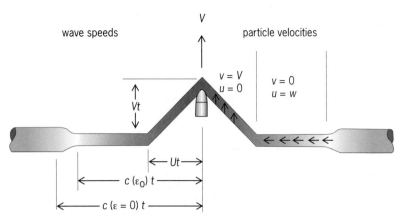

Fig. 21. Stress and transverse waves in a single fiber subjected to projectile impact, where c is the longitudinal wave speed, ε is the increment of strain, t is time, u and v are the longitudinal and vertical particle velocities, respectively, V is the projectile velocity, the product $c(\varepsilon_0)t$ is the instantaneous distance over which the strain is nonzero, the product $c(\varepsilon = 0)t$ is the instantaneous distance to the zero-strain wave front, the product Ut is the transverse wave half-width, the product Vt is the amplitude of the transverse wave, and w is the constant particle velocity.

$$\int_{t_i}^{t_f} F\,dt = m\left(V_i - V_f\right) \qquad (3)$$

the impact force F produced by the projectile on the target as a function of the linear momentum change,

where F is the impact force, t represents time, m is the mass of the projectile, V_i is the initial velocity of the projectile, and V_f is the final velocity of the projectile ($V_f = 0$ for nonpenetrating impacts). Because energy is conserved, the energy balance expressed in Eq. (4) for a rigid (nondeforming) projectile governs

$$\tfrac{1}{2} m \left(V_i^2 - V_f^2 \right) = E_{\text{damping}} + E_{\text{elastic}} + E_{\text{plastic}} \\ + E_{\text{friction}} + E_{\text{kinetic}} \qquad (4)$$

the impact event if heat dissipation, acoustic energy, and any rotational kinetic energies of the projectile are neglected for simplification.

The left side of Eq. (4) represents the kinetic energy of the projectile. Each term on the right side of the equation represents a specific energy-absorption mechanism provided by the fabric target, where E_{damping} is the energy dissipated through viscous damping; E_{elastic} is the elastic strain energy (recoverable); E_{plastic} is the plastic (inelastic) strain energy; $X = E_{friction}$ is the energy dissipated through friction produced at the yarn crossover regions, yarn-projectile contact interfaces, and layer-to-layer interactions; and finally, E_{kinetic} is the kinetic energy. For deformable projectiles, such as lead handgun rounds, similar energy-absorption terms are added to equation (4) to include the elastic and plastic strain energies of the projectile.

Ballistic impact of woven fabrics. Many dynamic effects observed in ballistic impacts on soft woven body armors parallel what occurs when a baseball is caught in the webbing of a catcher's mitt. Consider cases in which a deformable projectile and a rigid projectile impact identical multilayered woven fabric armors having clamped edges, as shown in the half-symmetry models of **Fig. 22.** Both projectiles initially contact a minimal number of yarns; these are known as the primary yarns. The primary yarns begin to compress transversely, and stress waves initiate and propagate along both yarn directions dissipating energy away from the impact site.

The crossover regions, however, reflect some of the energy back toward the impact site, a negative characteristic of the woven construction in contrast to UD fabrics. The primary yarns begin to deflect out from the fabric plane in the direction of projectile travel. These dynamic deflections are the transverse waves and can lead to yarn pull-out, a deformation mode in which the primary yarns grossly displace out from the fabric plane. More primary yarns are gradually recruited (depending on the projectile shape, diameter, hardness, and yarn sizes) and

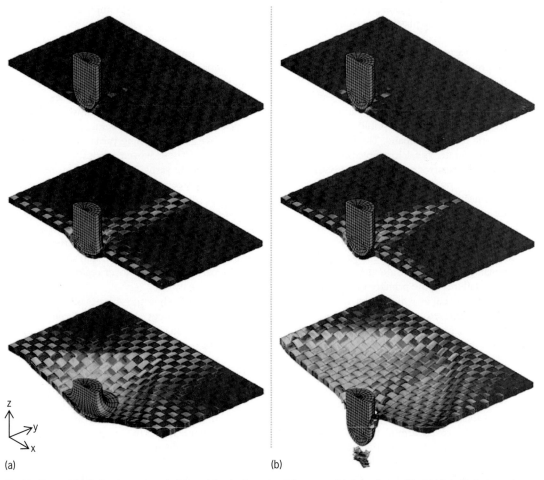

(a)　　　　　　　　(b)

Fig. 22. Numerical (finite element analysis) models of a four-ply, plain-woven fabric system subjected to ballistic impact, showing yarn stress-wave color contours, (a) projectile blunting (deformable projectile case), and (b) fabric penetration failure (rigid projectile case).

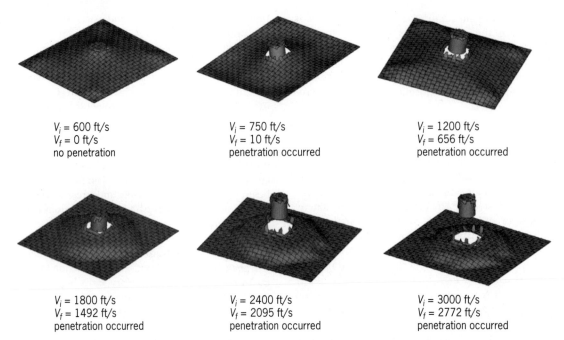

V_i = 600 ft/s
V_f = 0 ft/s
no penetration

V_i = 750 ft/s
V_f = 10 ft/s
penetration occurred

V_i = 1200 ft/s
V_f = 656 ft/s
penetration occurred

V_i = 1800 ft/s
V_f = 1492 ft/s
penetration occurred

V_i = 2400 ft/s
V_f = 2095 ft/s
penetration occurred

V_i = 3000 ft/s
V_f = 2772 ft/s
penetration occurred

Fig. 23. Numerical models of plain-woven fabrics subjected to two-grain FSP impacts showing transverse waves and yarn failures for different projectile velocities, where V_i is the initial velocity of the projectile, and V_f is the final velocity of the projectile.

attempt to straighten (crimp interchange). Friction develops at the crossover regions, and further deflection induces elastic and plastic (inelastic) stretching of the yarns.

Secondary yarns (those not directly in contact with the projectile) begin to participate because of friction developed at the crossover regions. Once the crossover friction is overcome, slip and shearing between yarn families occur, and the interstices (regions of oblique susceptibility shown in the woven fabric of Fig. 2) expand. The projectile then begins to decelerate. The deformable projectile plastically deforms (referred to as blunting) with possible fracture sites produced. Blunting often develops a mushroom-shaped appearance that increases the diameter of the projectile's tip, causing an increase in the number of primary yarns and an expanded distribution of the impact force. The rigid projectile does not deform; the distribution of the impact force remains localized, causing increased probability of stress failures in the primary yarns. A peak deflection is produced, at which point the projectile is either fully arrested or allowed to penetrate if a sufficient number of primary yarns have failed. Yarn failures and penetrations are shown in **Fig. 23** for FSP impacts on single-layer woven fabrics.

The proper levels of yarn mobility within the weave (that is, crimp interchange, stretching, shearing rotation, and pull-out) enable the fabric to dissipate ballistic-impact energy. For the stacked multilayer, soft woven armor, the ballistic-energy absorbability does not necessarily scale with the energy-absorption capacities of its individual layers. Layer-to-layer interactions may prevent the stacked layers from achieving their individual energy-absorption capacities because compressive stresses of the primary yarns in stacked fabrics can exceed those observed in single-layer impacts as reported by P. M. Cunniff.

Yarn pullout is an observable deformation mechanism occurring in woven fabrics subject to ballistic impact. Therefore, yarn pull-out is often tested to determine the fabric's frictional characteristics. Studies by Y. Duan and colleagues and P. V. Cavallaro and A. M. Sadegh have shown that the dynamic energy-absorption capacities of woven fabrics increase with an increasing yarn-to-yarn coefficient of friction μ. These woven fabrics were tested by extracting single yarns from them and monitoring the force-displacement response using a fixture such as the one shown in **Fig. 24** developed by K. M. Kirkwood and colleagues. The initial extraction force produces a peak resistance. When yarn slippage starts, the resistance force decreases as the extracted yarn is pulled from the weave, and the number of actively participating crossover regions sequentially decreases one by one.

Outlook. Soft body armors have evolved into highly sophisticated protective devices delivering unprecedented protection levels against some of the harshest physical threats. Yet ballistic and fragment threats remain a primary concern for the military and law enforcement communities. Continued effective protection of these communities requires further evolution of body armor; that is, the development of improved fiber materials, manufacturing processes, and relevant mechanics that outpace future increases in weapon effectiveness levels.

As demonstrated here, the research required to advance soft body armor protection levels demands a deeper and more thorough understanding of the material behaviors across many dimensional scales.

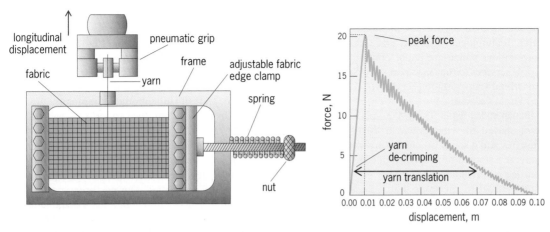

Fig. 24. Examples of yarn pull-out test fixture and force versus displacement plot.

Further investigation of the complex dynamics at each scale will increasingly incorporate the virtual environment through robust, physics-based, numerical modeling tools using, for example, explicit finite-element-analysis techniques coupled with experimental validation testing. Advanced numerical models that unite the armor and human body elements will be aggressively pursued for developing new methods and materials for defeating ballistic and fragment threats, while mitigating back-face signatures and behind armor blunt-trauma injuries.

For background information *See* BALLISTICS; FINITE ELEMENT METHOD; IMPACT; MANUFACTURED FIBER; MATERIALS SCIENCE; NANOTECHNOLOGY; NUMERICAL ANALYSIS; POLYESTER RESINS; POLYMER; POLYOLEFIN RESINS; SPINNING (TEXTILES); STRESS AND STRAIN; TEXTILE; TEXTILE CHEMISTRY in the McGraw-Hill Encyclopedia of Science & Technology.

Paul V. Cavallaro

Bibliography. *Ballistic Resistance of Body Armor NIJ Standard-0101.06*, National Institute of Justice, July 2008; *Ballistic Test Method for Personal Armour Materials and Combat Clothing, NATO Standardization Agreement (STANAG) 2920*, 2d ed., North Atlantic Treaty Organization, July 31, 2003; L. Cannon, Behind armour blunt trauma: An emerging problem, *J R Army Med Corps*, 147:87–96, 2001; P. V. Cavallaro and A. M. Sadegh, *Crimp-imbalanced protective (CRIMP) fabrics*, NUWC-NPT Technical Report 11,957, Naval Undersea Warfare Center Division Newport, Newport, RI, March 31, 2010; P. M. Cunniff, An analysis of the system effects in woven fabrics under ballistic impact, *Textil. Res. J.*, 62:495–509, 1992; *Department of Defense Military Standard, V$_{50}$Ballistic Test for Armor, MIL-STD-662F*, December 1997; Y. Duan et al., A numerical investigation of the influence of friction on energy absorption by a high-strength fabric subjected to ballistic impact, *Int. J. Impact Eng.*, 32:1299–1312, 2006; J. W. S. Hearle, P. Grosberg, and S. Backer, *Structural Mechanics of Fibers, Yarns and Fabrics*, Wiley, New York, 1969; W. Johnson, C. Collins, and F. Kindred, *A Mathematical Model for Predicting Residual Velocities of Fragments After Perforating Helmets and Body Armor*, Technical Note 1705, Army Ballistic Research Laboratories, October 1968; K. M. Kirkwood et al., *Yarn Pull-out as a Mechanism for Dissipation of Ballistic Impact Energy in Kevlar KM-2 Fabric, Part I: Quasi-Static Characterization of Yarn Pull-Out*, Army Research Laboratory ARL-CR-537, May 2004; E. Merkle et al., Assessing behind armor blunt trauma (BABT) under NIJ Standard-0101.04 conditions using human torso models, *J. Trauma Inj. Infect. Crit. Care*, 64(6):1555–1561, 2008; R. M. Peetz et al., Polymerization, *McGraw-Hill Encyclopedia of Science & Technology*, 10th ed., vol. 14, pp. 181–191, McGraw-Hill, New York, 2007; F. T. Pierce, The geometry of cloth structure, *J. Textil. Inst.*, 28(3):T45–T96, 1937; D. Roylance and S. Wang, *Penetration Mechanics of Textile Structures, TR-80/021*, U.S. Army Natick Research & Development Command, June 1979; H. H. Yang, *Kevlar Aramid Fiber*, Wiley, New York, 1993.

Space flight, 2010

2010 was a year of change in space flight, and a preview of the possible future of space exploration. The space shuttle program, which was to be retired in 2011, launched humans into space three times. The National Aeronautics and Space Administration (NASA) cancelled its Constellation program, which, among other missions, would have returned astronauts to the Moon. Instead, U.S. President Barack Obama committed NASA to a series of developmental goals leading to new spacecraft for reaching low-Earth orbit and new technology for potential missions beyond the Moon. The *International Space Station (ISS)* celebrated 10 years of habitation and gained approval for on-orbit operations through at least 2020. Mars exploration using robotic explorers continued, as did the discovery of planets around other stars via the Kepler mission. The *Hubble Space Telescope* continued to deliver incredible images and scientific information about the universe. And, using radar measurements from a NASA instrument aboard India's *Chandrayaan-1* spacecraft, scientists found more evidence for water, in quantities never before expected, on the Moon.

Commercial spaceflight companies also had several successes in 2010. These included Space Exploration Technologies (SpaceX) and Orbital Sciences Corporation, which are under contract to develop supply services to the *ISS* as part of NASA's Commercial Orbital Transportation Services (COTS) program, and Virgin Galactic.

Human space flight. The NASA space shuttle program started to wind down in 2010, with three launches and a scrub of a fourth toward the end of the year followed by extensive delays to repair the shuttle's external fuel tank.

On February 8, the space shuttle *Endeavour* lifted off from NASA's Kennedy Space Center on the first space shuttle mission of 2010 (mission STS-130). The launch followed a smooth countdown and relatively few technical problems. *Endeavour* traveled to the *ISS* carrying two European modules: Node-3 (Tranquility) and the Cupola, a bay window–like observation area (**Fig. 1**). Their installation completed the non-Russian part of the *ISS*, with more than a third of the pressurized station elements designed and built in Europe. As the Cupola was commissioned, a moon rock from the Apollo 11 mission was placed inside. The crew consisted of Commander George Zamka, Pilot Terry Virts, and Mission Specialists Robert Behnken, Nicholas Patrick, Kathryn Hire, and Stephen Robinson. The shuttle landed back at Kennedy Space Center on February 21.

Expedition 22 Commander Jeff Williams and Flight Engineer Max Suraev landed in a *Soyuz TMA-16* spacecraft in Kazakhstan on March 18, 2010, after a $5^1/_2$-month stay aboard the *ISS*.

Tracy Caldwell Dyson, a NASA astronaut, and Russian cosmonauts Alexander Skvortsov and Mikhail Kornienko launched in a *Soyuz TMA-18* spacecraft from the Baikonur Cosmodrome in Kazakhstan on April 2 at 12:04 AM EDT. They docked at the *ISS* on April 4.

Space shuttle *Discovery* lit up Florida's Space Coast sky about 45 min before sunrise at 6:21 AM on April 5, with its launch from the Kennedy Space Center. The shuttle and its crew of Commander Alan Poindexter, Pilot James P. Dutton, Jr., and Mission Specialists Rick Mastracchio, Clayton Anderson, Dorothy Metcalf-Lindenburger, Stephanie Wilson, and Japan's Naoko Yamazaki ended a 14-day journey of more than 6.2×10^6 mi (10×10^6 km) with a 9:08 AM landing back at Kennedy Space Center on April 20. This mission (STS-131) had delivered science racks, new crew sleeping quarters, equipment, and supplies to the *ISS*. During three spacewalks, the crew installed a new ammonia storage tank for the station's cooling system, replaced a gyroscope for the station's navigation system, and retrieved a Japanese experiment from outside the Kibo laboratory for examination on Earth.

Space shuttle *Atlantis* launched from Kennedy Space Center on May 15 at 2:20 PM. The crew on this mission (STS-132) included Mission Specialists Piers Sellers, Steve Bowen, Michael Good, and Garrett Reisman, Pilot Tony Antonelli, and Commander Ken Ham. The mission delivered the Russian-built Mini Research Module-1 to the *ISS*, and ended its 12-day journey of more than 4.8×10^6 mi (7.7×10^6 km) with an 8:48 AM May 26 landing at Kennedy Space Center.

Expedition 23 Commander Oleg Kotov and Flight Engineers T. J. Creamer and Soichi Noguchi landed their *Soyuz-17* spacecraft in Kazakhstan on June 1, after $5^1/_2$-month stay aboard the *ISS*.

The launch of the *Soyuz TMA-19* on June 15 from the Baikonur Cosmodrome in Kazakhstan sent NASA astronauts Doug Wheelock and Shannon Walker and Russian cosmonaut Fyodor Yurchikhin to the *ISS*. They docked for their long-duration mission two days later. On July 31, an *ISS* RPC1 (Remote Power Controller 1), essentially a circuit breaker, tripped open, powering off an ammonia pump and resulting in the loss of one half of the cooling to the *ISS*. On August 16 a spacewalk by Wheelock and Dyson was completed successfully in 7h 20min, fully accomplishing its objective of installing the spare pump module plus additional tasks. This contingency spacewalk, the third required after the July 31 failure, finished restoring cooling capability to the station and demonstrated once again our ability to adapt to hardware failures in space.

Expedition 24 Commander Skvortsov and Flight Engineers Dyson and Kornienko landed their *Soyuz TMA-18* spacecraft in Kazakhstan on Saturday, September 25, wrapping up a 6-month stay aboard the *ISS*. The spacecraft's undocking from the *ISS* and its subsequent landing in Kazakhstan occurred a day later than planned because of a hatch sensor problem that had delayed the craft's separation from the Russian Poisk module's docking port on the station's Zvezda module. The problem had prevented hooks on the Poisk side of the docking mechanism from opening. Station crew members installed a series of jumper cables, bypassing the sensor, and the Poisk module hooks retracted. Following undocking and

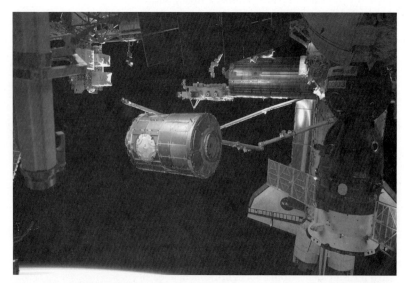

Fig. 1. The Canadian "Candarm2" robot arm moves the European module Tranquility out of the cargo bay of the American space shuttle *Endeavour* and past Japan's Kibo module while a Russian Soyuz sits docked in the foreground, illustrating the international aspects of the *International Space Station*. (*Photo courtesy of NASA*)

a normal descent, the crew landed at 1:23 AM near Arkalyk, Kazakhstan.

NASA astronaut Scott Kelly and Russian cosmonauts Oleg Skripochka and Alexander Kaleri launched aboard the *Soyuz TMA-01M* spacecraft from the Baikonur Cosmodrome in Kazakhstan at 7:10 PM EDT on October 7.

On November 5, NASA managers made the decision to scrub the launch of space shuttle *Discovery* (mission STS-133) due to a hydrogen leak at the ground umbilical carrier. It was later found that the external tank had a series of cracks in its foam as well as its underlying structural metal. Repairs and analysis on the tank required the rest of the 2010 calendar year.

Expedition 25 Commander Wheelock and Flight Engineers Walker and Yurchikhin landed safely in their *Soyuz TMA-19* spacecraft in Kazakhstan on November 25, after 5-month stay aboard the *ISS*.

At 2:09 PM EST on December 15, a *Soyuz TMA-20* spacecraft lifted off from the Baikonur Cosmodrome with the Expedition 26 crew onboard, headed to the *ISS*. The crew included Russian cosmonaut Dmitry Kondratyev, American astronaut Catherine Coleman, and European Space Agency (ESA) astronaut Paolo Nespoli of Italy.

Robotic solar system exploration. On January 26, after six years of roaming the Red Planet, NASA's Mars Exploration rover *Spirit* became a stationary robot. NASA designated the once-mobile explorer a stationary science platform after efforts over several months failed to free it from a sand trap. *Spirit's* sister spacecraft, *Opportunity*, continued roving.

On March 1, using data from NASA's lightweight Mini-SAR (miniature synthetic aperture radar) instrument aboard India's *Chandrayaan-1* spacecraft, scientists detected water ice deposits near the Moon's north pole. The Mini-SAR found more than 40 small lunar craters containing ice. The craters range in size from 2 to 15 km (1 to 9 mi) in diameter. Although the total amount of ice depends on its thickness in each crater, which could not be measured by the radar, it is estimated that there could be at least 6×10^8 metric tons (1.3×10^{12} lb) of water ice at the pole.

In May, NASA's *MErcury Surface, Space ENvironment, GEochemistry, and Ranging* (*MESSENGER*) spacecraft took an image of the Earth and Moon at a distance of 183 million km (114 million mi) [**Fig. 2**]. This image was acquired as part of *MESSENGER's* campaign to search for vulcanoids, small rocky objects that are believed to exist in orbits between Mercury and the Sun. No vulcanoids were detected, but the *MESSENGER* spacecraft was in a unique position to look for them. *MESSENGER's* vulcanoid searches occurred near perihelion passages, when the spacecraft's orbit brought it closest to the Sun on its path to orbit the planet Mercury.

On June 13, Japan's *Hayabasa* asteroid probe entered Earth's atmosphere at over 43,500 km/h (27,000 mi/h), the second fastest such entry on record. *Hayabusa* was the first craft ever to retrieve asteroidal material directly from its source and re-

Fig. 2. The *MESSENGER* spacecraft captured this image looking back at the Earth-Moon system. In the lower left portion of this image, the Earth can be seen, as well as the much smaller Moon to Earth's immediate right. (*Courtesy of NASA/Johns Hopkins University Applied Physics Laboratory/Carnegie Institution of Washington*)

turn it to Earth, having picked up tiny grains of material from the near-Earth asteroid Itokawa in mid-September 2005. The fiery heat of reentry, which exceeded 2760°C (5000°F), destroyed most of the spacecraft, but fortunately a heat shield protected the capsule containing the samples.

On June 14, NASA scientists announced an estimate, based on testing of lunar samples combined with computer modeling of mineral crystallization, that the volume of water molecules locked inside minerals in the Moon's interior could exceed the amount of water in North America's Great Lakes. Scientists at the Carnegie Institution's Geophysical Laboratory in Washington, along with other scientists, determined that the water was likely present very early in the Moon's formation as hot magma started to cool and crystallize. This finding means water is native to the Moon, rather than having been transported there (for example, by meteor impacts) as previously theorized.

On July 1, the Mini-RF (miniature radio frequency) instrument, a synthetic aperture radar onboard NASA's *Lunar Reconnaissance Orbiter* (*LRO*), returned the first high-resolution view of a potentially ice-rich crater near the north pole of the Moon. With a resolution 10 times better than the Mini-SAR aboard the *Chandrayaan-1* spacecraft, Mini-RF enabled the observation of details of the crater's interior.

On July 10, the *Rosetta* spacecraft of the European Space Agency (ESA) achieved a milestone on its journey to the comet Churyumov-Gerasimenko. *Rosetta* flew past asteroid Lutetia on its second and final pass of the asteroid belt between the orbits of Mars and Jupiter, while traveling at about 15 km/s (9 mi/s) [54,000 km/h (33,600 mi/h)] passing merely 3162 km (1965 mi) from the asteroid. In addition to returning spectacular close-up images of the asteroid, *Rosetta's* close flyby of Lutetia and eventual landing on the mission's primary target, Churyumov-Gerasimenko, is expected to gather crucial data not just on asteroids and comets but on the evolution of the solar system itself.

On August 19, NASA announced that newly discovered cliffs in the lunar crust indicate that the Moon shrank globally in the recent geologic past and might still be shrinking today. This was according to a team analyzing new images from the *Lunar Reconnaissance Orbiter*, which provided important clues to the Moon's recent geologic and tectonic evolution.

In September, scientists presented indications that the Martian moon Phobos formed relatively near its current location via re-accretion of material blasted into Mars's orbit by some catastrophic event. Two independent approaches of compositional analyses of thermal infrared spectra, from ESA's *Mars Express* and NASA's *Mars Global Surveyor* missions, yielded very similar conclusions. The re-accretion scenario was further strengthened by measurements of Phobos's high porosity from the Mars Radio Science Experiment (MaRS) onboard *Mars Express*.

On October 28, NASA determined that the area where the Mars rover *Spirit* got stuck contained evidence that water, perhaps as snow melt, has trickled into the subsurface fairly recently and on a continuing basis. Researchers took advantage of *Spirit's* stationary situation to examine in great detail soil layers that the wheels had exposed, as well as neighboring surfaces. *Spirit* had made 33 cm (13 in.) of progress in its last 10 backward drives before its energy levels fell too low for further driving in February 2010.

On November 4, as part of NASA's EPOXI mission, the *Deep Impact* spacecraft successfully flew by comet Hartley 2 and returned images. [The name EPOXI is a combination of the names of two mission components: Extrasolar Planet Observations and Characterization (EPOCh) and Deep Impact Extended Investigation (DIXI).] Hartley 2 is the fifth comet nucleus visited by a spacecraft.

On November 10, the China National Space Administration (CNSA) released the first photos from its recently launched *Chang'e-2* lunar orbiter. The images were of the Bay of Rainbows (Sinus Iridium), which China had scheduled to be the potential landing location of its *Chang'e-3* rover mission.

In December, 33 years after its launch, NASA's *Voyager 1* spacecraft reached a distant point at the edge of the solar system where there is no outward motion of solar wind. Heading toward interstellar space some 17.4×10^9 km (10.8×10^9 mi) from the Sun, *Voyager 1* had crossed into an area where the velocity of the hot ionized gas, or plasma, emanating directly outward from the Sun had slowed to zero. Scientists suspect the solar wind had been turned sideways by pressure from interstellar wind in the region between stars. The event was a major milestone in *Voyager 1's* passage through the heliosheath, the turbulent outer shell of the Sun's sphere of influence, and the spacecraft's upcoming departure from the solar system.

Also in December, NASA's *Mars Odyssey*, which launched in 2001, broke the record for longest-serving spacecraft at Mars. The probe began its 3,340th day in Martian orbit at 5:55 PM PST (8:55 PM EST) on December 15 to break the record set by NASA's *Mars Global Surveyor*, which had orbited Mars from 1997 to 2006.

Commercial space flight. On March 22, the commercial space flight company Virgin Galactic announced that its spaceship *VSS Enterprise* had completed its inaugural captive carry flight (that is, it was attached to a mothership which provided propulsion) from Mojave Air and Space Port in California. The *VSS Enterprise* test flight program, intended to continue through 2011, was to progress from captive carry to independent glide and then powered flight, prior to the start of commercial operations carrying space tourism passengers to the edge of space. The *VSS Enterprise* later made its first crewed flight, still carried by a mothership, on July 15.

On June 4, at Cape Canaveral, Florida, SpaceX achieved the maiden launch of its *Falcon 9* rocket carrying a mockup of the company's *Dragon* capsule, intended to carry cargo to space as part of NASA's Commercial Orbital Transportation Services (COTS) program. With the retirement of NASA's space shuttles, SpaceX is expected to launch at least 12 missions to carry cargo to and from the *ISS*. The *Falcon 9* rocket and *Dragon* spacecraft also are designed to eventually carry astronauts into low-Earth orbit.

On October 10, at Mojave Air and Space Port, Virgin Galactic completed the first piloted free flight of the *VSS Enterprise*. The spaceship was released from its mothership at an altitude of 13,700 m (45,000 ft). During its first flight, the spaceship was piloted by Pete Siebold and co-pilot Mike Alsbury. The two main goals accomplished by the flight were the clean release of the spaceship from its mothership, and the free flight, glide back, and landing at Mojave Air and Space Port.

In November, Orbital Sciences Corporation test-fired the first-stage rocket engine for its *Taurus II* rocket and opened the mission control center that will support the company's COTS program missions for NASA. The company shipped the *Taurus II* stage-one core in December to NASA's Wallops Flight Facility in Virginia for assembly. Orbital is under contract with NASA to fly eight cargo missions to the *ISS*.

On December 8, SpaceX successfully launched its *Falcon 9* rocket carrying a fully functional *Dragon* capsule. The crewless flight was the first for NASA's COTS program, and the launch was the first commercial launch and return of a spacecraft from low-Earth orbit.

Other activities. On January 4, NASA's *Kepler* space telescope discovered its first five new exoplanets, named Kepler 4b, 5b, 6b, 7b, and 8b. Launched on March 6, 2009, from Cape Canaveral Air Force Station in Florida, the *Kepler* mission continuously and simultaneously observes more than 150,000 stars. *Kepler's* photometer, which detects small shifts in stars' brightness as orbiting planets pass in front of them, already has measured hundreds of possible planet signatures that are being analyzed.

On February 4, NASA released the most detailed and dramatic images ever taken of the dwarf planet Pluto. The images from NASA's *Hubble Space Telescope* show an icy, mottled, dark, molasses-colored world undergoing seasonal surface color and brightness changes. The observations indicate that Pluto has become significantly redder (compared with images taken by *Hubble* in 1994), while its illuminated northern hemisphere is getting brighter. These changes are most likely consequences of surface ice melting on the pole facing the Sun and then refreezing on the other pole, as the planet moves along its 248-year-long orbital period.

Launched on February 11, NASA's *Solar Dynamics Observatory*, or *SDO*, is the most advanced spacecraft ever designed to study the Sun. The spacecraft provides 16-Mpixel images of the Sun (compared with 2 Mpixels for high-definition television) in a broad range of ultraviolet wavelengths. The initial images returned by *SDO* confirmed its new capability of allowing scientists to better understand the Sun's dynamic processes. Some images from the spacecraft showed details of material streaming outward and away from sunspots. Others showed extreme close-ups of activity on the Sun's surface. These solar events can greatly affect Earth.

On April 22, the United States Air Force launched its first robotic *X-37B* space plane. Development of the *X-37B* began in the late 1990s, with NASA starting the project. It was later picked up by the Defense Advanced Research Projects Agency (DARPA) and ultimately completed by the Air Force. The project is now under the Air Force Rapid Capabilities Office. The robotic space plane orbited the Earth on a secret military mission until it landed at Vandenberg Air Force Base in California on December 3.

On May 6, NASA successfully tested a launch pad abort system developed for the *Orion* crew module. (The module was originally part of the canceled Constellation program that was to return humans to the Moon, and is now expected to be used to ferry astronauts to the *ISS*.) The information gathered from the test was to be used to improve future launch abort systems, resulting in safer and more reliable crew escape capability during rocket launch emergencies.

Also on May 6, the first scientific results from the *Herschel Infrared Space Observatory* were announced, revealing hidden details of star formation. New images showed thousands of distant galaxies creating new stars, and star-forming clouds in the Milky Way. *Herschel's* observation of the star-forming cloud RCW 120 revealed an embryonic star which appears to be destined to turn into one of the biggest and brightest stars in our galaxy within the next few hundred thousand years. The developing star already

Fig. 3. *Hubble Space Telescope* view of the Lagoon Nebula. (*Courtesy of NASA, ESA*)

contains 8 to 10 times the mass of the Sun and is still surrounded by an additional 2000 solar masses of gas and dust from which it can feed further. *Herschel* is an ESA space observatory with science instruments provided by the European-led Principal Investigator consortia and with important participation from NASA. *See* HERSCHEL SPACE OBSERVATORY.

On April 8, Europe's first mission dedicated to studying the Earth's ice was launched from Kazakhstan. ESA's *CryoSat-2* satellite was put into a polar orbit, from which it was to send back data on how ice is responding to climate change and the role it plays in the Earth's climate.

A Delta 4 rocket roared into space on May 27 carrying the first in a powerful new series of Global Positioning System (GPS) satellites for the U.S. Air Force. The *GPS IIF SV-1* satellite that launched from Pad 37 at Cape Canaveral Air Force Station in Florida marked the 349th launch in the Delta program's 50-year history.

On September 23, a spectacular new NASA/ESA *Hubble Space Telescope* image was released that revealed the Lagoon Nebula (also known as Messier 8). The dramatic view, of a massive cloud of gas and dust sculpted by intense radiation from hot young stars deep in the heart of the nebula, was captured by the telescope's Advanced Camera for Surveys (ACS) [**Fig. 3**].

On October 26, an international group of scientists used data from NASA's *Kepler* spacecraft to detect stellar oscillations, or "starquakes," that yield new insights about the size, age, and evolution of stars.

Launch summary. Globally there were 74 space launches in 2010, with 23 being commercial. Four launches were failures: two Indian, one Korean, and one Russian.

Russia once again led in launches, with 31 (see **table**). Thirteen of these launches were commercial. Of the 18 non-commercial launches, nine were devoted to the *International Space Station*: five crewless Progress modules launched on Soyuz launch vehicles on *ISS* supply missions, and four crewed Soyuz missions. Eight Russian launches were for military purposes; one of these was a civil mission using the Rockot vehicle to launch the *Gonets M2* satellite. The

Space launches in 2010		
Country of launch	Attempts	Successful
Russia	31	30
United States	15	15
China	15	15
Europe	6	6
Japan	2	2
India	3	1
Israel	1	1
South Korea	1	0
Total	74	70

one Russian launch failure was of a Proton M vehicle with three Russian Glonass M navigation satellites intended for medium Earth orbit. Launch vehicles used by Russia were Proton M (12), Soyuz (9), Dnepr M (3), Rockot (2), Soyuz 2 (2), Kosmos 3M (1), Soyuz U (1), and Molniya (1).

The United States and China tied for the second-most launches in 2010 with 15 each. This marked a dramatic increase for China, up from six launches in 2009, and a large drop for the United States, down from 25 launches in 2009.

Four U.S. launches were commercial. Of the 11 noncommercial U.S. launches, three were space shuttle missions, two brought classified payloads into space for the National Reconnaissance Office, five carried Department of Defense (DOD) payloads or were sponsored by the DOD, and one was a NASA civil government mission. The three space shuttle launches were the most for any U.S. launch vehicle. Other vehicles used included Atlas V 501 (2), Atlas V 531 (1), Atlas V 401 (1), Delta II (1), Delta IV medium (1), Delta IV Medium + (1), Delta IV Heavy (1), Falcon 9 (2), and Minotaur IV (2). All launches were successful.

All Chinese launches were with variants of the Long March vehicle. The Long March 4C was used four times; the 2D, 3A, and 4C were used three times each; and the 3B and 4B were used once each. All Chinese launches were noncommercial and crewless. Nine launches carried governmental civil missions (communications, meteorological, remote sensing, and scientific).

Europe successfully used the Ariane 5 for all six of their launches in 2010. The Ariane 5 launched 12 geostationary orbit (GEO) communications satellites. Eleven of the satellites were commercial, and one was a military communications satellite.

Japan had two successful launches of its H-IIA rocket. The first launch carried four low Earth orbit (LEO) and two interplanetary spacecraft. The second launch carried a geostationary (GEO) satellite. The Japanese launch site at Tanegashima was opened to use throughout the year by an agreement between the Japanese government and the local fishing industry. Previously the launch site could only be used for half of the year, not during fishing season.

India had three launches and two were unsuccessful. Both failures were with Geosynchronous Satellite Launch Vehicle (GSLV) spacecraft carrying GEO

communications and navigation satellites. A Polar Satellite Launch Vehicle (PSLV) successfully carried five satellites for Indian and international missions to LEO and Sun synchronous orbit (SSO).

Israel's Shavit launch vehicle was successful in placing the Israeli Ministry of Defense satellite *Ofeq 9* into orbit. South Korea's Korea Space Launch Vehicle (KSLV)-1 vehicle failed to launch the Korean LEO atmospheric research satellite *STSAT 2B* into orbit. This was the second KSLV launch attempt and the second failure.

For background information *see* ASTEROID; ASTRONAUTICS; COMET; EXTRASOLAR PLANETS; GALAXY, EXTERNAL; HUBBLE SPACE TELESCOPE; MARS; MOON; NEBULA; PLUTO; SCIENTIFIC AND APPLICATIONS SATELLITES; SOLAR SYSTEM; SOLAR WIND; SPACE FLIGHT; SPACE STATION; SPACE TECHNOLOGY; SPACE SHUTTLE; STAR; SUN; SYNTHETIC APERTURE RADAR (SAR); UNIVERSE in the McGraw-Hill Encyclopedia of Science & Technology.   Donald Platt

Bibliography. *Aviation Week & Space Technology*, various 2010 issues; *Commercial Space Transportation: 2010 Year In Review*, Federal Aviation Administration, January 2011; ESA Press Releases, 2010; NASA Public Affairs Office, News Releases, 2010.

Squidworm

The vast expanse of the oceanic water column (all the water below the surface and above the deep seafloor) supports a wide and fantastic array of animals. In many cases, little or nothing is known about these animals. The majority of life in the water column is dependent on productivity derived from sunlight in the surface waters; therefore, animal abundance and biomass generally decrease with increasing water depth. However, as one descends to within a few hundred meters of the seafloor, animal abundance and biomass increase again. This bottom few hundred meters is called the benthic boundary layer, the demersal zone, or the benthopelagic zone and is home to a diverse suite of animals who take advantage of the concentration of organic matter and other animals near the seafloor. This habitat is particularly difficult to study because it cannot be effectively sampled with either traditional water column or seafloor sampling equipment. Additionally, many of the animals found in the demersal zone, like the animals swimming and floating farther up in the water column, are fragile or gelatinous and thus are easily damaged by sampling gear. The development of vehicles capable of free operation in the deep water column has facilitated the discovery of many unusual animals in this habitat and has allowed a better understanding of the community found there because they allow one to directly observe the community and to selectively sample it. A fantastic example of these discoveries is *Teuthidodrilus samae*, known as the squidworm (see **illustration**), which is a large and flamboyant segmented worm that is common in the broad benthic boundary layer of the deep Celebes Sea (in the western Pacific Ocean).

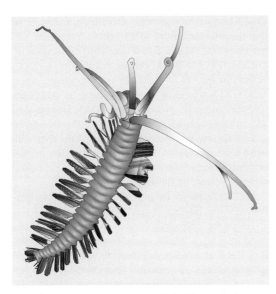

Line drawing of a squidworm (*Teuthidodrilus samae*).

Discovery. In October 2007, an international team of researchers led by Laurence Madin of the Woods Hole Oceanographic Institution was exploring the water column of the Celebes Sea with a remotely operated vehicle (ROV). The ROV is controlled from aboard a research vessel and is equipped with high-definition cameras and sampling devices that are specifically designed to allow gentle collection of delicate swimming animals. On several ROV dives, the team observed large, actively swimming worms with unique head appendages and collected seven individuals.

Habitat. The Celebes Sea is located between Indonesia and the Philippines. It is considered a pocket basin because its deep central portion [approximately 6200 m (3.85 mi) in depth] is completely surrounded by shallow sills. The shallow sills and density differences between the cold, deep water and the warm, shallow water effectively isolate the deep water of this basin from that of neighboring basins and the Pacific Ocean. Evolutionary biology has shown that a common mechanism of speciation (diversification or the divergence of a single species into two) is isolation. Truly isolated communities are relatively unknown in the deep sea, so this area is particularly interesting for studies of biodiversity. Additionally, the Celebes Sea is at the center of the "Coral Triangle," an area considered to be a conservation hot spot because of the high diversity and endemism of shallow-water corals and fishes. This area is also the center of geographical distributions and diversity of deep-sea fish groups such as hatchetfish and anglerfish. The exploratory expedition that discovered the squidworm (and several other new species) was motivated by the hypothesis that the deep-water fauna would be equivalently diverse because of the unusual geology.

Ecology. Squidworms were collected at water depths of 2000–2800 m (1.2–1.7 mi) and were observed 1–100 m (3–330 ft) above the seafloor. No individuals were observed on the seafloor, nor were they seen to interact with it in any way. Their guts were found to contain fine sediment particles, indicating the possibility that these creatures had been feeding on the seafloor; however, the particles could also have been obtained from the water column.

Numerous squidworms were observed within single ROV dives. This is unusual for any animal living in the water column, especially considering the vast three-dimensional space available to live in and the very limited reach of the lights and cameras of the ROV. Based on this information, it is hypothesized that squidworms are abundant in the deep waters of the Celebes Sea. This probable abundance and the large size of individuals suggest that this species has a significant ecological effect on other members of the deep, demersal community of the Celebes Sea. However, much remains to be learned about their ecology—for example, who eats them, who or what they eat, why they are often seen in large numbers, and what their reproductive behaviors are.

Description. Squidworms are translucent, dark-brown, segmented worms that bear bands of iridescent muscles running the length of their ventral and dorsal surfaces. The sides of their bodies are lined with expansive fans of flattened, glasslike bristles that sparkle with reflected light in the glare of the powerful ROV lights. Squidworms are among the largest members of the polychaete worm family Acrocirridae, reaching nearly 10 cm (4 in.) in length and 1 cm (0.4 in.) in width. They also have ten unique head appendages which can double the length of the animal, eight of which are long and extend out in all directions and two that remain coiled under the head. It is these head appendages that give the organism its name because they resemble the arms of a squid (*Teuthidodrilus* means squidworm). The eight head appendages taper to a fine extensible tip that is highly innervated and likely supports a suite of mechanoreceptors (touch) or chemoreceptors (smell/taste). These receptors allow the animals to sense the water at distances of more than a body's length away, which could be important for avoiding predators and finding food and mates in this vast habitat. When floating or swimming, the animals hold these long appendages (which are also used to breathe and are called branchiae) in very specific postures that maximize the volume of water that they inhabit. The other two (coiled) appendages are used for feeding and have a groove running their entire length that is filled with millions of cilia (hairs) that collect and pass particles to the mouth. Many small, highly branched, upright structures that support more chemosensory receptors are tucked below the base of the branchiae. These upright structures are not known in any other worms of this group (Cirratuliformia) and thus are particularly interesting because they represent an evolutionary innovation.

Additional observations. *Teuthidodrilus* is currently a monotypic genus (containing only one described species), but that is not expected to be the case as exploration of deep water continues with ROVs and submersibles. A related animal was observed, although not collected, off western India at a depth of 1000 m (3300 ft) in 2004 and was recorded by the SERPENT (ROV) project. Differences in depth and posture suggest that it could be an additional species; however, until a specimen can be obtained, the exact determination remains uncertain. More recently, an expedition to explore seamounts near an Indonesian marine protected area observed numerous squidworms at depths of approximately 800 m (2620 ft). All of these animals were observed on a single ROV dive within a few meters of the seafloor. However, these observations were not accompanied by collection of the animals, so it cannot be determined if they are members of *T. samae* or another species belonging to the genus *Teuthidodrilus*.

For background information *see* ANNELIDA; BIODIVERSITY; BIOGEOGRAPHY; DEEP-SEA FAUNA; MARINE BIOLOGICAL SAMPLING; MARINE ECOLOGY; OCEANOGRAPHY; POLYCHAETA; UNDERWATER PHOTOGRAPHY; UNDERWATER VEHICLES in the McGraw-Hill Encyclopedia of Science & Technology.   Karen J. Osborn

Bibliography. T. Koslow, *The Silent Deep: The Discovery, Ecology, and Conservation of the Deep Sea*, University of Chicago Press, Chicago, 2007; K. J. Osborn, L. P. Madin, and G. W. Rouse, The remarkable squidworm is an example of discoveries that await in deep-pelagic habitats, *Biol. Lett.*, 7:449–453, 2011; K. J. Osborn et al., Deep-sea swimming worms with luminescent "bombs," *Science*, 325:964, 2009; B. H. Robison, Deep pelagic biology, *J. Exp. Mar. Biol. Ecol.*, 300:253–272, 2004; B. H. Robison and W. M. Hamner, Pocket basins and deep-sea speciation, pp. 755–757, in R. Gillespie and D. A. Clague (eds.), *Encyclopedia of Islands*, University of California Press, Berkeley, 2009.

STAT3

Signal transducer and activator of transcription 3 (STAT3) is a latent cytosolic transcription factor that signals directly from cell surface receptors to the nucleus, coupling the activation of these receptors to gene activation. Originally discovered as an acute-phase response factor that is activated after stimulation by interleukin-6 (IL-6), STAT3 was subsequently shown to participate in signaling by polypeptide growth factors and oncoproteins, linking its activation to cancer. STAT3 is encoded by an evolutionarily conserved gene and is one of a family of seven mammalian proteins (STAT1, STAT2, STAT3, STAT4, STAT5A, STAT5B, and STAT6) that vary in size between 750 and 850 amino acids and share a 20–50% sequence identity. The major structural features of all STAT proteins include an N-terminal STAT dimerization domain, a coiled-coil domain involved in protein–protein interactions, a central DNA-binding domain, an Src homology (SH2) domain, a conserved tyrosine residue at position 705 (Tyr^{705}), and a C-terminal transcriptional activation domain. Upon activation of cell surface receptors, monomeric STAT3 proteins bind to phosphorylated receptors, forming homodimers through their SH2 domains.

STAT3 homodimer proteins can then translocate to the nucleus and bind DNA within promoter target sites that share a 9-base-pair (bp) consensus sequence, TTCCGGGAA. Precise regulation of STAT3 activation is critical; if STAT3 is deregulated, aberrant STAT3 signaling may contribute to malignant transformation by promoting cell-cycle progression and/or cell survival.

This review will detail the mechanisms that control STAT3 regulation, its aberrant regulation in cancer and inflammatory syndromes, and a proposed means to target its inhibition therapeutically.

STAT3 regulation by phosphorylation and acetylation. During normal signal transduction, STAT3 is activated by cell surface receptors that are stimulated by a large number of cytokines [for example, IL-6, IL-10, and tumor necrosis factor-α (TNFα)] and growth factors [for example, epidermal growth factor (EGF), platelet-derived growth factor (PDGF), and granulocyte colony-stimulating factor (G-CSF)], leading to receptor aggregation (**Fig. 1**). Multiple activated growth factor receptors with intrinsic tyrosine kinase (TK) activity, including EGF, PDGF, CSF-1R, G-protein-coupled receptors, and the T-cell receptor complex and CD40 receptors, can directly phosphorylate STAT3 proteins. In contrast, cytokine receptors lacking intrinsic TK activity, such as the interferon or IL-6 receptors, recruit the Janus kinase (JAK) family of tyrosine kinases to facilitate STAT3 activation. In this way, the receptor-associated JAKs become activated by autophosphorylation following ligand engagement and subsequently phosphorylate tyrosine residues within the cytoplasmic tails of the cytokine receptors. In a third nonreceptor tyrosine kinase–mediated mechanism, cytoplasmic TKs, including the oncogenic proteins Src and Abl kinase, can directly phosphorylate STAT3. In any of these scenarios, the receptor phosphotyrosines serve as docking sites for the recruitment of inactive cytoplasmic STAT3 monomers. Phosphorylation of a single tyrosine residue (Tyr^{705}) on the receptor-bound STAT3 monomer induces homodimerization of STAT3 monomers through a combination of phosphotyrosine and SH2 domain interactions (Fig. 1). In addition to tyrosine phosphorylation, STAT3 also undergoes serine (Ser) phosphorylation at Ser^{727} by several serine kinases; this is a reported requirement for its transcriptional activation.

Other than phosphorylation on tyrosine and serine sites within the carboxyl-terminal region, STAT3 is also modified by acetylation at a lysine residue (Lys^{685}) by CREB-binding protein (CBP/p300), which is a histone acetyltransferase protein. STAT3 acetylation is required for stable dimerization and thus cytokine-stimulated DNA binding and transcriptional regulation.

STAT3 responsive genes. STAT3 transcriptionally activates a variety of genes involved in cell survival, angiogenesis (the origin and development of blood vessels), and cellular migration. For example, these genes activate apoptosis (programmed cell death) inhibitors such as Bcl-xL, Mcl-1, and survivin; cell-cycle regulators such as cyclin D1, Myc, and the cyclin-dependent kinase inhibitor p21CIP/WAF; and angiogenesis inducers such as vascular endothelial growth factor (VEGF). A large number of genes that are induced early after STAT3 activation are transcription factors themselves, including junB, egr1, KLF4, bcl-6, and NFIL3, suggesting that STAT3 can regulate broad programs of gene expression. Microarray analysis of STAT3-mediated transcriptional changes after longer periods of activation identified other genes involved in oncogenic pathways, with several genes mediating cell migration and invasion. Interestingly, in addition to transcriptional activation, STAT3 also represses transcription of the p53 tumor suppressor gene, thereby blocking downstream activation of p53-regulated gene programs. This latter mechanism may account for the deregulated growth and survival of primary tumor cells harboring wild-type p53 genes that express high levels of STAT3 protein.

STAT3 regulation by GTPases. In addition to receptor- and non-receptor-mediated activation by growth factors and oncogenes, a novel pathway of STAT3 activation has recently emerged from observations of increased STAT3 phosphorylation associated with increases in cell density. Both normal and transformed cells that have been grown to postconfluence or forced to form multicellular aggregates display a dramatic increase in STAT3–Tyr^{705} phosphorylation, DNA binding, and transcriptional activity. E-cadherin, a plasma membrane glycoprotein that controls the organization and dynamics of cell adhesion, was shown to be involved in this process when its activity was disrupted by chemical and genetic methods. A definitive demonstration of STAT3 activation as a direct effect of cadherin engagement, rather than a secondary effect of cell-to-cell proximity, was determined through E-cadherin mutational analyses. Further examination of this mechanism revealed a role for Rac1 and Cdc42, which belong to the Rho family of GTPases [a group of monomeric guanosine triphosphate (GTP)–binding proteins]. Rac1 and Cdc42 were found to be required for STAT3 activation following cadherin engagement; this was shown by the negative effects on STAT3 activation following downregulation of Rac1 and Cdc42 through small hairpin RNA (shRNA) expression or pharmacological inhibition. The current model for density-dependent STAT3 activation is that Rac1/Cdc42 activation increases IL-6 family cytokine production, which in turn activates the gp130 family of cytokine receptors (**Fig. 2**).

Role of STAT3 in normal development and the immune system. Conventional deletion of *stat3* in embryonic stem cells of mice results in early embryonic lethality, suggesting that STAT3 alone, unlike its other family members, is essential for embryonic growth and survival. Conditional deletion of *stat3* in hematopoietic tissues, including T cells, neutrophils, and macrophages, supports the role of STAT3 in cytokine-mediated signaling pathways. Specifically, conditional deficiency of *stat3* in CD4+ T cells impairs IL-17 production and limits IL-17-associated pathology, whereas retroviral overexpres-

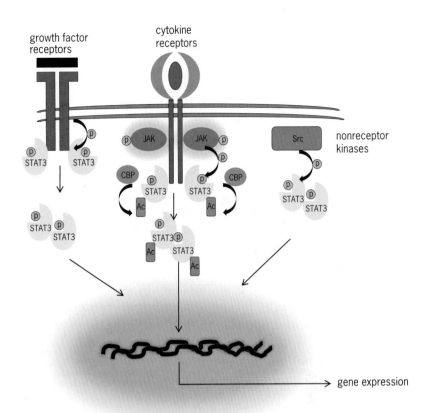

Fig. 1. Receptor- and non-receptor-mediated model of STAT3 activation. In receptor-mediated STAT3 activation, growth factors or cytokines bind to their respective receptors and then either undergo autophosphorylation (growth factor receptors) or recruit tyrosine kinases (cytokine receptors), including the Janus-activated kinases (JAK) that phosphorylate the cytoplasmic tails of the receptors. Once phosphorylated, the receptors recruit STAT3 monomers, which bind to the receptors through their SH2 domains. STAT3 monomers are then phosphorylated at Tyr^{705} and form dimers through their SH2 domains. In another mechanism of activation, STAT3 is phosphorylated by cytoplasmic tyrosine kinases (such as Src and Abl) or acetylated at Lys^{685} by a histone acetyltransferase protein [CREB-binding protein (CBP)]. Phosphorylation (p) or acetylation (Ac) allows STAT3 dimer formation and translocation of the complex to the nucleus, where it can activate or repress gene transcription from site-specific promoters.

sion of constitutively active STAT3 induces IL-17 production. The requirement for STAT3 in human IL-17 production was revealed in patients with Hyperimmunoglobulin E syndrome (Hyper IgE syndrome, HIES; also known as Job syndrome), a rare immunodeficiency disorder characterized by skin inflammation, bacterial pneumonias, elevated serum IgE levels, and tooth and bone abnormalities. Large family gene linkage studies uncovered germline *STAT3* mutations in probands with the disease that primarily localized to regions within the conserved SH2 and DNA-binding domains (note that probands are clinically affected individuals through whom a family is found that can be used to study the genetics of a particular disorder). Patients with HIES have a severely impaired ability to produce T helper 17 (Th17) cells, which probably results in their clinical symptoms. Based on several consistent findings in humans and animal models, a role for STAT3 in regulating mucosal innate immunity has been proposed. Pathologic studies of mice with a conditional deletion of *stat3* in hematopoietic cells showed evidence

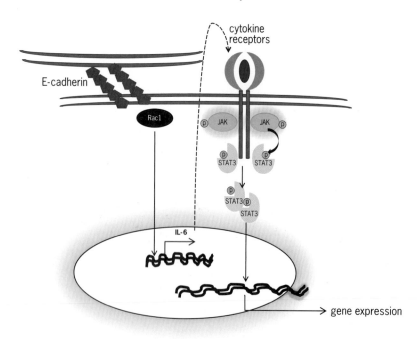

Fig. 2. Model of STAT3 activation by cadherin and Rac1/Cdc42. In an alternative mode of activation, cell–cell contact through cadherin engagement increases Rac1 and Cdc42 proteins, which induce IL-6 transcription. An increase in IL-6 production stimulates cytokine receptors, which then activate STAT3 through receptor-mediated activation.

of a Crohn's disease–like colitis, whereas immunohistochemical studies showed activated STAT3 expression in the intestines of patients with inflammatory bowel disease. Genome-wide association studies (GWAS) uncovered the single-nucleotide polymorphism, rs12948909, within the *STAT3* gene in patients with Crohn's disease and ulcerative colitis.

Role of STAT3 in cancer. STAT3 is constitutively activated in cancers of the breast, prostate, lung, head, and neck, as well as in multiple myeloma and leukemia. A constitutively active form of STAT3 alone is sufficient to transform cultured fibroblasts to anchorage independence (the ability of cells to proliferate in the absence of adhesion to extracellular matrix proteins) and points to a causal role for STAT3 in cancer development. Most of the described oncogenic functions of STAT3 depend on its Tyr^{705} phosphorylation status; however, an additional oncogenic role for STAT3 has been recently described that is dependent on serine phosphorylation and takes place in mitochondria. Recent evidence also suggests a crucial role for STAT3 in selectively inducing and maintaining a procarcinogenic inflammatory microenvironment, both at the initiation of malignant transformation and during cancer progression. STAT3 is linked to inflammation-associated tumorigenesis that is initiated by genetic alterations in malignant cells, as well as by many environmental factors, including chemical carcinogens, sunlight, infection, cigarette smoking, and stress.

Drug targets for STAT3 inhibition. The large body of data validating STAT3 as a target for cancer therapy, and the tolerance of normal cells for the loss of STAT3

function, has driven the effort to identify molecules that inhibit STAT3. In cell lines, inhibition of STAT3 activity using genetic or pharmacologic approaches reduces cancer cell growth and induces apoptosis. A few of the strategies to inhibit STAT3 that are currently under exploration are based on targeting the different structural domains of the protein and indirect targeting of the upstream components of the pathway, including the inhibition of tyrosine kinases that block aberrant STAT3 signaling. As STAT3 is negatively regulated through numerous mechanisms [for example, suppressors of cytokine signaling (SOCS), protein inhibitor of activated STAT (PIAS), protein phosphatases, and ubiquitin-dependent proteasomal degradation], additional strategies targeting STAT3 are also being actively investigated. Although few clinical trials targeting STAT3 are in place in 2011, it is likely to become an important targeting tool in the future as more is understood about the significance of this molecule in cancer and other diseases.

For background information *see* CANCER (MEDICINE); CELL (BIOLOGY); CELL BIOLOGY; CELLULAR IMMUNOLOGY; CYTOKINE; GENE; ONCOGENES; ONCOLOGY; PROTEIN; PROTEIN KINASE; SIGNAL TRANSDUCTION; TRANSCRIPTION; TUMOR SUPPRESSOR GENES in the McGraw-Hill Encyclopedia of Science & Technology.   Rachel A. Altura

Bibliography. B. B. Aggarwal et al., Signal transducer and activator of transcription-3, inflammation, and cancer: How intimate is the relationship?, *Ann. N. Y. Acad. Sci.*, 1171:59–76, 2009; S. M. Holland et al., STAT3 mutations in the hyper-IgE syndrome, *N. Engl. J. Med.*, 357:1608–1619, 2007; G. Niu et al., Role of Stat3 in regulating p53 expression and function, *Mol. Cell. Biol.*, 25(17):7432–7440, 2005; H. Yu, D. Pardoll, and R. Jove, STATs in cancer inflammation and immunity: A leading role for STAT3, *Nat. Rev. Cancer*, 9:798–809, 2009; P. Yue and J. Turkson, Targeting STAT3 in cancer: How successful are we?, *Expert Opin. Investig. Drugs*, 18(1):45–56, 2009.

Stem cell maintenance in embryos and adults

Stem cells are capable of self-renewal. Provided they have the ability to differentiate into various types of mature cells, they are classified as pluripotent stem cells. Examples of well-studied stem cell types include embryonic stem cells, induced pluripotent stem cells, and tissue-specific stem cells, which enable the corresponding tissues to sustain cell-based homeostasis and which contribute to regenerative mechanisms. These tissue-specific stem cells have been found in blood, nervous tissue, mesenchyme, and skin, and they are regulated both intrinsically and extrinsically. Intrinsic regulation is programmed by genes and transcription factors that are active in the stem cells themselves, whereas extrinsic regulation is accomplished by cytokines and matrices secreted by niche cells surrounding stem cells. Among tissue-specific stem cells, hematopoietic stem cells (HSCs) are the stem cells in the blood lineage and

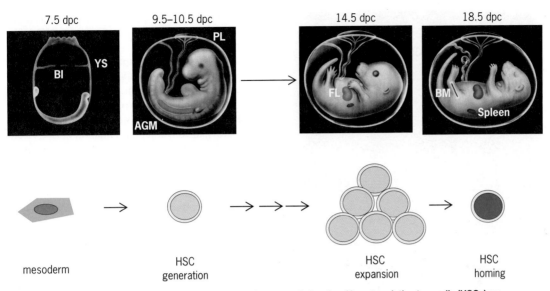

Fig. 1. Schema of hematopoietic development during embryogenesis in mice. Hematopoietic stem cells (HSCs) are generated in the yolk sac (YS), aorta-gonad-mesonephros (AGM) region, and placenta (PL) from early- to mid-gestation periods [7.0–11.0 days postcoitum (dpc)]. They expand to the fetal liver (FL) at midgestation and later migrate (home) temporarily to the fetal spleen and finally to fetal bone marrow (BM). The blood island (BI) is the first site of hematopoiesis.

they have been used in patients for transplantation therapy in the fields of hematology and regenerative medicine. Most sources of HSCs rely on healthy donors and umbilical cord blood. However, problems remain in HSC transplantation, including shortages of donors and risks for donors [for example, transplant rejection and graft-versus-host disease (GVHD)]. Understanding of HSC regulation, especially with regard to niche cells, will enable us to improve HSC therapy through novel stem cell engineering.

HSC maintenance in embryos. In the mouse embryo, the site for blood production changes during the 20-day gestation period. HSCs are initially generated in the yolk sac, aorta-gonad-mesonephros region, and placenta. This initial stage is followed by expansion of HSCs into the fetal liver. Then, later in gestation, after residing temporarily in the fetal spleen, HSCs finally reside in the fetal bone marrow (**Fig. 1**).

The yolk sac is a membranous sac attaching to an embryo and contains the blood islands, which are the first site of hematopoiesis (the process by which the cellular elements of the blood are formed), particularly erythropoiesis (the process by which erythrocytes are formed), and development of the circulatory system after 7.0 days postcoitum (dpc, which is a designation of embryonic age). Among the three germ layers (endoderm, mesoderm, and ectoderm), mesodermal cells differentiate into both hematopoietic cells inside and endothelial cells (ECs) outside of the blood islands. Endodermal cells differentiate into unclassified mesenchymal cells (MCs) that fill out the other spaces of the yolk sac. Therefore, both ECs and MCs likely comprise a niche for yolk sac hematopoiesis. Members of the Hedgehog family of proteins, which are important regulators of many developmental processes, reportedly function in yolk sac hematopoiesis, but the

niche regulation of yolk sac hematopoiesis remains unclear.

After yolk sac hematopoiesis, HSCs are generated in the aorta-gonad-mesonephros region, where aggregates of the cells expressing HSC surface protein markers (for example, c-Kit, CD31, and CD34) are observed and display HSC activity. These aggregates are named hematopoietic clusters or aortic clusters. At 10.5 dpc, these hematopoietic clusters are frequently observed attaching to the walls of large arteries (as if they are generated from EC layers). Because the aorta-gonad-mesonephros region consists of ECs, gonad cells, mesonephros cells, and unclassified MCs, the cells likely comprise a niche for HSC generation. MCs of the aorta-gonad-mesonephros region reportedly increase the number of HSCs at 10.5 dpc through the Hedgehog signaling pathway. In addition, ECs of the aorta-gonad-mesonephros region at 10.5 dpc likely have a role in HSC regulation through secretion of a cytokine known as stem cell factor (SCF). Taken together, several cell components comprise a niche for HSC generation.

The placenta connects the fetus to the mother and is highly vascularized in structure. It functions not only in gas exchange and fetal nutrition but also in hematopoiesis at 8.5–13.5 dpc. In particular, HSCs are generated in the placenta at 11.5 dpc, with their numbers dramatically increasing at 11.5–12.5 dpc. Similar to the aorta-gonad-mesonephros region, hematopoietic clusters are observed in the vasculature of the placenta (**Fig. 2**). To examine the roles of niche cells surrounding the clusters, they were isolated by both laser-capture microdissection and flow cytometry methods. Among several cytokine genes, only the *SCF* gene was expressed particularly in the niche cells. Administration of blocking antibody to c-Kit, the receptor for SCF, clearly demonstrated that SCF/c-Kit signaling is pivotal in HSC regulation in the placenta.

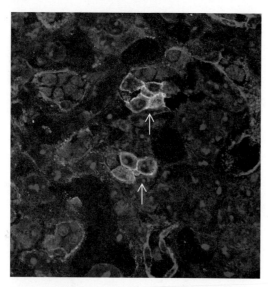

Fig. 2. Confocal image of hematopoietic clusters in the aorta of the placenta (PL) at 11.5 days postcoitum (dpc). Sections of the placenta were made from ICR mouse strain embryos at 11.5 dpc, stained with antibodies (CD34 and c-Kit). CD34/c-Kit double-positive cells indicate hematopoietic clusters (arrows), equivalent to hematopoietic stem cells (HSCs). TOTO-3 was used as a stain for nuclei.

After HSCs are generated in the yolk sac, aorta-gonad-mesonephros region, and placenta, they migrate to the fetal liver, where the numbers of HSCs dramatically increase at 12–16 dpc and they differentiate particularly into erythrocytes. The fetal liver consists of hepatoblasts (hepatocyte precursors), ECs, and hematopoietic cells (as well as other cells). Among them, hepatoblasts have a role in HSC differentiation, particularly erythropoiesis through secretion of mainly erythropoietin (EPO), a cytokine that regulates erythropoiesis, and SCF. In addition, hepatoblasts increase the number of HSCs through production of several other cytokines, including angiopoietin-like 3 (Angptl3), insulin-like growth factor-2 (IGF2), SCF, and thrombopoietin (TPO). Taken together, hepatoblasts comprise a niche for HSC regulation through cytokine secretion. However, it remains unclear how HSCs maintain their self-renewal capability under extensive cytokine exposures that stimulate HSC differentiation. Further investigation is necessary to resolve how niche cells regulate HSC potency in the fetal liver.

After the numbers of HSCs increase in the fetal liver, they migrate to the fetal spleen. At this location, HSCs differentiate particularly into macrophages at 13.5–14.5 dpc. Fetal spleen-derived stromal cells enable HSCs to differentiate into macrophages, but not lymphocytes. It is not known how HSC differentiation is regulated in this tissue. After transient hematopoiesis in the fetal spleen, HSCs gradually migrate to the bone marrow after 16.5 dpc, and they remain there for the duration of their lives. Here, the stromal cells and MCs secrete a chemokine, called CXC-chemokine ligand 12 (CXCL12), which attracts the HSCs expressing the CXCL12 receptor CXCR-4 and results in HSCs migrating to the bone marrow.

Aside from this homing (migration) mechanism, little is known about niche regulation in fetal bone marrow.

During embryogenesis, spatiotemporal regulation of HSCs occurs extrinsically by several niche cell components. Accumulation of evidence about embryonic HSC regulation by niche cells will facilitate the development of novel therapies of in vitro generation, expansion, and differentiation of HSCs from embryonic stem (ES) cells and induced pluripotent stem (iPS) cells in the future.

HSC maintenance in adults. In adults, HSCs residing in the bone marrow self-renew slowly and are able to differentiate into leukocytes, erythrocytes, and platelets, thereby maintaining homeostasis in peripheral blood. Compared to embryonic HSCs, one important characteristic of adult HSCs is their cell cycle status. The cell cycle is the series of events for cell division and duplication. It consists of five phases: G0, G1, S, G2, and M phases. The G0 phase is the state in which cells stop cell division and is called quiescence. The cell cycle status of most adult HSCs is G0 phase, whereas that of embryonic HSCs is a non-G0 phase. This quiescent status is essential in interrupting the exhaustion of HSCs and is tightly regulated by surrounding niche cells. In response to blood cell loss, HSCs exit from quiescence, divide, and differentiate into mature blood cells.

Bone marrow consists of several kinds of cells, including osteoblasts (which generate bone), ECs, neural cells, fibroblasts (stellate connective tissue cells found in fibrous tissue), hematopoietic cells, and MCs. Several reports suggest that both osteoblasts and ECs comprise niches for HSC regulation. Osteoblastic niches are located at the endosteal region (the inner surface of bone), where HSCs bind through their expression of adhesion molecules, including neural cadherin (N-cadherin) and very late antigen 4 (VLA4) integrin, which bind to the same and/or other adhesion molecules expressed on osteoblasts. Moreover, N-cadherin interacts with the extracellular matrix, including an acidic glycoprotein (osteopoietin) and a hyaluronic acid secreted from osteoblasts. These adhesion molecules are important for retaining HSCs in the osteoblastic niches. To maintain their quiescent status in the osteoblastic niches, osteoblasts secrete several molecules: (1) vascular endothelial growth factor (angiopoietin-1), (2) TPO, (3) SCF, and (4) CXCL12. Angiopoietin-1, TPO, SCF, and CXCL12 respectively bind to their corresponding receptors of Tie2, MPL, c-Kit, and CXCR4 expressed on HSCs. The bindings of these molecules result in inhibition of HSC division to maintain the quiescent status of HSCs.

In contrast to the osteoblastic niches, HSCs residing in endothelial niches are not quiescent. The endothelial niches are located in the center of the bone marrow cavity and are composed of sinusoid ECs expressing vascular endothelial cadherin (VE-cadherin). In response to blood cell loss, sinusoid ECs secrete CXCL12 to attract HSCs from osteoblastic niches to endothelial niches, resulting in HSC proliferation and differentiation to replenish blood cells.

The endothelial niches are also important for regulating HSC migration from the osteoblastic niches to blood circulation. Administration of granulocyte colony-stimulating factor (G-CSF) causes endothelial niches to secrete neutrophil protease, which degrades adhesion molecules and keeps HSCs in the osteoblastic niches. Using this mechanism, HSCs are harvested from peripheral blood by G-CSF administration and are clinically used for HSC transplantation therapy.

Recently, it has been reported that mesenchymal stem cells (MSCs) expressing nestin, an intermediate filament protein, constitute an essential niche in the bone marrow. Loss of nestin-expressing MSCs reduces the number and homing capacity of HSCs. However, it remains unclear how MSCs cooperate with osteoblastic and endothelial niches.

The quiescent status of HSCs is regulated extrinsically by several niche cell components. With this in mind, stem cells in leukemia are considered to be an important target for therapy because leukemic stem cells may be quiescent in osteoblastic niches, causing relapse of leukemia after complete remission. Thus, understanding of adult HSC regulation by niche cells will facilitate the development of novel technologies to freely control the quiescent status of HSCs and leukemic stem cells, ultimately leading to therapies targeting leukemic stem cells as well as HSC expansion and differentiation in the future.

For background information *see* CELL (BIOLOGY); CELL CYCLE; CELL DIFFERENTIATION; CELL LINEAGE; CYTOKINE; EMBRYOLOGY; EMBRYONIC DIFFERENTIATION; GENETIC ENGINEERING; HEMATOPOIESIS; REGENERATIVE BIOLOGY; STEM CELLS; TRANSPLANTATION BIOLOGY in the McGraw-Hill Encyclopedia of Science & Technology.

Daisuke Sugiyama; Tomoko Inoue; Kasem Kulkeaw

Bibliography. M. J. Kiel and S. Morrison, Uncertainty in the niches that maintain haematopoietic stem cells, *Nat. Rev. Immunol.*, 8:290–300, 2008; Z. Li and L. Li, Understanding hematopoietic stem-cell microenvironments, *Trends Biochem. Sci.*, 31:589–595, 2006; S. Méndez-Ferrer et al., Mesenchymal and haematopoietic stem cells form a unique bone marrow niche, *Nature*, 466:829–834, 2010; H. K. Mikkola et al., Placenta as a site for hematopoietic stem cell development, *Exp. Hematol.*, 33:1048–1054, 2005; T. Sasaki et al., Regulation of hematopoietic cell clusters in the placental niche through SCF/Kit signaling in embryonic mouse, *Development*, 137:3941–3952, 2010; D. Sugiyama and K. Tsuji, Definitive hematopoiesis from endothelial cells in the mouse embryo; a simple guide, *Trends Cardiovasc. Med.*, 16:45–49, 2006; D. Sugiyama et al., Hepatoblasts comprise a niche for fetal liver erythropoiesis through cytokine production, *Biochem. Biophys. Res. Commun.*, 410:301–306, 2011; A. Trumpp, M. Essers, and A. Wilson, Awakening dormant haematopoietic stem cells, *Nat. Rev. Immunol.*, 10:201–209, 2010; A. Wilson and A. Trumpp, Bone-marrow haematopoietic-stem-cell niches, *Nat. Rev. Immunol.*, 6:93–106, 2006.

Strategic decision making

Strategic decisions are decisions that critically influence the performance and survival of a firm. These strategic decisions are typically made by the senior managers of a firm, and generally involve high levels of uncertainty and complexity, occur in dynamic contexts, and include many different stakeholders, often with conflicting interests. Examples of strategic decisions include deciding whether a firm should collaborate with another firm to develop an innovative new product or go it alone, and whether a company should diversify into a new industry or market as opposed to concentrating on its core businesses.

A common theme—risk—underlies many of the theories and perspectives used to examine strategic decision making. Consistent with common usage, managers use the term risk to refer to the possibility of outcomes that are worse than their expected levels, and the degree to which this poor performance could hurt managers and their firms. This usage of risk differs from the usage in many academic fields that associate risk with variability of outcomes or choices involving well-specified potential outcomes, where each outcome is associated with a probability of occurrence. Researchers of strategic decision making generally use the term risk when referring to decisions that involve uncertainty or ambiguity.

Three theoretical perspectives, namely, the behavioral theory of the firm (BTOF), behavioral decision theory (BDT), and agency theory, dominate the work on strategic decision making. These theories are described in the following text.

Behavioral theory of the firm (BTOF). According to this theory, firms consist of coalitions of stakeholders. As such (and in contrast to traditional economic conceptualizations of firms as entities with a single goal or a well-defined set of many goals), firms have aspiration levels on many dimensions of performance. That is, firms seek to achieve desired levels of performance related to areas such as sales, profits, and so on. Various factors influence aspiration, including a firm's performance relative to its peers and the firm's historical performance. The theory predicts that the firm will make few changes to its existing routines and will operate as usual when the performance of a firm exceeds its aspiration level. However, when the performance of a firm falls below its aspiration level, the firm will seek to raise its performance by making changes that typically increase firm risk (see **illustration**). In this theory (and again in contrast to traditional economic theory), firms do not seek optimal results; rather, they seek to "satisfice," or produce good-enough results; in this context, the term satisfice refers to aiming to achieve satisfactory results. Empirical work on the behavioral theory of the firm has used field or archival data on firms.

Although the above description captures the essence of the theory, it should be noted that the theory also takes into account contingent factors that influence the likelihood that firms will change their routines in response to performance below

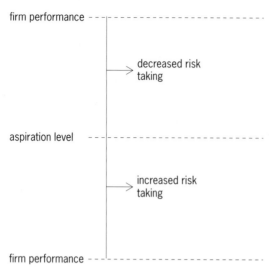

Organizational level prediction of the behavioral theory of the firm (BTOF). When the performance of a firm exceeds its aspiration level, the firm makes few changes to its existing routines and operates as usual. When the performance of a firm falls below its aspiration level, the firm seeks to raise its performance by making changes that typically increase firm risk.

aspiration levels. One critical factor is the amount of slack, or excess resources, held by the firm. Firms can use their slack resources to buffer themselves from changes in the external environment, freeing them from having to change routines every time that performance falls even minimally below aspirations.

Empirical research on the behavioral theory of the firm strongly supports parts of the theory. Many studies find that firms performing below their aspiration levels make more changes and take more risks than firms performing above their aspiration levels. The behavioral theory of the firm has influenced many other organizational theories, including institutional theory (which explains why firms tend to resemble each other), population ecology (which explains variations in organizational forms), and organizational learning.

Behavioral decision theory (BDT). In contrast to the organization-level perspective adopted by the behavioral theory of the firm, behavioral decision theory examines individual decision making under conditions of uncertainty, almost exclusively based on experiments. Early work consisted of an inventory of systematic ways (termed *heuristics*) in which individual choice deviated from expected utility maximization. These heuristics derived from such things as associating the frequency of an event with how easy it is to remember it, or overinterpreting the implications of small samples. Possibly the most influential theory in this area of research is prospect theory, which was proposed initially by Daniel Kahneman and Amos Tversky in the late 1970s. Prospect theory states that individuals judge outcomes with reference to a reference point. Decision behavior varies depending on whether the potential outcomes exceed or fall below the reference point. Unlike the behavioral theory of the firm, prospect theory does not include a full theory of the determination of the

reference point; in the literature, researchers' experimental designs imposed the reference points. However, discussions of the reference point suggest that it depends on various experiential, psychological, and social factors.

Prospect theory addresses situations where individuals face well-defined choices involving uncertainty. In this theory, the value of an alternative action equals a weighted sum of the outcomes that could result from that action. The weighted sum has two components. First, the value of the outcome itself is determined by a value function that has different curvatures for outcomes above or below the reference point. Second, these outcome values are then weighted by a function of their probabilities. This weighting function underweights outcomes with mid-range probabilities and overweights outcomes with extremely low probabilities.

Much of the strategic decision-making work on prospect theory claims that it predicts risk aversion for outcomes above the reference point and risk seeking for outcomes below the reference point. However, these implications have two difficulties. First, they assume that all of a gamble's potential outcomes are either above or below the reference point. Although this is easy to do in an experiment, most strategic choices involve both positive and negative potential outcomes. The value function is extremely risk averse for gambles involving both positive and negative outcomes. The second difficulty is that they completely ignore the impact of the probability weighting function.

Agency theory. Deriving from economics' expected utility theory of rational choice, agency theory distinguishes between principals (a firm's owners, shareholders, or board of directors) and agents (managers and employees of the firm). The theory assumes an honest, risk-neutral principal, and an amoral, risk-averse agent. The principal wants the agent to act in the principal's interests. The theory assumes that the principal cannot tell if the agent is acting in the principal's interests, or that it is expensive to do so. The theory then relates the form of agent compensation to agent behavior. For example, agency theorists have claimed that compensating agents solely by salary (which often depends on firm size) results in excess growth of the firm, and insufficient risk-taking. Generally, agency theorists have favored compensating managers based on returns to the principal. Researchers have used agency theory to examine strategic decisions including franchising, outsourcing, alliance formation, and so on.

The rational choice assumptions of agency theory are often not essential to its predictions. For example, although agency theory would imply that salary compensation, where salary depends on firm size, gives managers an incentive to grow the firm excessively, any sensible theory of managerial behavior would make this prediction. The distinction here is that agency theory assumes that the agent optimally responds to the incentives, whereas few (if any) of the empirical applications of the theory test this optimality prediction.

Other approaches to strategic decision making. In addition to the aforementioned three major approaches, researchers also have examined strategic decision making from a variety of other perspectives. This work often focuses on the influence of one or more of the following factors: context (for example, the uncertainty of the environment, the size of the organization, and demographics of the decision makers); process (for example, the level of systematic analysis and political activity); content (for example, a corporate merger or a choice of business level strategy); implementation (for example, resources available for implementation); and outcomes (including outcomes associated with the decision examined and aggregate firm outcomes based on average firm decision processes). One major perspective focuses on cognition, both at the level of the individual manager and at the level of the top management team, as a critical factor explaining strategic decisions. Studies in this area examine how individuals' cognitive maps or team characteristics such as demographics and mental models influence strategic decisions. Other research examines the effects of decision framing and of individual heuristics and biases. Still other studies examine political behavior around strategic choices. Several studies focus on the risk inherent in strategic decisions; these examine the determinants of individuals' and organizations' risk propensities and perceptions, the impact of risk on organizational outcomes, and the different types of risk in strategic decisions.

Recent advances in strategic decision making. Recent work on strategic decision making has attempted to integrate different theories in this area, such as the behavioral theory of the firm and prospect theory, and prospect theory and agency theory (the "behavioral agency model"). Other studies attempt to advance the field by examining the validity of the assumptions underlying the theories across different national contexts. As always, the true test of these theories lies in how well they can help us understand real-life events, such as the recent financial crises around the world.

For background information *see* COGNITION; DECISION ANALYSIS; DECISION SUPPORT SYSTEM; DECISION THEORY; MOTIVATION; PROBABILITY; PROBLEM SOLVING (PSYCHOLOGY); PSYCHOLOGY; RISK ASSESSMENT AND MANAGEMENT; SOCIOBIOLOGY in the McGraw-Hill Encyclopedia of Science & Technology.

Devaki Rau; Philip Bromiley

Bibliography. P. Bromiley and D. Rau, Risk taking and strategic decision making, pp. 307–326, in P. Nutt and D. Wilson (eds.), *Handbook of Decision Making*, Wiley, Chichester, U.K., 2010; J. Child, S. Elbanna, and S. Rogrigues, The political aspects of strategic decision making, pp. 105–138, in P. Nutt and D. Wilson (eds.), *Handbook of Decision Making*, Wiley, Chichester, U.K., 2010; R. M. Cyert and J. G. March, *A Behavioral Theory of the Firm*, Prentice-Hall, Englewood Cliffs, NJ, 1963; A. Fiegenbaum and H. Thomas, Attitudes toward risk and the risk-return paradox: Prospect theory explanations, *Acad. Manage. J.*, 31:85–106, 1988; R. M. Holmes, Jr., et al., Management theory applications of prospect theory: Accomplishments, challenges, and opportunities, *J. Management*, 37:1069–1107, 2011; M. Jensen and W. Meckling, Theory of the firm: Managerial behavior, agency costs and ownership structure, *J. Financial Econ.*, 3:305–360, 1976; D. Kahneman and A. Tversky, Prospect theory: An analysis of decision under risk, *Econometrica*, 47:263–291, 1979; J. G. March and Z. Shapira, Managerial perspectives on risk and risk taking, *Management Sci.*, 33:1404–1418, 1987; J. G. March and H. Simon, *Organizations*, Wiley, New York, 1958; V. Papadakis, I. Thanos, and P. Barwise, Research on strategic decisions: Taking stock and looking ahead, pp. 31–70, in P. Nutt and D. Wilson (eds.), *Handbook of Decision Making*, Wiley, Chichester, U.K., 2010; J. P. Walsh, Managerial and organizational cognition: Notes from a trip down memory lane, *Organ. Sci.*, 6:280–321, 1995.

Submarine power cables

The demand for inexpensive, reliable, and clean electric power has been increasing steadily in recent years. This is particularly true for remote locations, such as islands and offshore platforms. Submarine power cables (SPC) can connect those places to the onshore electric network. Islands that have been powered by local, small, inefficient power plants can now be supplied from shore with less expensive and more reliable electric power. Also, offshore oil and gas production platforms can draw benefits from a shore-based power supply. Onboard power generation occupies precious space and requires maintenance. A reliable maintenance-free power supply from shore cuts costs and increases platform safety.

Demand from the offshore oil and gas industry is only one of the driving forces for the remarkable technical development of submarine power cables and their installation. Another application with a rapidly increasing demand is bulk power transmission between cities, regions, and countries. This is often done with high-voltage direct current (HVDC) technology, using a new breed of submarine power cables. Today, a single submarine power cable system can transmit the power of a nuclear plant over hundreds of miles.

Design of submarine power cables. All submarine power cables have three vital components: the conductor, electrical insulation, and the armoring. The choice of conductor configuration and insulation material is determined by the application of the cable, while the design of the armoring responds to the submarine site conditions and the installation procedure. Traditionally, copper is used as the conductor material, but aluminum is taking market share because it is less expensive. Since the conductor can have a live potential far higher than 100,000 V, the electrical insulation of the cable conductor against the grounded cable sheath is a crucial function. Most modern submarine power cables have cross-linked polyethylene (XLPE) insulation, which has

Fig. 1. Mass-impregnated HVDC cables. (*a*) First generation: 100 kV, 20 MW (1954). (*b*) 450 kV, 600 MW (1993).

superseded the oil-filled cables of earlier times. This polymeric insulation material is extruded over the inner conductor, together with shielding layers. After the extrusion, the entangled molecules are crosslinked (locked in their position), creating a tough, void-free, temperature-resistant insulation layer. Another insulation material with a long track record is mass-impregnated paper insulation, used today for HVDC submarine cables with the highest voltage ratings (**Fig. 1**). Since humidity is the prime enemy of any electrical insulation, high-voltage submarine power cables usually have a lead sheath to keep humidity out.

From the earliest days, it was evident that submarine cables must be protected against fishing gear, anchors, and other subsea threats. The vast majority of submarine power cables are wrapped with one or two layers of heavy steel wires, creating a sturdy armoring. This provides both tensional strength for the installation process and impact protection over the entire lifetime of the cable. In order to reduce eddy-current losses in ac cables, armoring wires of copper, aluminum, or stainless steel are used occasionally. An external coating made from polypropylene strings can sport a few white or orange stripes, providing better optical contrast for the cameras of underwater robots.

Grid interconnectors. The largest submarine power cables connect regional or national grids. The Trans-

bay Cable (200 kV dc) in San Francisco and the Neptune Cable (500 kV dc) between New Jersey and Long Island feed bulk power to urban areas. The connection of autonomous electric power grids by long-distance submarine power cables has become popular in Europe. Submarine power cables connect the United Kingdom to France; Sweden to Germany, Poland, and Denmark; Norway to the Netherlands; and so forth. Demand and supply on each side change hour by hour, and cable operators earn a profit from the differences in the electricity prices on the two sides. Nine HVDC interconnectors span more than 100 mi (161 km), and more long-haul systems are under construction.

These cable systems are operated with HVDC, since ac transmission over distances of 40 mi (64 km) or more would require enormous charging currents for the cable capacitance. HVDC eliminates these charging currents. A pair of single–core cables carries direct current and is connected to the land-based ac grid by means of voltage converters. An HVDC cable system requires much less conductor material, insulation, and armoring than an ac system with the same transmission power, and has much less losses. The disadvantage is the cost of the voltage converter stations; still, an HVDC transmission system can be less costly over its lifetime.

The grand old man of submarine power cables, the mass-impregnated cable, is suitable for the largest HVDC interconnectors. The electric current is carried in copper or aluminum conductors of up to 2.5 in. diameter (a cross-sectional area of 3000 mm²). Figure 1 depicts two vintages of mass-impregnated cables. Both have a copper conductor, insulation made of high-density paper impregnated with a high-viscous compound, and a lead sheath. Distance is no issue for mass-impregnated cables. The first cable of this kind carried a humble power of 20 MW over a submarine distance of 63 miles (100 km) in 1954. Today, the NorNed cable transmits 700 MW over 360 mi (580 km) [see **table**]. The commercial limit for a pair of mass-impregnated cables is about 2000 MW at 500 kV dc.

In recent years, HVDC cables with extruded dc-XLPE insulation also have been developed for submarine use. This material is expected to oust mass-impregnated insulation in the future, since it is easier to manufacture. Extruded HVDC cables are available for voltage ratings up to 320 kV.

Grid interconnections or city infeeds over shorter distances can also be made by ac cable systems, such as the 345-kV Bayonne cable system between New Jersey and Brooklyn [distance 7 mi (11 km)]. There

Some remarkable submarine power cables					
Project name	Voltage	Submarine cable length	Location	Cable type	Remarks
NorNed	450 kV dc	360 mi (580 km)	Norway–Netherlands	Mass-impregnated	Longest submarine power cable
SAPEI	500 kV dc	264 mi (425 km)	Italy–Sardinia	Mass-impregnated	Largest water depth 5315 ft. (1620 m)
Goliath	123 kV ac	66 mi (106 km)	Norway	3-core XLPE	Longest ac cable
Bayonne	345 kV ac	7 mi (11 km)	United States	Single–core XLPE	Highest submarine ac voltage in United States
Neptune	500 kV dc	51 mi (82 km)	United States	Mass-impregnated	Highest submarine dc voltage in United States

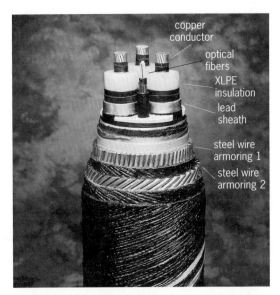

Fig. 2. Three-core XLPE submarine cable for 123 kV. The outer armoring layer with short lay length (also known as "rock armor") provides extra lateral impact protection.

are three single–core cables making up an ac system, with copper wire armoring in lieu of steel in order to reduce the eddy-current losses in the armoring. The insulation of large ac systems is made from XLPE or, in a few cases, from oil-filled paper insulation.

Cables for offshore wind farms. There is an enormous boom in offshore wind farm construction, with high growth rates in Europe and, to some extent, in China. A farm can have up to 100 individual turbines. The turbines are interconnected with three-core medium-voltage submarine power cables (currently up to 36 kV). The power is collected in an offshore transformer platform, which is connected to shore by a larger subsea export cable. **Figure 2** demonstrates such a three-phase XLPE export cable with copper conductors, each surrounded by an XLPE insulation layer. Extruded filler profiles in the interstices between the three cores ensure that the cable is round rather than triangular, and accommodate fiber-optic cables for data transmission. Wind farm power export is most often accomplished by heavy three-core XLPE cables at 150–170 kV, and in a few cases at 245 kV. The cables can have conductor sizes up to 1.75 in. diameter (1600 mm^2) and more than 10 in. (250 mm) outer diameter.

Some of the wind farm projects in the European North Sea are located more than 120 mi (193 km) from the nearest onshore substation, making ac cable systems not viable. These remote wind farm clusters are connected to the shore grid by a pair of extruded single-core HVDC cables. One cable has positive voltage, the other negative. A single–layer steel wire armoring is sufficient for the relatively shallow water [with a depth of less than 300 ft (100 m)]. The first submarine application of this cable type, however, was not for wind power, but was the Cross Sound cable in 2002, which brings power from Connecticut to Long Island. The tech-

nology has made progress since, and now some extruded HVDC cables at 300–320 kV are being constructed in the North Sea to connect remote offshore wind farms. They are always installed in pairs. With proper sizing, each pair can transmit up to 1000 MW of power, the equivalent of a nuclear reactor.

Installation. The installation of submarine power cables requires highly specialized marine equipment. Worldwide, a few dedicated cable-laying vessels (CLV) exist, with large turntables to carry 7000 tons of cable each. Depending on the cable dimension, this translates into an uninterrupted length of 100 mi (160 km) of cable. This is enough for most applications. If necessary, cable lengths can be jointed at sea. Today, only a handful of companies worldwide are able to perform this specialized job.

Basically, the cable-laying vessel steams ahead at a controlled speed, paying out the cable, which goes down in a catenary line and arrives neatly on the seafloor (**Fig. 3**). The reality, however, is different. The permitted cable corridor is often narrow; the seafloor can be rocky or littered with boulders, wreckage, or dumped ammunition. The cable-laying vessel needs to keep on course despite changing winds, and the vertical movements caused by waves generate dynamic forces in the suspended cable. Satellite-based dynamic positioning systems in combination with advanced propulsion systems enable modern cable-laying vessels to maneuver with high precision even in difficult weather.

The tensional forces during laying (and eventually recovering) of the cables limit the water depth in which the cables can be installed. Three-core cables have been laid in more than 1500 ft (500 m) of water, and single-core cables in 5000 ft (1600 m) of water.

Since damage can cause a massive loss of income to the operator and high repair costs, submarine power cables are protected by embedding them into the seafloor using powerful ploughs or water-jetting

Fig. 3. Offshore supply vessel laying a pair of extruded HVDC cables.

devices. A burial depth of 3–6 ft (1–2 m) provides a good protection against the most prevalent threats: fishing gear and ship anchors.

Outlook. Although submarine power cables have achieved an impressive performance, there are challenges ahead. These can be put into four words: more power, greater depth. More power can be achieved by higher transmission voltage. The manufacturers, however, do not speak openly about their development goals.

The offshore industry and transmission system operators are asking for submarine power cables for locations with greater depth. Connecting the islands of Hawaii would require the installation of submarine power cables in more than 6000 ft (2000 m) of water, which has not yet been done. Also, deep-water floating oil and gas production platforms need to be connected to power from shore, requiring a larger working depth for submarine power cables.

Today's submarine cable design can be stretched to some more power and depth. In the end, new concepts of electrical insulation and mechanical strength will help to string the most powerful lines through the largest depths.

For background information *see* ELECTRIC POWER TRANSMISSION; SUBMARINE CABLE in the McGraw-Hill Encyclopedia of Science & Technology.

Thomas Worzyk

Bibliography.   C. C. Barnes, *Submarine Telecommunication and Power Cables*, Peregrinus, Stevenage, U.K., 1977; O. I. Gilbertson, *Electrical Cables for Power and Signal Transmission*, Wiley, New York, 2000; T. Worzyk, *Submarine Power Cables: Design, Installation, Repair, Environmental Aspects*, Springer, Berlin, 2009.

The shape of the universe

What is the shape of space? While this question may have once seemed more philosophical than scientific, modern cosmology has the chance to answer it using the oldest observable light in the universe, the cosmic microwave background radiation (CMB). NASA's *Wilkinson Microwave Anisotropy Probe* (*WMAP*) has made a detailed map of the CMB sky that has been used to provide answers to many age-old questions about the nature of the universe.

The CMB is the oldest light in the universe reaching us, from 379,000 years after the big bang. The surface of last scattering marks the time that the universe had cooled enough for light to propagate freely through space. The CMB photons have thus all traveled the same distance to reach our detectors and form a two-dimensional sphere (a 2-sphere) surrounding the Earth. A map of the CMB sky is usually shown as a flattened sphere with different colors or shades showing temperature variations in the CMB photons (**Fig. 1**). The temperature of the CMB photons reaching us today is 2.725 K, and the total temperature variation is only parts in a million (10^{-6}).

While it is certainly possible that the universe extends infinitely in each spatial direction, the *WMAP* data contain a hint that the universe is not infinite in extent. *WMAP* shows a limit to the size of large-scale structures in the universe, with none extending more than $60°$ across the sky. This was first noted by the *COsmic Background Explorer* (*COBE*) and was later confirmed by *WMAP*. The missing structure on global scales could be explained by a universe that is fundamentally not big enough to sustain them and hints at a special distance scale in the universe.

It is possible that the three spatial dimensions of our universe have a finite volume without having an edge, just as the two-dimensional surface of the Earth is finite but has no edge. In such a universe, a straight path in one direction could eventually lead back to where it started and would be considered a closed path. For a short enough closed path, we expect to be able to detect an observational signature revealing the specific shape, or topology, of our universe.

Geometry and topology. An important question answered by the *WMAP* mission is that of the curvature of space (**Fig. 2**). The matter and energy density of the universe indicate that space is very nearly flat. Even if space is not quite flat, the *WMAP* data tell us that the radius of curvature of the universe is at least of the order of the size of the observable universe, and space can be considered to be nearly flat.

The *WMAP* sky also provides clues about the topology of the universe. The mathematical field of topology is sometimes described as rubber-sheet geometry in that stretching a surface or space does not change its topological properties. For example, the surface of a doughnut is topologically equivalent to the surface of a coffee cup (**Fig. 3**). In topology, this surface is known as a two-dimensional torus or 2-torus. A 2-torus can also be represented by a square with opposite sides identified or "glued" together. A three-dimensional torus can similarly be

Fig. 1. The *Wilkinson Microwave Anisotropy Probe* (*WMAP*) 7-year cosmic microwave background radiation (CMB) data. An all-sky map is used with the Milky Way Galaxy providing the horizontal diameter and the galactic poles at top and bottom. Dark patches are slightly cooler and light patches are slightly hotter. (*Courtesy of NASA/WMAP Science Team*)

represented by a cube with opposite faces identified. In such a space, light traveling in one direction would eventually come back to where it started. It is possible that our universe has a similar multiply connected global topology and that there exist closed paths in one or more spatial dimensions. For example, if our universe is a three-dimensional torus, then there exist closed paths in all three spatial dimensions.

Observable topologies for our universe would have a characteristic fingerprint in the CMB sky. The set of possible topologies for our universe is determined by the curvature of space. In a flat universe, the allowable topologies are restricted to a set of 18 possibilities. It has been shown that a nearly flat universe would have an observational fingerprint very similar to that of an exactly flat universe. Neil J. Cornish, David N. Spergel, and Glenn D. Starkman have described the CMB signature revealing the shape of space as "circles in the sky."

Circles in the sky. In a universe with a nontrivial topology, there are closed paths in one or more spatial dimensions. This means that, if we could see far enough in one direction, we would see a copy of ourselves. A copy of our own galaxy, solar system, and planet could possibly be observed many light-years away. Due to the finite speed of light, if we were to observe a distant copy of our galaxy we would see it as it existed in the past. One way to think about a multiply connected space is to picture multiple copies of the fundamental domain. The fundamental domain of a three-torus universe would be a cube or a rectangular prism. We can think of infinite space as tiled with copies of the fundamental domain. In a multiply connected universe, each copy of the fundamental domain contains a copy of the Earth, and surrounding each copy of the Earth is a copy of the 2-sphere of the surface of last scattering. Since, in a homogeneous and isotropic geometry, the intersection of two 2-spheres is a circle, and these circles in the CMB are physically the same place in space, we expect to find matching circles of temperature patterns in the CMB when looking in two different directions in the sky (**Fig. 4**).

Matching circles of temperature patterns in the CMB sky would be the observational signature we would need to find the shape of our universe. The *WMAP* data has been searched for pairs of circles with matching temperature patterns but no statistically significant matches have been found. This suggests that, if we do live in a universe with a nontrivial topology, the fundamental domain is larger than the observable universe and that copies of the CMB do not intersect one another to produce matching circles.

There is a specific nontrivial topology for the universe that has received some attention from researchers. A dodecahedral topology would be allowed in a slightly positively curved space. The fundamental domain would be 12-sided and we would expect to find 6 pairs of matching circles in the CMB sky (**Fig. 5**). There is still no conclusive evidence

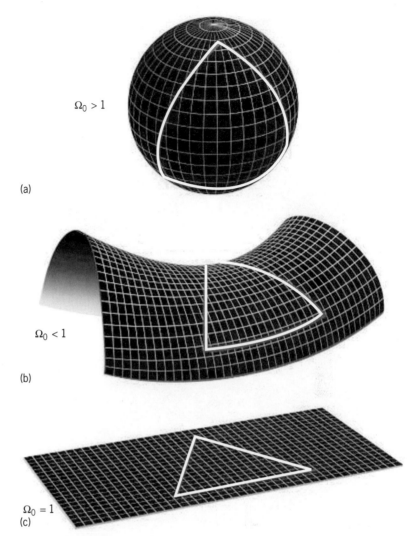

$\Omega_0 > 1$

(a)

$\Omega_0 < 1$

(b)

$\Omega_0 = 1$
(c)

Fig. 2. Two-dimensional surfaces with (*a*) positive curvature, (*b*) negative curvature, and (*c*) zero curvature. Observations suggest that the three spatial dimensions of our universe have a flat or nearly flat geometry. The density parameter Ω_0 is the ratio of the density of the universe to the critical density. By this definition $\Omega_0 = 1$ for a flat universe. (*Courtesy of NASA/WMAP Science Team*)

Fig. 3. The surface of a doughnut is topologically equivalent to the surface of a coffee cup and is known as a torus.

that we live in a dodecahedral space, but it remains an interesting possibility.

While the shape of our universe remains a mystery, the matching-circles test can be used to place a lower bound on the size of the universe. The fact that we do not observe any circles with matching temperature patterns in the CMB means that the fundamental domain of our universe must be at least 24×10^9 parsecs across ($\approx 78 \times 10^9$ light-years).

For background information, *see* BIG BANG THEORY; COSMIC BACKGROUND RADIATION;

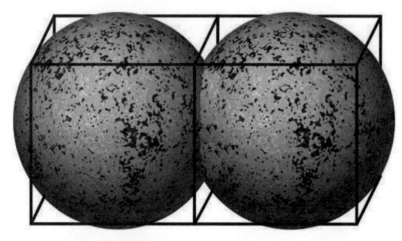

Fig. 4. In a multiply connected universe, all of space must be tiled with copies of the fundamental domain. Here the fundamental domain is a cube, and associated with each domain is a copy of the surface of last scattering. The intersections are physically the same place in space, producing matching circles of temperature patterns in two different directions as observed from the Earth.

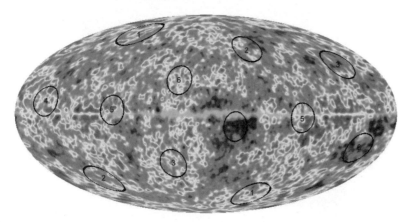

Fig. 5. An all-sky CMB map with the locations indicated in black for 6 sets of circles with matching temperature patterns that are expected for a certain orientation of a universe with a dodecahedral topology.

COSMOLOGY; TOPOLOGY; TORUS; UNIVERSE in the McGraw-Hill Encyclopedia of Science & Technology.   Joey Shapiro Key

Bibliography. F. Levin, *Calibrating the Cosmos: How Cosmology Explains Our Big Bang Universe*, Springer, New York, 2007; J. Levin, *How the Universe Got Its Spots: Diary of a Finite Time in a Finite Space*, Weidenfeld & Nicolson, London, 2002, Anchor, New York, paper 2003; J. P. Luminet, *The Wraparound Universe*, AK Peters, Ltd., Wellesley, MA, 2008.

The star that changed the universe

On the night of October 5/6, 1923, Edwin Powell Hubble took an astronomical photograph that is now famous (**Fig. 1**). He used the 100-in. (2.5-m) Hooker telescope on Mount Wilson, above Pasadena, CA, to expose a 4 × 5-in. (102 × 127-mm) glass plate for 45 min. He had pointed the telescope at a field in M31, the great nebula in Andromeda, and he was engaged in a long-term effort to first find Cepheid variable stars in M31 and to then measure their light curves. Doing this would enable Hubble to determine the distance to M31, a matter much in dispute at that time.

Hubble's achievement. Hubble was doing exactly what he had been hired to do by the Director of Mount Wilson, George Ellery Hale, and that was to apply the power of the world's largest telescope—then six years old—to address big questions such as the size and structure of our universe. In the early 1920s, little to nothing was known about very basic cosmological questions, such as: How big is our Galaxy, and where are we in it? Are there other galaxies too, or is the Milky Way the only one? How big is the universe, and how old? This was literally the subject of debate between Curtis and Shapely in early 1920. The Mount Wilson Observatory was supported primarily by Andrew Carnegie, and it later became part of the Carnegie Observatories.

Hubble set out to break through this uncertainty and he had the tools to do it, tools like the 100-in. telescope that nobody else could hope to approximate. The other crucial tool was the Cepheid, a type of variable star that Henrietta Leavitt, working at the Harvard College Observatory in 1912, had found to have a unique and vital property, namely a relationship between the period of its variation and the intrinsic brightness of the star. She had found this relation by examining photographic plates taken of the Magellanic Clouds. The distance to the Clouds was not known, but by assuming (reasonably) that all their

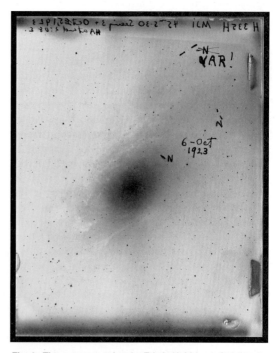

Fig. 1. The exposure taken by Edwin Hubble on October 6 (Universal Time), 1923. Written on the emulsion side (that is, in reverse) is the plate number (H335H for Hooker telescope and Hubble), exposure time, and so forth. Written on the glass side in India ink are later notations, namely the "N" marks to note novae, with one crossed out and replaced by "VAR!" when Hubble determined that it was a Cepheid. That star is Hubble's variable number 1. (*Reproduced with permission of The Carnegie Observatories, Pasadena, CA*)

stars were at the same distance she could verify the relationship. In 1913, Ejnar Hertzsprung provided an absolute calibration by studying Cepheids in the Milky Way; by the 1920s, all of this was well known. In addition to having an intrinsic period-luminosity relation, Cepheids were also ideal for determining the distance to M31 because they are inherently luminous and so can be detected as individual stars at large distances. Moreover, they vary by a magnitude or more, meaning the way they vary was detectable even using photographic plates.

But the first critical step to success was finding any Cepheids at all in M31. Hubble, in that era long before electronic imaging, used photographic plates that had to be chemically developed and then put into a different chemical, a "fixer," to preserve the image. Hubble exposed his plates and then impatiently pulled them out of the fixer before they were fully ready (his plates needed reprocessing in later years to ensure their survival). He looked for stars that got brighter by comparing each exposure to an earlier one. Many of these were novae, marked with an "N." Novae get bright very quickly and they then fade away, just as Cepheids do. It was later that he realized that that star near the top of the exposure in Fig. 1 had not only gotten fainter, it had come back on this later exposure and become bright again. It had to be a Cepheid, and he excitedly wrote that famous "VAR!" after he crossed out the "N." It took many more years of hard work before Hubble published his results in 1929. He knew before then, of course, that M31 was another galaxy beyond our Milky Way, but he needed measurements of many Cepheids to build a convincing case for this profound discovery (although a first, quick note was published by Hubble in 1925).

Copies of Hubble's plate. Hubble's 1923 plate is an important historical artifact that was taken with his own hands and embodies his primary scientific goals and achievements. It also represents one of the main reasons that the *Hubble Space Telescope* (*HST*) was built in the 1980s and launched in 1990. For all these reasons, ten film copies of the plate were flown on the *HST* deployment mission, STS-31, in April, 1990, as part of the Official Flight Kit (OFK) of the Marshall Space Flight Center (MSFC). A NASA center that launches its spacecraft on the shuttle is entitled to include items of its choice in its OFK. These are frequently crew patches or the like that can be given out to those who worked on the spacecraft later, but the HST Project at MSFC also included the 10 film copies. (Allan Sandage first suggested the 1923 plate as an appropriate artifact to commemorate the launch of the *HST* and arranged for its loan to the Space Telescope Science Institute through the cooperation of the Carnegie Observatories. A number of people at MFSC helped to see that the ten copies flew on STS-31.)

The intent at the time was to get the copies back, to have a print made from each, and to present the negative and framed print as a gift to the organizations that played key roles in the construction and launch of the *HST*. An astronomer would go out to

those groups to present the gift and to show what astronomers were doing with this new, marvelous telescope.

The impetus for doing that all fell apart after the deployment, once the flawed mirror was revealed. The *HST* was no longer a source of pride; it was an embarrassment. In addition, attempts to locate and retrieve the copies from MSFC failed.

Years elapsed, and, finally, Servicing Mission 4 (SM4) was nearly becoming a reality at long last. SM4 was planned to be the last visit of the space shuttle to the *HST*, and it would be especially fitting if those ten film copies could be found and reflown on SM4. New attempts were made to follow every possible lead, but all turned into dead ends; they could not be found. But a few film copies remained of the 1923 plate, and two were flown on SM4; both were recovered.

HST observations of Hubble's variable. The original intent is still to be fulfilled, but an obvious and natural question is whether the *HST* itself ever observed that Cepheid that Hubble found in 1923. The first step was to determine the name and celestial coordinates of that star marked with the "VAR!": it turns out to be Hubble's variable number 1, fittingly. A search of the *HST* archive then showed there were some exposures that came very close to the star, but none included it. The Hubble Heritage team at the Space Telescope Science Institute volunteered to obtain some exposures of variable number 1 with the new cameras on the *HST*, but to use that scarce time to its best they needed to know how the star is varying today, and, in particular, the phase of the light curve so that it would be possible to catch the star near both its maximum and minimum light levels.

To do that, the Hubble Heritage team enlisted the help of the American Association of Variable Star Observers (AAVSO). It is possible for citizen-scientists in the twenty-first century to own a telescope and

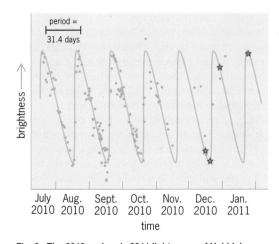

Fig. 2. The 2010 and early 2011 light curve of Hubble's variable number 1, assembled from observations of the American Association of Variable Star Observers (AAVSO), with the large stars marking the times when the *Hubble Space Telescope* obtained an exposure. Note that Edwin Hubble determined that this star had a photographic magnitude of 18.4 at its brightest and 19.3 when faint.

Fig. 3. A montage put together at the Space Telescope Science Institute by Z. Levay. (*a*) A modern image of M31. Even this image shows only a small portion of this huge galaxy. The footprint of Edwin Hubble's 1923 plate is shown as a large white rectangle. Within that large footprint one can see the very small footprint of the *HST* image taken with Wide Field Camera 3 (WFC3) (*photograph courtesy of Robert Gendler*). (*b*) The WFC3 image, enlarged, with Hubble's variable number 1 shown circled. Even this short exposure with the *HST* reveals thousands to millions more stars in M31 than Hubble could detect in the 1920s. (*c*) Hubble's 1923 plate.

instrument that can do better on faint stars than Hubble could with the world's largest telescope in his time. By putting together the observations of these amateurs (an effort in which the AAVSO, directed by Arne Henden, provided critical and enthusiastic support) the Cepheid's light curve and its phase were determined. The Hubble Heritage Team (especially Lisa Frattare, Keith Noll, and Zolt Levay) then carried out observations in which the *HST* took four separate exposures (**Fig. 2**). It is also helpful to see Hubble's exposure in comparison to the *HST*'s (**Fig. 3**). Even though Hubble's 1923 exposure samples much more of M31 than we can now with the *HST*, his 4×5-in. plate sampled only a tiny portion of what is (at the 100-in. telescope) an image about 1.5 m (60 in.) across in its full extent (at the Newtonian focus of that telescope, which is what Hubble used). Hubble had the equivalent of binoculars; the *HST* is like a microscope.

Hubble's 1923 observation changed the course of astronomy forever. We are now, unambiguously, in a very large universe, one that is expanding, also thanks to Hubble and Lemaitre. Cepheids remain as useful mileposts in the twenty-first century and still are observed by *Hubble* (the telescope) to establish

the distances to the gallaxies far beyond M31 (most of which were much too far away to be observed as more than smudges in 1923). *Hubble* the telescope has the capability that Hubble the man could only dream of (and I doubt he could truly have conceived of the *HST*'s power), and the telescope has been aptly named.

For background information *See* ANDROMEDA GALAXY; ASTRONOMICAL IMAGING; ASTRONOMICAL PHOTOGRAPHY; CEPHEIDS; COSMOLOGY; HUBBLE CONSTANT; HUBBLE SPACE TELESCOPE; LIGHT CURVES; MAGELLANIC CLOUDS; NOVA; UNIVERSE; VARIABLE STAR in the McGraaw-Hill Encyclopedia of Science & Technology.   David R. Soderblom

Bibliography. E. Hertzsprung, Über die räumliche Verteilung der Veränderlichen vom δ Cephei-Typus, *Astr. Nach.*, 196:201, 1913; E. P. Hubble, Cepheids in spiral nebulae, *The Observatory*, 48:139–142, 1925; E. P. Hubble, A spiral nebula as a stellar system, Messier 31, *Astrophys. J.*, 69:103–158, 1929; H. S. Leavitt and E. C. Pickering, Period of 25 stars in the Small Magellanic Cloud, *Harvard Coll. Obs. Circ.*, 173:1–3, 1912; V. Trimble, The 1920 Shapley-Curtis discussion: background, issues, and aftermath, *Pub. Astron. Soc. Pac.*, 107:1133–1144, 1995.

3D graphic displays

Three-dimensional (3D) graphic display systems are currently attracting unprecedented levels of interest. This has given rise to a great diversity of approaches from displays intended for use with hand-held devices to large virtual-reality installations. Irrespective of the nature of the technology, in order to be successful, 3D displays must meet the expectations of the complex human visual system. Here, various cues to depth are summarized and this forms the basis for discussion on a number of diverse exemplar approaches to display implementation. This is structured around a simple classification scheme.

Cues to depth. When viewing our surroundings, the physical separation of our eyes gives rise to small disparities in the images cast onto the retina. This occurs because each eye captures the scene from a different vantage point and on the basis of processes that are far from understood. These disparities generate a strong impression of 3D relief. This binocular parallax cue to depth provides a better insight into the relative and absolute spatial positions of objects comprising the visual scene.

Our visual impression of the natural world is based on a range of monocular and binocular cues to depth. Linear perspective, height in the visual field, and shading underpin image depiction on conventional (monocular) flat-screen display systems. Other important cues are not supported by the traditional implementation of this display modality—specifically, binocular parallax, convergence (the visual axes of the eyes intersecting at the point of fixation), accommodation (the eyes focusing on the point of fixation), and motion parallax (the relative motion of objects located at different distances). This latter cue can manifest itself in two ways. Firstly, relative motion may be based on the movement of components within the image scene [motion parallax caused by image dynamics (PID)]. Alternatively, information on spatial relationships may be more easily discerned by changes in vantage point [motion parallax caused by observer dynamics (POD)]. Some 3D displays limit support for POD to horizontal head movements, whereas other approaches more closely mimic the real-world experience by supporting both horizontal and vertical changes in vantage point.

Some forms of 3D graphic displays extend the capabilities of the traditional display screen by also supporting binocular parallax, while other approaches endeavor to incorporate the additional cues associated with the visual impression of the physical world. However, the relative reliance that we subconsciously place upon cues varies according to the nature of an image scene and the forms of information that we are attempting to extract. Consequently, the fact that a display supports a larger number of cues is not necessarily advantageous. A key requirement is to ensure that supported cues conform to our real-world expectations.

Display classification. A broad range of approaches may be adopted in the implementation of 3D graphic displays. When considered from the standpoint of cues that they are able to support, they can be conveniently grouped into four categories. These are summarized in the **table** and are briefly discussed below.

Head-coupled perspective displays. These are of particular relevance to hand-held devices such as iPhones. They exhibit the visual characteristics of the monocular display and support horizontal and vertical forms of POD. In this latter respect, the inbuilt camera is used to track a user's vantage point and, on the basis of this information, the 3D geometry of the image scene is computed. The effect is akin to viewing the world through a single eye. Such systems capitalize on the often undervalued significance that motion parallax has as a depth cue and on this basis provide an impression of three-dimensionality.

Stereoscopic displays. Both the stereoscopic and autostereoscopic (Type I) classes of display are fundamentally based on the stereoscope, which was pioneered by Charles Wheatstone and David Brewster in the first half of the nineteenth century. At the present time, these classes of display are the focus of extensive research and development activity.

In the case of the stereoscopic class of display, two views onto an image scene are formed (the vantage points differing by a distance approximately equal to the inter-ocular separation). One view is presented to the left eye and the other to the right. Disparities in the retinal images provide support for the binocular parallax depth cue.

Stereoscopic displays differ in terms of the way in which the left and right views are presented to the visual system. In the case of the anaglyph approach, the views are displayed simultaneously and are distinguished by the range of colors employed in their depiction. Viewing glasses (comprising colored filters, usually red and blue or red and green) ensure that only the appropriate view is visible to each eye. This approach imposes limitations on color content.

A simple display classification scheme		
Class	Examples	General characteristics
Head-Coupled Perspective	Conventional flat-screen display + camera, Monocular multiview	Pictorial cues + H + V,PID + H + V,POD
Stereoscopic	Anaglyph, Lenticular, Parallax barrier	Pictorial cues + H + V,PID + Binocular parallax
Autostereoscopic Type I	VR and IVR, Multiview	Pictorial cues + H + V,PID + Binocular parallax + H,POD + freedom in vantage point
Autostereoscopic Type II	Volumetric, Holographic	Pictorial cues + H + V,PID + Binocular parallax + H + V,POD + oculomotor cues + freedom in vantage point

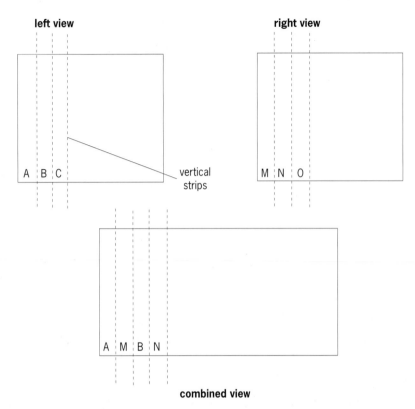

Fig. 1. The left and right views are divided into vertical strips and then interleaved.

Alternative glasses-based approaches often rely on the left and right views being depicted sequentially, that is, as alternate image frames. In one scenario, liquid-crystal-based glasses are used. Each eyepiece

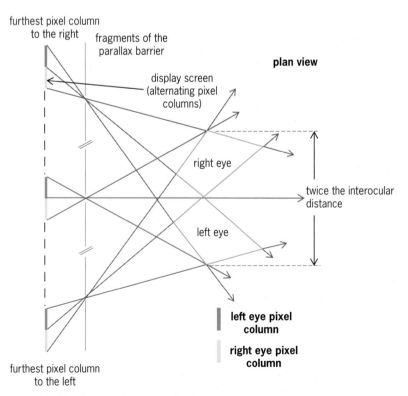

Fig. 2. The general principle of operation of the parallax barrier technique.

is switched between transmissive and opaque states (switching being synchronized to the display's frame-refresh frequency) in such a way that only one eye-piece is transmissive at any time. Thus, the left and right views are directed to the appropriate eye. Since only alternate frames are visible to each eye, flicker (both perceived and subliminal) can be problematic and necessitates the use of a refresh frequency of about 120 Hz.

A related approach incorporates a polarizing filter that forms a faceplate to the display. The polarization characteristics of this filter are switched between two orthogonal states in synchronism with the frame-refresh frequency. Left and right views are depicted as alternate frames. Viewing glasses typically contain orthogonal polarized filters, ensuring that the appropriate view is directed to the intended eye.

Large-screen systems employing this technique commonly make use of two digital light processing (DLP) projectors that simultaneously output image frames to a single diffusive screen. Polarizing filters attached to each DLP projector ensure that the frames from the projectors are orthogonally polarized. Again, orthogonally polarized viewing glasses are used and, since the left and right views are depicted simultaneously, a standard frame-refresh frequency can be employed.

The parallax barrier and lenticular techniques support glasses-free 3D. In the case of the former, a mask comprising a set of opaque vertical strips is placed in front of a conventional display screen. In the simplest case, the left and right views of an image scene are sectioned into vertical strips and interleaved (**Fig. 1**). As indicated in **Fig. 2**, the barrier ensures that only one set of image strips are visible to each eye. In the case of the lenticular approach, the image again comprises a set of interleaved vertical strips, but in this case a sheet of vertical column lenses are fitted to the display and ensure that only one set of strips are visible to each eye.

Both approaches result in only half of the displayed pixels being visible to each eye and the presence of the parallax barrier reduces image brightness.

Autostereoscopic Type I displays. This class of display extends the stereoscopic approach by providing support for horizontal POD. Thus, a user is able to move from side to side and view an image scene from different perspectives. One approach employs parallax barrier or lenticular techniques coupled with a camera-based head-tracking technology. Thus, the content depicted on the interleaved image strips is updated according to the viewing position. Latency (lag between movement and image update) can be an issue; however, there are continual improvements in this area.

Recent developments in implementing dynamic barriers (in which the barrier takes the form of a liquid-crystal panel) offer exciting opportunities by enabling barrier parameters (for example, the barrier pitch) to be adjusted dynamically in accordance with viewing position. Liquid-crystal panels are also being used in the implementation of dynamic lenticular panels. In this case, lens elements are formed by applying nonuniform fields across the liquid crystal material.

Immersive virtual reality displays (IVR) allow users to don a special viewing apparatus and gain a strong sense of immersion in a synthetic world. Here, the headgear employs two small display screens, from which the left and right views are projected into the eyes.

Considerable developments are taking place in forming tiled arrays of displays. These may be arranged as a planar surface or, as in the case of the Varrier display, they may be arranged in a semicircle around the user to create a semi-immersive environment. Head tracking is used to support horizontal POD. Until recently, the extent of the bezel (rim) around each display panel has led to significant inter-panel gaps. This problem is gradually being resolved.

Accommodation/convergence breakdown. As indicated above, both stereoscopic and autostereoscopic (Type I) classes of display are fundamentally based on the principles of the stereoscope. Invariably, displays of this type require the eyes to focus on the display screen(s) on which the left and right views are depicted. However, their visual axes converge on the feature of interest within the 3D scene. Thus, the two cues become decoupled. This is referred to as accommodation/convergence (A/C) breakdown and is contrary to the real-world experience, where the eyes focus and converge on the point of fixation. Unless handled carefully, this can lead to visual strain and headaches. Important considerations are the degree and rapidity of decoupling.

Autostereoscopic Type II displays. In principle, these displays extend the cues supported by Type I systems and incorporate vertical POD together with natural support for accommodation and convergence.

In the case of the volumetric approach, images are depicted in a transparent 3D physical volume (image space). Since these images are able to occupy three physical dimensions, a broad range of depth cues (including linear perspective) are supported in a natural manner. Volumetric displays often allow considerable freedom in viewing position; for example, viewing from any position around a spherical image space.

Numerous approaches may be adopted in the implementation of volumetric displays. By way of a simple example, an image space may be formed using a rapidly rotating planar screen (the axis of rotation being in the plane of the screen), comprising a rectangular array of light-emitting elements. If an element is briefly turned on, each time it passes through a certain spatial location, and if the screen is rotating with sufficient rapidity, a continuous point of light in space (voxel) will be seen. This forms the basis for 3D image construction.

Other forms of volumetric display place no reliance on mechanical motion. For example, an image space may comprise a gaseous medium, with voxels being formed at the intersection of two non-visible laser beams (utilizing the stepwise excitation of fluorescence).

Important implementation goals include support for voxel formation at regular locations as defined by a 3D lattice, invariance in image quality with viewing direction, and the eradication of dead zones (regions in which one or more image quality parameters fall below a minimum threshold).

Usually volumetric systems give rise to translucent images. However, recent work includes the creation of voxels possessing anisotropic light-emission characteristics and this forms the basis for the creation of opaque images. Unfortunately, displays of this type are unable to provide natural support for vertical POD.

Image space. 3D graphic displays enhance the visualization process by allowing 3D content to occupy (or appear to occupy) 3D space. This assists in the interpretation of spatial relationships and object dynamics. In addition, such systems are able to advance interaction processes, allowing tasks that are inherently 3D to be carried out within a 3D space. This provides opportunities to employ alternative and synergistic interaction tools, including haptic probes that allow image components to exhibit material properties such as solidity.

When comparing the visualization and interaction opportunities offered by a display modality, it is helpful to consider the nature of the image space within which visible images are formed. Four indicative forms of image space are summarized below.

Apparent image space. This is associated with display technologies that are fundamentally based on the principle of the stereoscope (stereoscopic and autostereoscopic Type I). The visible image is depicted on a planar screen and the impression of three-dimensionality is based on disparities between the left and right views. Consequently, this form of image space has no physical basis. Insertion of physical interaction tools (such as haptic probes) into this apparent image space can disturb the image formation and cause visual conflict (the accommodation and convergence cues coupling and decoupling as a user switches gaze between the probe and the image scene).

Virtual image space. This appears to exist behind some form of optical component such as a mirror or glass plate. The 3D display marketed by Reachin Technologies (and which is based on the nineteenth century Pepper's Ghost theatrical illusion) provides an example of such an image space. This supports the use of haptic probes and ensures that the presence of the probe does not interfere with image formation.

Physical image space. This is associated with the volumetric approach and takes the form of a transparent physical image space, which occupies three spatial dimensions. Usually, the underlying techniques used for the formation of the image space preclude the insertion of physical interaction tools. Thus, the image and interaction spaces are physically separated.

Ethereal image space. This is also known as "free" image space and is characterized by images that appear to be suspended in midair. In one form, the image space comprises a cloud of light-scattering particles into which image content is projected. A key issue concerns the need to address particles when they pass through the appropriate spatial coordinates. This process is greatly simplified by substituting the particle cloud for a set of stacked surfaces, with each formed from discrete light-scattering particles. This approach is adopted in the implementation of the AquaLux 3D display developed at Carnegie Mellon University.

This form of image space permits the insertion of physical interaction tools, although the presence of such tools may impact on local image formation.

Ongoing activity. Despite the tremendous progress that has been made in the implementation of the stereoscopic and autostereoscopic (Type I) classes of display, accommodation/convergence breakdown remains problematic. Unless properly handled, this can lead to serious usability issues. Extending parallax barrier techniques to support horizontal POD for multiple simultaneous users is difficult, not only from the perspective of display technology but also in terms of data throughput. Here, for certain forms of application, volumetric displays offer a practical solution. Although it is a simple matter to construct an operational volumetric system, the development of a high-performance display is a most challenging proposition.

Electroholography remains the long sought-after goal of 3D graphic-display developers and there can be little doubt that general-purpose holographic displays will ultimately become a practical proposition. However, the computational requirements remain daunting. For example, a 100 mm^2 hologram with a maximum viewing angle of 30° (both horizontally and vertically) requires about 33×10^9 data samples per frame. If each sample is represented by eight bits then, assuming a frame-refresh frequency of 30 Hz, the data throughput is around 10^{12} bytes/s. This throughput increases linearly with the dimensions of the hologram.

For background information *see* ELECTRONIC DISPLAY; EYE (VERTEBRATE); HOLOGRAPHY; PERCEPTION; STEREOSCOPY; VIRTUAL REALITY; VISION in the McGraw-Hill Encyclopedia of Science & Technology.

Barry G Blundell

Bibliography. P. C. Barnum, S. G. Narasimhan, and T. Kanade, A Multi-Layered Display with Water Drops, *ACM SIGGRAPH 2010 (Emerging Technologies section)*, Los Angeles, CA, 2010; B. G. Blundell, *3D Displays and Spatial Interaction: Exploring the Science, Art and Evolution and Use of 3D Technologies*, vol. 1: *From Perception to Technology*, Walker & Wood Limited, 2011; B. G. Blundell, *About 3D Volumetric Displays*, Walker & Wood Limited, 2011; J. Napoli et al., Radiation therapy planning using a volumetric 3-D display: PerspectaRAD, *Proc. SPIE*, 6803, 2008; T. Peterka et al., Advances in the Dynallax solid-state dynamic parallax barrier autostereoscopic visualization display system, *IEEE Trans. Visualiz. Comp. Graphics*, 14(3):487–499, 2008; D. J. Sandin et al., The Varrier™ autostereoscopic virtual reality display, *ACM Trans. On Graphics, Proc. SIGGRAPH 2005*, 24(3):894–903, 2005.

Thunderstorm-generated turbulence

Turbulence is a well-known hazard to the aviation sector. It is responsible for numerous injuries each year, with occasional fatalities, and it is the underlying cause of many people's fear of air travel. Not only are encounters with turbulence a safety issue, but they are the source of millions of dollars of operational costs to airlines, with the increased costs being passed on to the consumer. For these reasons, pilots, dispatchers, and air-traffic controllers attempt to avoid turbulence wherever possible. A common method of avoidance involves circumventing regions or altitudes where turbulence was recently encountered by other aircraft. Empirical rules are also used to identify weather patterns known to be conducive to the generation of turbulence; pilots employ these rules during flight, and operational weather forecasters on the ground also provide guidance. However, these methods are imprecise, and recent research into turbulence generation processes, by thunderstorms especially, is laying the foundation for improving methods of strategic turbulence forecasting and tactical turbulence avoidance.

It has been known for some time that turbulence at the cruising levels of commercial aircraft is caused by a variety of meteorological processes. These include jet streams, strong wind shear, frontal systems, mountain-induced flows, and thunderstorms. Turbulence that is not associated with clouds and thunderstorms is usually invisible and is referred to as clear-air turbulence (CAT). Regions within clouds and thunderstorms are usually turbulent, but thunderstorm-generated turbulence can extend well outside of the cloudy air. Such near-cloud turbulence (NCT) has been recorded thousands of feet above the tops of clouds and at least 100 miles away from active thunderstorm regions. It follows that CAT and NCT are particularly hazardous because they are impossible for pilots to see and cannot be detected using standard onboard radars. NCT has been poorly understood for some time, and traditional reports of turbulence made by pilots (pilot reports; PIREPS) are too imprecise and sparse to elucidate the governing processes. Nonetheless, recent work using high-resolution numerical modeling of thunderstorms has exposed many of the important processes underlying NCT, including the important role of atmospheric gravity waves, which has only recently become appreciated.

Gravity waves. Developing convective clouds and thunderstorms are known to generate atmospheric gravity waves. Gravity waves are formed by the vertical displacement of stably stratified air induced by the thunderstorm circulation; buoyancy acts to restore the parcels to their original altitude, which sets up an oscillation. The gravity waves generated by thunderstorms can typically have periods and

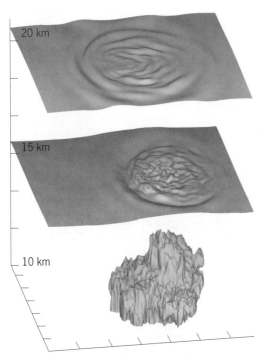

Fig. 1. Three-dimensional rendering of a high-resolution thunderstorm simulation. Lower surface shows the cloud outline above 10 km (33,000 ft). Upper surfaces show vertical displacement at 15 km and 20 km (49,000 and 66,000 ft).

horizontal wavelengths as small as 10 min and 5 mi (8 km), respectively.

Gravity waves are generated throughout the entire cloud life cycle and propagate away from the storm, both horizontally and vertically. An example three-dimensional visualization of a high-resolution thunderstorm simulation shows that the vertical displacements associated with the waves closely resemble ripples caused by a pebble thrown into a pond (**Fig. 1**). Like water waves, atmospheric gravity waves can break. It is the breaking of the gravity waves that is important for NCT. In particular, the horizontal wavelengths of gravity waves are usually too long to elicit a strong aircraft response, but the process of breaking creates smaller-scale motions (turbulence) that strongly influence aircraft.

Even though there is an established body of literature dealing with gravity-wave instability processes, recent numerical simulations have elucidated important information about the role of gravity waves as a turbulence hazard and the properties of the atmosphere that control that hazard. For example, it has recently been shown that the vertical gradient in wind speed (that is, wind shear) above the cloud can influence the vertical extent and intensity of above-cloud NCT. The stability of the atmosphere, which can be represented by the vertical gradient in temperature, also has a strong influence on the turbulence. These are important new findings because current Federal Aviation Administration (FAA) guidelines for avoiding thunderstorm-generated turbulence do not take wind shear or stability into account.

An example turbulence event. Much of our recent understanding of NCT was motivated by a severe turbulence encounter above a developing thunderstorm over North Dakota in 1997. During this event, 22 passengers and crew members were injured. This event was not unique, and recent events over the United States and other parts of the world show common characteristics. The primary tool used for studying these events is high-resolution numerical modeling. These models are similar in many respects to the models used to make operational weather forecasts, except that they employ much finer grids that can resolve thunderstorm and turbulence production processes more realistically. The benefit of numerical modeling is that the three-dimensional structure of the evolving flow can be examined to determine the governing processes without the ambiguity caused by sparse observational datasets.

A high-resolution simulation of the North Dakota thunderstorm shows turbulence extending well above the cloud top (**Fig. 2**). In this case, the turbulence is found above the developing convective updrafts, that is, the columns of air within the thunderstorm that can move upward at speeds greater than 60 mi/h (100 km/h). The turbulence extends up to 10,000 ft (3 km) above the cloud, and an aircraft flying in the clear air above the cloud top would be subject to both gravity-wave perturbations and turbulence. In particular, the turbulence above the cloud coincides with the gravity-wave-induced perturbations in potential temperature; the potential temperature is the temperature that the air would be if it were compressed adiabatically to a pressure of 1000 hPa. The isopleths of potential temperature provide an indication of the air's vertical displacement, and the regions where they steepen and overturn identify breaking and likely turbulence production.

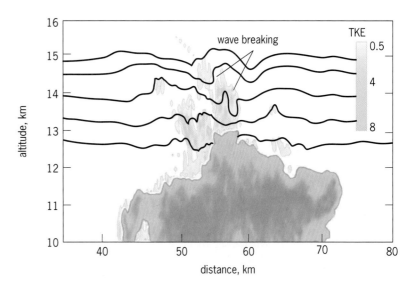

Fig. 2. Examples of turbulence due to gravity wave breaking. Cross section through a simulated cloud showing the concentration of condensed water (that is, cloud) with gray shading, isopleths of potential temperature (solid black lines), and regions of turbulence outside of the cloud (color shading) measured in terms of turbulence kinetic energy (TKE, $m^2 s^{-2}$).

Observing the hazard. Precise observations of aircraft turbulence encounters are also crucial for improved understanding and avoidance of thunderstorm-generated turbulence. Not only can reliable observations validate numerical models, but they can provide important information about the frequency of occurrence and the predominant spatial locations of the hazard. For example, extensive datasets of historical PIREPS have demonstrated a propensity for turbulence in regions known for thunderstorm activity. However, traditional PIREPS have relatively sparse coverage and spatial and temporal uncertainties that are too large to permit the determination of the precise location of turbulence relative to cloud boundaries. Therefore, using PIREPS, it is impossible to determine whether the turbulence was encountered inside or outside a cloud, a prerequisite for studying thunderstorm-generated turbulence.

In recent years, however, a new source of data for studying turbulence has emerged. An automated in-situ turbulence reporting system has been installed on many commercial aircraft in the United States. The system employs an algorithm that uses the existing aircraft accelerometer measurements and knowledge of the aircraft response function to determine an aircraft-independent atmospheric turbulence metric known as the eddy dissipation rate (EDR). Median and peak EDRs are recorded along with the time and location of the measurements, providing extensive data coverage of both smooth and turbulent air, which is quantified with an objective intensity measure. These data have proved extremely valuable for case studies and are providing many new opportunities for turbulence research. An example of in-situ EDR measurements from several aircraft on August 5, 2005, identifies moderate to severe turbulence in the clear air just outside the cloud boundary (**Fig. 3**).

Ongoing research. Near-cloud turbulence is not limited to regions above the cloud, and cases of turbulence adjacent to clouds are the focus of active research as well (Fig. 3). Studies have shown that horizontally propagating gravity waves can break adjacent to clouds, or can simply perturb an environment that is close to the threshold for turbulence production. There are also some indications that clouds might act like obstacles to the flow, forming turbulent wakes in their lee. In many cases, especially when the cloud tops are high in the summer or at lower latitudes, turbulence adjacent to clouds is probably more likely to be encountered by aircraft, since most aircraft are not able to fly high enough to go over the storms, and is therefore a future research priority.

Despite ongoing advances in numerical weather prediction, explicit predictions of turbulence for the aviation sector are years away. The turbulent scales of motion are simply too small to be resolved by standard weather models, and the chaotic nature of the processes that are at play necessitates an ensemble of forecasts to be constructed. Improvements to existing approaches using current capabilities are also being explored. For example, a variety of

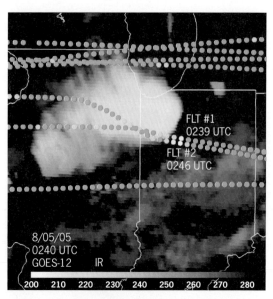

Fig. 3. In-situ measurements of near-cloud turbulence. Turbulence measurements from EDR-equipped commercial aircraft superposed on a GOES-12 infrared satellite image. Each dot is a peak EDR measurement, and colors represent values categorized as smooth (darkest shading), light (lighter shading), moderate (lightest shading), and moderate-to-severe (white) turbulence.

high-resolution model experiments are being used to identify the governing processes to help formulate improved avoidance guidelines for thunderstorms. Other topics of active research include using modern satellite and ground-based radar instruments as tools for diagnosing turbulence in real time, as well as developing methods for extracting (unresolved) turbulence information from operational weather prediction models.

The recent enhancements in numerical modeling and observational capabilities have provided a wealth of opportunities for thunderstorm-generated turbulence research. In recent years, gravity waves have been identified as an important turbulence generation process, but it is likely that other undiscovered processes are also at play, and continued research is essential to improve turbulence forecasting and avoidance systems.

For background information *see* ACCELEROMETER; AERONAUTICAL METEOROLOGY; CLEAR-AIR TURBULENCE; CLOUD; MIDDLE-ATMOSPHERE DYNAMICS; STORM DETECTION; THUNDERSTORM; WEATHER FORECASTING AND PREDICTION in the McGraw-Hill Encyclopedia of Science & Technology.

Todd Lane; Robert Sharman

Bibliography. T. P. Lane et al., An investigation of turbulence generation mechanisms above deep convection, *J. Atmos. Sci.*, 60:1297–1321, 2003; P. F. Lester, *Turbulence: A New Perspective for Pilots*, Jeppesen Sanderson, Englewood, CO, 1994; C. J. Nappo, *An Introduction to Atmospheric Gravity Waves*, Academic Press, San Diego, CA, 2002; J. K. Wolff and R. D. Sharman, Climatology of upper-level turbulence over the contiguous United States, *J. Appl. Meteorol. Clim.*, 47:2198–2214, 2008.

Top quark at the Tevatron

The fundamental building blocks of matter are leptons and quarks. Leptons, consisting of charged leptons and neutrinos, come in three flavors: electron, muon, and tau. Quarks, on the other hand, come in six flavors: up, down, strange, charm, bottom, and top. They range in mass from a few thousandths of a gigaelectronvolt (GeV) to more than 100 GeV. Normal matter is made up of up and down quarks and electrons, while the other flavors occur very rarely in the nature. It took nearly 100 years after the discovery of electrons to progress to the most recent discovery of a fundamental particle, that of the top quark. Protons and neutrons, on the other hand, are composite particles made up of quarks. A proton is made up of two up quarks and one down quark, while a neutron is made up of one up quark and two down quarks. An antiproton, the antiparticle of proton, is made up of two antiup quarks and one antidown quark.

The discovery of the top quark, the sixth quark, at the Tevatron in 1995 completed the identification of the content of matter and marked the triumphal success of the standard model of particle physics. The Tevatron, at Fermilab in Illinois, is the most powerful machine colliding proton and antiproton beams, at an energy 1.96 TeV (1 TeV = 10^{12} eV). An imaginary axis can be associated with the proton such that forward is the direction of the proton beam while backward is the direction of the antiproton beam. This terminology will be useful in the following discussion of the forward and backward directions.

Forward-backward asymmetry in top-quark production. The standard model consists of a set of matter particles called leptons and quarks described above, and a set of force carriers responsible for the electromagnetic, weak, and strong interactions. Each quark or lepton has four degrees of freedom, of which two form the left-handed component and the other two form the right-handed component. These two components behave very differently under the weak interaction, which was used by Enrico Fermi to explain nuclear beta decay. However, these two components behave identically under the interactions of quantum chromodynamics (QCD).

The top quark is produced mainly by quark-antiquark fusion through the strong interaction, also called quantum chromodynamics. The Feynman diagram that describes this interaction is depicted in **Fig.1**. The coil-like structure in the middle is the gluon, the force carrier for quantum chromodynamics. Since quantum chromodynamics does not distinguish between left-handed and right-handed components of quarks, the proton-antiproton collisions at the Tevatron should produce equal amounts of top quarks in the forward and backward directions. An example of a forward-backward difference would be the production of muons in electron-positron collisions, because the force carrier of the weak interaction, the Z boson, couples differently to left-handed and right-handed leptons. In the case of top-quark production at the Tevatron, certainly

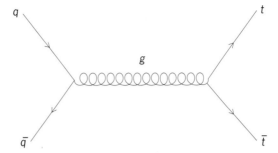

Fig. 1. Quark-antiquark (q-q̄) fusion into a pair of top (t) and antitop (t̄) quarks. The force carrier in the middle is the gluon (g).

no difference is expected between the forward and backward directions. Even with higher-order corrections, the forward-backward difference or asymmetry is only a few percent, and it is not easy to measure such a small difference experimentally.

One of the experiments at the Tevatron, the Collider Detector at Fermilab (CDF), detected a forward-backward asymmetry as large as 20% in the summer of 2008 while the standard model predicted a level of only about 5%, as explained above. Statistically, such a large asymmetry could be the result of experimental fluctuations at a probability of a few percent, which is a small number but cannot be ruled out completely. More data have been accumulated since then, and new results were announced at the beginning of 2011. The overall asymmetry was a little smaller but was still quite different from the prediction of the standard model. However, the most surprising aspect of the new results was that the asymmetry showed very peculiar behavior when taking into account the invariant mass of the top and antitop system. The invariant mass is a characteristic of the total energy and momentum of a system of objects that is the same in all frames of reference related by Lorentz transformations. Mathematically, the invariant mass M of two particles (1 and 2), with energies E_1 and E_2 and momenta $\vec{p}_1$ and $\vec{p}_2$, is given by the equation below. The

$$M = [(E_1 + E_2)^2 - (\vec{p}_1 + \vec{p}_2)^2]^{1/2}$$

asymmetry was small and consistent with zero in the small invariant mass region but as large as 47% in the large invariant mass region of the top-antitop system. Statistically, the chance of experimental fluctuation was further reduced to 0.1%.

W′ boson. Scientists began to take the asymmetry very seriously. It could be an effect due to existence of new particles or new interactions of the top quark, but still statistical fluctuation cannot be completely ruled out. A number of possibilities have been put forward to explain the phenomena. They can be categorized into two types. One is a new particle propagating in a virtual manner in the t channel, as shown in **Fig. 2**. The incoming quark scatters with the antiquark, except for head-on collisions, to give easily the top quark to the right-hand side and the antitop to the left-hand side. Also, since the proton contains more quarks than antiquarks, the diagram naturally explains why the top quark probably goes in

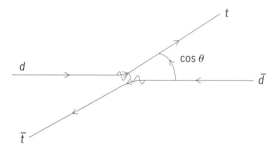

Fig. 2. Feynman diagram showing the scattering of a down (*d*) quark with an antidown ($\bar{d}$) quark into a pair of top (*t*) and antitop ($\bar{t}$) quarks.

the forward direction while the antitop goes in the backward direction, as long as the collision is not head-on.

So far, the flavor behavior in the interactions has not been discussed. Flavor is conserved in the strong interactions but changes only mildly in the weak interaction. In Fig. 2, the incoming quark is a light quark, since the chance of finding a heavy quark inside a proton is substantially less than that of finding a light quark. As the scattering proceeds, the light quark changes into the heaviest quark, the top quark. In the standard model, the probability of this transition is extremely small, less than 0.1%, which is way below that suggested by the current data. The interpretation of the data therefore requires a new type of force. An extraordinary example of such a force was given in K. Cheung, W. Y. Keung, and T. C. Yuan in 2009. This new type of force is carried by a singly-charged spin-one boson, denoted by W', which has properties that are similar to those of the W boson in the standard model but is heavier and different in the way it couples to the standard-model particles. It connects the down quark to the top quark or the antidown to the antitop. The strength of the coupling is required to be as strong as the strong interactions, and predictions of the mass of the W' boson range from 200 to 600 GeV.

An immediate test of this explanation can be performed at the Tevatron and at the Large Hadron Collider (LHC), which is a high-energy proton-on-proton machine at CERN in Geneva, Switzerland which is operated at a center-mass-energy 3.5 times as high as the Tevatron. Since this W' boson is relatively light compared to the energy of the LHC, it can be produced readily in the collisions there, and will decay subsequently into top and antidown quarks. The final state consists of a top-antitop pair and a light quark. The top quark and the light quark will give rise to a bump in the invariant-mass spectrum. This is a clean and obvious signature if the W' boson is the real reason behind the data.

Another possibility is an electrically-neutral spin-one Z' boson, similar to the standard-model Z boson. The Z' boson connects an up quark to the top quark or antiup to antitop in the modification of Fig. 2. Its other behaviors are similar to those of the W' boson described above. Furthermore, this Z' boson offers another interesting signature: like-sign top-quark pair production at the LHC. Same sign means two

top quarks instead of one top quark and one antitop quark. However, some preliminary results from the LHC do not support this interpretation.

Axigluon-type explanations. The other type of explanation involves a diagram similar to Fig. 1, but with a new type of force carrier in the middle. Similar to the Z boson that gives forward-backward asymmetry to muons, as described above, the new force carrier is required to have different couplings to the left- and right-handed light and heavy quarks; even the signs are different. It is unnatural, if not impossible, to build such a model.

An immediate test of this explanation can also be performed at the LHC. It will produce, if the new force carrier is lighter than 1.5 TeV, a visible bump or peak in the invariant-mass spectrum of the top-antitop system. It will take a longer time to accumulate enough events in the high-mass region to carry out this test than that of the explanation based on the W' boson.

Prospects. Particle physics is entering a very exciting era, with the processing of the large data set that has already been collected at the Tevatron and with the LHC running at very high energy and high luminosity. Both theorists and experimenters are looking forward to any sign of new physics. The forward-backward asymmetry in top-pair production described here is one of the most recent exciting developments in high-energy physics. Interpretations have been put forward by theorists to explain the data and experimenters are actively testing these new ideas. Hopefully, in a few years we will be clear about what kinds of new physics beyond the standard model should be investigated. *See* LARGE HADRON COLLIDER: FIRST YEAR.

For background information *see* ELEMENTARY PARTICLE; FEYNMAN DIAGRAM; GLUONS; LEPTON; PARTICLE ACCELERATOR; QUANTUM CHROMODYNAMICS; QUARKS; RELATIVITY; STANDARD MODEL; WEAK NUCLEAR INTERACTIONS in the McGraw-Hill Encyclopedia of Science & Technology.   Kingman Cheung

Bibliography. T. Aaltonen et al., Forward-backward asymmetry in top-quark production in $p\bar{p}$ collisions at $\sqrt{s} = 1.96$ TeV, *Phys. Rev. Lett.*, 101:202001 (8 pp.), 2008; T. Aaltonen et al., Evidence for a mass dependent forward-backward asymmetry in top quark pair production, *Phys. Rev.*, D83:112003 (23 pp.), 2011; K. Cheung, W. Y. Keung, and T. C. Yuan, Top quark forward-backward asymmetry, *Phys. Lett.*, B682:287–290, 2009; K. Cheung and T. C. Yuan, Top quark forward-backward asymmetry in the large invariant mass region, *Phys. Rev.*, D83:074006 (7 pp.), 2011; V. Barger, W. Y. Keung, and C. T. Yu, Asymmetric left-right model and the top pair forward-backward asymmetry, *Phys. Rev.*, D81:113009(7 pp.), 2010.

Transparent light-harvesting materials

Solar cells, or photovoltaics (PVs), have been developed as a promising technology for renewable energy. Among solar cells, organic photovoltaic solar cells (OPVCs), or the technology to convert sunlight into electricity using thin films of organic semiconductors, such as conjugated polymers, have been under intense research over the past decades as potential cost-effective replacements for the more expensive but currently more efficient silicon-based devices. OPVCs provide several attractive features, including the use of cheap, lightweight plastic materials (conjugated polymers) with good solubility in common organic solvents, cost-effective processing methods (such as printing and dip- and spin-casting techniques), and mechanical flexibility. In polymer-based OPVCs, the conjugated polymer, usually an electron donating (p-type) semiconducting material, harvests the light to create an excited state known as exciton (an electron and a hole bound together). Excitons can be separated into electrical charges by using appropriate electrical fields, for example by the use of electrodes with different work functions, such as indium tin oxide (ITO) [with high work function] and aluminum (Al) or gold (Au) [low work function]. There are several known OPVC device architectures, including single-layer (conjugated polymer) OPVCs, double-layer OPVCs (electron donor-acceptor layers using polymer plus dye, p-type plus n-type conjugated polymers, or polymer plus fullerene combinations), multilayer PVs (alternations of donor and acceptor materials), or bulk heterojunction PVs (blended polymer-fullerene composites). The addition of electron-accepting materials [dyes, fullerenes (C_{60} and its derivatives) or n-type conjugated polymers] enhances exciton dissociation into separated charges, thus improving the power conversion efficiency of OPVCs. Bulk heterojunction OPVCs based on blends of conjugated polymers and fullerenes are technologically the most promising, with current reports claiming power conversion efficiencies as high as 8%.

Regardless of their architecture, conventional polymer-based OPVCs require an active layer (conjugated polymer) at least 100 nm thick for light to be efficiently absorbed in order to generate electricity. At this thickness, conventional polymer-based PVs are opaque, preventing their application in technologies where transparency of the film or device is sought. One important example is power-generating (or photovoltaic) windows, where the device needs to preserve some transparency while still generating electricity.

In this article, we describe a conjugated polymer-based transparent light-harvesting thin film with ordered microporous structure which, when doped with electron-accepting materials (such as fullerenes), exhibits highly efficient charge separation and charge transport when illuminated by visible light. These properties make it a promising candidate for the development of highly transparent OPVCs.

Fabrication. The composition of a transparent light-harvesting thin film with an ordered microporous honeycomb-like structure is given in **Fig. 1**a, and a scanning electron microscopy (SEM) image of this particular film is shown in

Fig. 1b. The conjugated polymer used for film fabrication is a poly(phenylene vinylene) derivative, poly{2,5-bis[3-(N,N-diethylamino)-1-oxapropyl]-1,4-phenylenevinylene}, hereafter named P1. P1 efficiently absorbs blue-to-green light (absorption peak at 460 nm) and emits green-yellowish photoluminescence (emission peak at 550 nm). The P1 polymer thin films with honeycomb-like structures are fabricated by a self-assembly method known as the breath-figure technique (BFT), which is schematically shown in Fig. 1c. The BFT method exploits the property of water droplets to condense onto cold surfaces in hexagonal arrays known as breath figures. P1 is dissolved in a nonpolar volatile solvent, such as chlorobenzene or chloroform, and a small amount of the polymer solution is drop-casted on a water surface under conditions of controlled humidity. Deposition on the water surface allows easy transfer of the self-assembled film on a solid substrate, such as silicon. Solvent evaporation induces a cooling effect, causing a severe drop in temperature that initiates condensation of the water droplets onto the polymer surface in breath figures. As the organic solvent evaporates, the void space created by the waterdroplets allows formation of regularly packed conjugated polymer honeycomb structures driven by capillary forces. Without BFT, the as-cast film deposited on a silicon substrate from the same solution of P1 in chlorobenzene is structurally featureless with respect to any framework, as can be seen in Fig. 1d. We also fabricated transparent thin films with honeycomb-like structures from blends of P1 conjugated polymer and a fullerene derivative, emerald green fullerene (EG-C_{60}) [and schematically shown in Fig. 1b], by starting BFT from a P1/EG-C_{60} blend in chlorobenzene. The choice of these particular electron donor-acceptor materials was based on previous studies showing that photoexcited P1 undergoes efficient charge (electron) transfer when complexed with EG-C_{60}, for example, when mixed in solvents such as chloroform or chlorobenzene.

Structural properties. Conjugated polymer thin films with honeycomb-like structures, cast either from P1 polymer in chlorobenzene (Fig. 1b and e) or from P1/EG-C_{60} blended in chlorobenzene (Fig. 1f), exhibit structural regularity over large areas (up to 1 mm^2, Fig. 1a), consisting of hexagons of 3-4 μm in size and with node and frame thicknesses of 0.6 and 0.25 μm, respectively. The polymer film located at the center of the hexagons is extremely thin, down to molecular layer thickness (a few nanometers), making the honeycomb-like thin film highly transparent as most of the polymer material is concentrated within the hexagonal frame. We have found that the polymer concentration used for BFT plays a dominant role in determining the final morphology and transparency of the honeycomb structure, with an ideal concentration of 4 mg/mL of P1 in chlorobenzene, resulting in a highly uniform honeycomb structure. Higher polymer concentrations (>6 mg/mL) resulted in thick films with rounded holes and decreased transparency, whereas lower concentrations (<2 mg/mL) resulted in disconnected rings.

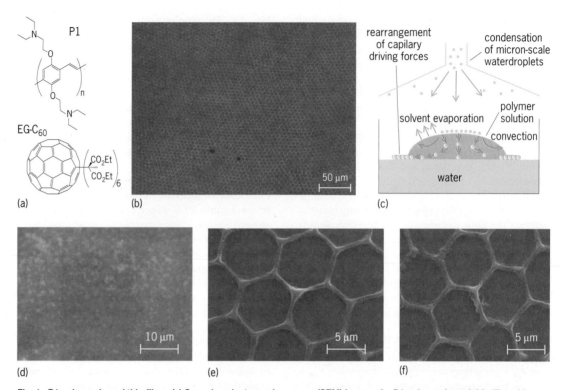

Fig. 1. P1 polymer-based thin films. (*a*) Scanning electron microscopy (SEM) image of a P1 polymer-based thin film with a honeycomb-like structure deposited from chlorobenzene. (*b*) Chemical structures of conjugated polymer P1 and emerald fullerene (EG-C$_{60}$) used to fabricate transparent thin films with honeycomb-like structure. (*c*) Scheme of the breath figure technique (BFT) method used to produce large-area conjugated polymer-based thin films with honeycomb-like structures. (*d*) SEM image of a P1 film drop-cast without BFT. (*e*) SEM image of P1 only honeycomb-like film by BFT. (*f*) SEM image of blended P1/EG-C$_{60}$ self assembled by BFT and exhibiting honeycomb-like structures with highly regular hexagonal structures. (*Reprinted with permission from H. Tsai et al., Structural dynamics and charge transfer via complexation with fullerene in large area conjugated polymer honeycomb thin films, Chem. Mater., 23(3):759–761, 2011. Copyright © 2010 American Chemical Society.*)

Optical properties of polymer-only thin films with honeycomb-like structure. P1 polymer-only thin films with honeycomb-like structure are highly transparent (>80% estimated transparency) at the center of the hexagon rings when observed by optical transmission microscopy (as shown in **Fig. 2a**, shadowed regions). At the same time, these same thin films are highly fluorescent (photoluminescent), for example, when observed under blue light, with the most intense fluorescence observed at the nodes (intersection points) and frame of the hexagon rings (Fig. 2b), thus demonstrating their transparent and light-harvesting character. A nanoscopic investigation of the P1 polymer thin films with honeycomb-like structure by confocal fluorescence microspectroscopy reveals different spectral properties when compared to those of the as-cast, structurally homogeneous films deposited without BFT (films like in Fig. 1d). In particular, photoluminescence intensity signals recorded from various regions of a hexagonal ring, namely, nodes [region (2), **Fig. 3a** and b] and frame [region (3) in Fig. 3a and b], although very strong (see intensity image in Fig. 3a), are characterized by quenched fluorescence lifetimes (see color-coded lifetime image in Fig. 3b, nodes and frame featuring lifetimes of approximately 0.4 ns). In contrast, the as-casted structurally homogeneous film and a chlorobenzene solution of P1 feature lifetimes approximately 1.5 ns. The same regions of the hexagon

[nodes and frames but not center, that is, region (1) in Fig. 3a and b] feature red-shifted and structured photoluminescence spectra, opposite to the rather broad and structureless spectra observed for the as-cast film. The spectroscopy observed for the center regions of the hexagons [region (1), Fig. 3a and b] is very similar to that of the as-cast P1 film and of the P1 dissolved in chloroform. Thus, various parts of the hexagon structure (center, frame, and node) exhibit various spectral behaviors caused by differences in polymer packing–polymer chain interaction induced by the time scale of solvent evaporation. The center region of the hexagon consists of an extremely thin polymer layer that forms directly underneath the waterdroplet. Rapid formation of this extremely thin film, which is structurally and spectroscopically similar to the as-cast film, is because the polymer solution becomes quite thin as evaporation occurs. In contrast, the frame and nodes of the hexagon are situated between water droplets, and the thickness of the polymer solution at these regions is hundreds of nanometers. Hence, formation of the fully evaporated film will be slower, with the possibility of forming more structured assemblies or aggregates of the polymer. Indeed, the observation of high photoluminescence intensity signals, quenched photoluminescence lifetimes and red shifted, structured spectra from hexagon nodes and frames indicate strong packing of the polymer

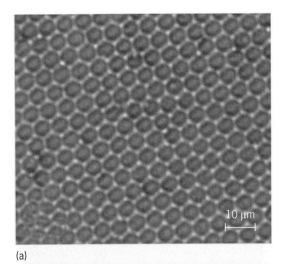

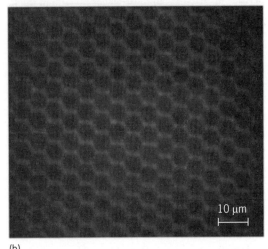

Fig. 2. Images of a P1-based light-harvesting thin film with honeycomb-like structure by (a) optical transmission and (b) fluorescence microscopy. (*Reprinted with permission from H. Tsai et al., Structural dynamics and charge transfer via complexation with fullerene in large area conjugated polymer honeycomb thin films, Chem. Mater., 23(3):759–761, 2011. Copyright © 2010. American Chemical Society*)

chains, resulting in highly ordered regions (aggregates). In such ordered regions, the polymer can exhibit extended chain conformations, leading to enhanced interchain interaction known as $\pi - \pi$ stacking. This particular property of the transparent, thin honeycomb framework aids charge mobility when charge transfer is present, similar to observations for bulk heterojunction OPVCs based on polythiophenes and fullerenes.

Charge transfer in polymer-fullerene blended thin films with honeycomb-like structure. To demonstrate the potential applicability of the transparent thin films with honeycomb-like structure here for building integrated photovoltaic devices such as power generating windows, we investigated the magnitude of charge transfer by doping such films with fullerene, either forming blended films (P1/EG-C$_{60}$) or double-layer films [P1 plus tris-malonic acid fullerene (TM-C$_{60}$)]. The latter sample was obtained by adsorbing the fullerene derivative TM-C$_{60}$ on

top of the polymer film created by BFT to mimic double-layer OPVC architectures. TM-C$_{60}$ is known to exhibit efficient charge transfer when complexed with P1 polymer.

Confocal fluorescence microspectroscopy shows the magnitude of polymer photoluminescence quenching, in this case by charge transfer to fullerene, thus quantifying the magnitude of charge transfer as a function of thin-film architecture. For double-layer P1/TM-C$_{60}$ films with honeycomb-like structure, quenching by charge transfer is observed mainly at the center of the hexagon rings (**Fig. 4a**), presumably because of the rather thin polymer layer in this region experiencing good interfacial contact with TM-C$_{60}$. For the hexagonal rings, due to the relatively thick polymer layer of these structures (few hundred of nanometers), quenching by charge transfer takes place only at the surface where good interfacial contact exists between the polymer and fullerene. Thus, the photoluminescence signals from these regions exhibit only slightly quenched lifetimes compared to undoped P1 polymer thin films with honeycomb-like structure (Fig. 3b). The quenching by charge transfer is much more dramatic in the case of P1/EG-C$_{60}$ blended film with honeycomb-like structure in which nearly complete quenching is observed both at the frame and nodes (Fig. 4b). These observations demonstrate efficient change transfer within the entire honeycomb framework when the composite polymer-fullerene thin film with honeycomb-like structure is made from a blend.

In summary, we have developed a conjugated-polymer-based thin film with highly regular microporous structure, exhibiting high transparency and efficient charge-transfer when blended with fullerenes. The unique structure adopted by the polymer within the honeycomb framework, with ordered and strongly interacting polymer chains should facilitate charge transport, a property that influences the power conversion efficiency of solar cells. Combined with the cost-effective method used for film deposition, resulting in a large-area framework, the

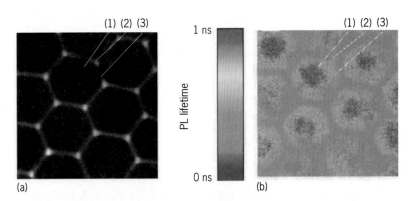

Fig. 3. The same 12 × 12 μm area of a P1 polymer thin film with honeycomb-like structure as seen by (a) confocal fluorescence intensity imaging and (b) confocal fluorescence intensity lifetime imaging. Regions (1), (2), and (3) denote parts of the hexagonal structure, namely, the center, node, and frame, respectively. (*Reprinted with permission from H. Tsai et al., Structural dynamics and charge transfer via complexation with fullerene in large area conjugated polymer honeycomb thin films, Chem. Mater., 23(3):759–761, 2011. Copyright © 2010 American Chemical Society*)

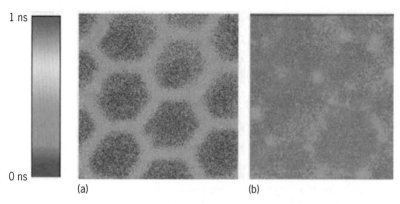

Fig. 4. Confocal fluorescence lifetime images of (a) double layer polymer-fullerene (P1/TM-C$_{60}$) thin film with honeycomb-like structure and (b) blended polymer-fullerene (P1/EG-C$_{60}$) thin film with honeycomb-like structure. (*Reprinted with permission from H. Tsai et al., Structural dynamics and charge transfer via complexation with fullerene in large area conjugated polymer honeycomb thin films, Chem. Mater., 23(3):759–761, 2011. Copyright © 2010 American Chemical Society*)

transparent thin films described here are promising active-layer candidates for the development of a new class of transparent OPVCs. Current efforts in our group are focused on scaling-up the deposition method and on implementing such active layers in photovoltaic devices.

[This research was supported by the U.S. Department of Energy (DOE), Office of Basic Energy Sciences, Division of Materials Science and Engineering. Research carried out in part at the Center for Functional Nanomaterials of Brookhaven National Laboratory (US-DOE contract DE-AC02-98CH10886) and at the Center for Integrated Nanotechnologies at Los Alamos National Laboratory (US-DOE contract DE-AC52-06NA25396).]

For background information *see* COMPOSITE MATERIALS; CONJUGATION AND HYPERCONJUGATION; ELECTRON-HOLE RECOMBINATION; FLUORESCENCE; FLUORESCENCE MICROSCOPE; FULLERENE; HOLE STATES IN SOLIDS; ORGANIC CONDUCTOR; PHOTOVOLTAIC CELL; PHOTOVOLTAIC EFFECT; SEMICONDUCTOR; SOLAR CELL; SOLAR ENERGY; WORK FUNCTION (ELECTRONICS) in the McGraw-Hill Encyclopedia of Science & Technology.

Mircea Cotlet; Zhihua Xu; Hsing-Lin Wang; Hsinhan Tsai

Bibliography.   U. H. F. Bunz, Breath figures as a dynamic templating method for polymers and nanomaterials, *Adv. Mater.*, 18:973–989, 2006; B. Kippelen and J. L. Bredas, Organic photovoltaics, *Energ. Environ. Sci.*, 2(3):251–261, 2009; H. Tsai et al., Structural dynamics and charge transfer via complexation with fullerene in large area conjugated polymer honeycomb thin films, *Chem. Mater.*, 23(3):759–761, 2011.

2010 eruption of Eyjafjallajökull, Iceland

The explosive eruption of the Icelandic volcano Eyjafjallajökull in April–May 2010 caused unparalleled disruption to air travel, with about 100,000 flights cancelled in Europe during April 15–21. Compared to many other eruptions, the volcanic plume was neither very high nor was the discharge of ash by the volcano high. The large impact of this event was caused by a combination of the volcano's location, the long duration (39 days) of the more or less continuous explosive eruption, and the persistent jet-stream flow of northwesterly winds carrying eruption clouds toward Europe. A change in European aviation regulations on April 21 regarding permissible ash concentration for operation of passenger jets was needed to get commercial aircraft in Europe back in the air.

Geological setting. Eyjafjallajökull is a large volcano on the south coast of Iceland (**Fig. 1**). At its base, it is 25 km long (east–west) and 15 km wide (north–south) and rises to 1650 m above sea level. The upper part is covered by an 80 km^2 ice cap from which the volcano takes its name [jökull (glacier/ice cap)]. The glacier is in most places 50–100 m thick, except in the 2–3-km-wide summit caldera where it is 200–300 m thick. The volcano has gradually been constructed over the last 800,000 years in a large number of small- to medium-sized eruptions. Together with the neighboring and much larger Katla caldera to its east and the partly submarine Vestmannaeyjar central volcano to its southwest, Eyjafjallajökull lies at the tip of the southwards propagating eastern volcanic zone of Iceland. The volcanic zone forms part of the boundary between the major tectonic plates of North America to the west and Eurasia to the east. Eyjafjallajökull is not very active, with only four eruptions known prior to 2010 over the last 1500 years. Three of these were explosive summit eruptions, while one was a fissure eruption on the southwest flank. All were modest in size and smaller than the 2010 eruption. The products of Eyjafjallajökull are predominantly basalts and basaltic andesites, whereas more evolved rocks, such as trachyandesites and trachytes, are found on the upper slopes.

Flooding of meltwater in jökulhlaups (glacial outburst floods) is an integral part of volcanic activity in ice-covered volcanoes because of rapid melting when the hot magma interacts with the ice. The lowlands to the south and west of Eyjafjallajökull are important agricultural areas, and many farms lie within 15 km of the summit on the south side. Hazard assessment and risk mitigation measures, aimed at reducing the danger to people caused by flooding during volcanic eruptions, were implemented in the years before the 2010 eruption. These plans were put to good use during the eruption, when 800 people had to be evacuated repeatedly from their homes because of flooding danger.

2010 eruption. The first known sign of unrest preceding the 2010 eruption was an increase in earthquake activity in the crust under Eyjafjallajökull in 1992. This was followed by inflation of the volcano, as magma flowed into the crust at 5-6 km depth in 1994 and again in 1999. A small inflation event occurred in 2009, and a steady increase in seismicity and inflation began at the end of the year. In late February 2010, seismicity and the rate of inflation increased a great deal, culminating in the outbreak of a small basaltic fissure eruption on the northeast slope on March 20. A small lava field was formed over a period of 3 weeks. The flank eruption came to an end on April 12. The main eruption occurred in the summit caldera, beginning at about 1:30 a.m. on April 14. It was preceded by an intense swarm of earthquakes. The eruption was at first subglacial, but after a few hours a small eruption plume had emerged through the cloud cover, and a jökulhlaup flowed northward, down the outlet glacier Gígjökull. The outburst flood lead to the closure of the main road along the south coast while the eruption plume gradually grew, having reached an elevation of 10 km by evening. The dark and heavily tephra-loaded eruption plume was diverted by strong westerly winds toward the east and then southeast. The drift of the plume could be tracked using satellite images. By April 16, a dilute cloud had reached mainland Europe (**Fig. 2**). Minute fallout of fine ash was reported in many places, including the Faroe Islands, Shetland, Norway, Denmark, Scotland, England, and Germany.

The explosive eruption was continuous for 39 days between April 14 and May 22. The intensity varied a great deal with the most powerful phases occurring from April 14-18 and May 5-17 (**Fig. 3**). The peak discharge observed during the eruption was about 1000 tonnes per second (10^6 kg/s), late on April 14 (Fig. 1). For most of the duration of the eruption, winds carried the ash plume toward the east, southeast, or south. From April 20–May 4, when explosive activity was relatively weak, a lava flow formed that extended about 3 km down the Gígjökull outlet glacier. The lava progressed along the base of the glacier, melting the ice along its way and formed a deep ice canyon that cut the glacier into two parts along most of its length. By May 19 the eruption was clearly declining, and the last day of continuous activity was on May 22. In the months that followed, a steam cloud issued from the vent, and explosions involving small amounts of magma occurred in early June.

The magma from the April–May summit eruption is trachyandesite, with SiO_2 content of 58–60%. Studies of the ground deformation from Global Positioning System (GPS) point surveys and satellite-based interferometric synthetic aperture radar (InSAR) measurements indicate inflow of new basaltic magma into the roots of the volcano at 4–6 km depth from January into May. This magma was erupted in the flank eruption but, apparently, it heated and mobilized a stagnant body of older, evolved magma and triggered its eruption at the summit. It is likely that the older magma resided in a sill-like body, and repeated intrusion of new magma into the base of this magma body caused the several peaks observed in the magma discharge and ash generation during the eruption.

Once the subglacial eruption had melted its way through the 200-m-thick ice, the explosive, tephra-producing eruption occurred through ice cauldrons with vertical walls that quickly grew to a diameter of 400-600 m. The magma interacted with the

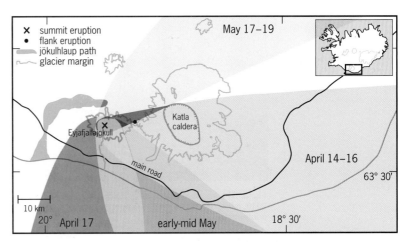

Fig. 1. The Eyjafjallajökull volcano and the main fallout sectors in the 2010 eruption. The ice cap and flood paths of meltwater in jökulhlaup are indicated. The floods occurred on April 14–15. After that time, melting was relatively slow compared to the first two days of the eruption.

meltwater and its presence affected the fragmentation of the magma and the eruption plume. Compounds, such as fluorine, were effectively scrubbed from the surfaces of tephra particles in the meltwater and the steam-rich plume. This changed after a few days, when water access to the vent was much reduced, leading to a large increase in the amount of fluorine released. Grazing animals were either kept inside or moved away from the most affected areas, to avoid fluorine poisoning. These measures proved successful in protecting the livestock.

The tephra erupted in Eyjafjallajökull was unusually fine-grained. Up to 20% of the material erupted in the first few days were classified as aerosols (diameter $<10~\mu$m) and up to 50% of the mass was very

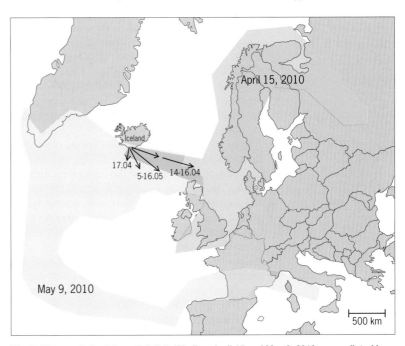

Fig. 2. Dispersal of ash from Eyjafjallajökull on April 15 and May 9, 2010, as predicted by the London Volcanic Ash Advisory Center. Ash was eventually dispersed in all directions around the volcano, but the sectors to the east and southeast were most affected because of the persistent northwesterly upper tropospheric winds during the eruption. Dates indicate the direction of dispersal during the main periods of activity.

Fig. 3. The eruption plume of Eyjafjallajökull on May 11, 2010, a time of medium discharge of ash toward the southeast.

fine ash (diameter ≤ 63 μm). This high abundance of small particles, with very long settling times in the atmosphere, contributed to the dispersal of clouds containing relatively small amounts of ash over distances well in excess of 2000 km. These dilute ash clouds were observed by ground-based lidar and research aircraft in many places in northern and central Europe, following peaks in explosive activity in April and May.

The eruption produced an estimated 0.27 km^3 of tephra transported by the eruption plume, with about half being deposited on land and the other half mostly in the sea south and east of Iceland. The amount transported to Europe was very minor. About one-fourth of the mass erupted was transported by meltwater or flowed as lava. The total amount of magma erupted (airborne, water-transported, and lava flows) is equivalent to 0.18 km^3 of dense rock. The eruption is classified as moderate in size, and if classified on the basis of maximum plume height (10 km) and magma discharge, its Volcanic Explosivity Index (VEI; scale = 0–8) is 3.

In the three previous eruptions of Eyjafjallajökull, in 920 AD, 1500 AD, and 1821–1823 AD, an eruption in the Katla caldera followed or took place at about the same time. This correlation has not been explained with a physical model, but it suggests that the likelihood of Katla erupting has increased after the 2010 eruption.

Many volcanoes in Iceland are capable of producing volcanic ash clouds that will disperse ash toward Europe, depending on the upper tropospheric winds. The air flow usually has an easterly trend, with flow toward the northeast, east, and southeast being the most common. Basaltic eruptions from ice-covered volcanoes in Iceland are usually explosive because of magma-water interaction, and often lead to 8–12-km-high plumes that may persist for some days. Shorter (duration of hours), but slightly more powerful eruptions of intermediate magma have taken place in the Hekla volcano. Seven explosive eruptions occurred in Iceland between 1970 and 2004, but the plumes were all diverted toward the northeast, avoiding mainland Europe. It is therefore a coincidence that appreciable closures of European air space caused by an Icelandic eruption did not take place before 2010.

Impact. After the eruption, the thickness of ash cover on the lower slopes of Eyjafjallajökull was 5–15 cm, and lahars (flows of mixture of volcanic ash and water) caused difficulties in some places in the aftermath of the eruption. Grazing of sheep on parts of the lower slopes will have to be abandoned for some time because of the damage to vegetation. Drifting ash can also be a problem. However, these difficulties are limited to a small area on the south side of the volcano. Minor fallout was observed in most parts of Iceland, but it was so small that it did not affect agriculture or other economic activity.

The closure of air space over extensive areas from April 15–21 resulted in the largest aviation shutdown in Europe since World War II. The closure was based on the principle that the encounter of aircraft and volcanic ash should be avoided at all costs. It soon became clear that such a strict criteria could lead to extended closures while the volcano remained active. A more practical approach, based on recommendations from jet engine manufacturers, was adopted after April 21. An upper limit of permissible concentrations of ash in the air was specified at 2 milligram (mg)/m^3. Later in the eruption, an additional time-limited zone with concentration of 2–4 mg/m^3 was defined. The concentration estimates for Eyjafjallajökull were based on computer models of atmospheric flow and dispersal made by the London Volcanic Ash Advisory Center (VAAC). Considerable uncertainty exists in such models. Satellite observations coupled with improved estimates of vent discharge from radars and other instrumentation may in the future be important in providing better constraints on the actual ash concentrations in eruption cloud in the event of an eruption.

For background information *See* AEROSOL; ANDESITE; BASALT; EARTHQUAKE; EUROPE; GEODESY; GLACIOLOGY; JET STREAM; LIDAR; MAGMA; PLATE TECTONICS; PYROCLASTIC ROCKS; TRACHYTE; VOLCANO; VOLCANOLOGY in the McGraw-Hill Encyclopedia of Science & Technology.   Magnús T. Gudmundsson

Bibliography. M. T. Gudmundsson et al., Eruptions in Eyjafjallajökull Volcano, Iceland, *Eos*, 91:90–191, 2010; M. T. Gudmundsson et al., Volcanic hazards in Iceland, *Jökull*, 58:251–268, 2008; U. Schumann et al., Airborne observations of the Eyjafjalla volcano ash cloud over Europe during air space closure in April and May 2010, *Atmos. Chem. Phys.*, 11:2245–2279, 2011; F. Sigmundsson et al., Intrusion triggering of the 2010 Eyjafjallajökull eruption, *Nature*, 426:426–430, 2010; T. Thordarson and G. Larsen, Volcanism in Iceland in historical time: Volcano types, eruption styles and eruptive history, *J. Geodyn.*, 43:118–152, 2007.

Ultrahigh-temperature ceramics

Ultrahigh-temperature ceramics (UHTCs) are refractory, high thermal conductivity materials that have significant potential for use in aerospace applications, especially sharp leading edges of wings. The materials have been under development for many

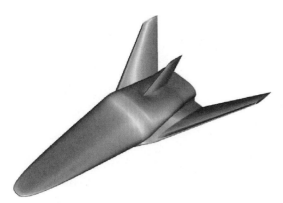

Fig. 1. Concept vehicle showing leading edges and nose.

years, but interest has increased dramatically over the last 10 years. Progress has been made in processing these materials and characterizing and improving their properties. The two major issues are low fracture toughness and insufficient oxidation resistance. Approaches to improve these properties include the development of composites to improve fracture toughness and the use of additives and compositional changes to improve oxidation.

UHTCs are a family of materials that includes the borides, carbides, and nitrides of transition elements such as hafnium, zirconium, tantalum, and titanium. UHTCs possess some of the highest melting points of known materials. In addition, they are very hard, have good wear resistance and mechanical strength, and relatively high thermal conductivities (compared to other ceramic materials). Strong covalent bonding is responsible for the high melting points, moduli, and hardness of the UHTC family of materials. High negative Gibbs energies of formation also give UHTCs excellent chemical and thermal stability under many conditions. In comparison to carbides and nitrides, the diborides tend to have high thermal conductivity, which gives them good thermal shock resistance and makes them ideal for many high-temperature thermal applications. The diborides of hafnium and zirconium are of particular interest to the aerospace industry for sharp leading-edge applications (**Fig. 1**), which require chemical and structural stability at extremely high operating temperatures.

Hafnium diboride and zirconium diboride, with melting points of 3380°C and 3247°C, respectively, are the focus of this paper. UHTCs have the potential to enable the development of sharp hypersonic vehicles or atmospheric entry probes capable of the most extreme entry conditions.

Sharp leading edges can significantly enhance the aerodynamic performance of a vehicle, improving cross range (the ability of a vehicle to deviate from the most direct or planned entry path, thereby accessing a greater range of landing opportunities) and maneuverability and contributing to accuracy and improved safety, especially for crewed vehicles. The amount of heat generated on entry and thus the temperature that a material must withstand depends upon the energy that reaches the material from the atmosphere during entry and the way in which the material can dissipate heat. Blunt bodies, such as capsules or non-lifting bodies, have a boundary layer that forms around the surface during entry, significantly reducing friction and thus heating. This boundary layer is very thin in very sharp bodies, exposing the material to much higher temperatures. In addition, blunt bodies have a large surface area from which to reradiate thermal energy; sharp leading edges have less surface area and thus must handle the energy differently. **Figure 2** illustrates the energy balance in a sharp leading edge. Most of the energy at the tip must be conducted away through the material and then reradiated out the sides. Thus, a material with a high thermal conductivity is required.

Materials for sharp leading edges can be reusable but need specific properties that are different from more conventional thermal protection materials used in acreage or blunt configurations because of geometry and very high temperatures. A material used for a sharp leading edge must have both very high-temperature capability (although for short times) and high thermal conductivity, while being resistant to the oxidation occurring during reentry. As an example, materials used on the leading edges of the space shuttle (reinforced carbon-carbon) saw temperatures around 1650°C, while materials for vehicles with sharp leading edges will experience temperatures in excess of 2000°C. However, it is important to note that the high temperatures are only seen at the very tip of the leading edge (**Fig. 3**),

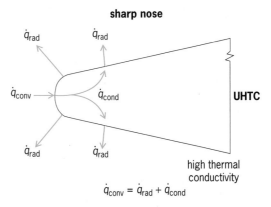

Fig. 2. Energy balance in a sharp leading edge, where $\dot{q}$ is heat flux (convective, radiative, conduction).

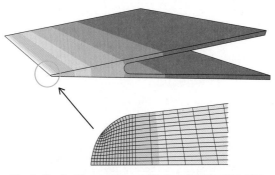

Fig. 3. Nominal temperature profile in a sharp leading edge (darker color indicates highest temperature).

meaning that there can be quite substantial thermal stresses which high thermal conductivity can alleviate.

Background. Hafnium diboride (HfB_2) and zirconium diboride (ZrB_2) materials were first investigated in the early 1950s for nuclear reactor applications. There was extensive work in the 1960s and 1970s by ManLabs for the U.S. Air Force that showed the potential of HfB_2 and ZrB_2 as nosecone and leading-edge materials.

There was a substantial gap in sustained development during the 1980s and most of the 1990s, although the Air Force Research Laboratory (AFRL) did consider UHTCs for long-life, human-rated (certified for human transportation) turbine engines in the 1980s.

In the late 1990s, NASA Ames teamed with Sandia National Laboratories New Mexico, Air Force Space Command, and TRW and revived interest in HfB_2/SiC, ZrB_2/SiC ceramics for sharp leading edges. Materials were made by vendors for two ballistic flight slender hypervelocity aerothermodynamic research probe (SHARP) experiments: SHARP-B1 in 1997, which had a UHTC nosetip, and SHARP-B2 in 2000, which had a UHTC strake assembly. The material was not recovered from SHARP-B1 but the strakes were recovered from SHARP-B2. Subsequent examination of the material indicated that the processing was very poor and that improvements were needed to assess the potential of the UHTCs. This led to a number of other programs sponsored by NASA, including the Space Launch Initiative (SLI), Next Generation Launch Technology (NGLT), and Ultra Efficient Engine Technology (UEET) programs from 2001–2005, NASA's Fundamental Aeronautics

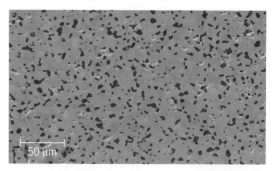

Fig. 5. Microstructure of HfB_2-20 vol % SiC (dark phase) made by hot-pressing.

Program to 2009, and a substantial interest both nationally and internationally in these materials. Most recently, NASA and the Air Force have been jointly funding the National Hypersonic Science Center for Hypersonic Materials and Structures led by Teledyne Scientific & Imaging LLC. **Table 1** lists some of the recent and current efforts in UHTCS at universities, government agencies, and international laboratories.

Processing. The refractory nature of UHTCs means that processing them into dense shapes requires high temperatures and often high pressures. Bulk UHTCs are most often fabricated through hot pressing in either resistance- or induction-heated furnaces, using graphite dies, at temperatures ranging from 1900–2100°C and pressures of 60–100 MPa, in processes that have not changed much since the 1960s. High melting temperatures make it extremely difficult to consolidate pure samples using conventional hot pressing. The phase diagram for the Hf-B system, shown in **Fig. 4**, illustrates the high melting point and very narrow compositional range over which Hf B_2 is stable. Zr-B has a lower melting temperature.

Work by Man Labs found that additives could eliminate billet cracking and make dense, fine-grained microstructures achievable. In particular, adding SiC from 5–30 vol % improved UHTC densification and oxidation resistance. SiC is the most common additive used today to control grain growth and provide oxidation resistance, although there are substantial research efforts exploring other additives such as W compounds, $MoSi_2$ and $TaSi_2$ to improve strength, microstructure, toughness and oxidation resistance. A typical microstructure of a hot-pressed HfB_2-20 vol % SiC material is shown in **Fig. 5**. Other ways of fabricating UHTCs include pressureless sintering, spark plasma sintering (SPS) or field assisted sintering (FAS), reaction sintering, and vapor deposition techniques for coatings. The fine-grained microstructure shown in **Fig. 6** is typical of that obtained by SPS.

An issue with powder techniques is mixing of the diboride and silicon carbide powders that have very different densities (11.2 g/cm³ for HfB_2 versus 3.2 g/cm³ for SiC). Drying techniques such as rotary evaporation or freeze-drying can be used to prevent separation of the powders.

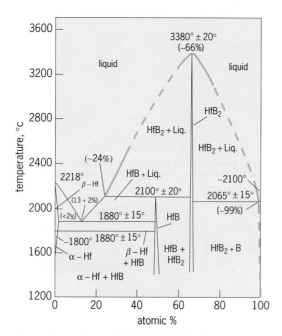

Fig. 4. Phase diagram for Hf-B system. (A. McHale, ed., *Phase Equilibria Diagrams: Phase diagrams for Ceramists,* **vol. X, American Ceramic Society, 1994. Reprinted with permission of The American Ceramic Society,** *http://www.ceramics.org; all rights reserved)*

TABLE 1. Recent UHTC research efforts and institutions

ZrB2 Based Ceramics	Catalytic Properties of UHTCs
Missouri University of Science & Technology	PROMES-CNRS Laboratory, France
US Air Force Research Lab (AFRL)	CNR-ISTEC
NASA Ames & NASA Glenn Research Centers	CIRA, Capua, Italy
University of Illinois at Urbana-Champaign	SRI International, California
Harbin Institute of Technology, China	Imaging and Analysis (Modeling)
Naval Surface Warfare Center (NSWC)	University of Connecticut
NIMS, Tsukuba, Japan	AFRL
Imperial College, London, UK	NASA Ames Research Center
Korea Institute of Materials Science	Teledyne (NHSC-Materials and Structures)
CNR-ISTEC	Oxidation of UHTCs
HfB$_2$ Based Ceramics	AFRL
NASA Ames Research Center	NASA Glenn Research Center
NSWC—Carderock Division	Georgia Institute of Technology
Universidad de Extramdura, Badajoz, Spain	Missouri University of Science & Technology
CNR-ISTEC, Italy	Texas A & M University
Fiber Reinforced UHTCs	CNR-ISTEC, Italy
Chinese Academy of Sciences, Shenyang	University of Michigan, Ann Arbor, Michigan
University of Arizona	NSWC – Carderock
MATECH/GSM Inc., California	Harbin Institute of Technology, China
AFRL	University of Illinois at Urbana-Champaign

Alternative sources of SiC include preceramic polymers and various coating techniques. The microstructures obtained can be quite different and have shown evidence of improved fracture toughness. An example of a microstructure with acicular grains formed by the addition of a preceramic polymer source of SiC is shown in **Fig. 7**.

Properties. The properties of UHTCs depend on composition and processing, and much progress has been made in improving strength and reliability of these materials. Some typical properties are given in **Table 2**. Typical strengths for HfB$_2$ materials are in the range of 300 to 500MPa, with much higher strengths for reported for some ZrB$_2$ materials with WC or MoSi$_2$ additives. The room-temperature ther-

mal conductivity of pure HfB$_2$ can be as high as 120 W/mK, but is reduced by the addition of SiC and is quite dependent on impurities and processing. Fracture toughness is generally reported to be 4–5 MPa m$^{1/2}$.

Evaluation. The evaluation of the performance of these materials for thermal protection requires that they be tested in an environment that simulates reentry conditions. No test can simulate all the parts of the trajectory where the heat flux and pressure are changing as a vehicle descends through the atmosphere. However, arc jets can be used to evaluate performance at selected conditions that can cover some of those experienced during reentry. Arc jets generate a plasma through the discharge of an

TABLE 2. Typical UHTC properties

Property	HfB$_2$/20 vol % SiC	ZrB$_2$/20 vol % SiC
Density (g/cc)	9.57	5.57
Strength (MPa) 21°C 1400°C	356 ± 97[*] 137 ± 15[*]	552 ± 73[*] 240 ± 79[*]
Modulus (GPa) 21°C 1400°C	524 ± 45 178 ± 22	518 ± 20 280 ± 33
Coefficient of thermal Expansion (x10^{-6}/K) RT	5.9	7.6
Thermal conductivity (W/mK)[#]RT	80	99

[*] Flexural strength
[#] R. P. Tye and E. V. Clougherty, The thermal and electrical conductivities of some electrically conducting compounds, *Proceedings of the Fifth Symposium on Thermophysical Properties*, C. F. Bonilla (ed.), The American Society of Mechanical Engineers, Sept. 30–Oct. 2, 1970, pp 396–401, 1970.

electric arc. The conditions are determined by a variety of operating factors. The main differences between exposure to a furnace environment and to an arc jet environment are the directionality of the heating and the pressure. Furnaces typically provide heat from three directions from radiating elements. During reentry the heating is primarily from one direction and can be both convective and radiative in nature. In addition, the pressure is substantially less

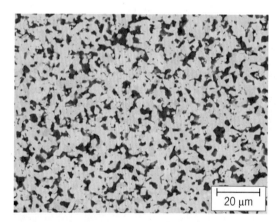

Fig. 6. Microstructure of HfB$_2$-20 vol % SiC (dark phase) made by spark plasma.

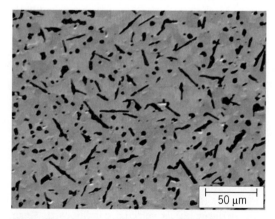

Fig. 7. Microstructure of HfB$_2$-15 vol % SiC made with a preceramic polymer source for SiC showing elongated grains of SiC.

than 1 atmosphere, depending upon the trajectory. Arc jets provide only convective heating.

Typical UHTC arc jet samples are shown in **Fig. 8**.

The behavior of a UHTC in the arc jet depends on the heat flux, pressure, and time. However, a typical response is the formation of an oxide layer on the surface with a depletion zone below, as shown in **Figs. 9** and **10** described by Levine et al. The oxide layer thickness is related to whether the conditions are such that there is active oxidation of SiC, that is, the pressure is sufficiently low that SiO is formed and vaporizes rather than forming a silica layer. Hafnia or zirconia will also form, depending upon the composition of the base material. Additives to the material can alter the composition and the properties of the oxide that is formed. The formation of a white or light colored oxide reduces the emissivity of the material and thus its ability to reradiate energy, resulting in higher temperatures in the material; thus a dark oxide is desirable. The conduct and interpretation of arc jet testing is complex, and more detail can be found in the references cited above.

Computational modeling. The fundamental properties of ZrB$_2$ and HfB$_2$ were computationally modeled since 2009. Ab initio computations detailed the different bonding motifs in these materials, showing the relationship between electronic structure and their unusual mechanical and thermal properties. Interatomic potentials were developed to enable atomistic simulations of these materials for the first time. These potentials were used to compute the lattice thermal conductivity of single-crystal materials from atomic-scale simulations. Experimental agreement was very good. Further, the structure, energetics, and thermal resistance of idealized grain boundaries have also been calculated.

Outlook. UHTCs have properties of great interest and applicability for aerospace applications, especially for vehicle sharp leading edges. The high thermal conductivity and refractory properties of these materials are particularly attractive. The high density is not an issue if they are to be used in small quantities on the forward part a vehicle, as they can help to move the center of gravity forward. These materials do have two major issues, neither of which is

Fig. 8. Typical UHTC samples for testing in an arc jet.

Fig. 9. Post-test arc-jet nosecone model after a total of 80 min of exposure. Total exposure is the sum of multiple 5- and 10-min exposures at heat fluxes from 200 W/cm².

insoluble but both of which are the focus of research efforts.

First, although these materials have good oxidation resistance, they do in general contain SiC and are subject to oxidation and material loss. Approaches to overcome oxidation include the reduction in the amount of SiC, and the addition of materials that can alter the oxidation behavior by changing the composition of the oxide or inhibit its formation.

The second issue with UHTCs is their low fracture toughness of around 4 MPa m$^{1/2}$. The best approach to improve fracture toughness is to make a composite material. This avenue is being explored by a number of research groups (Table 1).

Other issues involved with developing a composite are the availability of high-temperature fibers that are compatible with the UHTC matrix (currently only SiC and C-based fibers are being used, although there are efforts to develop UHTC-based fibers), fiber coatings, and processing techniques. Finally, there have been developments in forming a tougher microstructure by the in-situ formation of elongated grains that can deflect or bridge cracks.

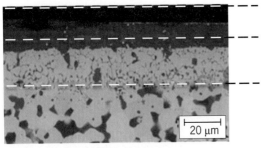

Fig. 10. Microstructure of UHTC tested in arc jet showing, from the top, oxide later, depleted zone, and base material.

For background information *see* ATMOSPHERIC ENTRY; CERAMICS; COMPOSITE MATERIAL; FREE ENERGY; REFRACTORY; SINTERING; SPACECRAFT STRUCTURE; STRAKE; THERMAL STRESS in the McGraw-Hill Encyclopedia of Science & Technology.

Sylvia M. Johnson

Bibliography. E. V. Clougherty, D. Kalish, and E. T. Peters, Research and development of refractory oxidation resistant diborides, *AFML-TR-68-190*, Man-Labs Inc., Cambridge, MA, 1968; J. Cotton, Ultrahigh-temperature ceramics, *Advanced Materials & Processes*, 168(06):26–28, 2010; E. L. Courtright et al., Ultra high temperature assessment study: Ceramic matrix composites, *AFWAL-TR-91-4061*, Wright Patterson Air Force Base, Ohio, 1992; R. A. Cutler, Engineering properties of borides, in *ASM Engineered Materials Handbook*, vol. 4: *Ceramics and Glasses*, S. J. Schneider (ed.), ASM International, pp. 787–803, 1991; M. S. Daw, J. W. Lawson, and C.W. Bauschlicher, Interatomic potentials for zirconium and hafnium diboride, *Comp. Mat. Sci.*, in press, 2011; M. Gasch et al., Processing, properties and arc jet oxidation of hafnium diboride/silicon carbide ultra high temperature ceramics, *J. Mate. Sci.*, 39:5925–5937, 2004; M. Gasch and S. Johnson, Physical characterization and arcjet oxidation of hafnium-based ultra high temperature ceramics fabricated by hot pressing and field-assisted sintering, *J. Eur. Ceram. Soc.*, 30(11):2337–2344, 2010; M. Gasch, S. M. Johnson, and J. Marschall, Oxidation characterization of hafnium-based ceramics fabricated by hot-pressing and electric field-assisted sintering, *Proc. 7th International Conference on High Temperature Ceramic Matrix Composites (HT-CMC7)*, Bayreuth, Bavaria, Germany, pp. 846–853, 2010; A. F. Guillermet and G. Grimvall, Phase stability properties of transition metal diborides, *Am. Inst. Phy. Conf. Proc.*, 231:423–431, 1991; S. Johnson et al., Recent developments in ultrahigh temperature ceramics, *AIAA-2009-7219, 16th AIAA/DLR/DGLR International Space Planes and Hypersonic Systems and Technologies Conference*, Bremen, Germany, Oct. 19–22, 2009; S. M. Johnson et al., Ultra high temperature ceramics: issues and prospects, *Proc. 7th International Conference on High Temperature Ceramic Matrix Composites (HT-CMC7)*, Bayreuth, Bavaria, Germany, pp. 819–831, 2010; L. Kaufman and E. V. Clougherty, Investigation of boride compounds for very high temperature applications, *RTD-TRD-N63-4096*, Part III, ManLabs Inc., Cambridge, MA, March 1966; D. Kontinos, K. Gee and D. Prabhu. Temperature constraints at the sharp leading edge of a crew transfer vehicle, *AIAA 2001-2886 35th AIAA Thermophysics Conference*, Anaheim, CA, June 11–14, 2001; J. W. Lawson, C. W. Bauschlicher, and M. S. Daw, Ab initio computations of electronic, mechanical and thermal properties of ZrB$_2$ and HfB$_2$, *J. Am. Ceram. Soc.*, in press, 2011; S . Levine et al., Evaluation of ultra-high temperature ceramics for aeropropulsion use, *J. Europ. Ceram. Soc.*, 22(14-15):2757–2767, 2002; A. McHale (ed.), *Phase Equilibria Diagrams: Phase diagrams for Ceramists*, vol. X, American Ceramic Society, 1994; D. Sciti et al., Properties of a pressureless sintered ZrB2-MoSi2 ceramic composite, *Journal of the American Ceramic Society*, 89(7): 2320–2322, 2006; D. M. Smith et al., Arc-heated facilities, in *Advanced Hypersonic Test Facilities*, vol. 198, *Progress in Astronautics and Aeronautics*, F. K. Lu and D. E. Marren (eds.), AIAA, 2002.

Ultrashort laser pulse measurements

Observing the evolution of an event in time requires a device that is faster than the event. To examine how a balloon pops, for example, we can take a sequence of images with a strobe light whose flashes are shorter than the duration of the popping. And if we then want to know how short the strobe-light flash is, we must use a photodiode that is even faster to measure it. But how do we then measure the response of a photodiode? Clearly, this problem goes on and on to shorter and shorter times. When we reach the time scale of a millionth of a billionth of a second (or a femtosecond, 1 fs = 10^{-15}s), we reach the frontier, the shortest technological events ever generated: ultrashort laser pulses, which last for only a few femtoseconds. How do we measure them when there is no shorter event available?

This dilemma is the essence of the field of ultrashort laser pulse measurement. These laser pulses are thousands of times shorter than the response of the fastest detectors available. So what can we do? The solution to this seemingly hopeless problem begins with a clever intuition: if there is no shorter event available, how about using the event to measure itself?

Autocorrelation. The idea of using an ultrashort laser pulse to measure itself is the basis of a method called autocorrelation (**Fig. 1**), invented in the 1960s. In an autocorrelator, the pulses to be mea-sured are first split into two replicas. Then with a few mirrors and a translation stage, the two pulses are re-combined with a controllable time delay between them. (Since we know the speed of light, we can re-late the variable beam path difference to the relative delay.) At the point where the two beams are over-lapped, we place a nonlinear optical medium, usually a second-harmonic generation (SHG) crystal, which generates light of twice the frequency if both beams are present at the same time. In the arrangement shown in Fig. 1, the second harmonic is emitted in a direction between the two beams and has an inten-sity proportional to the product of the two incoming light pulse intensities at their times of arrival. As a re-sult, when the two pulses overlap perfectly in time, we will have maximum second harmonic. If one of the pulses arrives earlier or later than the other, the second harmonic is weaker. If the pulses do not over-lap at all, there will be no second harmonic. As a result, measuring the second harmonic energy as a function of the relative delay gives a rough measure of the pulse duration. The shorter the pulse duration, the thinner the autocorrelation width. The resulting trace is called the intensity autocorrelation. The de-tector used to measure the signal intensity can be very slow, since the temporal resolution is achieved by the fine adjustment of the relative delay.

The fact that an intensity autocorrelator uses an event only as short as the event itself brings some handicaps, however. First, it yields only a rough mea-sure of the pulse width, yielding no information about the actual pulse shape. Pulses with very dif-ferent intensity profiles can have the same autocor-relation (Fig. 1). So researchers must assume a pulse shape (such as a Gaussian) to obtain a laser-pulse duration. Worse, ultrashort laser pulses also neces-sarily have broad frequency spectra, so it would be good to know the evolution of frequencies (or the phase or color) in time, but autocorrelation gives no

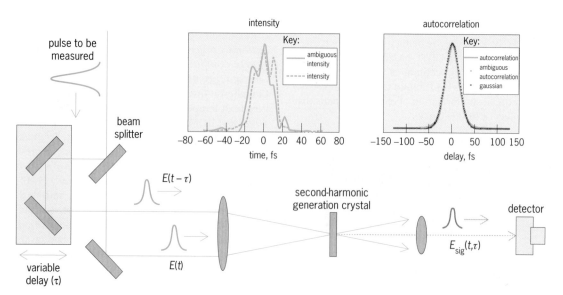

Fig. 1. Typical setup of an intensity autocorrelator. The inset shows two possible plots of pulse intensity versus time that yield the same autocorrelation traces.

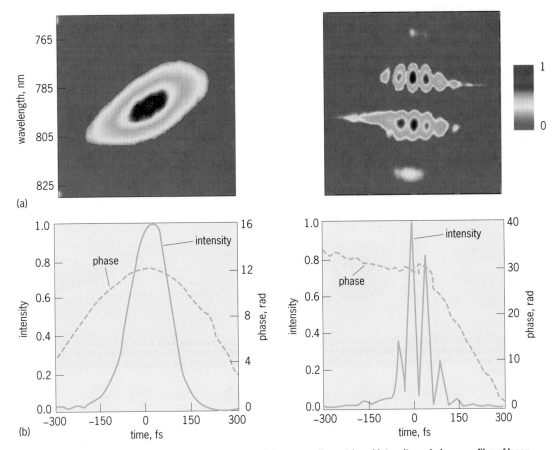

(a)

(b)

Fig. 2. FROG technique. (*a*) Experimental FROG traces. (*b*) Corresponding retrieved intensity and phase profiles of laser pulses.

information at all about the pulse phase. Clearly, scientific research needed a much better pulse measurement method.

FROG. A significant step forward occurred in 1991. Rick Trebino and coworkers realized that measuring the spectrum, not the energy, of the autocorrelator's output light at each delay yielded sufficient information to completely determine the pulse intensity and phase, despite the use of the pulse to measure itself. Their experimental setup is essentially the same as that of an autocorrelator, with the exception that a spectrometer is used measure the second-harmonic spectrum. This yields a data trace of intensity versus both delay and frequency—effectively a spectrogram (or a musical score) of the pulse (**Fig. 2***a*). Next, an iterative algorithm retrieves both the intensity and the phase of the pulse from the measured data, without any a priori assumptions about the pulse (Fig. 2*b*). Named frequency-resolved optical gating (FROG), this method became an immensely powerful tool in ultrashort laser research. It has been successfully used over a wide range of laser wavelengths, pulse durations, and complexities, and it remains the method of choice for accurate and reliable ultrashort pulse measurements.

GRENOUILLE. Recently Trebino and coworkers realized that FROG could be significantly simplified (**Fig. 3**). Their first simplification was to replace the beam splitter, delay stage, and all alignment mirrors with a single simple optic called a Fresnel biprism.

This approach is based on a consequence of the geometry of two beams that cross at an angle, namely, that the time delay between them is mapped onto the transverse position (**Fig. 4***a*). Even better, this approach can measure a single pulse; while the measurements in Fig. 1 must be performed over a train

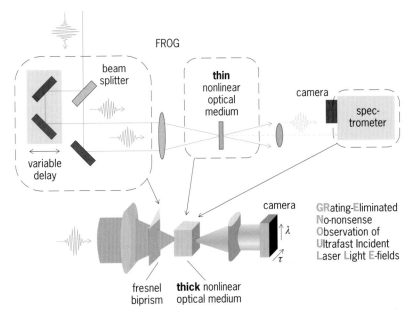

Fig. 3. Experimental schemes of FROG and its simplification to GRENOUILLE.

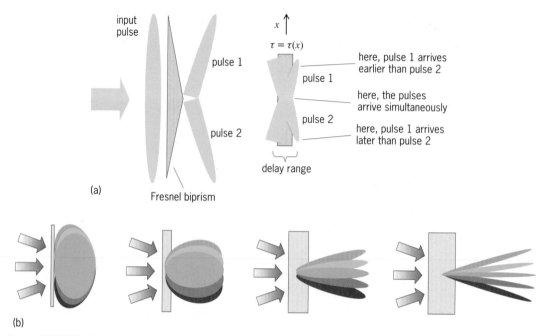

(a)

Fresnel biprism

(b)

Fig. 4. GRENOUILLE concepts. (*a*) Crossing beams at an angle maps the relative delay onto transverse position. (*b*) The second-harmonic light generated by the crystal depends on angle. Also, the angular width of a given color depends on the crystal thickness. The thicker the crystal, the narrower the angular width of a given color, allowing a thick crystal to also act as a spectrometer.

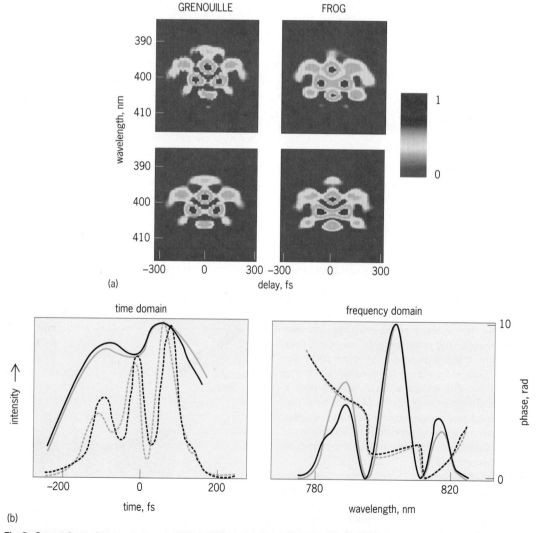

(a)

(b)

Fig. 5. Comparison of measurements with both FROG and GRENOUILLE. (*a*) Experimental traces. (*b*) Retrieved pulses. Colored lines show FROG measurements; black lines show GRENOUILLE measurements.

of pulses (at best achieving one delay per pulse), the Fresnel biprism achieves all delays on each pulse.

A second clever trick eliminates the spectrometer. This approach works successfully because nonlinear optical interactions generate different colors in different directions (Fig. 4b), and these color distributions also depend on the crystal thickness. For autocorrelators and FROGs, the crystal must normally be very thin (typically 0.1 mm or less) so that, at the beam's angle of incidence, all the colors in the pulse are converted to the second harmonic. On the other hand, with a very thick crystal, only a small range of second harmonic frequencies occur at a given angle. So, if the incoming beam is focused tightly into the crystal, a wide range of beam angles occurs, and every frequency component in the pulse will be converted to the second harmonic at a different angle in the output beam. But angularly resolving the beam's colors in this manner is precisely what spectrometers do. As a result, a thick crystal and a simple spatially resolving detector, like a charge-coupled device (CCD) camera, can replace the thin crystal and spectrometer used in a FROG setup.

These simplifications turn FROG into a compact, simple, and user-friendly device, which requires almost no alignment. It is called GRating-Eliminated No-nonsense Observation of Ultrafast Incident Laser Light E-fields (abbreviated as GRENOUILLE, French for frog).

Despite all these simplifications, GRENOUILLE performs almost as well as a regular FROG with no functional compromise. It can also measure complex pulses, as FROG does (**Fig. 5**). In addition, the highly symmetrical structure of the device also allows the measurement of even more properties of the ultrashort pulse, such as space-time couplings. For example, an ultrashort pulse can propagate with its intensity front tilted with respect to the direction of propagation. Minute tilts of this sort can be detected with GRENOUILLE, but not with a standard autocorrelator or FROG.

These advances in ultrashort laser pulse measurements have made crucial contributions to the research and applications of laser science. Indeed, these powerful characterization tools are available even to nonspecialists. More recent research and development have extended such methods even to the full evolution of pulses in space and time and also with extreme complexities.

For background information *see* INTERFERENCE OF WAVES; LASER; NONLINEAR OPTICAL DEVICES; NONLINEAR OPTICS; OPTICAL PULSES; ULTRAFAST MOLECULAR PROCESSES; WAVEFORM DETERMINATION in the McGraw-Hill Encyclopedia of Science & Technology.

Selcuk Akturk; Xun Gu

Bibliography. S. Akturk et al., Measuring pulse-front tilt in ultrashort pulses using GRENOUILLE, *Opt Express*, 11:491–501, 2003; J. A. Giordmaine et al., Two-photon excitation of fluorescence by picosecond light pulses, *Appl. Phys. Lett.*, 11:216–218, 1967; P. O'Shea et al., Highly simplified device for ultrashort-pulse measurement, *Opt. Lett.*, 26:932–934, 2001; R. Trebino et al., Measuring ultrashort laser pulses in the time-frequency domain using frequency-resolved optical gating, *Rev. Sci. Instrum.*, 68:3277–3295, 1997.

Unducted fans for aircraft propulsion

The application of the gas turbine toward aviation propulsion, leading to the turbojet engines of Frank Whittle and Hans Von Ohain in the late 1930s, redefined aviation transportation for generations to come. The turbojet supplanted the then prevalent internal combustion engine technology with a simple, lightweight, and low-cost alternative. It allowed aircraft to climb to altitudes well beyond those attainable for a conventional internal combustion engine. It also enabled flight speeds to advance from the low subsonic arena of propeller-driven airplanes into the high subsonic and supersonic flight regimes. However, the benefit of increased flight speed offered by the turbojet came at the expense of reduced propulsive efficiency. Propulsive efficiency is a measure of how well jet exhaust power is transferred to the airplane. A slow-turning propeller imparting a small change in velocity over a large stream of air had been replaced with a small but high-velocity airflow to produce thrust.

In the late 1950s, aviation propulsion underwent another great leap forward with the implementation of the ducted turbofan engine architecture. This architecture recaptured a portion of the propulsive efficiency previously foregone in the transition from propeller to turbojet, while retaining capability for high subsonic and supersonic flight. This was accomplished by expanding the turbojet (now termed the engine core) exhaust through a low-pressure turbine and driving a large, relatively low-pressure-ratio ducted fan. Subsequent decades of commercial engine design and innovation were dedicated to realizing additional propulsive efficiency through evolutionary reductions to fan pressure ratio and corresponding increases in engine bypass ratio (the ratio of fan bypass to core airflow.)

The next leap forward in aviation propulsion took place in 1986 with the first flight of a B727-100 powered by a GE36 unducted fan (UDF) engine. The UDF engine architecture is a unique form of turbofan engine architecture that removes the fan cowl and ducting and replaces a moderately low pressure ratio with a very-low-pressure-ratio counterrotating fan. Extensive flight testing of the UDF was performed on a McDonnell Douglas MD-80 proof-of-concept vehicle (**Fig. 1**).

While the architecture offered a substantial performance advantage over a ducted turbofan with equivalent core technology, concerns over noise and complexity in conjunction with a fall in the price of jet fuel kept the engine from entering into production at the time. However, engine makers continued to mature the architecture using internal and government funding. The combination of high fuel prices and volatility in fuel prices in the 2010s have sparked renewed interest in the innovative UDF engine architecture. More than 20 years after flight

Fig. 1. GE36 unducted fan proof-of-concept vehicles.

test, the architecture remains unmatched in demonstrated efficiency at high subsonic flight speeds.

Performance advantage. Small changes in airplane fuel burn drive significant changes in a commercial airline's profitability. Reductions in fuel burn also provide a handle for mitigating aircraft engine emissions in an era of increasing concern over transportation-related air pollution. The high-level characteristics available to the propulsion system designer for influencing the airplane fuel burn are propulsive efficiency, thermal efficiency, propulsion system weight, and propulsion system drag.

The propulsive efficiency of the UDF positively differentiates it from the ducted turbofan more than any other high-level characteristic. The propulsive efficiency (η_{PR}) is defined by Eq. (1), where $P_{A/C}$ is

$$\eta_{PR} = \frac{P_{A/C}}{\Delta KP_{\text{Jet}}} \qquad (1)$$

the aircraft power, and ΔKP_{Jet} is the change in jet power across the engine. Simplifying the above expression by neglecting the presence of fuel in the exhaust jet yields a reasonable approximation, Eq. (2), for propulsive efficiency in terms of flight

$$\eta_{PR} = \frac{2v_0}{v_0 + v_{\text{Jet}}} \qquad (2)$$

velocity v_0 and exhaust jet velocity v_{Jet}. The exhaust jet velocity is predominantly a function of fan pressure ratio and flight Mach number. The relationship between fan pressure ratio and propulsive efficiency is shown in **Fig. 2** for a cruise condition.

On the right side of Fig. 2, a state-of-the-art moderate-bypass-ratio engine is shown at a fan pressure ratio of 1.7. From this baseline the UDF holds a propulsive efficiency advantage of ~24%. As the fan pressure ratio of advanced ducted fans trends downward toward the UDF, the propulsive benefits available for transitioning to a UDF will also diminish. However, for the next generation of ducted engines a gap of 12–16% is anticipated to remain.

A detailed analysis is required to properly account for how much of this advantage can be translated into fuel burn reduction. Items such as the additional weight of the large unducted fan module, cooling flows for sections of the module exposed to hot core exhaust gas, and potentially pylon blowing for acoustic considerations will consume a portion of the benefit. Typical estimates for the fuel burn advantage of an advanced UDF entering into service in 2025 relative to an advanced high bypass ducted turbofan with similar core thermodynamic characteristics are in the range of 7–11%. The adverse impact of fan nacelle weight and fan duct pressure drop on fuel burn for lower-fan-pressure-ratio ducted engines will make it challenging to continue to close this performance gap going forward, likely giving the UDF a long term performance advantage.

The UDF aerodynamic and acoustic challenges associated with achieving these levels of performance will be discussed, as well as some of the innovative solutions to these challenges.

Fan aerodynamic innovation. At cruise flight Mach numbers, the outer portion of the UDF fan blades operate at transonic Mach numbers, where shock waves can add significant losses. To mitigate the strength of shocks the blades are highly swept in a manner similar to transonic wings. The high degree of blade sweep also serves to dephase the noise generated by front rotor wakes interacting with the rear rotors.

In a conventional ducted fan, the axial flow velocity through the engine at a given fan rotational speed does not vary significantly with aircraft flight velocity. On a UDF, axial flow velocity through the fan is a strong function of aircraft flight velocity.

Therefore, between takeoff operation and cruise operation, UDF fan blades see significant swings in relative flow angle, operating Mach number, and lift coefficients. An aircraft wing copes with these variations by employing variable-camber features such as flaps and slats. A UDF employs variable-pitch fan blades, but variable-camber airfoils are currently mechanically too complex. Therefore, new airfoil shapes are being designed that operate more efficiently over a wide range of operating conditions.

Because the UDF has no casing around the fan, the fan blades generate tip vortices analogous to those

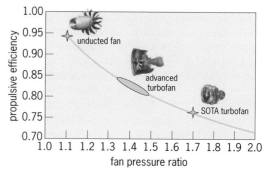

Fig. 2. Relationship between propulsive efficiency and fan pressure ratio.

from aircraft wing tips. As on aircraft, these vortices are strongest at takeoff conditions. On the UDF, the vortices from the front rotor interact with the aft-rotor aerodynamics to generate significant noise. A considerable amount of aerodynamic design effort is concentrated on developing configurations that control the strength of front-rotor vortices and reduce the sensitivity of the aft rotor to these vortices.

Modern blade design. Modern computational tools provide the ability to simulate open fan flow physics and blade response using three-dimensional (3D) computational fluid dynamics (CFD) and computational aeroacoustics. These tools were used to optimize the modern blade designs, reducing interaction noise, while maintaining the fuel burn advantage of the UDF architecture. Optimization of the forward blade trailing wake and tip vortex is illustrated in **Fig. 3**.

Unducted fan noise physics. Aircraft noise is regulated by governmental authorities. Prior to entering service, each aircraft must demonstrate compliance with community noise statutes to achieve certification status.

The primary sources of aircraft community noise are the airframe and the engines. For a conventional ducted turbofan engine, the sources, in order of importance, are jet noise, fan noise, and core noise (compressor and turbine). The noise sources for a UDF engine are essentially the same, but their relative contributions differ, as shown in **Fig. 4**. The jet and core (compressor and turbine) noise sources are referenced in the bottom half of the figure because their contribution to the aircraft system noise is small compared to the fan noise. However, the complex counterrotating open fan gives rise to multiple sources of "fan noise," as illustrated in the upper portion of Fig. 4.

The dominant sources of community noise have been highlighted in color in Fig. 4. As shown in the figure, these sources are associated with rotor blade response to ingested flow distortion, due to the incidence angle, the wake behind the mounting pylon, or the interaction of the aft rotor with the trailing wake or vortex from the forward rotor.

Acoustic progress and challenges. Recent testing of several modern and historical blade designs provided

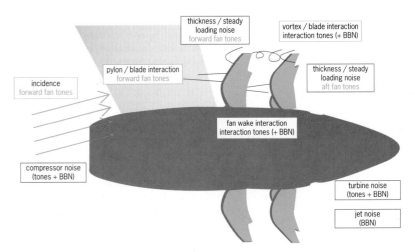

Fig. 4. UDF engine noise sources.

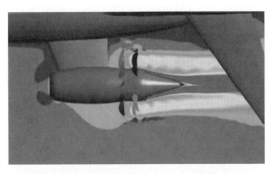

Fig. 5. Cut plane through a UDF engine installed on a commercial aircraft with velocity contours, showing impact of installation on the UDF.

opportunities to validate the CFD/CAA based design tools. However, a critical aspect of the UDF engine concept is integration with the airframe. The strong influence of the pylon and airframe is clearly illustrated in **Fig. 5**, which shows velocity contours on a cut plane through an engine installed on a typical commercial aircraft. Installed performance optimization is one of the primary challenges to developing a modern UDF product.

In conclusion, advances in analytical capability since the flight testing of the first UDF concepts and innovative experimental work have enabled significant strides toward a UDF-powered aircraft.

For background information *see* AERODYNAMIC SOUND; AIRCRAFT DESIGN; AIRCRAFT ENGINE PERFORMANCE; COMPUTATIONAL FLUID DYNAMICS; TRANSONIC FLIGHT; TURBOFAN; TURBOJET in the McGraw-Hill Encyclopedia of Science & Technology.

Kurt Murrow; Andrew Breeze-Stringfellow; John Wojno

Bibliography. A. Breeze-Stringfellow, Open rotors: benefits and challenges, Ultra-high bypass ratio panel discussion, *ASME Turbo Expo*, June 14–18, 2010, Glasgow, Scotland; G. Hoff et al., Experimental performance and acoustic investigation of modern, counterrotating blade concepts, NASA rep. NASA-CR0185158, 1990; B. A. Janardan and P. R. Gliebe, Acoustic characteristics of counterrotating unducted fans from model scale tests, *J. Aircraft*, 27(3):268–275, March 1990; L. H. Smith, Unducted fan

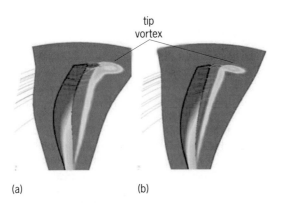

Fig. 3. Modern blade design optimization. (a) Baseline blade. (b) Improved blade.

aerodynamic design, *J. Turbomachinery*, 109:313–324, 1987; J. Wojno, NASA/General Electric Open Rotor Research Test Campaign, Ultra-high bypass ratio panel discussion, *ASME Turbo Expo*, June 6-10, 2011, Vancouver, Canada.

Ventilation and climate control of deep mines

As underground mines reach greater depths, higher temperatures, higher humidity, and more gas emissions are encountered. The topic is of interest to various audiences with very different viewpoints. Workers in the mines expect healthy and safe working conditions with an acceptable comfort level for a typically heavy workload. Mining operators wish to comply with regulations while minimizing the cost of investment and operation. Mining industry regulatory organizations demand compliance with standards for safety and health. Environmental and workers' advocates wish to see improvements in working conditions as well as lower emission standards. Mining engineers try to design a mine within the constraints of all of the above. Profitable exploitation must be secured amid the factors of ever-increasing energy costs, stricter conditions, and lower tolerance levels to health risks and accidents. In addition, ventilation engineers and technicians labor to meet the demands of all parties and to provide safe and healthy working conditions underground at a minimum cost. This balancing act requires intimate familiarity with the nature of geology and the ore deposit as well as their interaction with the engineered mining operation. It is helpful to review some aspects of the parameters that influence the task as well as the underlying physical processes that a mine ventilation engineer must encounter and consider. This article focuses on the issues in metal mines.

Effects of the geologic formation on mine ventilation. The ore body and the rock mass, known as the strata, around a mine opening are the prime elements of the boundary conditions for a mine ventilation and air conditioning task. The strata represent sources of (1) contaminant gas species, (2) moisture and water, and (3) heat.

Contaminant gases. Gas species that may be liberated from the strata are the natural contaminants that affect both safety and health. Such gases include carbon dioxide (CO_2), carbon monoxide (CO), hydrogen sulfide (H_2S), sulfur dioxide (SO_2), and methane (CH_4). To keep the working area safe and healthy, adequate air flow must be provided by ventilation to dilute contaminant gases.

Gas liberation from the strata to the air space occurs by diffusion and convection. The strata are formed of porous and fractured geologic deposits and give off stored gas or generate new species after exposure to oxygen supplied by ventilation. The transport of gaseous species in geologic media is subject to advanced theories in geophysics. According to the basic governing laws introduced by A. Fick

and H. Darcy as shown in Eq. (1), the mass flux den-

$$q_m = \rho D_m \, \text{grad}\,(c)|_w + k/\nu \, \text{grad}\,(P)|_w \qquad (1)$$

sity, q_m [in kg/(m²s)] units, may be written as the sum of the diffusive and convective flux components across a unit wall surface area of the working opening, where ρ (in kg/m³), D_m (in m²/s), c (in kg/kg), k (in m²), and ν (in m²/s) are the mixture density, species diffusion coefficient, concentration, species permeability of the strata, and species kinematic viscosity, respectively. The grad(c) and grad(P) terms denote, respectively, the gradient (which, in the one-dimensional case, is a simple differential with respect to distance from the wall) of the concentration and the gradient of the barometric pressure of gas species in the strata pores and fractures in close vicinity to the wall of the air space. Note that these gradients are time-dependent. Changes in gas concentration in the air space by ventilation generates a variable gas influx component by ordinary diffusion. Changes in barometric pressure due to weather or modulation of ventilation pressure likewise induce a Darcy-type, pressure diffusion component. These coupled transport-process effects must be accounted for when advanced mine ventilation control is considered.

Equation (1) is widely used in advanced Earth science and engineering model calculations and numerical simulations, and excellent transport codes are available commercially, such as the TOUGH simulation program. Recent detailed mine ventilation calculations, especially for coal mines with methane liberation, increasingly use similar governing equations that take into consideration the variation of strata permeability with the changing water saturation level as well as with mechanical stress. However, engineers of metal mines rarely use such detailed models, resorting instead to simple, empirical correlations based on mine field measurements. Empirical correlations, however, are blind to coupled transport process effects and risk overlooking the potential of increased gas liberation that may be caused by automatic control of the ventilation system.

A new generalized and time-dependent transport model for gas species, vapor mass, and heat-flow prediction in the strata has been developed and applied to ventilation simulations in underground openings. According to the Numerical Transport Code Functionalization (NTCF) model, the time-dependent species flux density, q_m, in Eq. (1) may be expressed in matrix-vector equation as in Eq. (2),

$$[q_m] = [q_m^c] + [[M_c]][c - c^c] + [[M_P]][P - P^c] \qquad (2)$$

where the bracketed variables are vectors composed of the sampled values of the variables taken at preselected time instants. Vectors $[q_m^c]$, $[c^c]$, and $[P^c]$ are "central" values around which the linearized model in Eq. (2) is valid and defined during model identification. Matrices $[[M_c]]$ and $[[M_P]]$ are dynamic admittance operators of constant coefficients that are identified by the NTCF procedure. This procedure employs system identification of the process of liberation of the gas species from the strata using a numerical transport model (that is, a code such as TOUGH) with a set of predetermined test boundary conditions as input histories with time. Equation (2) may be regarded as an algebraic, general representation of the differential model given in Eq. (1). The gradients, as driving forces of the diffusive and convective flux components, are replaced with finite differences, varying directly with time through the time series of the sampled boundary values given by vectors $[c]$ and $[P]$. The obvious advantage of using Eq. (2) is that the boundary values $[c]$ and $[P]$ can be calculated or modeled from the ventilation air pressure and concentration, and can even be measured for verification. In contrast, the grad $(c)|_w$ and grad$(P)|_w$ gradients in Eq. (1) are unknown and difficult-to-capture variables, dependent on time and the boundary conditions of c and P on the bounding wall. To evaluate these gradients, the complete, coupled solution of gas liberation from the strata wall must be described with a transport model (program code), such as TOUGH2, but that can be used only for a given boundary condition for c and P, both of which are dependent on the ventilation problem itself. Therefore, the solution to the gradients in Eq. (1) must be iterated directly with the solution to the ventilation model for the air space, whereas the dynamic admittance matrices in Eq. (2) can be predetermined and captured for a mine from preselected model runs or even from field measurements for given strata.

Moisture and water. Moisture and/or water vapor transported from the porous and fractured strata are natural sources of humidity in the ventilating air and represent a significant component of the heat stress or comfort index of workers underground. The vapor mass flux density q_v that enters the air space from the strata may be described with a similar model as for a gas species and thus with an equation similar to Eq. (1), but now with material properties, diffusivity, and permeability all referring to water vapor in the porous and fractured strata. Concentration of vapor in the model may be substituted as mass fraction, ω, or even partial vapor pressure, P_v, which can be converted between each other, or from absolute or relative humidity, if one knows the temperature of the air–vapor mixture. In the coupled processes of simultaneous diffusive, convective, as well as evaporative vapor transport due to heat from the strata, an NTCF model may be used that accounts for the driving forces of vapor mass fraction ω, barometric pressure P, and temperature T, as in Eq. (3), where

$$[q_v] = [q_v^c] + [[M_\omega]][\omega - \omega^c] + [[M_P]][P - P^c]$$
$$+ [[M_T]][T - T^c] \qquad (3)$$

the dynamic admittance matrices $[[M_\omega]]$, $[[M_P]]$, and $[[M_T]]$ may be determined against model identification runs using the hydrologic model of the strata, or by direct field measurements during time-variable disturbances. In the majority of current mine ventilation models, the vapor transport from the strata

assumes pool (unconstrained) evaporation from a partially wet surface. With the mass-transport coefficient, β, and the vapor mass fraction–driving force, $\omega_s - \omega$, from a wall element at saturated temperature, the vapor flux density from a unit surface area at any time instant is given as in Eq. (4). The fraction of the

$$q_v = uf\beta(\omega_s - \omega)|_w \qquad (4)$$

contact surface between the strata and the air space from which evaporation takes place is expressed by a wetness factor, wf in Eq. (4), an empirical number that can only be inverse-evaluated by trial and error during the calibration of the ventilation and climate model against field measurements. The wetness factor can hardly ever be verified by direct measurement or visual observation even if the wet surface portions are darker in appearance, because evaporation may take place from deeper layers of the strata that are hidden from the naked eye. However, a simple evaporation model with a calibrated wetness factor works surprisingly well for mine ventilation applications when the strata are situated under the water table, and the natural hydrology must be engineered with drainage, grouting, and sealing anyway—a situation that is difficult to model with a porous and fractured rock mass model.

Heat. The heat from the rock strata in deep underground mines is a significant portion of the thermal load on the working environment. The strata heat flux density may be described as the sum of conduction and convection components by the boundary equation (5), where ρ (in kg/m^3), c_p [in [J/(kg K)], a

$$q_b = \rho c_p a \operatorname{grad}(T)|_w + q_{bc}|_w \qquad (5)$$

(in m^2/s) are density, specific heat, and thermal diffusivity, respectively all multiplying the temperature gradient at the wall, grad $(T)|_w$. The last term, $q_{bc}|_w$, is the convective component of heat flux density due to all gas, water, and vapor mass fluxes entering the air space from the rock strata. The corresponding NTCF model to Eq. (5) expresses the temperature and the pressure differences explicitly as driving forces in the conduction and convection terms as in Eq. (6),

$$q_b] = [q_b^c] + [[H_T]][T - T^c] + [[H_P]][P - P^c] \qquad (6)$$

where the dynamic admittance matrices $[[H_T]]$, and $[[H_P]]$ may be determined against model identification runs using the thermal-hydrologic model of the strata, for example, by applying a TOUGH2 model. The advantage of using a strata heat transport model in the form of Eq. (6) is its simplicity and accuracy, as well as its applicability to any geometric configuration, such as an arbitrary shape in the cross section of the opening. In simple cases, when only dry heat conduction is considered in the strata, such as in most mine ventilation and climate-control applications, the convective term in Eq. (6) is eliminated.

Effects of the engineered mine workings on mine ventilation. Mining machinery and work activities add contaminant gases, water, and vapor, as well as heat, to the air space. Electrical machinery dissipates heat into the air-flow network. Diesel engines emit CO_2, CO, water vapor, heat, and nitrous fumes (that is, a mixture of oxides of nitrogen including NO_2, N_2O_4, and NO). Process water used for dust control contributes to humidity in the mine. Workers underground add heat by metabolism, as well as vapor and CO_2 by perspiration and breathing. Even the fans that are used to drive ventilation dissipate heat into the air flow as a result of compression work as well as the viscous and kinetic energy loss in the machinery. Compressed air used as mine energy transmission is the lone exception, because it adds air and cools as much as it heats, giving a net zero balance in the energy budget of the mine underground, provided that the compressor is operated on the surface.

The resulting concentration of gas species, humidity, and temperature in the mine air space are the dynamic combination of the geologic effects defining the boundary conditions and engineered effects defining the sources as well as the sinks (that is, the fresh air intake and the air coolers and dehumidifiers employed in the ventilation system). To evaluate the final outcome, careful modeling and simulation must be done for reliable prediction of the air quality underground. Four sets of equations must be solved in a coupled way.

The gas species concentration model in the air space may be described by Fick's second law as in Eq. (7), where ρ is the density of the air and gas

$$\rho \frac{\partial c}{\partial t} + \rho v_i \frac{\partial c}{\partial x} = \rho D \frac{\partial^2 c}{\partial x^2} + \rho D \frac{\partial^2 c}{\partial y^2} + \rho D \frac{\partial^2 c}{\partial z^2} + q \tag{7}$$

mixture in the air space; x, y, and z are Cartesian coordinates; t is time; c, D, and q are respectively concentration, diffusion coefficient, and mass flux source of a given gas species; and v_i is air velocity in flow channel i in a discretized flow network model.

The water vapor concentration model in the air space also follows Fick's second law and Eq. (7) as applied to humidity mass fraction in Eq. (8), where

$$\rho \frac{\partial \omega}{\partial t} + \rho v_i \frac{\partial \omega}{\partial x} = \rho D \frac{\partial^2 \omega}{\partial x^2} + \rho D \frac{\partial^2 \omega}{\partial y^2} + \rho D \frac{\partial^2 \omega}{\partial z^2}$$
$$+ qc + qs + qm \tag{8}$$

ω and D are the mass fraction and diffusion coefficient for water vapor; qc, qs, and qm are sources or sinks of vapor fluxes due to condensation (a negative value), superheated steam sources such as in diesel exhaust, and evaporated process and strata water, respectively.

The heat balance in the air space is governed by Fourier's second law as in Eq. (9), where q_b is the

$$\rho c \frac{\partial T}{\partial t} + \rho c_p v_i \frac{\partial T}{\partial x} = \rho c_p a \frac{\partial^2 T}{\partial x^2} + \rho c_p a \frac{\partial^2 T}{\partial y^2}$$
$$+ \rho c_p a \frac{\partial^2 T}{\partial z^2} + q_b \tag{9}$$

source or sink of heat flux due to direct heat dissipation in the dx, dy, dz grid block of the air space.

The air flow in the ventilation network is governed by the Navier-Stokes equation. In its simplified form for flow channels along given grid lines, it reads as Eqs. (10)–(12), where v_x, v_y, v_z are ve-

$$\rho \left(\frac{\partial v_x}{\partial t} + \mathbf{v} \cdot \nabla v_x \right) = \rho g_x - \frac{\partial P}{\partial x} + F_x \tag{10}$$

$$\rho \left(\frac{\partial v_y}{\partial t} + \mathbf{v} \cdot \nabla v_y \right) = \rho g_y - \frac{\partial P}{\partial y} + F_y \tag{11}$$

$$\rho \left(\frac{\partial v_z}{\partial t} + \mathbf{v} \cdot \nabla v_z \right) = \rho g_z - \frac{\partial P}{\partial z} + F_z, \tag{12}$$

locity components of vector v; g_x, g_y, g_z are gravitational forces, which include buoyancy in the x, y, and z directions; and F_x, F_y, F_z are viscous and kinetic dissipation terms. The viscous dissipation terms are calculated from Moody's friction resistance coefficient, while the kinetic dissipation terms are evaluated using fitting-loss coefficients in simple mine ventilation network calculations.

The governing equations for gas, humidity, heat, and air flow constitute the coupled transport model for the air space. A computational fluid dynamic (CFD) model may be created by the coupled solution of these governing equations, together with the constitutive equations for conduction, diffusion, and radiation. The differential equations may be integrated over a finite element, resulting in an integrated-parameter CFD solution. This approach, used in the coupled ventilation and climate simulation model, MULTIFLUX, allows for reducing the number of discretization elements in the computational domain.

Targets and control of ventilation and climate conditions. For the most common contaminant gas species, mine ventilation textbooks tabulate the allowable threshold limit values (TLV) for safety and health, specified and defined in three different ways: (1) by the time-weighted-average (TWA) value; (2) by the short-term exposure limit (STEL) value; and (3) by giving the ceiling limit (CL) value. For temperature and humidity limits, the combined effects on the human body are specified in the form of heat stress indices or an effective/equivalent temperature, TE. These parameters include the wet-bulb air temperature (related to humidity concentration and temperature together), the dry-bulb air temperature (that is, the "real" temperature), and the air velocity; and they may even depend on the workers' clothing as well as the level of physical work load. The effective temperature is often close to the wet-bulb temperature, and its specified value is $TE_{max} = 28°C$ in most countries. It must be noted, however, that the core (rectal) body temperature and the heart rate together are the ultimate measure of the heat and humidity stress on the human body, and physical discomfort rapidly decreases productivity and may cause a decline in work care and attention.

Careful design of the mine ventilation system with gas species concentration, humidity, temperature, and air flow control is necessary to maintain a safe and healthy working environment underground that supports adequate and economic production. Modern concepts in instrumentation, monitoring,

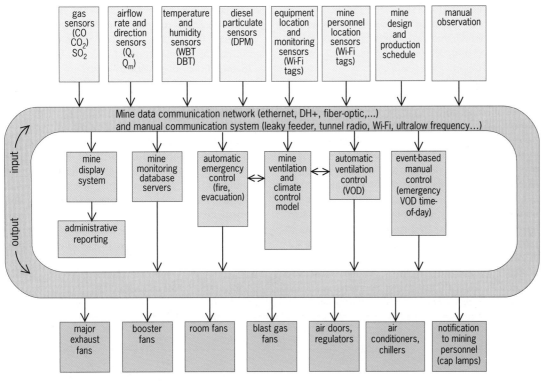

Components of a mine ventilation and climate-control system.

communication, and control have given rise to new concepts in ventilation and climate control as well. The **illustration** summarizes the components of such a control system, which includes an option for ventilation on demand (VOD).

For background information *see* COMPUTATIONAL FLUID DYNAMICS; INDUSTRIAL HEALTH AND SAFETY; MINING; NAVIER-STOKES EQUATION; UNDERGROUND MINING; VENTILATION in the McGraw-Hill Encyclopedia of Science & Technology. George L. Danko

Bibliography. R. B. Bird, W. E. Stewart, and E. N. Lightfoot, *Transport Phenomena*, John Wiley & Sons, 1960; G. Danko, Functional or operator representation of numerical heat and mass transport models, *J. Heat Transfer*, 128:162–175, 2006; G. Danko, D. Bahrami, and J. Birkholzer, A turbulent transport network model in MULTIFLUX coupled with TOUGH2, *J. Nucl. Technol.*, 174:1–26, 2011; L. H. Hartman et al., *Mine Ventilation and Air Conditioning*, 2d ed., Krieger Publishing Company, 1991; M. J. McPherson, *Subsurface Ventilation Engineering*, 2d ed., Mine Ventilation Services, 2009; K. Pruess, C. Oldenburg, and G. Moridis, *TOUGH2 User's Guide, Version 2.0*, Report LBNL-43134, Lawrence Berkeley National Laboratory, Earth Sciences Division, Berkeley, CA, 1999; J. R. Welty, C. E. Wicks, and R. E. Wilson, *Fundamentals of Momentum, Heat and Mass Transfer*, 3d ed., John Wiley & Sons, 1984.

Vitamins in plants

Vitamins are organic molecules that are essential for metabolism in both animals and plants. Although plants are able to synthesize their own vitamins, animals do not form or insufficiently form these micronutrients, which therefore must be ingested in the diet. Vitamin deficiencies lead to diseases that, in some cases, can be severe and eventually cause death. **Table 1** lists a number of human disorders caused by vitamin deficiencies. Although vitamin deficiencies (such as vitamin A deficiency) are normally associated with restricted food availability in low-income countries, there is a surprisingly inadequate intake of vitamins (in particular, vitamins B_9 and C) among industrialized countries because of unbalanced diets. Fruits and vegetables are rich sources of many of the vitamins in the human diet. Besides having a high concentration of vitamins, these foods

TABLE 1. Human disorders caused by vitamin deficiencies

Disorder	Deficiency
Blindness	Vitamin A deficiency
Beriberi	Vitamin B_1 deficiency
Ariboflavinosis	Vitamin B_2 deficiency
Pellagra	Vitamin B_3 deficiency
Abdominal distress and burning feet syndrome	Vitamin B_5 deficiency
Anemia and risk of cardiovascular disease	Vitamin B_6 deficiency
Basal ganglia disease	Vitamin B_7 deficiency
Neural tube defects in the fetus	Vitamin B_9 deficiency in pregnant women
Pernicious anemia	Vitamin B_9 and vitamin B_{12} deficiency
Scurvy	Vitamin C deficiency
Rickets and osteomalacia	Vitamin D deficiency
Nerve damage and reproductive failure	Vitamin E deficiency
Bleeding disorders	Vitamin K deficiency

can be ingested in a fresh state, thus preserving their vitamin levels; this is in contrast to foods of animal origin, which undergo cooking or thermal treatments that destroy thermolabile vitamins to a variable extent.

Water-soluble vitamins. Vitamins can be classified by their water solubility. Water-soluble vitamins include vitamin C (ascorbate) and the B-group vitamins—B_1 (thiamine), B_2 (riboflavin), B_3 (niacin), B_5 (pantothenic acid), B_6 (pyridoxine, pyridoxal, pyridoxamine, and their phosphorylated derivatives), B_7 (biotin, also known as vitamin H), B_9 (tetrahydrofolate and its one-carbon derivatives), and B_{12} (cobalamin). Vitamin C and most B vitamins are ubiquitous in plants; only vitamin B_{12}, which is exclusively synthesized by prokaryotes, is lacking in plants. Humans can synthesize vitamin B_3 from the

essential amino acid tryptophan, although 60 mg of this amino acid is required to make 1 mg of vitamin. Vitamin C is the most abundant vitamin in plants, reaching the highest concentrations in fruits (up to 228 mg/100 g in guava), followed by some flowers (broccoli and cauliflower) and leaves (cabbage) [**Table 2**]. In contrast, B vitamins range from nearly 1 mg to a few mg per 100 g in the most concentrated plant tissues, with the notable exception of vitamin B_7, which is in the range of several μg/100 g. B vitamins mainly accumulate in seeds. The bran of certain seeds contains remarkable concentrations of vitamins B_1 and B_3. For example, rice bran contains 39- and 24-fold higher amounts of vitamins B_1 and B_3, respectively, than polished rice, and wheat bran contains a 5-fold higher concentration of vitamin B_3 than wheat germ. In the case of vitamins B_2, B_3, B_5,

TABLE 2. Vitamin contents of the 10 richest plant foods (according to the Danish Food Composition Databank)*

Provitamin A† 0.8 mg	B₁ 1.4 mg	B₂ 1.6 mg	B₃ 18 mg	B₅ 6 mg	B₆ 2 mg	B₇ 150 μg	B₉ 0.2 mg	C 60 mg	E‡ 10 mg	K₁ —$
Rose hip 11.4 mg/100 g	Rice bran 2.75 mg/100 g	Tea 0.95 mg/100 g	Rice bran 34 mg/100 g	Peanuts, dried 2.8 mg/100 g	Dill, dried 1.71 mg/100 g	Soybeans, dried 60 μg/100 g	Mung beans, dried 0.625 mg/100 g	Guava 228 mg/100 g	Sunflower seeds, dried 49.5 mg/100 g	Parsley 0.79 mg/100 g
Carrot 9.07–11 mg/100 g	Sunflower seeds, dried 2.1 mg/100 g	Soybeans, dried 0.87 mg/100 g	Wheat bran 29.6 mg/100 g	Wheat bran 2.4 mg/100 g	Pistachio nuts, dried 1.7 mg/100 g	Kale 36 μg/100 g	Soybeans, dried 0.28 mg/100 g	Red sweet pepper 187 mg/100 g	Almonds 24 mg/100 g	Spinach 0.56 mg/100 g
Dandelion leaves 9 mg/100 g	Wheat germ 1.45 mg/100 g	Almonds 0.8 mg/100 g	Peanuts, dried 20 mg/100 g	Peas, dry 2 mg/100 g	Wheat germ 1.42 mg/100 g	Peanuts, dried 34 μg/100 g	Broccoli 0.239 mg/100 g	Red hot chili pepper 166 mg/100 g	Pine nuts, dried 24 mg/100 g	Coriander 0.31 mg/100 g
Dill 5.98 mg/100 g	Brazil nuts 1.13 mg/100 g	Wheat germ 0.61 mg/100 g	Tea 7.5 mg/100 g	Soybeans, dry 1.92 mg/100 g	Wheat bran 1.38 mg/100 g	Wheat bran 24 μg/100 g	White beans, dried 0.226 mg/100 g	Lemon peel 129 mg/100 g	Hazelnuts, dried 21 mg/100 g	Chives 0.31 mg/100 g
Parsley 5.69 mg/100 g	Alfalfa seeds 1.08 mg/100 g	Alfalfa seeds 0.58 mg/100 g	Sunflower seeds, dried 6.8 mg/100 g	Buckwheat groats 1.45 mg/100 g	Garlic 1.24 mg/100 g	Walnuts 19 μg/100 g	Chervil 0.22 mg/100 g	Broccoli 114 mg/100 g	Wheat germ 11 mg/100 g	Chickpeas, dried 0.264 mg/100 g
Kale 5.05 mg/100 g	Peanuts, dried 0.91 mg/100 g	Dill 0.43 mg/100 g	Fennel seeds 6.05 mg/100 g	Lentils, dry 1.36 mg/100 g	Linseeds 0.79 mg/100 g	Banana 5.5 μg/100 g	Spinach 0.22 mg/100 g	Green sweet pepper 99.1 mg/100 g	Linseeds 8.2 mg/100 g	Broccoli 0.26 mg/100 g
Sweet potato 4.47 mg/100 g	Wheat bran 0.89 mg/100 g	Wheat bran 0.36 mg/100 g	Wheat germ 5.8 mg/100 g	Tea 1.3 mg/100 g	Sesame seeds 0.79 mg/100 g	Avocado 3.6 μg/100 g	Wheat germ 0.19 mg/100 g	Kiwi fruit 92.7 mg/100 g	Peanuts, dried 8.2 mg/100 g	Kale 0.25 mg/100 g
Hot chili pepper 4.39 mg/100 g	Soybeans, dried 0.874 mg/100 g	Fennel seeds 0.353 mg/100 g	Wheat kernels 5.6 mg/100 g	Wheat germ 1.2 mg/100 g	Millet 0.75 mg/100 g	Carrot 3.4 μg/100 g	Chickpeas, dried 0.18 mg/100 g	Strawberry 76 mg/100 g	Blackberry 5.5 mg/100 g	Brussels sprouts 0.25 mg/100 g
Spinach 4.19 mg/100 g	Pistachio nuts, dried 0.87 mg/100 g	Celery 0.34 mg/100 g	Barley groats 5.5 mg/100 g	Hazelnuts 1.15 mg/100 g	Walnuts 0.74 mg/100 g	Rice 3 μg/100 g	Parsley root 0.18 mg/100 g	Lychee 71.5 mg/100 g	Kale 5.4 mg/100 g	Mung beans, dried 0.17 mg/100 g
Watercress 4.01 mg/100 g	Poppy seeds 0.85 mg/100 g	Chervil 0.34 mg/100 g	Brown rice 5.09 mg/100 g	Avocado 1.1 mg/100 g	Lentils, dried 0.556 mg/100 g	Peas 3 μg/100 g	Cauliflower 0.165 mg/100 g	Cauliflower 68.1 mg/100 g	Brazil nuts 5 mg/100 g	Spring cabbage 0.17 mg/100 g

*European Union recommended daily amounts (RDA) are shown below each vitamin, and vitamin concentrations are shown under each food item. Also note that requirements of certain vitamins may vary depending on sex, age, pregnancy, and intake of other nutrients.
† β-Carotene equivalents.
‡ α-Tocopherol equivalents.
$ There is no RDA for vitamin K₁.

B_7, and B_9, other organs of specific plants can reach similar concentrations as seeds; these other organs include leaves (kale and chervil) and fruits (avocado and banana).

Water-soluble vitamins act as cofactors in enzyme-catalyzed reactions in central metabolism in both plants and animals. In the case of most B vitamins, the biologically active cofactors are phosphorylated (thiamine pyrophosphate from B_1, and pyridoxal phosphate from B_6) or associated with nucleotides [flavin mononucleotide (FMN) and flavin adenine dinucleotide (FAD) from B_2, and nicotinamide adenine dinucleotide (phosphate) or NAD(P)$^+$ from B_3]; they may also be other derived forms [coenzyme A and acyl carrier protein (ACP) derived from B_5, di- and tetrahydrofolate polyglutamates from B_9, and adenosylcobalamin and methylcobalamin from B_{12}]. An important metabolic process dependent on an adequate vitamin B_{12} supply is myelin biosynthesis. Land plants, though, do not require cobalamins and therefore lack this vitamin; unlike animals, they have cobalamin-independent methionine synthase. However, more than half of all algal species need cobalamin as a cofactor for methionine synthase but do not have the capacity to synthesize this vitamin; instead, they obtain it through a close association with bacteria in exchange for fixed carbon. This is the case with nori (*Porphyra yezoensis*), which is widely used to make sushi.

In particular, vitamin C has an outstanding role in plants, acting in the same way as in animals, namely as a cofactor of a high number of 2-oxoacid-dependent dioxygenases, which catalyze the incorporation of O_2 into an organic substrate. In animals, ascorbate is involved as an enzyme cofactor in the synthesis of hydroxyproline-rich proteins, which are essential components of collagen, the main structural protein in skin, bones, tendons, and ligaments; an analogous function is developed in plants as hydroxyproline-rich proteins are components of the plant cell wall. Furthermore, ascorbate acts as a cofactor in plants in (1) the de-epoxidation of violaxanthin in the photoprotective xanthophyll cycle, (2) the biosynthesis of the hormones abscisic acid, ethylene, and gibberellins, and (3) the hydrolysis of glucosinolates. Interestingly, certain B vitamins and vitamin C are required as cofactors in the biosynthetic pathways of other vitamins in plants. In this way, methyltetrahydrofolate is involved in the biosynthesis of several vitamins, including B_1 and B_2 (through participation in the purine biosynthetic pathway), as well as B_5 and K_1. In addition, S-adenosylmethionine, intimately linked to folates, is required for the biosynthesis of vitamins E and K_1. Pyridoxal phosphate is also a cofactor in the biosynthetic pathway of vitamin B_7, whereas thiamine pyrophosphate participates in the route leading to geranylgeranyl diphosphate, a common precursor of the lipid-soluble provitamin A and vitamins E and K_1. Finally, ascorbate is involved in the biosynthesis of homogentisate, which is a precursor of vitamin E (see **illustration**).

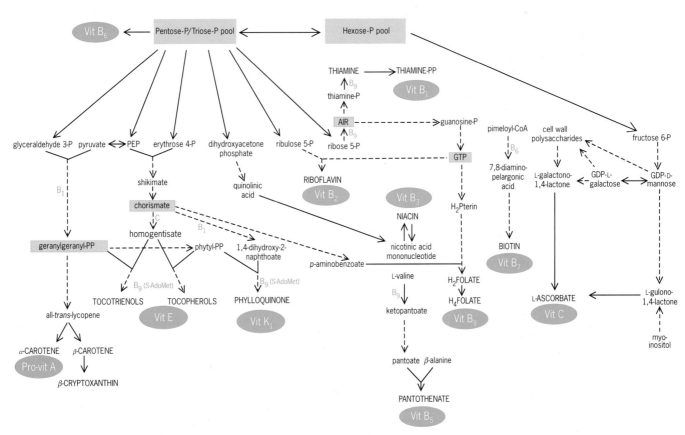

Overview of vitamin biosynthesis in plants. Note that several vitamins share common precursors and that some vitamins act as cofactors for the biosynthesis of others.

Besides its role as a cofactor, ascorbate has a prominent antioxidant function in both animals and plants. Ascorbate is the specific substrate for ascorbate peroxidase, which scavenges hydrogen peroxide through the ascorbate–glutathione cycle. Ascorbate also can act as an antioxidant in a nonenzymatic manner, donating electrons to a wide range of substrates; in this manner, an important function of ascorbate is the regeneration of tocopherol (vitamin E) radicals. Ascorbate can modulate (through its redox state) cell proliferation and elongation, as well as other developmental and environmental responses, including programmed cell death (apoptosis), senescence, and pathogen responses. New functions for some B vitamins are now emerging. Vitamins B_1 and B_6 play a role in resistance of both plants and animals to oxidative stress; however, it is unknown whether this protective effect is exerted through a direct antioxidant action or indirectly through their role as redox cofactors.

Lipid-soluble vitamins. Lipid-soluble vitamins include vitamin A (retinoids), vitamin D (mainly ergocalciferol or vitamin D_2 and cholecalciferol or vitamin D_3, collectively known as calciferol), vitamin E (tocochromanols, which include four tocopherols and four tocotrienols), and vitamin K (K_1 or phylloquinone is the major dietary form found in plants, K_3 or menadione is also present in plants, and K_2 or menaquinone-4 is the active form found in animal tissues). Vitamin A is not found in plants; instead, plants have large amounts of provitamin A, a group of carotenoids (mainly the carotenes, α- and β-carotene, and the xanthophyll β-cryptoxanthin) from which herbivores and omnivores are able to synthesize vitamin A. Humans and vertebrates in general are also able to synthesize cholecalciferol from cholesterol in the skin after UV-B light exposure from the sun or artificial sources. However, sun exposure can be insufficient during winter at latitudes with weak sunlight, in individuals with dark skin, or when skin is covered by clothing; therefore, diet becomes an important source of vitamin D. Because vitamin D has only been recently detected in a few plant families (Solanaceae, Cucurbitaceae, Fabaceae, and Poaceae), and cholesterol is a minor sterol in most plant species, plants are not considered as a source of either vitamin D or provitamin D.

Although β-carotene is present in all photosynthetic organisms, α-carotene and β-cryptoxanthin are not uniformly distributed within the plant kingdom. Similarly, a heterogeneous distribution is found among vitamin E compounds: tocopherols are present in all plant species studied so far, whereas tocotrienols only appear in a range of unrelated plant species. Leaves are generally the main source of β-carotene, with a concentration in the range of 1–11 mg/100 g. α-Carotene and β-cryptoxanthin are less abundant and are mainly found in fruits such as pumpkins; carrot is a remarkable root vegetable that contains higher amounts of α- and β-carotene than leaf vegetables; and β-cryptoxanthin, which is mostly in an esterified form, is also found in seeds and flowers. Vitamin E accumulates in seeds, particularly

oil seeds, with concentrations close to 50 mg/100 g in sunflower seeds. Tocotrienols are found in high amounts in fruits and seeds, but only in some species. It is noteworthy that vitamin E is highly concentrated in oils obtained from these fruits and seeds; one example is palm oil, which is rich in tocotrienols. Vitamin E in the form of tocopherols is also present in photosynthetic tissues such as leaves, but with remarkable 10- to 20-fold lower levels compared to seeds. The highest vitamin K_1 concentrations are found in most green vegetables (for example, in the range of several hundred μg per 100 g for parsley), although seeds (peas, beans, and wheat), fruits (strawberries and cucumber), tubers (potato), and roots (carrot) contain this vitamin at lower concentrations.

Vitamin A comprises a group of compounds known as the retinoids, which have many physiological functions [for example, regulation of growth, development, and reproduction; low-light and color vision; adaptive immunity; and maintenance of epithelial structure (including the skin)]. The antioxidant properties of carotenoids seem to play a role in the prevention of chronic diseases in humans. Carotenoids are a large group of more than 700 compounds that give the particular red, orange, and yellow pigmentation to fruits and flowers to attract pollinators and feeders, thus playing an important role in plant reproduction and seed dispersal. In addition, carotenoids in chloroplasts (cell plastids) contribute to light harvesting and protect the photosynthetic apparatus from photooxidative damage. Vitamin D has an essential function in regulating calcium and phosphorus homeostasis in vertebrates and hence aids in bone formation and maintenance; new roles have been recently assigned to this vitamin, including regulation of muscle contractility, cell growth and differentiation, immunomodulation, and vascular, endocrine, and reproductive functions. Vitamin E compounds are chain-breaking antioxidants that prevent lipid peroxidation. The antioxidant action of vitamin E seems to be involved in prevention of atherosclerosis, nerve damage, and hemolysis, and observational studies indicate a role of this vitamin in the prevention of cardiovascular disease. Furthermore, vitamin E is essential for reproduction and modulates immune functions. Tocochromanols also have an essential antioxidant function in plants. α-Tocopherol, the major form found in chloroplasts, has a relevant protective role in preventing photooxidative damage under high light intensities, having a complementary function to that of carotenoids. In contrast, an antioxidant function of tocotrienols in fruits and seeds where they accumulate has not yet been clearly established. Finally, vitamin K is a cofactor for γ-glutamyl carboxylase, which is involved in the synthesis of proteins that participate in blood coagulation, bone metabolism, prevention of vascular calcification, and vascular repair. In plants, phylloquinone constitutes the second step in the photosystem I redox chain, whereas menadione seems to be involved in redox reactions at the plasma membrane.

Outlook. Vitamins are essential not only for animals but also for plants. Plants can synthesize their own vitamins through complex and interrelated metabolic networks. Humans then can benefit from the richness of vitamins present in plants, which deserve a central place in a balanced diet.

For background information *see* ANTIOXIDANT; ASCORBIC ACID; BIOTIN; CAROTENOIDS; NIACIN; NUTRITION; PLANT PHYSIOLOGY; RIBOFLAVIN; THIAMINE; VITAMIN; VITAMIN A; VITAMIN B_6; VITAMIN B_{12}; VITAMIN D; VITAMIN E; VITAMIN K in the McGraw-Hill Encyclopedia of Science & Technology.

María Amparo Asensi-Fabado; Sergi Munné-Bosch

Bibliography. M. W. Davey et al., Plant L-ascorbic acid: Chemistry, function, metabolism, bioavailability and effects of processing, *J. Sci. Food Agric.*, 80:825–860, 2000; S. Mooney et al., Vitamin B_6: A long known compound of surprising complexity, *Molecules*, 14:329–351, 2009; F. Rébeillé et al., Roles of vitamins B_5, B_8, B_9, B_{12}, and molybdenum cofactor at cellular and organismal levels, *Nat. Prod. Rep.*, 24:949–962, 2007; A. G. Smith et al., Plants need their vitamins too, *Curr. Opin. Plant Biol.*, 10:266–275, 2007.

Wastewater treatment for deployed military forces and disaster response

Forward operating bases (FOBs) are transitory facilities used by the various armed forces of the United States and its allies to provide secure bases of operations for military missions in overseas environments. FOBs can be and are used by all the military services, but given the nature of their missions, the Army and the Marine Corps use these facilities the most. FOBs typically are constructed by the military engineering units of the Army Corps of Engineers (USACE) or the Civil Engineer Corps (CEC) of the United States Navy (USN), or contractors under the guidance of USACE or CEC.

Prior to the recent wars in Iraq and Afghanistan, the doctrine of the combined military forces of the United States was based on fast-moving offensive operations. This model did not envision the establishment of long-term operational basecamps, as it assumed that units would be moving quickly. Indeed, the early stages of Operation Enduring Freedom and Operation Iraqi Freedom (the Second Gulf War) were characterized by rapid offensive movements throughout Afghanistan and Iraq. Rapid movement precluded the need for extensive and long-term wastewater treatment.

However, as these efforts shifted from operations against established military units to combating counterinsurgency, the development of relatively long-term (months to years) FOBs became necessary. The 2010 documentary by National Geographic, *Camp Leatherneck*, provides some insight into the environment and terrain in southern Afghanistan as well as issues faced by basecamps [in this case, operated by the United States Marine Corps (USMC) in this area]. Camp Leatherneck had a military force of approximately 7000 and was expected to get larger. However, other bases in the area were much smaller,

indicating that bases vary quite a bit in size. The soldiers were housed in tents and mobile buildings, indicating that investment in expensive fixed systems would not be warranted. Camp Leatherneck also supported other USMC basecamps in southern Afghanistan. Transportation of supplies was challenging because of poor roads and insurgent activity, particularly involving the use of improvised explosive devices (IEDs). A second 2010 documentary, *Restrepo*, coverered a smaller camp operated by the U.S. Army in a mountainous area of northeastern Afghanistan. Human waste was collected and periodically burned using gasoline or diesel fuel. These and other sources make it clear that the development and operation of FOBs offer many challenges in terms of wastewater treatment.

A 2009 manual recommended the use of deployable wastewater treatment systems or the use of lagoon and wetland systems to achieve adequate sanitary treatment. In January 2011, one of the authors (Victor Medina) met with members of the Army Engineer School and the Directorate of Environmental Integration, who indicated that the development of effective wastewater treatment for basecamps is a critical need for the Army.

Deployable wastewater treatment is also needed for disaster response. An illustrative example is the aftermath of the earthquake in Haiti. The disaster damaged wastewater treatment operations. As a result, a terrible cholera epidemic occurred months after the earthquake, resulting in more than 200,000 infections and 4000 deaths. Although the exact source of the infection is not known, it is clear that improperly treated wastewater was a factor.

This article describes one effort to address the need for wastewater treatment for military operations and for disaster response. The first prototype of a deployable wastewater treatment reactor was developed with funding from the Air Force Institute for Operational Health through 2007. Since then, design changes and system improvements have been tested and incorporated, resulting in the development of a deployable wastewater treatment reactor called the deployable aerobic aqueous bioreactor by the Texas Research Institute for Environmental Studies at Sam Houston State University, Lamar University, Sul Ross University, and the Army Engineer Research and Development Center (ERDC).

Desirable attributes for deployable treatment systems. The Strategic Environmental Research and Development Program (SERDP) is funded by the Department of Defense, the Department of Energy, and the U.S. Environmental Protection Agency (EPA) to support promising research projects on sustainable environmental practices by the various entities of the Department of Defense. In December 2010, SERDP announced its statements of needs for the fiscal year 2012 program, and one of these (ERSON-12-01) requested the development of innovative approaches to sustainably treat wastewater for FOBs. Five desirable attributes were given. The first was the ability to treat a range of wastewater types that may be generated in an FOB, including graywater (from showers, laundry, and so on) and blackwater (from toilet activities). A second attribute was that the system be scalable in size for application to a wide range of camp sizes and needs. As discussed earlier, camps can vary in size from thousands to less than 100. The third attribute was that the technology be capable of operating in an energy-neutral configuration. Transportation of fuel is costly and, more important, dangerous, as it exposes drivers to conventional and IED attacks. The fourth attribute was to develop a system that was capable of water reuse. Even partial reuse, such as to allow for skin contact, toilet flushing, or vehicle washing, could be beneficial. Another form of reuse could be to allow the water to be safely used by indigenous populations for irrigation. Finally, the fifth attribute indicates that it is desirable for the system to have low operator involvement. These attributes are focused on basecamps, but they are applicable for disaster response as well.

Deployable aerobic aqueous bioreactor. The deployable aerobic aqueous bioreactor (DAAB) is a portable semiautomated wastewater treatment facility for rapid deployment in military, disaster relief, and humanitarian mission theaters. Personnel requirements for DAAB operation are minimal. The DAAB is shipped and operates in two 20-ft (6-m) ISO (International Organization for Standardization) containers, and the system can be operational and achieve EPA effluent standards within 48 h of arrival.

The basic DAAB consists of two units: a biological treatment unit (BTU) capable of treating 20,000 gallons of municipal wastewater (approximately 500 persons) per day, and a control and power unit (CPU) capable of providing control and power to 4 BTUs at a single location (**Fig. 1**). Each BTU contains a primary sedimentation tank, two attached-growth aeration tanks, and blowers. The CPU contains a diesel generator, a filtration unit, an ultraviolet disinfection system, cultivation units, and a ruggedized touchscreen computer controller.

The DAAB is designed to have minimal power needs. However, as it is currently configured, supplementary power is needed. The average energy demand of one BTU is 15 kW, and the 30-kW onboard generator requires 200 gal of diesel or jet fuel per week to operate at maximum load. The majority of the power (>98%) is consumed by the blowers for aeration, the grinder pump for bringing wastewater into the system, and the effluent pump for filtration. The system also typically requires 5 gallons of SAE 30 motor oil for 1 to 3 weeks of operation. The air filter should be replaced every 1 to 3 weeks as well. Alternatively, the DAAB unit can operate on external, 220-V, three-phase power. It has a 65-A maximum draw.

Previous efforts to deploy portable wastewater treatment systems in Bosnia have been reported to have suffered from weather effects. The small reactor volumes were subject to changes in temperature, and even freezing. The DAAB uses a closed reactor system, which should minimize these effects. This should also protect the unit from dust and assist in minimizing odors from the unit.

These units are designed to be shipped and operated in ISO containers, which are used by the U.S. Armed Forces. The units can be shipped via air. **Figure 2** shows the shipping containers holding a DAAB unit for transport to Afghanistan.

Rapid system start-up. A critical factor for the successful use of a deployable system is the need

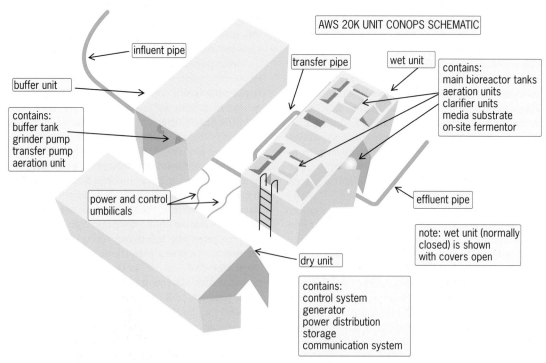

Fig. 1. Typical setup for the DAAB unit.

Fig. 2. DAAB packaged for shipping to Afghanistan.

for rapid system start-up. This has been achieved through the development of a novel method of preserving bacterial cultures in a metabolically active state that was developed at Sam Houston State University. The DAAB utilizes a combination of six bacterial strains preserved in pellets to cultivate a biofilm on a fixed media in the aerobic treatment tanks. These bacterial strains are cultivated during the first 24 h in wastewater from the site while the BTU is filling. The bacterial cultures are then injected into the BTU, which is operated in a recirculation mode for an additional 24 h to cultivate the media. This allows the unit to become fully operational in 48 hours.

Novel disinfection system and reverse osmosis. In some cases, it may be desirable to provide disinfected effluent. This is particularly desirable for most types of water reuse, particularly those that may allow skin contact. Typically, disinfection is accomplished by the use of chlorination or ozone treatment. Chlorination requires the transport of cylinders of chlorine gas or salts, which is costly and creates hazards for the soldiers or contractors who must transport the materials. Ozone can be generated on-site, but doing so requires substantial amounts of fuel. The current configuration uses UV disinfection. A novel solution has been the development of an electrochemical cell to achieve disinfection. Recent research has shown that mild electric fields can generate zones of oxidation and reduction that can be harnessed for chemical reactions. Studies to support the DAAB reactor development have resulted in an effective electrochemical cell for disinfection of reactor effluent. The cell uses substantially less energy than ozone generation systems. An optional reverse osmosis unit has also been developed. Coupling reverse osmosis with disinfection allows the unit to completely meet EPA standards for potable water.

Commercialization. Active Water Sciences is a company that was formed to produce commercial units based on the DAAB design. Six units have been produced by AWS and sold to the U.S. Army through EnviroTech for use in Afghanistan. While the results of these deployments will be incorporated into the fact sheet when available, these commercial units do not have disinfection systems, since they were not required in the contract specifications.

Energy-neutral systems. Developing an energy-neutral system is the next logical step in reactor design. Transport of fuel to basecamps can be costly and dangerous. A system that did not require supplementary fuel, or at least minimized the need for it, would be better received.

As it is currently configured, the DAAB is designed to be energy efficient. However, as discussed previously, fuel is required for its operation. One possibility for reducing fuel use is to develop a microbial fuel cell that could replace one of the aeration tanks. This would provide preliminary treatment of the wastewater while generating energy. Other possibilities could be reactions that generate hydrogen, which can be used as a fuel, or even the use of bacteria that generate fatty acids that could be distilled as a biodiesel. The problem with each of these approaches is that human waste tends to have low energy content per unit of volume. The estimated energy output possible from human waste sources for a 20,000-gal treatment system is on the order of 10.5 kW maximum, less than the 15-kW demand of the system. Other waste streams could increase energy production, particularly wastes produced from food service activities.

Outlook. Deployable wastewater systems could be valuable for reducing the environmental impact of military operations and improving responses to natural disasters. The DAAB system is a viable deployable system that is available for use. Other systems could perform similarly. Further development of effective systems is warranted.

For background information *see* DISINFECTION; FILTRATION; SEWAGE TREATMENT; WASTEWATER REUSE; WATER TREATMENT in the McGraw-Hill Encyclopedia of Science & Technology.

Scott A. Waisner; Victor F. Medina

Bibliography. P. Gupta, Bioethanol production from biomass of *Saccharum spontaneum*, *Journal of Biofuels*, 1(1), 2010; V. F. Medina and S. A. Waisner, Military solid and hazardous wastes: Assessment of issues at military facilities and base camps, in T. Letcher and D. Vallero, *Waste: A Handbook for Management*, Elsevier, Amsterdam, 2011; K. P. Nevin et al., Power output and columbic efficiencies from biofilms of *Geobacter sulfurreducens* comparable to mixed community microbial fuel cells, *Environ. Microbiol.*, 10(10):2505–2514, 2008; U.S. Army Corps of Engineers (USACE), *AED Design Requirements: Package Wastewater Treatment Plants and Lagoons*, Afghanistan Engineering District, 2009.

Wave energy converters

Since Stephen Salter's seminal paper in 1974, it has been recognized that ocean waves provide a very rich renewable resource from which to generate electricity. The energy resource, which is normally measured in terms of the available wave power per meter of crest width, exceeds 77 kW/m in the North Atlantic. The zones of highest wave power are found between the latitudes of 30° and 60° in both

Fig. 1. LIMPET Oscillating Water Column, a 500-kW plant built and grid-connected by Voith Hydro Wavegen in 2000, located at Claddach Farm on the Rhinns of Islay, Scotland.

Fig. 2. Test section of a 600-kW machine installed by Wave Star Energy at Hantsholm, Denmark, in 2009 and grid-connected since 2010.

hemispheres. Waves are generated by winds blowing across the surface of the ocean, providing for the natural storage (and concentration) of wind energy in the water near the surface. Once they have been created, surface waves travel thousands of kilometers with little energy loss. As they approach the coastline, however, energy intensity decreases as a result of interaction with the seabed, although reflection and refraction of waves often leads to energy "hot spots."

Wave energy converters seek to convert the kinetic energy of the waves into electricity. The wave energy sector is at an exciting stage, with precommercial deployments of devices now occurring in numerous regions around the world. However, there has been no trend toward a dominant technology, particularly for wave energy devices, and, with new devices being proposed continually, evaluation and comparison of the technologies and their potential for large-scale deployment is difficult. Since 2000, around 10 different converters have been demonstrated at either large or full scale, and two small farms of devices have been deployed. Currently, over 4000 different energy-conversion techniques have been patented worldwide.

Figures 1 through 4 show examples of the diversity of devices currently deployed in the oceans. **Figure 1** shows the front face of LIMPET, an oscillating water column that uses wave motion inside the chamber to drive reciprocating air flow through a

Wells turbine. The air spring and turbine are located behind the concrete wall, which has an opening just below the water to allow wave energy to excite the air spring. Wave Star (**Fig. 2**) uses the heaving motion of two large floats to move arms attached to the fixed platform; the arms drive hydraulic motors, which in turn power the generator. Pelamis (**Fig. 3**) also uses a hydraulic system to generate electricity, but in this case the relative motion of pairs of the five floats is used to drive hydraulic rams located at the hinges. Finally, **Fig. 4** shows OYSTER, an oscillating flap fixed to the seabed, which is used to pump water to a pelton wheel located in an on-shore generator.

EquiMar project and classification system. To assist in comparing these disparate technologies, the European Commission funded the EquiMar project to develop a suite of protocols for the "Equitable Assessment of Marine Energy Converters." This 3-year project, involving 22 organizations from 11 countries, published its final suite of protocols in April 2011. The protocols cover the environmental, engineering, and economic assessment of wave (and tidal) energy converters.

In addition to the protocols, a large number of technical reports, including a device classification template and a sea trials manual, are available from the project website. Previous attempts to classify wave energy converters have relied on visual descriptions of the basic device form (as, for example,

Fig. 3. The 750-kW, E-ON, P2 wave energy converter at the European Marine Energy Centre, at Billia Croo, Orkney, Scotland, installed and grid-connected by Pelamis Wave Power in October 2010. (*Photograph Courtesy of Pelamis Wave Power*)

Fig. 4. Oyster 1, a 315-kW hydroelectric wave energy converter under test at the European Marine Energy Centre at Billia Croo, Orkney, Scotland, installed and grid-connected by Aquamarine Power in the Autumn of 2009. (*Photograph Courtesy of Aquamarine Power*)

heaving buoy, oscillating water column, surge wave converter, overtopping device, attenuator, and so forth). EquiMar provides a more detailed, and flexible, classification system, which not only describes accurately how devices operate but also defines key aspects of device subsystems and device performance metrics. EquiMar recommends this "layered" structure to describe various elements of the device. The top layer includes information that will allow the user to verify the basic form of the device, providing information on the method of energy extraction and characterization of the physical form and motion paths of the hydrodynamic subsystem. Layer 2 offers information concerning the power take-off system, while layer 3 addresses how the device is kept in place in the marine environment and how key aspects of the device are controlled.

The classification characterizes the device in a progressive and compartmented manner in order to provide a complete and logically flowing description. Wave and energy converters are split into four discrete subsystems as shown in **Fig. 5**. The components shown within each subsystem are examples and are not indicative of any particular device. The classification template defines the device by the way it captures energy; through the shape/trajectory of any component motion paths and the physical principles involved. In this manner, devices that are similar in appearance can be differentiated. All four subsystems are defined using the classification, and the output parameter such as electrical power is also specified. The classification is for use by all marine energy stakeholders and is of most use to those wishing to compare devices in an equitable manner.

Energy capture. Wave energy converters have theoretical limitations on the amount of power they can absorb, similar to the Betz limits that apply to conventional wind turbines. A useful principle is that a good wave radiator is also a good wave absorber when placed in an incident wave field. If the wave radiated by a moving body is of the same amplitude and frequency as the incident wave, is traveling in the same direction, and is in antiphase with it, then the two waves interfere destructively, the energy being absorbed by the body and leaving still water behind it.

In two dimensions, a small heaving or surging body will radiate waves of equal size that propagate fore and aft of the body, symmetrically for a heaving body, antisymmetrically for a surging body. Maximum absorption occurs when the body moves to create waves propagating upstream and downstream that are half the amplitude of the incoming wave. Under these conditions, an absorbing body that is free to move in a single degree of freedom has a maximum efficiency of 50%.

A body able to move in two degrees of freedom (for example, heaving and surging, or heaving and pitching) can combine the symmetric and antisymmetric modes to radiate a wave in one direction only, the same as that of the incident wave. It can then cancel it with no reflection, thus achieving 100% absorption.

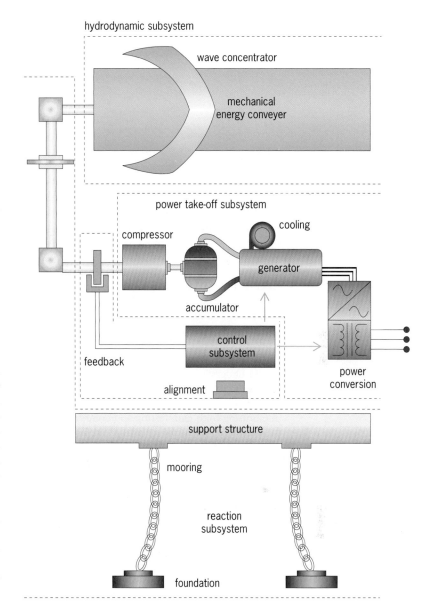

Fig. 5. The four subsystems comprising a generic wave energy converter.

In three dimensions, the theoretical limits are associated with the shape of the converter's far-field radiation pattern and how it interacts with the incident wave, with a greater focused radiated wave field giving a larger theoretical limit. Consequently, in three dimensions, the heaving and surging bodies no longer have the same efficiency.

The theoretical limits are attained when power is extracted under the optimal control of the power take-off system. The optimal control is that which both (1) induces resonance in the motion of the wave energy converter by providing a reactive power that matches and cancels out its external reactance (that is, stiffness and inertia terms), and (2) matches the external wave radiation damping of the converter. This is termed complex conjugate control, and its implementation requires foreknowledge of the incoming wave.

Project development. It is important to consider the effect of developing farms of devices for large-scale

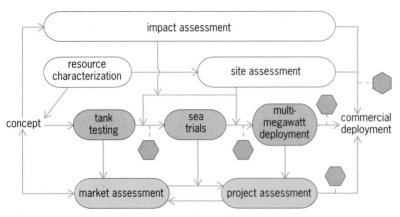

Fig. 6. Development stages of a marine energy project from initial concept to commercial deployment, showing the environment considerations (white), engineering (gray), and economic assessment (color).

energy exploitation. Because of physical constraints, it is unlikely that a single wave energy converter will have a rated capacity much larger than 1 MW. EquiMar recommends that the decision process outlined in **Fig. 6** be followed by a technology developer, starting with an initial concept based in part on an understanding of the available resource and the likely future market. This initial concept should then be tested at small scale in a controlled environment using both tank testing and computer simulations to understand the behavior and limitations of the primary interface of the energy converter. Such small-scale tests allow mooring geometries and likely wake and radiated wave effects to be explored, and detailed survival and operating envelope tests to be performed. All the successful wave and tidal energy concepts (to date) have taken this initial, cautious approach, which takes advantage of the relatively low costs of learning lessons at small scale in the laboratory before moving up to a larger laboratory scale. At the end of this stage, a detailed analysis of the laboratory test results can be combined with information about the conditions at test locations to decide whether sea trials should be commissioned.

The program of sea trials represents the first occasion under which the device is subjected to an uncontrolled environment. Although best practice would be to deploy the device into the ocean at circa 1:4 scale, there are only three (out of the 16 current and proposed) European test sites (Nissum Bredding in Denmark, Galway Bay in Ireland, and EMEC in the United Kingdom) that operate at this scale. Sea trials allow the technology developer to learn how the device performs in the open ocean and critically to understand the processes, vessels, and equipment needed for installation, operation, maintenance, and recovery of the device for the first time. The cost of performing such trials is at least an order of magnitude more than that associated with even large-scale laboratory tests. The view taken by EquiMar was that such sea trials should extend to small, precommercial, array deployments and may require several generations of full-scale devices to be deployed at several test locations until sufficient information is collected about the operability and reliability of a particular technology to allow for commercial deployment.

Site assessment is the key starting point for the commercial deployment of a multimegawatt farm. The selection of a site must be based not only on the available local resource but also on other marine spatial planning considerations, such as the location of shipping lanes, fishing grounds, and underwater pipes and cables, as well as on the strength of the local electricity grid. Following the selection of a site, there will be a significant amount of work in obtaining the required consents, and this should be the first time the devices are to be deployed outside a recognized sea-trial area. After the selection and characterization of the site, the specific operating envelopes of the machines to be deployed will be designed and full-scale deployment can begin.

The go/no-go decision on a commercial deployment will be extremely complex, and information about the installation, operating, and removal costs of the technology will be critical. Improved technology pricing models drawing on the engineering information available from model, scale, and prototype tests are critical to this stage, as they will allow legislators and possible investors to make informed decisions about revenue streams.

Resource assessment. In determining the deployment location and test conditions for a wave energy converter, it is critical to have a good understanding of the local wave climate and associated bathymetric and meteorological conditions. The wave climate is normally parameterized from observations made by wave measurement instruments (buoys, satellite data, and pressure and acoustic instruments can be used) deployed at a specific site. The proposed International Electrotechnical Commission (IEC) technical specification (62600-100) recommends that the wave climate be measured for at least a year to account for any seasonal variations. From these measurements the joint distribution of wave height and wave period are derived from spectral analysis of the data. By analyzing wave spectra over a long period at a particular location, the discrete joint distribution of wave power (as a function of wave height and wave period) can be estimated using the method of bins. The average power performance of a particular device at the site can then be estimated by multiplying the frequency in each bin by power generated by the machine under the wave height and period associated with that bin. The required output power is normally summarized in a power matrix (**Fig. 7**).

A number of additional parameters are also of interest, including spectral shape, directionality of waves, directional frequency spectrum, tidal changes of the mean water depth, tidal current velocities and direction, marine current direction and velocity, and wind speed and direction. If the site is to be used for deployment of either full-scale devices or large-scale demonstration machines, then a detailed bathymetric survey must be undertaken.

The importance of detailed knowledge of conditions at a deployment site cannot be underestimated. This information can be used to help plan deployment, recovery, and maintenance operations, and will allow a detailed risk assessment to be completed.

	power period (T_{pow}), s																
	5.0	5.5	6.0	6.5	7.0	7.5	8.0	8.5	9.0	9.5	10.0	10.5	11.0	11.5	12.0	12.5	13.0
0.5	idle	idle	idle	idle	idle	idle	idle	idle	idle	idle	idle	idle	idle	idle	idle	idle	idle
1.0	idle	22	29	34	37	38	38	37	35	32	29	26	23	21	idle	idle	idle
1.5	32	50	65	76	83	86	86	83	78	72	65	59	53	47	42	37	33
2.0	57	88	115	136	148	158	152	147	138	127	116	104	93	83	74	66	59
2.5	89	138	180	212	231	238	238	230	216	199	181	163	146	130	116	103	92
3.0	129	198	260	305	332	340	332	315	292	266	240	219	210	188	167	149	132
3.5	–	270	354	415	438	440	424	404	377	362	326	292	260	230	215	202	180
4.0	–	–	462	502	540	546	530	499	475	429	384	366	339	301	267	237	213
4.5	–	–	544	635	642	648	628	590	562	528	473	432	382	356	338	300	266
5.0	–	–	–	739	726	731	707	687	670	607	557	521	472	417	369	348	328
5.5	–	–	–	750	750	750	750	750	737	667	658	586	530	496	446	395	355
6.0	–	–	–	–	750	750	750	750	750	750	711	633	619	558	512	470	415
6.5	–	–	–	–	750	750	750	750	750	750	750	743	658	621	579	512	481
7.0	–	–	–	–	–	750	750	750	750	750	750	750	750	676	613	584	525
7.5	–	–	–	–	–	–	750	750	750	750	750	750	750	750	686	622	593
8.0	–	–	–	–	–	–	–	750	750	750	750	750	750	750	750	690	625

significant wave height (H_{sig}), m

Fig. 7. Power matrix for the Pelamis (P-750) wave energy converter. (*Reproduced from the P750 brochure, http://www.pelamiswave.com/media/pelamisbrochure.pdf*)

The combined meteorological and oceanic climatic conditions are commonly referred to as met-ocean conditions. The EquiMar Sea-Trial manual provides detailed information about the assessment and presentation of met-ocean data.

Hydrodynamic subsystems. To extract energy from a wave, an appropriate damping force must be applied by a component of the wave energy converter in contact with, and providing resistance to, the wave. This is called the primary conversion component of the converter, which must in turn be referenced, via a power take-off mechanism, to the seabed, the shoreline, or another component that will not simply move along with the primary conversion component. It is possible for two primary components to react against each other so long as the wave forces acting on them are out of phase. Floats and flaps are the most common types of primary conversion components. In the case of oscillating water column (OWC) devices, the primary conversion component is the pressurized air chamber that is referenced to atmospheric air pressure via a turbine for power take off.

The hydrodynamic subsystem is classified according to the exciting force and the axis of motion of the primary component. Exciting forces are described in terms of buoyancy, inertia, seabed, shoreline, lift, and potential. **Figure 8** shows six of the most common excitation modes.

The axis of motion is described in terms of heave (vertical motion), surge (horizontal motion in the wave travel direction), pitch (rotation about the horizontal axis orthogonal to the direction of wave travel), and fixed (devices that are fixed in position). The final point is to record the number of moving members comprising the hydrodynamic subsystem.

In this way we can categorize LIMPET (Fig. 1) as a buoyancy-shoreline-heave/surge, Wave Star (Fig. 2) as buoyancy-seabed-heave, Pelamis (Fig. 3) as buoyancy-boyancy-heave/sway, and Oyster (Fig. 4) as inertia-seabed-surge.

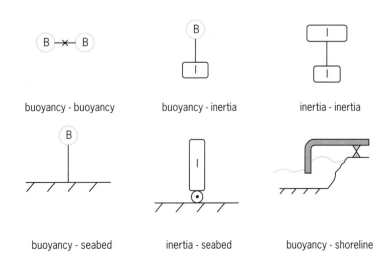

Fig. 8. Classification of hydrodynamic subsystems in terms of the exciting force on the primary component.

Economics and policy. Wave energy is an emerging technology, and for it to become competitive with more established renewable energy technologies (for example, off-shore wind), both economic and policy support are needed. The International Energy Agency's implementing agreement on ocean energy (Ocean Energy Systems Implementing Agreement, OES-IA) reports regularly on national and multinational support mechanisms.

The U.S. Department of Energy has recently applied the system of Technology Readiness Levels (TRLs) to ocean energy, and provides financial support for the development of new technologies and demonstration projects. In Europe the Commission has recently announced that the 7th Framework program will include funding for the demonstration of a small wave farm. Other initiatives include the announcement by the Scottish government of the Saltire Prize (a 10 million prize that will be awarded to the team that can demonstrate, in Scottish waters, a commercially viable wave or tidal stream energy technology that achieves the greatest volume of electrical output over the set minimum hurdle of 100 GWh over a continuous 2-year period using only the power of the sea).

Macroeconomic analysis by the Fraser of Allander Institute has shown that wave energy projects will provide significant economic benefit to local communities of all energy projects, with each gigawatt installed creating around 7000 direct jobs. The European Ocean Energy association estimates that each gigawatt of installed wave energy converter offsets 725 kilotons per year of carbon dioxide.

If the significant deployment targets set for 2020 and 2050 by national governments around the world are to be met, there is an exciting future ahead for wave energy.

This research was supported by the European Commission's 7th Framework program under Grant Agreement FP7-213380, and by the United Kingdom Engineering and Physical Sciences Research Council as part of the Supergen Marine program (Grant No EP/E040136/1).

For background information *see* ENERGY SOURCES; OCEAN WAVES in the McGraw-Hill Encyclopedia of Science & Technology.   David M. Ingram

Bibliography. J. Cruz, *Ocean Wave Energy*, Springer, Berlin, 2008; B. Holmes et al., *Equimar: Sea Trial Manual*, University of Edinburgh School of Engineering, 2011; D. M. Ingram et al., *Protocols for the Equitable Assessment of Marine Energy Converters*, University of Edinburgh School of Engineering, 2011.

Web 2.0 technologies

The term Web 2.0 is used to describe the second generation of the World Wide Web (WWW), which is more dynamic and interactive than its predecessor, Web 1.0. Although the term Web 2.0 follows the trend to number the new releases or versions of software products, in reality the term is nothing more than a name for a set of Internet-based tools and how to use these tools to encourage users to participate and share opinions and resources, thus creating a colossal content online for other users. In contrast with Web 1.0 where Internet users were "passive" in the sense that most of what the users did was to read static Web pages or pages that were rarely updated, Web 2.0 is a new paradigm where a new set of Internet-based tools and their use emphasize peers' social interaction and collective intelligence. This way, Web 2.0 not only helps users contribute to a Web site's content but also follow up with the Web site's latest content without even opening the actual Web page. Based on these features, Web 2.0 is commonly known as the wisdom Web, people-centric Web, participate Web, and read/write Web. Most Web 2.0 sites include a search engine that helps users locate the content other users have created. This technology has a more democratic approach than its predecessor, allowing all the citizens of the Web (webizens) to share and distribute their ideas and information. Although there are concerns about intellectual property, proprietary information, privacy, security, and control, Web 2.0 technologies can also help improve the collaboration and communication within the corporate environment and also across the multiple vertical industries.

Advantages of Web 2.0 technologies. As mentioned earlier, Web 2.0 encourages Internet users to actively participate as contributors and to help customize the media and technology as they see fit either for their own purposes or for the benefit of the Web community in which they participate. In this sense, there are a number of advantages regarding Web 2.0, and the following are among the most important.

Knowledge management. This technology facilitates flexible Web design, a responsive user interface, and collaborative content creation and modification by users. Knowledge management enhances the creation of new applications by reusing content and by establishing social and support networks. In other words, knowledge management can help in sharing, retrieving, organizing, and leveraging knowledge. The content on the Web is not just limited to viewing by the user, it also enables the user to actively contribute.

Rapid application development. The three principal approaches used to create the Web 2.0 applications are asynchronous JavaScript and XML (AJAX™), Adobe Flex™, and the Google Web Toolkit™. AJAX enhances the user interface and makes the applications highly interactive and more responsive. Responsiveness is achieved by exchanging a small amount of data with the server so that the entire Web page does not have to be reloaded when a change is made by a user request. Another application development solution is Adobe Flex, which creates and delivers cross-platform rich Internet applications (RIAs) on the Web. Adobe Flex creates applications that are engaging, interactive, and radically enhance the user experience by increasing user interactivity. The Google Web Toolkit can help the developers create applications using the Java development tools.

A compiler helps translate the Java application to a browser-compliant script and the toolkit widgets helps construct the user interface elements. This approach overcomes the limitations of writing AJAX applications, as it uses a union of technologies that, at the same time, provides users with a dynamic and standard-compliant experience.

Customer relationship management. By data mining of the customer information, more customers can be reached. The technology also adds the ability to communicate with customers more effectively and receive from them insights and concerns.

Collaboration and communication. Discussions can be co-ordinated, projects and tasks can be synchronized, and communication streams can be audited when using this technology. Web 2.0 technologies can improve the collaboration and communications across multiple industries, but the technology is not well accepted yet.

Web 2.0 technologies and services. The majority of Web 2.0 technologies are in the area of collaboration and communication, including several new technologies such as blogs, really simple syndication (RSS), wikis, mashups, podcasts, tags, crowd-sourcing, and social networks. The main yield of these technologies is knowledge management as they create content that improves over time. The **table** shows the mapping between some of the Web 1.0 and Web 2.0 technologies. It depicts how the technologies have become more dynamic in Web 2.0, as compared to Web 1.0.

Blogs. A blog is a two-way Web-based communication tool that allows users to enter their thoughts, ideas, and comments freely through a Web site. A blog entry created by the user can contain text, images, or links to other blogs, that is, whatever information the user would like to share publicly (open to anyone) or privately (open to restricted users). This gives rise to a blogosphere, which contains all the blogs in a community or social network. Any Internet user can post a blog online and the readers can place a comment as a blog post or create discussions about a topic. This technology demonstrates the ease of Web publishing using a collection of tools that makes the blogging process easier. The Web content published online in a blog can be aggregated for a particular use.

Wikis. A wiki is a Web-based and collaborative authoring system to create and edit content on-line through a Web browser. Wikis use server-side processing to convert content into HTML instantaneously. Thus, it helps users discuss evolving issues with higher communication efficiency and productivity. The other advantages of a wiki are that it creates knowledge repositories that are centralized and shared, brings thoughts of diverse individual together, and the content posted online by the users can evolve over time. Wikis have adopted the philosophy and use of open-source software, along with its advantage of collaborative software development.

Really simple syndication (RSS). As opposed to a wiki, decontextualization is central to RSS. This is a family of Web-feed formats that are used to organize content from blogs or Web pages. It is in the form of a condensed XML file that contains information items and links to the information sources. This technology (using the RSS reader or aggregator) helps inform users about any updates to the blogs or Web sites of interest.

Mashups. A Web mashup is a Web page or Web site that combines information and services from multiple sources on the Web. These can be grouped into seven categories: mapping, search, mobile, messaging, sports, shopping, and movies. The most commonly used mashups are the mapping mashups. Mashup technology is gradually replacing rapid application development (RAD) technology.

Tags, folksonomy, and tag clouds. Tags can be viewed as keywords added to the content in a blog or wiki using a social page tag tool. It makes the Web page searching process easier. Folksonomy are the user-created taxonomies of information. Users categorize the content they find or post online using tags or labels. The main advantage of this technology is that it is open ended and does not have a controlled vocabulary. A tag cloud lists the content tags on a Web page using a visual representation. The visual illustration differentiates between each tag's popularity levels. Tags are very useful in social bookmarking, by which users bookmark their favorite pages and assign tags to the content. These tags can be shared with other online users.

Podcasts. Podcasts are time and location independent digital files. Users can subscribe to the podcast feeds (audio/video RSS feeds), download them automatically, and transfer them to a portable device for later use. The podcasts can make the information easily available when traveling. Interviews and lectures can be recorded and viewed later. This technology is being widely used in the education community for student lectures.

Crowdsourcing. Web 2.0 technologies rely more on the "crowd," that is, the people to contribute to the online content. The collective and massive content created by a group of people has far more value, in general, than that created by a single individual. One example of this type of content is Wikipedia®.

Social networks. Social networks can consist of individuals or groups. Some of the popular social networks are Facebook, Orkut, LinkedIn®, and Flickr®. Each individual or group has a profile linked to another individual or group. There are various

Mapping of Web 1.0 and Web 2.0 technologies

Web 1.0	Web 2.0
Britannica Online	Wikipedia
Personal Web sites	Blogging
Domain name speculation	Search engine optimization
Page views	Cost per click
Publishing	Participation
Content management systems	Wikis
Directories	Tagging
Stickiness	Syndication

collaboration tools included in these networks that help in sharing audio, video, and documents. Other applications available in some of these social networks are playing games and chatting. Some of them, such as LinkedIn and Twitter are used just for sharing information with contacts.

Social applications. Web 2.0 is commonly associated with Web applications that facilitate interactive information sharing and collaboration. Most of the traditional applications focus on the delivery of the content but Web 2.0 applications focus on social connectivity. Applications such as collaborative wikis, blogging, social bookmarking, and online social networks have become increasingly popular in users' personal and professional lives. Examples of such portals that create new applications that were previous unthinkable are MySpace™, Flickr, and YouTube. These communities and portals are growing.

Business. Web 2.0 is changing how the world understands the status of information and knowledge. It is also altering the ways in which a user contributes to the information applications. The main advantage of this technology for corporate is its potential to benefit from the knowledge and information within an organization and recognize its socially reliant status. Web 2.0 technology can be adapted to the business environment and responsiveness of the changing business information needs. One of the advantages of Web 2.0 is in improving the customer relations, by making it is easier to reach customers and receive feedback. In addition, using technologies such a wikis and blogs, it becomes easier for the employees to communicate effectively and promptly.

Education. Technologies such as blogs and wikis help the learning environment by improving users' learning and writing ability and by increasing the interaction among learners and providers. The applications support social and active learning. E-learning 2.0 is an application that facilitates the creation and delivery of personalized, social, and lifelong learning through Web platforms that use Web 2.0 technologies. Two examples of wiki resources for e-learning 2.0 are Edublogs.org and wikispaces.com. Another e-learning 2.0 application that uses RSS and blogging, ChinesePod®, is for learning Mandarin Chinese. An example of open-source social networking software designed for education is known as Elgg®.

Web 2.0 and e-government. All the new technology developments, such mobile applications, multimedia applications, and geographical information systems, have societal implications. They have changed the way the citizens interact with the government as well as with each other. These developments also affect the services provided by the government. The government is providing more and more services online now. Thus, there is interweaving of technological and societal developments that bring e-government and Web 2.0 closer. The important aspects on Web 2.0 that focus on the services provided by the government are virtual networks, sharing of information, active users that influence the products, and the dynamic content of information.

For background information *see* DATA MINING; INTERNET; WORLD WIDE WEB in the McGraw-Hill Encyclopedia of Science & Technology.

Ramon A. Mata-Toledo; Pranshu Gupta

Bibliography. S. J. Andriole, Business impact of Web 2.0 technologies, *ACM Communications*, 53(12): 67–79, 2010; D. D. Kool and J. V Wamelen, Web 2.0: A new basis for e-government? *Information and Communication Technologies: From Theory to Applications: 3rd International Conference*, Damascus, Syria, pp.1–7, April 2008; K. J. Lin, Building Web 2.0, *Computer*, 40(5):101–102, 2007; R. McLean, B. H. Richards, and J. I Wardman, The effect of Web 2.0 on the future of medical practice and education: Darwinian evolution or folksonomic revolution? *Med J Aust.*, 187(3):174–177, 2007; S. Murugesan, Understanding Web 2.0, *IT Professional*, 9(4):34–41, 2007; T. O. Reilly and J. Battelle, Web squared: Web 2.0 five years on, Web 2.0 Summit, Oct 2009; H. Shaohua and W. Peilin, Web 2.0 and social learning in a digital economy, *Knowledge Acquisition and Modeling Workshop: IEEE International Symposium*, Wuhan, China, pp.1121–1124, December 2008; L. Tredinnick, Web 2.0 and business: A pointer to the intranets of the future? *Business Information Review*, 23:228–234, 2006.

Web to print

Web to print (also referred to as web2print or W2P) is a ubiquitous term most broadly defined as a class of software tools that are customized for print-centric e-commerce or e-procurement. Web to print facilitates the interaction between those who buy print and related products with those who sell such products, using the Internet as a medium for the exchange. Although the term Web to print is used generically, the hyphenated phrase (Web-to-print®) is a trademark of Belmark, Inc., a label printing company in De Pere, Wisconsin. Belmark was initially granted this trademark in 1999 and renewed it in January 2011. Despite this fact, Web to print has become the de facto term used to describe print-centric e-commerce, as evidenced by the millions of citations returned in the Web search results for the phrase "Web to print." Nearly all of the vendors in the space refer to their software solutions as Web to print.

Web to print developed as a concept during the late 1990s when software development companies began to build e-commerce platforms that catered to the unique needs of the print community. Among the product offerings and business models were auction sites that let print buyers anonymously submit jobs for which printers could bid, and e-commerce storefronts offering automated order entry, estimating, job tracking, and file delivery. Many of the print e-commerce solutions were offered through the Application Service Provider (ASP) model, with the vendor hosting a printer's branded e-commerce site and often claiming a percentage of any sale transaction through the system. Some offered licensed solutions that the print provider could self-host and manage independently. The number of software providers grew quickly, and by the year 2000, there were more than 40 print e-commerce solutions on the market.

As the number of print e-commerce products grew, the integration between client-facing Web portals and the production management information systems (MIS) used by most print manufacturers became a point of concern. In 2000, a consortium of MIS and print e-commerce companies established PrintTalk®, a cooperative community to address the issue. Today, PrintTalk is a specification maintained by the International Cooperation for the Integration of Processes in Prepress, Press, and Postpress Organization or CIP4, which is a nonprofit standards association. PrintTalk, as defined by CIP4, is "an XML standard that provides a single format for print providers to collaboratively communicate business transactions and specifications of print product both with their print buyers and among themselves."

Many of those first print e-commerce software companies folded during the dot-com bust of 2000–2001. While this delayed the widespread adoption of Web to print solutions in the print service provider community for a couple of years, it rebounded strongly by the mid-2000s. A 2008 study by the market research firm InfoTrends found that more than half of printing firms offered some sort of Web to print solution, and more than two-thirds of print buyers purchased print over the Internet. Today, a majority of graphic media companies offer some type of Web to print solution.

There are three overarching functions of a Web to print solution. These include automation, quality assurance, and integration with the end-user's internal business processes. A customer-facing e-commerce website helps streamline the sometimes complicated process of purchasing printed products by automating many tasks that traditionally fell to customer service and estimating departments in a commercial printing business. Some of these include job ticketing, electronic file upload and verification (preflight), paper selection, size and quantity determinations, database linking, and job costing. These Web to print solutions offer automation of these tasks and provide an audit trail to ensure that the projects are produced to the end-users' specifications. Most importantly, Web to print solutions provide a bridge between the customer's workflow and procurement process and the print service provider's production workflow process. This integration allows print service providers to grow beyond being suppliers of printed products to becoming valued marketing and supply-chain managers for their corporate clients.

The business models for Web to print systems tend to fall into four broad categories that cater to different types of customer solutions and products: print procurement, marketing and brand management, document management, and workflow automation.

Print procurement. This includes e-commerce-style storefronts, where users can buy printed or related products. These storefronts may allow for the ad-hoc upload of client files or customization of pre-loaded templates for all types of printed products for businesses (business cards, envelopes, brochures, and promotional items) and consumers (invitations, calendars, greeting cards, and personalized apparel). Most offer instant quotes and a shopping cart interface for payment, with credit-card and sales-tax support. Online "do-it-yourself" book publishers fall into this category, as do sites selling specific printed products, such as personalized labels, posters, buttons, and even household items like pillows, shower curtains, and sheets. Some print-service providers offer a Web to print interface as an added way for customers to place orders for products while still maintaining a traditional face-to-face sales force or physical store. Others, such as VistaPrint, are pure e-commerce companies, doing all of their business via their web storefronts.

Marketing and brand management. While these also employ e-commerce-style storefronts, the distinction in this category is that they are built and branded for a specific organization or purpose while maintained by the graphic media company. Brand management means that sites usually provide templates in which digital assets (such as layout file and images), color charts, and approved use of branded collateral are administered, helping corporate marketing departments with a diverse user base manage their brand from a central repository. Centralized billing and reporting on customer print-spending histories are major benefits of these types of solutions to corporate brand managers. Cross-media marketing campaign management is a large and growing component of web solutions that support marketing programs. Such systems allow users to manage all facets of a marketing campaign, including email, two-dimensional barcode links [Quick Response (QR) codes], personalized web micro-sites, social-media engagement, and mobile device communication from a web-based dashboard.

Document management. A self-service web storefront can be used for document management, which includes administering stored product inventory and shipping. There is a great deal of crossover between this category and branded storefronts. But as a general rule, document-management solutions offer true fulfillment capabilities, wherein an authorized customer can manage both its digital assets and physical stocked inventory based on preestablished rules and conditions. Document management systems integrate warehouse inventories with management information systems, automating processes such as placing orders for reprints, pick and pack, product delivery, client invoicing, and reporting. These systems help reduce waste caused by over-ordering product and reduce warehousing and inventory costs for the client, while creating a dedicated customer for the graphic media company.

Workflow automation. Integrating the client-facing web storefront with "back-end" print production has become one of the most critical considerations in building a Web to print solution, especially in the digital printing arena. Most digital press vendors offer Web to print software created to tie directly into the press workflow system. This means that projects can be delivered to the print service provider through the website with the client-created job ticket and estimate already in place, checked for accuracy, and placed directly into a print queue. Depending on the trust relationship between the print provider and the client, the project can then printed without any press operator intervention, for a truly "hands-off" and end-to-end workflow. Prepress-centric solutions, aimed at content-creators and offering preflight, file delivery, soft proofing, and approval, have been in use for more than a decade. These solutions are starting to be married to e-procurement storefronts to create a direct workflow chain from the customer to the printing press. One such solution is the Kodak INSITE® Storefront, a Web to print solution that integrates with the Kodak PRINERGY® prepress (print production) workflow solution.

Some graphic media companies have built their own solutions using hired or in-house information technology (IT) and Web-development talent, particularly those with requirements and security needs that are so stringent that no preexisting solution will meet their needs, such as those that support financial-service organizations. Many, however, use one of the dozens of Web to print software solutions on the market today. These products have evolved significantly from their late 1990s predecessors. With the trend toward cloud computing, a model that enables on-demand network access to a shared pool of computing resources (networks, servers, apps, and services), many Web to print solutions are offered through software as a service (SaaS). In this model, the Web to print software provider manages the entire infrastructure of the system for the graphic media company, including responsibility and management of IT components. Users typically access these applications over the Internet through a Web browser. Widely used software, such as Google's Gmail email system, is an example of an SaaS application, as is the print service provider Mimeo.com.

Web to print solutions are critical to the success of graphic media companies today and into the future because human communication and business engagement over the Internet is widespread and growing. Worldwide Internet access grew by 480% from 2000 to 2011, with more than 2 billion worldwide users; today more than 78% of the U.S. population uses the Internet. According to research by the Pew Internet and American Life project, 66% of Americans bought something over the Internet in 2010. By 2009, businesses-to-business (B-to-B) activity, of which print e-commerce is a part, accounted for 91% of e-commerce in the United States, with e-commerce accounting for 42% of total shipments in the manufacturing sector. Both consumers and businesses have grown accustomed to buying and selling services of all kinds via the Internet. Graphic media companies must therefore offer products using the e-procurement tools buyers have become accustomed to.

Web to print solutions offer significant benefits to both the graphic media provider and the print buyer or customer. For print service providers, Web to print solutions offer access to a far larger customer base than they could achieve otherwise, and potentially worldwide. Building and managing branded solutions for corporate customers allows graphic companies to be seen as strategic partners in their customers' marketing efforts, in that their value is greater than the value of the products the service provider sells. Sixty percent of print service providers believe that web-enabled services improve production efficiency, customer acquisition, and retention rates.

Like any e-commerce site, Web to print portals offer the convenience of a 24-hour-a-day, 7-day-a-week, self-service access to a graphic media company's wares. For the print buyer, online document management can result in reduced static inventory, lower overall printing costs, and better overall project tracking. For marketers, an online management dashboard provides real-time results for many components of a marketing campaign, including email views, barcode links, website views, and postal deliveries. Services such as this, in addition to print procurement, are the cornerstone of the future of Web to print solutions.

For background information *See* INTERNET; PRINTING; SOFTWARE; WORLD WIDE WEB in the McGraw-Hill Encyclopedia of Science & Technology.

Julie Shaffer

Bibliography. S. McKibben and J. Shaffer, *Web-to-Print Primer*, Printing Industries of America Press, 2007; PrintTalk Version 1.3 2007-10-30, copyright © 2000–2007, International Cooperation for the Integration of Processes in Prepress, Press and Postpress (CIP4) with registered office in Zurich, Switzerland; U.S. Census Bureau, *2009 E-commerce multi-sector e-stats report*, released May 26, 2011; U.S. Trademark and Patent Office, Serial Number: 75747533, Registration Number: 2491305 (http://tarr.uspto.gov/servlet/tarr?regser=serial&entry=75747533).

Whooping crane restoration

Whooping cranes (*Grus americana*) [**Figs. 1** and **2**] are the rarest, most endangered cranes in the world. They numbered only 22 birds in 1937, and their breeding grounds were not discovered until

Fig. 1. Whooping cranes (*Grus americana*) at Aransas National Wildlife Refuge, Texas. (*Photo courtesy of Steve Hillebrand, U.S. Fish and Wildlife Service*)

Fig. 2. Whooping crane from the breeding program at the International Crane Foundation in Baraboo, Wisconsin. (*Photo courtesy of Glenn H. Olsen*)

1954. The status of the whooping crane population was instrumental in the enactment of the Endangered Species Act of 1973 in the United States. From these low numbers in the 1930s and 1940s, conservation and captive breeding efforts have increased their numbers to more than 550 whooping cranes today. Half of these whooping cranes are in the historic wild flock that migrates between the Texas Gulf Coast (in the United States) and northern Alberta (in Canada).

History. Even in pre-European settlement times, whooping cranes were never considered abundant, numbering an estimated 10,000 birds. In 1722, Iroquois Native Americans presented a dead whooping crane to English artist–naturalist Mark Catesby, who wrote the first scientific description of the species. In 1770, whooping cranes were seen in pairs, near what would later be discovered as their last breeding grounds, when Samuel Hearn first explored the Great Slave Lake area (now part of Canada's Northwest Territory). In 1805, Lewis and Clark reported sighting migrating whooping cranes at the mouth of the Little Missouri River.

Whooping crane numbers declined from about 10,000 in pre-settlement days to about 1000 in the 1870s. Subsistence hunting by settlers and draining of wetland breeding sites both contributed to these declines. Whooping cranes once bred from northern Illinois through southern Wisconsin, Iowa, Minnesota, and the eastern Dakotas in the United States; however, these areas became subject to settlement and intensive crop farming. Nests were also found in the Canadian prairie provinces of Manitoba, Saskatchewan, and Alberta, but these regions also came under the plow. The last known whooping crane nest was found in western Saskatchewan in 1922. For over 3 decades, no one knew where the remnant whooping crane flock went to breed.

Whooping crane protection. Aransas National Wildlife Refuge, on the Texas Gulf Coast near Corpus Christi, was established in 1937 to protect the wintering grounds of the whooping crane. This was one of the first steps to protect the species, helping to start it on the long slow road to recovery. Although the cranes were protected by law, illegal hunting continued to be a problem, with as many as

40 birds killed in the following decade. Any increase as a result of natural reproduction was erased by these hunting losses.

After establishment of the Aransas National Wildlife Refuge, no specific programs for whooping crane restoration were begun because of the lengthening shadows of World War II. Once the war was over, more energy was focused on whooping crane restoration. In 1946, the Cooperative Whooping Crane Project was started with joint support from the National Audubon Society and the United States Fish and Wildlife Service (USFWS). Robert Porter Allen, a National Audubon Society wildlife biologist, studied the whooping cranes at Aransas for two winters. The National Audubon Society paid for and distributed pamphlets to hunters describing the whooping crane and its status as a protected bird in Canada and the United States under the Migratory Bird Treaty Act of 1918. This effectively ended most shooting incidents.

In the early 1950s, Robert Porter Allen and USFWS biologist Bob Smith crossed northern Canada by air during the late spring and summer looking for the nesting grounds of the few remaining whooping cranes that wintered at Aransas. They found several lone birds in the Great Slave Lake area. In June 1954, a forester, who was fighting a fire in nearby Wood Buffalo National Park, about 50 km (31 mi) south of the lake, reported sighting (from a helicopter) a pair of whooping cranes on a nest. The area was very inaccessible. Several attempts to enter the area by boat and foot in 1954 proved unsuccessful. Finally, in 1955, Allen was able to gain access to the area along the Sass River by using a helicopter. He found nests for most of the birds that wintered at Aransas National Wildlife Refuge. The nesting ground was finally located. As it turned out, it was already protected as part of a large national park in Canada.

Captive breeding and reintroduction efforts. Whooping crane pairs mate at 4–5 years of age, and the females lay their first eggs. Each female lays only one or two eggs, with usually only one chick or "colt" being raised. However, whooping cranes are very long-lived for birds. Wild banded whooping cranes have lived into their thirties, and captive whooping cranes have lived into their forties. As the Canadian Wildlife Service, under the leadership of Ernie Kuyt, developed a plan to monitor and visit nests each spring, plans were also under way to establish the first captive breeding program in the United States. A lone whooping crane, injured on migration, became the first captive bird. He received the name Canus, which is derived from the initial letters of Canada ("CAN") and the abbreviation for the United States ("US"), because all whooping cranes are considered jointly owned by the two nations. Canus was moved in 1966 to the newly established breeding center at the Patuxent Wildlife Research Center [part of the USFWS (now part of the U.S. Geological Survey since 1996)] in Maryland. This whooping crane was joined by eggs removed from the nests in Canada. The Canadian Wildlife Service visited each nest to record data and monitor progress. If two viable eggs were present, one would be removed and either placed in another nest where there were no viable eggs or delivered to the captive facility. As the whooping crane pairs would only usually raise one colt, this resulted in more birds in the wild flock and additional whooping cranes to establish a captive breeding flock.

During the 1970s, the captive breeding flock at Patuxent Wildlife Research Center was slowly increased. Starting in the 1980s, emphasis was placed on establishing a new wild flock of whooping cranes. The first experiment involved putting whooping crane eggs from Patuxent and from the wild Wood Buffalo/Aransas flock in the nests of sandhill cranes (*Grus canadensis*) at Grays Lake National Wildlife Refuge in Idaho. The plan was to have the sandhill cranes raise the colts and teach them to migrate south to New Mexico. This plan worked to some extent. The sandhill cranes did rear and teach the whooping cranes the migratory route, but high mortality resulting from drought and predators greatly reduced the number of colts that survived each year. In addition, those whooping cranes that lived to maturity would not mate with each other. Young colts were found to imprint on their parents or other nearby birds. These colts were imprinted on gray sandhill cranes and did not associate with other white whooping cranes. The project ended in 1989, although some of the whooping cranes in the wild persisted into the late 1990s.

During the early 1990s, emphasis was placed on bolstering captive breeding of whooping cranes, and two new captive breeding centers were established—one at the International Crane Foundation in Baraboo, Wisconsin, and the other at the Calgary Zoo, in Calgary, Alberta. An additional captive breeding center has recently been established at the Audubon Zoo's Species Survival Center in New Orleans, Louisiana. Starting in 1994, whooping cranes costume-reared (that is, reared by humans in crane costumes) at the Patuxent Wildlife Research Center and the International Crane Foundation were released in central Florida to create a nonmigratory flock. All whooping cranes for this flock, as well as all cranes in subsequent efforts, were reared by humans in crane costumes to avoid the young birds imprinting on people. In addition, for proper imprinting, adult-age whooping cranes from the captive flock were housed next to the chicks being reared. The effort to establish a nonmigratory captive flock in Florida ceased in 2005 when a population viability analysis showed a less than 1% probability for success, primarily the result of continuing habitat loss and severe droughts leading to poor reproductive performance.

In 2001, after nearly a decade of testing with nonendangered sandhill cranes, 7 whooping cranes were led by costumed personnel from partner organization Operation Migration flying ultralight aircraft (**Figs. 3** and **4**) on a 2000-km (1240-mi) migration from central Wisconsin to the Florida Gulf Coast. Whooping cranes, similar to swans and geese, learn their migration route by following their parents on their first southward migration. There were

Fig. 3. Whooping cranes in flight behind an ultralight aircraft. (*Photo courtesy of the U.S. Fish and Wildlife Service*)

no whooping cranes left migrating in eastern North America, so researchers developed the technique of humans in costumes rearing the birds and flying south with the young birds following ultralight aircraft. Continued efforts to increase this flock have resulted in 110 whooping cranes in the wild, including 3 newly hatched chicks so far in 2011. Additional whooping cranes, hatched and reared at the International Crane Foundation, have been released each year since 2005 using a technique called direct autumn release. These whooping cranes are costume-reared by people until after fledging (that is, until the birds have developed wing feathers that are large enough for flight); then, they are taken out and released near some of the experienced adult whooping cranes who were part of the ultralight aircraft-led releases of previous years. The young whooping cranes follow the older birds on their first migration south and learn the route. These techniques have been successful, and reproduction by some of the released whooping cranes has occurred in the wild; however, there are still some problems associated with the released birds nesting in the wild.

The third reintroduction was started March 14, 2011, with the release of 10 whooping crane colts at White Marsh Lake, Louisiana. This area in Louisiana was once home to a small flock of nonmigratory whooping cranes, with the last one brought into captivity at the Audubon Zoo in New Orleans in 1951. Overall, the captive breeding centers are rearing an

Fig. 4. Teaching whooping cranes to fly behind an ultralight aircraft at Necedah National Wildlife Refuge in central Wisconsin. (*Photo courtesy of Glenn H. Olsen*)

additional 18 whooping crane chicks for release in Wisconsin (10 with ultralight aircraft technique and 8 for direct autumn release) and 17 for release in Louisiana in 2011.

For background information *see* AVES; BEHAVIORAL ECOLOGY; ECOLOGY; ECOSYSTEM; ENDANGERED SPECIES; ETHOLOGY; GRUIFORMES; MIGRATORY BEHAVIOR; PHYSIOLOGICAL ECOLOGY (ANIMAL); POPULATION DISPERSAL; POPULATION ECOLOGY; POPULATION VIABILITY; RESTORATION ECOLOGY in the McGraw-Hill Encyclopedia of Science & Technology.   Glenn H. Olsen

Bibliography. S. J. Converse et al., Evaluating propagation method performance over time with Bayesian updating: An application to incubator testing, pp. 110–117, in B. K. Hartup (ed.), *Proceedings of the Eleventh North American Crane Workshop*, International Crane Foundation, Baraboo, WI, 2008; P. A. Johnsgard, *Sandhill and Whooping Cranes: Ancient Voices over America's Wetlands*, University of Nebraska Press, Lincoln, NE, 2011; G. H. Olsen, The progress of restoring whooping crane populations, *Proceedings of the 32nd Annual Association of Avian Veterinarians*, Association of Avian Veterinarians, Denver, CO, 2011; D. Sakrison, *Chasing the Ghost Birds: Saving Swans and Cranes from Extinction*, International Crane Foundation, Baraboo, WI, 2007.

Wood preservative testing

Most wood species used in commercial and residential construction have little natural biological durability and will suffer from biodeterioration when exposed to moisture. Historically, this problem has been overcome by treating wood for outdoor use with toxic wood preservatives. As societal acceptance of chemical use changes, there is continual pressure to develop and market new types of durable wood products. In the last few years, several new types of wood preservatives have become available, and other new formulations are expected to appear on the near horizon. There is also increasing interest in nontraditional wood products, including wood–plastic composites (WPCs), chemically modified wood, and thermally modified wood, which are believed to be as durable as traditional wood products, yet are nontoxic. The rapid evolution of durable wood products has further highlighted an old problem in wood protection—namely, how do we evaluate long-term durability with short-term tests? This challenge is complicated by the wide range of exposure environments, types of structures, and service-life expectations.

Over the last century, numerous laboratory and field test methods have been developed to evaluate durability. Many of these methods have gained broad acceptance in Europe, Australia, Asia, and the United States. In the United States, the American Wood Protection Association (AWPA) has more than 20 preservative evaluation standard methods; other organizations, such as ASTM International (formerly known as the American Society for Testing and Materials), have applicable methods as well. Appendix A of the AWPA Standards provides detailed guidelines on the types of tests that may be needed to evaluate new wood preservatives. In almost all of these tests, the performance of the test product is compared to that of untreated wood and a well-known and commercially accepted durable product.

This article briefly discusses some of the most important tests used in evaluating the durability of new wood products. The tests include accelerated laboratory decay tests, aboveground and ground-contact field exposures, and tests that evaluate properties, such as resistance to leaching and effects on fastener corrosion. The tests mentioned in this article should be considered as only a minimum; other tests are necessary as well.

Accelerated laboratory decay testing. Laboratory tests are more rapid than field tests and are often the first step in evaluating durable wood products. The most widely used laboratory test in the United States is the soil-block (also called soil-bottle) decay test. The procedure for this test is described in ASTM Standard D1413 or AWPA Standard E10. In brief, a cube of the wood product is placed in a bottle that contains moist soil and a feeder strip that has been preinoculated with a specific decay fungus. The intent of the method is to provide the fungus with ideal conditions for colonizing the test material and to evaluate the ability of the wood product to resist colonization. The extent of protection is measured as weight loss of the cube after exposure, with weight loss of the test product compared to weight loss of unprotected wood or wood treated with a reference preservative. Weight losses of less than 3–4% are generally indicative of good protection. AWPA guidelines require the soil-block test for durable wood products intended for use in either aboveground or ground-contact applications (Use Categories 2, 3, and 4). However, the relationship between the results of the soil-block test and in-service durability is poorly understood. One of the drawbacks of the method is that the fungi evaluated may not be relevant for in-service conditions. The standard fungi have been selected for their known resistance to some of the conventional preservatives and are not necessarily those found degrading wood products. The vigor of the fungi also varies greatly between laboratories, and the results appear to depend on factors such as the age of the fungal culture, soil properties, and moisture content. Thus, although the soil-block test does provide some insight into the ability of a wood product to resist colonization by certain fungi, it does not offer great promise for predicting the service life of a wood product used either in ground contact or above the ground.

Ground-contact stake tests. Stake tests continue to be the primary method of evaluating products intended for in-ground use. AWPA guidelines require field stake tests for durable wood products intended for use in ground contact (Use Category 4), and they also recommend these tests for wood products used above the ground (Use Categories 2 and 3). In these tests, stakes [usually with dimensions of $19 \times 19 \times 457$ mm ($0.75 \times 0.75 \times 18$ in.), but other sizes are possible] are buried vertically to one-half their length in soil and then periodically removed and inspected. During inspection, the stakes are assigned a rating of 10 (unattacked), 9.5, 8, 7, 6, 4, or 0 (failure), depending on the extent of deterioration. AWPA Standard E7 details the test procedure. The durability of the test product is compared to that of untreated wood and a reference durable wood product.

Field stake evaluations are some of the most informative tests because they challenge the treated wood with a wide range of natural organisms under severe conditions. However, there are several factors that can interact to affect the results of these tests. Perhaps the most important of these factors are site conditions and duration of the test. Usually at least two different sites are used to account for differences in soil properties and types of organisms present, and at least one of the sites is in a region with a warm, moist climate. The AWPA Standards recognize that climate affects the rate of deterioration; for example, the minimum exposure time is 3 years in high-decay hazard areas such as southern Mississippi, whereas longer exposure times are required for low-decay hazard test sites such as Wisconsin. A test product intended for use in contact with the ground should be in nearly perfect condition after 3 years of exposure and should suffer only very minor attack after 5 years of exposure. However, results derived from northern climates can be misleading, even with longer exposures. For example, stakes that perform well for more than 5 years in Wisconsin can be virtually destroyed in fewer than 3 years in Mississippi. Products that perform well in ground-contact stake tests are generally very durable in aboveground applications, but the relationship between mediocre stake test performance and aboveground durability is not well understood.

Aboveground exposure tests. Aboveground field exposures are useful for treatments that will be used to protect wood above the ground. Although not as severe as field stake tests, aboveground tests do provide useful information on aboveground durability. There are many versions of aboveground tests, as described in AWPA Standards E9, E16, E18, E25, and E27. AWPA guidelines require one or more of these tests for wood products intended for use in aboveground applications (Use Categories 2 and 3), but they also are usually conducted for products intended for use in ground-contact application (Use Category 4). Specimens are exposed to the weather in an area with a warm, wet climate (usually either the southeastern United States or Hawaii). The specimens are designed to trap moisture and create ideal conditions for aboveground decay. The specimens are periodically inspected and given a rating for extent of deterioration using a scale (10, 9.5, 9, 8, 7, 6, 4, and 0) similar to that used in stake tests. Interpretation of the results of aboveground testing is challenging. The greatest source of difficulty is the wide variation in severity of exposure for wood used above the ground. The severity of aboveground exposure does vary with climate, but it also varies greatly with construction practices and localized site conditions (for example, moisture, temperature, and ultraviolet exposure). In areas where organic debris can collect in connections, the aboveground decay hazard may be higher than anticipated. Comparison to untreated wood specimens is especially important in these tests in order to assess the extent of the deterioration hazard.

Other common tests. Some types of tests do not involve biodeterioration. Instead, they evaluate properties that may indirectly affect biodeterioration or that provide information on other important properties. The laboratory leaching test helps to evaluate how rapidly the treatment will be depleted. A treatment needs leach resistance to provide long-term protection. In this test, small cubes of wood are immersed in water for 2 weeks. The methodology for this test is described in AWPA Standard E11. Laboratory corrosion testing is used to determine the compatibility of the treatment with metal fasteners, and the current procedures are described in AWPA Standards E12 and E17. Corrosion test methods continue to evolve to allow better correlation with in-service performance. Treatability testing is used to evaluate the ability of a treatment to penetrate deeply into the wood. Shallow surface treatments rarely provide long-term protection because degrading organisms can still attack the interior of the wood. Currently, there are no standard methods for evaluating treatability, but a recommended approach is provided in AWPA Appendix E. Strength testing compares the mechanical properties of treated wood with matched, untreated specimens. Treatment chemicals or processes have the potential to damage the wood, making it weak or brittle. ASTM Method 5664 is preferred for these strength-effect evaluations. AWPA guidelines require many of these tests for all intended applications, although the leaching test is not required for products intended for indoor use (Use Categories 1 and 2).

Adapting test methods to nontraditional wood products. With the commercial availability of nonconventional durable wood products such as WPCs and the emergence of chemically and thermally modified wood, there is an urgent need for modification of the current standards or development of entirely new standards. The current tests have been developed for more conventional wood protection systems and may not be appropriate for testing these newer materials. For example, WPCs (which, on average, consist of 50–65% wood fiber and 35–50% plastic) have slower moisture sorption compared to solid wood, and modified woods usually have a lower equilibrium moisture content compared to unmodified wood. As a result, these newer materials do not attain sufficiently high moisture content during the laboratory decay test period to facilitate fungal attack. Therefore, to increase the moisture content, WPCs can be preconditioned by soaking them in water at elevated or room temperature prior to insertion into the soil-block test. Simulating outdoor conditions by laboratory-accelerated weathering and then water-soaking the WPCs is another effective means of preconditioning prior to fungal durability testing. Many studies on these newer materials tend to modify the standards by using nonstandard protocols in order to understand the mechanism of decay resistance.

Field testing of wood-based materials, such as WPCs, provides additional valuable information on durability. In addition to moisture, fungi, and termite degradation, failure also can result from other environmental elements, including ultraviolet radiation, thermal cycling, and freeze-thaw cycling. These factors need to be considered in the testing of these materials, especially WPCs.

Review of test data and listing of new durable wood products. Once the appropriate tests of a wood product have been completed, the results are compiled and presented to one of two organizations for reviewing and listing of durable wood products. Traditionally, durable wood products have been reviewed by AWPA subcommittees, which are composed of representatives from industry, academia, and government agencies who have familiarity with conducting and interpreting durability evaluations. More recently, the International Code Council–Evaluation Service (ICC-ES) has evolved as an additional route for gaining building code acceptance of new types of pressure-treated wood. The ICC-ES does not standardize preservatives. Instead, it issues evaluation reports that provide evidence that a building product complies with the building codes. The tests required by ICC-ES are typically those developed by AWPA. It is important to note that separate toxicity evaluations by appropriate regulatory agencies (for example, the U.S. Environmental Protection Agency) are mandatory for any durable wood product that incorporates preservative pesticides.

For background information *see* CORROSION; ENVIRONMENTAL TOXICOLOGY; FUNGI; LEACHING; MOISTURE-CONTENT MEASUREMENT; WOOD ANATOMY; WOOD COMPOSITES; WOOD DEGRADATION; WOOD ENGINEERING DESIGN; WOOD PROCESSING; WOOD PRODUCTS; WOOD PROPERTIES in the McGraw-Hill Encyclopedia of Science & Technology.

Rebecca E. Ibach; Stan T. Lebow

Bibliography. ASTM International, *Annual Book of ASTM Standards*, Section Four: Construction, vol. 04.10: Wood, ASTM International, West Conshohocken, PA, 2010; AWPA, *AWPA Book of Standards*, American Wood Protection Association, Birmingham, AL, 2011; C. Hill, *Wood Modification: Chemical, Thermal and Other Processes*, Wiley, Chichester, West Sussex, England, 2006; S. T. Lebow, Chapter 15: Wood preservation, in *Wood Handbook: Wood as an Engineering Material*, Gen. Tech. Rep. FPL-GTR-113, Forest Products Laboratory, U.S. Department of Agriculture Forest Service, Madison, WI, 2010; T. P. Schultz et al., *Development of Commercial Wood Preservatives: Efficacy, Environmental, and Health Issues*, ACS Symposium Series 982, American Chemical Society, Washington, D.C., 2008.

Nobel Prizes for 2011

The Nobel prizes for 2011 included the following awards for scientific disciplines.

Chemistry. The chemistry prize was awarded to Dan Shechtman of the Technion (Israel Institute of Technology) for the discovery of quasicrystals.

Quasicrystals were discovered by Shechtman on April 8, 1982 at the National Bureau of Standards, Washington, D.C., which is now known as the National Institute of Standards and Technology (NIST), when he was studying the electron diffraction patterns of rapidly cooled alloys of aluminum and manganese and saw a sharp diffraction pattern of 10 spots arranged in a circle, indicating that it was an ordered crystal.

Diffraction patterns are seen as spots and describe the structure of crystals, which were then known as periodic arrangements of atoms or molecules with repeating symmetry, such that only diffraction patterns of 2, 3, 4, and 6 dots (2-, 3-, 4-, and 6-fold rotational symmetry) were allowed. Five- and 10-fold rotational symmetry could not have a periodic arrangement of atoms in the crystal.

Over the next 2 years, Shechtman teamed with a number of colleagues to figure out and explain how, for example, 5-fold symmetry was possible in a solid with crystalline properties and not an error, as it was largely considered by his peers. First he worked with I. Blech, who proposed a physical model for how quasicrystals formed, but their paper was rejected. (It was subsequently published in 1985.) In 1984, the journal *Physical Review Letters* published "Metallic phase with long-range orientational order and no translation symmetry," by D. Shechtman, I. Blech, D. Gratias, and J. W. Cahn, which generated a lot of controversy, interest, and finally success in the scientific community.

Shechtman's discovery of quasicrystals has resulted in interesting crystal structures made by unique materials preparation methods, and has also broadened the definition of what is a crystal by the International Union of Crystallography to include "any solid having an essentially discrete diffraction diagram."

Many practical applications of quasicrystals have resulted because of their favorable properties, including thermal stability (for heat protection), extreme hardness (for wear and scratch resistance), and low surface energy (for nonstick applications). Quasicrystals have also been shown to have good thermoelectric properties for refrigeration and power generation (low thermal conductivity). Quasicrystalline alloys, such as aluminum-copper-iron (AlCuFe), tend to be very brittle. However, their useful properties can be exploited by coating them as thin films on other substrates, such as aluminum or steel, by sputtering and other physical deposition techniques.

Quasicrystals also have been found to occur in nature in the mineral icosahedrite, a metal-rich mineral, containing aluminum, copper, and iron, from the Koryak Mountains, Russia.

For background information *see* CRYSTAL; CRYSTAL STRUCTURE; CRYSTALLOGRAPHY; QUASICRYSTAL; X-RAY CRYSTALLOGRAPHY; X-RAY DIFFRACTION in the McGraw-Hill Encyclopedia of Science & Technology.

Physiology or medicine. The prize in physiology or medicine was divided, with one half awarded jointly to Bruce A. Beutler of the University of Texas Southwestern Medical Center at Dallas, Texas, United States, and Jules A. Hoffmann of the Molecular and Cellular Biology Institute in Strasbourg, France, for their discoveries concerning the activation of innate immunity. The other half was awarded to Ralph M. Steinman of Rockefeller University, New York, United States, for his discovery of the dendritic cell and its role in adaptive immunity.

The immune system attempts to protect the body from harmful infections by pathogenic microorganisms. This involves neutralizing viruses; killing bacteria, fungi, and parasites; and dealing with potentially toxic proteins produced by some bacteria during the course of an infection. This year's Nobel Laureates have revolutionized our understanding of the body's immune system by discovering key principles for its activation.

The immune system has two main lines of defense against pathogens. The first is through a nonadaptive or innate response. If organisms manage to enter tissues, they are often recognized by molecules present in serum and by receptors on cells. For example, some receptors on cells, called Toll receptors, are related to proteins found in plants or insects that trigger the production of antimicrobial or antifungal substances. The innate response is not specific and is not improved by repeated encounters with the pathogen. Innate immunity leads to inflammation and to the destruction of invading pathogens.

If pathogens break through the innate response, the body responds with its second line of defense, the adaptive immune response. Adaptive immunity is brought about in part by lymphocytes, including T cells and B cells. It is characterized by specificity and also by the body's ability to respond more rapidly and more intensely when encountering a pathogen for a second time, a feature known as immunological memory. This permits successful vaccination and prevents reinfection with pathogens that have been successfully repelled by adaptive immunity.

This year's Nobel Prize winners in Physiology or Medicine helped elucidate the mechanisms triggering activation of innate immunity and mediating the communication between innate and adaptive immunity. Hoffman and Beutler discovered the sensors of innate immunity. Steinman discovered a new cell type, the dendritic cell, that initiates and controls adaptive immunity.

While investigating the immune system of fruit flies in 1996, Hoffman and his co-workers at the University of Strasbourg found that flies with mutations in the Toll gene died when infected with bacteria or fungi. Hoffman concluded that the product of the Toll gene, the Toll receptor, was responsible for sensing pathogens. He further concluded that activation of this gene was necessary to initiate the innate immune response and fight infections.

Around the same time that Hoffman was investigating fruit flies, Beutler was using mouse models to study human inflammatory diseases, which are caused by an overactive innate immune response. Beutler was looking for a receptor that could bind lipopolysaccharide (LPS), a potentially life-threatening endotoxin produced by bacteria. In 1998, Beutler and his colleagues at the University of Texas Southwestern Medical Center in Dallas found that mice resistant to LPS had a mutation in a gene that was similar to the Toll gene of the fruit fly. The protein encoded by this mammalian gene served as the LPS-binding receptor on cells. In discovering this Toll-like receptor (TLR), Beutler showed that insects and mammals use similar molecules to activate innate immunity. To date, scientists have identified at least ten different TLRs in humans that recognize and bind to a broad spectrum of pathogens.

When pathogens evade detection by innate immunity, adaptive immunity takes over. Until Steinman's research, scientists did not know the trigger that initiates adaptive immunity. Steinman discovered the dendritic cell—a specialized cell type—in 1973. At that time, Steinman and his colleagues at the Rockefeller University were investigating the induction of immune responses in mice. Through a series of later experiments, Steinman found that dendritic cells sense signals arising from the innate immune system, and then activate T cells (also called T lymphocytes). Once activated, T cells have a number of important functions, including helping B cells to produce antibodies, killing virus-infected cells, and regulating the level of the adaptive immune response. Steinman further demonstrated that the ability to activate T cells, and thus orchestrate the adaptive immune response, was unique among dendritic cells. Unfortunately, Steinman passed away before the news of his Nobel Prize reached him.

The fundamental and collective research of Hoffman, Beutler, and Steinman has made possible the development of new methods for preventing and treating disease. For example, today we have improved vaccines against infections, and scientists are working to stimulate the immune system to attack tumors. The research of these Nobel Laureates also helps us to understand why the immune system can attack the body's tissues, thus providing clues for new treatment of inflammatory diseases.

For background information *see* CELLULAR IMMUNOLOGY; CLINICAL IMMUNOLOGY; ENDOTOXIN; IMMUNITY; IMMUNOLOGICAL DEFICIENCY; IMMUNOLOGY; IMMUNOTHERAPY; INFECTION in the McGraw-Hill Encyclopedia of Science & Technology.

Physics. One half of the physics prize was awarded to Saul Perlmutter and the other half jointly to Brian P. Schmidt and Adam G. Riess for the discovery of the accelerating expansion of the universe through observations of distant supernovae. Perlmutter is a United States citizen working at the Lawrence Berkeley National Laboratory and the University of California at Berkeley. Schmidt is a citizen of the

United States and Australia, and works at the Australian National University, Weston Creek. Riess is a United States citizen working at The Johns Hopkins University and the Space Telescope Science Institute, Baltimore, Maryland.

In the time between the findings of Copernicus, Kepler, Galileo, and others in the sixteenth and seventeenth centuries and the early decades of the twentieth, the universe was seen as a fixed, static system, essentially identical with the Milky Way Galaxy. Albert Einstein took this static view into account when in 1916 he applied his theory of gravitation—the general theory of relativity—on a cosmological scale to the universe. He found that according to his theory the universe ought to be unstable unless he added a term to his equations to counterbalance the force of gravity, a term now called the cosmological constant—a kind of cosmic repulsion.

In the 1920s, other scientists, such as Alexander Friedmann and Georges Lemaître, revisited Einstein's calculations and cast doubt on his steady-state solution. There also had been hints of a nonstatic universe through early spectroscopic analyses of light emanating from distant stars. Then, in the 1920s, Edwin Hubble, using the new 100-in. (2.5-m) telescope at Mount Wilson, California, made discoveries showing that the Milky Way Galaxy was not the entire universe, and that the universe is not static—it is, in fact, expanding. Hubble observed a type of star known as a Cepheid variable in the Andromeda Nebula and other spiral nebulae. Cepheids, which had been discovered some years earlier by Harriet Leavitt, pulsate with a characteristic relation between the period of peak brightness and their luminosity that can be used to estimate their distance. Applying Leavitt's findings to a series of observations, Hubble concluded that these spiral nebulae were well beyond the Milky Way and were galaxies in their own right. He also measured the spectral composition of light coming from these distant stars and observed a so-called redshift, that is, a lengthening of the wavelength of light associated with certain elements toward the red end of the spectrum, suggesting that the observed star is moving away from the observer (radial Doppler effect). *See* THE STAR THAT CHANGED THE UNIVERSE.

Furthermore, the rate of recession appeared to increase with distance. This relationship is embodied in Hubble's law: a galaxy's distance is proportional to its radial recession velocity. These observations are now understood to mean that the universe is expanding in all directions.

Hubble's findings dramatically changed our view of the universe, and Einstein came to believe his cosmological constant was his greatest error. These conclusions raised a number of questions as to the shape, origins, and fate of the universe. Lemaître, for example, at about the same time Hubble was making his observations, held the view that the expansion of the universe would have started from a single quantum, a supposition that presaged the big bang theory.

The technologies of that era were inadequate to answer such questions. What was needed were stars much brighter than Cepheids that could be used as "standard candles" in the distant universe, highly sensitive imaging devices to replace photographic film, and modern computing and networking power. These became available in the 1980s and 1990s.

The object of the search by this year's winners of the Nobel Prize in physics was a so-called type Ia supernova. A supernova is the explosive death of a star, whereby the star's luminosity (intrinsic brightness) in the explosion can exceed that of an entire galaxy of stars or several billion (10^9) Suns. They are so bright that they can be detected in the far distant universe. Astronomers separate supernovae into two basic types, depending on the kind of star that explodes. Type I are small stars that explode in a runaway thermonuclear event that destroys the star. Type II are large stars that apparently run out of nuclear fuel and collapse under their own gravity in a catastrophic event, leaving behind a neutron star or black hole.

Type Ia supernovae occur in all types of galaxies, showing no preference for star-forming regions. They are end-of-life white dwarfs, similar in mass to the Sun, and typically occurring in a binary star arrangement. They are thought to steal mass from the companion star and become unstable, leading to the explosion. Type 1a supernovae are short-lived events, reaching peak luminosity in about 2 weeks and then declining in a quasi-exponential manner.

The brightest type Ia supernovae characteristically exhibit a slower decline in brightness than dimmer ones, which allows one to recognize them by the shape of their light curve. Thus they can be used as a "standard candle." That is, given a predictable intrinsic brightness, distance can be measured with good accuracy from apparent brightness. However, they are rare events. The last one believed to have been seen in our Milky Way Galaxy was observed by Tycho Brahe in 1572. Modern imaging and computer technologies have greatly facilitated the search for them. Charge-coupled devices (CCDs), similar in principle to those used in digital cameras but much larger and more sensitive, allow modern telescopes to image many thousands of galaxies in a single exposure. Repeated imaging of the same part of the night sky can provide the necessary data. High-speed computers and appropriate software can process the repeated images to extract the light of the supernovae from the background galactic light.

But this was challenging work. An appropriate type of supernova had to be found, and its brightness, redshift, and light curve needed to be measured. Given the short lifetime of a supernova, all this had to be done quickly. Two groups competed in the search for, and analysis of, supernovae. Perlmutter headed the Supernova Cosmology Project at Lawrence Berkeley Laboratory. Their effective techniques led to preliminary results based on seven supernovae found in 1994 and 1995 that suggested a slowing down of the expansion due to the decelerating effect of dark matter. However, the High-Z Supernova Search Team led by Schmidt, with

Riess playing a crucial role, came to the opposite conclusion in a paper published in 1998, that the universe was accelerating. A year later, Perlmutter's team published similar results as the High-Z Supernova Search Team. The two teams were in agreement: The rate of cosmic expansion is increasing as the universe grows older.

The mysterious force that counteracts gravity and drives expansion is called dark energy. Just as with dark matter, which fills much of the universe, its nature has not been determined. It appears that Einstein's cosmological constant may be one way of describing it. Whereas he introduced it to stabilize the universe, we now see it as driving its acceleration.

The current view is that the expansion of the universe after the big bang slowed (decelerated) with time because of the gravitational influence of dark matter, but about 5 billion (5×10^9) years ago, with an expanded, less dense universe, the effects of dark energy took over, changing deceleration to acceleration, a change known as the cosmic jerk.

For background information *see* ACCELERATING UNIVERSE; ASTRONOMICAL IMAGING; BIG BANG THEORY; CEPHEIDS; COSMOLOGICAL CONSTANT; COSMOLOGY; DARK ENERGY; DARK MATTER; HUBBLE CONSTANT; REDSHIFT; SUPERNOVA; UNIVERSE; WHITE DWARF STAR in the McGraw-Hill Encyclopedia of Science & Technology.

Contributors

Contributors

The affiliation of each Yearbook contributor is given, followed by the title of his or her article. An article title with the notation "coauthored" indicates that two or more authors jointly prepared an article or section.

A

Acton, Dr. Gary. *Department of Geology, University of California, Davis.* FIRST-ORDER REVERSAL CURVES (FORCs).

Agnarsson, Dr. Ingi. *Department of Biology, Faculty of Natural Sciences, University of Puerto Rico, San Juan.* DARWIN'S BARK SPIDER—coauthored.

Aktürk, Dr. Selçuk. *Engineering Physics Department, Faculty of Science and Letters, Istanbul Technical University, Turkey.* ULTRASHORT LASER PULSE MEASUREMENTS—coauthored.

Altura, Dr. Rachel A. *Department of Pediatrics, Division of Hematology-Oncology, Brown University and Hasbro Children's Hospital, Providence, Rhode Island.* STAT3.

Anaya, Dr. Paul. *Department of Medicine, University of Kentucky, Lexington.* ARSENIC-EATING BACTERIA—coauthored.

Andari, Elissar. *Center for Cognitive Neuroscience, Institute for Cognitive Science, Centre National de la Recherche Scientifique (CNRS), Bron, France.* OXYTOCIN AND AUTISM—coauthored.

Aprahamian, Prof. Ani. *Department of Physics, University of Notre Dame, Indiana.* ORIGIN OF THE HEAVY ELEMENTS.

Asensi-Fabado, Dr. María Amparo. *Departament de Biologia Vegetal, Universitat de Barcelona, Spain.* VITAMINS IN PLANTS—coauthored.

B

Baker, William F. *Partner, Skidmore, Owings & Merrill LLP, Chicago, Illinois.* CHARACTERISTICS OF SUPERTALL BUILDING STRUCTURES—coauthored.

Baker, Dr. William J. *Herbarium, Library, Art, and Archives, Royal Botanic Gardens, Kew, Richmond, Surrey, United Kingdom.* E-TAXONOMY.

Barry, Roger G. *Distinguished Professor Emeritus of Geography, National Snow and Ice Data Center, CIRES, University of Colorado, Boulder.* CRYOSPHERE AND CLIMATE.

Baschek, Dr. Burkard. *Department of Atmospheric and Oceanic Sciences, University of California, Los Angeles.* ROGUE WAVES—coauthored.

Baserga, Dr. Susan J. *Department of Molecular Biophysics and Biochemistry, Yale University School of Medicine, New Haven, Connecticut.* RIBOSOME BIOGENESIS AND DISEASE—coauthored.

Beran, Dr. Philip S. *Air Vehicles Directorate, Air Force Research Laboratory, Wright-Patterson Air Force Base, Dayton, Ohio.* QUANTITATIVE ASSESSMENT OF MAV TECHNOLOGIES—coauthored.

Berger, Prof. Lee R. *School of Geosciences and Institute for Human Evolution, University of the Witwatersrand, Johannesburg, South Africa.* AUSTRALOPITHECUS SEDIBA—coauthored.

Blackledge, Dr. Todd A. *Department of Biology, University of Akron, Ohio.* DARWIN'S BARK SPIDER—coauthored.

Blundell, Dr. Barry G. *School of Computing and Mathematical Sciences, Auckland University of Technology, New Zealand.* 3D GRAPHIC DISPLAYS.

Borden, Dr. Katherine L. B. *Institute for Research in Immunology and Cancer, University of Montreal, Quebec, Canada.* PML: PROMYELOCYTIC LEUKEMIA PROTEIN—coauthored.

Breeze-Stringfellow, Andrew. *Engineering Technologies, GE Aviation, Cincinnati, Ohio.* UNDUCTED FANS FOR AIRCRAFT PROPULSION—coauthored.

Briand, Dr. Danick. *The Sensors, Actuators and Microsystems Laboratory, Institute of Microengineering, Ecole Polytechnique Federale de Lausanne, Neuchatel, Switzerland.* HARVESTING WASTE ENERGY—coauthored.

Brigham, Prof. Lawson W. *UA Geography Program, University of Alaska, Fairbanks.* ARCTIC MARINE TRANSPORTATION.

Bromiley, Prof. Philip. *Dean's Professor of Strategic Management, Merage School of Business, University of California, Irvine.* STRATEGIC DECISION MAKING—coauthored.

C

Carrano, Dr. Matthew T. *Curator of Dinosauria, Department of Paleobiology, Smithsonian Institution, Washington, District of Columbia.* EVOLUTION OF THEROPOD DINOSAURS.

Cavallaro, Paul V. *Naval Undersea Warfare Center, Engineering, Test & Evaluation Department, Mechanics of*

Advanced Structures & Materials Team, Newport, Rhode Island. SOFT BODY ARMOR.

Centurion, Dr. Martin. *Department of Physics and Astronomy, University of Nebraska, Lincoln.* MOLECULAR DIFFRACTOGRAMS.

Chen, Dr. Z. Jeffrey. *Section of Molecular Cell and Developmental Biology, Center for Computational Biology and Bioinformatics, and Institute for Cellular and Molecular Biology, University of Texas at Austin.* PLANT HYBRID VIGOR.

Cheung, Prof. Kingman. *Department of Physics, National Tsing Hua University, Hsinchu, Taiwan, Province of China.* TOP QUARK AT THE TEVATRON.

Cheung, Dr. Peter C. K. *Food and Nutritional Sciences, School of Life Sciences, The Chinese University of Hong Kong, Shatin, New Territories, Hong Kong, China.* MUSHROOM SCLEROTIA.

Churchill, Dr. Steven E. *Department of Evolutionary Anthropology, Duke University, Durham, North Carolina, and Institute for Human Evolution, University of the Witwatersrand, Johannesburg, South Africa.* AUSTRALOPITHECUS SEDIBA—coauthored.

Coates, Prof. Andrew J. *Mullard Space Science Laboratory, University College London, Dorking, United Kingdom.* ENCELADUS.

Conroy, Dr. Gary. *Atkins and Pearce, Covington, Kentucky.* CARBON NANOTUBE RESPONSIVE MATERIALS AND APPLICATIONS—coauthored.

Cotlet, Dr. Mircea. *Center for Functional Nanomaterials, Brookhaven National Laboratory, Upton, New York.* TRANSPARENT LIGHT-HARVESTING MATERIALS—coauthored.

Crawford, Dr. Daniel. *Department of Ecology and Evolutionary Biology, University of Kansas, Lawrence.* OCEANIC ISLANDS: EVOLUTIONARY LABORATORIES.

Crist, Ms. Darlene Trew. *Director of Communications, Census of Marine Life, University of Rhode Island, Kingston.* CENSUS OF MARINE LIFE—coauthored.

Culjkovic-Kraljacic, Dr. Biljana. *Institute for Research in Immunology and Cancer, University of Montreal, Quebec, Canada.* PML: PROMYELOCYTIC LEUKEMIA PROTEIN—coauthored.

D

D'Agostino, Dr. Michelangelo. *High Energy Physics, Argonne National Laboratory, Illinois.* REACTOR NEUTRINO EXPERIMENTS—coauthored.

Dandino, Charles. *Mechanical Engineering, University of Cincinnati, Ohio.* CARBON NANOTUBE RESPONSIVE MATERIALS AND APPLICATIONS—coauthored.

Danko, Dr. George. *Department of Mining Engineering, University of Nevada, Reno.* VENTILATION AND CLIMATE CONTROL OF DEEP MINES.

Danovaro, Dr. Roberto. *Department of Marine Science, Polytechnic University of Marche, Ancona, Italy.* ANOXYPHILIC LORICIFERA—coauthored.

de Klerk, Bonita. *School of Geosciences and Institute for Human Evolution, University of the Witwatersrand, Johannesburg, South Africa.* AUSTRALOPITHECUS SEDIBA—coauthored.

de Rooij, Prof. Nico. *The Sensors, Actuators and Microsystems Laboratory, Institute of Microengineering, Ecole Polytechnique Federale de Lausanne, Neuchatel, Switzerland.* HARVESTING WASTE ENERGY—coauthored.

de Ruiter, Dr. Darryl J. *Department of Anthropology, Texas A&M University, College Station, Texas, and Institute for Human Evolution, University of the Witwatersrand, Johannesburg, South Africa.* AUSTRALOPITHECUS SEDIBA—coauthored.

Dean, Barry. *Department of Chemistry, Durham University, United Kingdom.* RING-OPENING METATHESIS POLYMERIZATION OF CYCLOOLEFINS—coauthored.

Dell'Anno, Dr. Antonio. *Department of Marine Science, Polytechnic University of Marche, Ancona, Italy.* ANOXYPHILIC LORICIFERA—coauthored.

Deville, Dr. Sylvain. *Laboratoire de Synthèse et Fonctionnalisation des Céramiques, CNRS, Cavaillon, France.* ICE-TEMPLATED CERAMICS.

Doepke, Amos. *Department of Chemistry, University of Cincinnati, Ohio.* CARBON NANOTUBE RESPONSIVE MATERIALS AND APPLICATIONS—coauthored.

Dror, Itiel. *Institute of Cognitive Neuroscience, University College London, United Kingdom.* COGNITIVE BIAS IN FORENSIC SCIENCE.

F

Fay, Dr. Michael F. *Royal Botanic Gardens, Kew, Richmond, Surrey, United Kingdom.* ANGIOSPERM PHYLOGENY GROUP (APG) CLASSIFICATION.

Fernández-Hernando, Dr. Carlos. *Department of Pathology, New York School of Medicine.* MICRORNA-33 (MIR-33)—coauthored.

Fogassi, Dr. Leonardo. *Istituto Italiano di Tecnologia, Brain Center for Social and Motor Cognition, and Dipartimento di Psicologia, University of Parma, Italy.* MIRROR NEURONS AND SOCIAL NEUROSCIENCE—coauthored.

Freed, Ms. Emily F. *Department of Genetics, Yale University School of Medicine, New Haven, Connecticut.* RIBOSOME BIOGENESIS AND DISEASE—coauthored.

Freeman, Dr. Daniel. *Professor of Clinical Psychology/MRC Senior Clinical Fellow, Department of Psychiatry, Oxford University, Warneford Hospital, Oxford, United Kingdom.* PARANOIA: MECHANISMS AND TREATMENT.

G

Gambi, Dr. Cristina. *Department of Marine Science, Polytechnic University of Marche, Ancona, Italy.* ANOXYPHILIC LORICIFERA—coauthored.

Gandini, Dr. Alessandro. *CICECO and Department of Chemistry, University of Aveiro, Portugal.* POLYMERS FROM RENEWABLE RESOURCES.

Gardner, Dr. Douglas J. *School of Forest Resources, University of Maine, Orono.* CELLULOSE NANOCOMPOSITES.

Garrett, Prof. Chris. *Department of Physics and Astronomy, University of Victoria, British Columbia, Canada.* ROGUE WAVES—coauthored.

Gemmrich, Dr. Johannes. *Department of Physics and Astronomy, University of Victoria, British Columbia, Canada.* Rogue waves—coauthored.

Goedeke, Ms. Leigh. *Department of Pathology, New York University School of Medicine.* MicroRNA-33 (miR-33)—coauthored.

Goldsmith, Dr. Michael R. *U.S. Environmental Protection Agency, National Exposure Research Laboratory, Research Triangle Park, North Carolina.* Nanoparticle risk informatics—coauthored.

Goodman, Dr. Maury. *High Energy Physics, Argonne National Laboratory, Illinois.* Reactor neutrino experiments—coauthored.

Green, Ms. Jill A. *Department of Pharmacology, Ohio State University, Columbus.* Primary cilia—coauthored.

Green, Dr. Richard E. *Department of Biomolecular Engineering, Baskin School of Engineering, University of California, Santa Cruz.* Neandertal genome.

Griffin, Prof. Matt J. *School of Physics and Astronomy, Cardiff University, United Kingdom.* Herschel Space Observatory.

Gu, Dr. Xun, *ABB Switzerland Corporate Research Center, Baden-Dättwil.* Ultrashort laser pulse measurements—coauthored.

Gudmundsson, Dr. Magnús T. *Institute of Earth Sciences, University of Iceland, Reykjavik, Iceland.* 2010 eruption of Eyjafjallajökull, Iceland.

Guo, Xuefei. *Department of Chemistry, University of Cincinnati, Ohio.* Carbon nanotube responsive materials and applications—coauthored.

Gupta, Pranshu. *Department of Computing and Information Sciences, Kansas State University, Manhattan.* Web 2.0 technologies—coauthored.

H

Hajj, Prof. Muhammad. *Department of Engineering Science and Mechanics, Virginia Polytechnic Institute and State University, Blacksburg.* Quantitative assessment of MAV technologies—coauthored.

Hamilton, Prof. Joseph H. *Department of Physics and Astronomy, Vanderbilt University, Nashville, Tennessee.* Discovery of element 117—coauthored.

Hopkins, Tracy. *Department of Cancer and Cell Biology, University of Cincinnati, Ohio.* Carbon nanotube responsive materials and applications—coauthored.

I

Ibach, Dr. Rebecca E. *Forest Products Laboratory, USDA Forest Service, Madison, Wisconsin.* Wood preservative testing—coauthored.

Ignatiadis, Dr. Michail. *Medical Oncology Department and Breast Cancer Translational Research Laboratory, Institut Jules Bordet, Brussels, Belgium.* Circulating cancer cells.

Ingram, Dr. David M. *Institute for Energy Systems, School of Engineering, The University of Edinburgh, United Kingdom.* Wave energy converters.

Inoue, Dr. Tomoko. *Division of Hematopoietic Stem Cells, Department of Advanced Medical Initiatives, Kyushu University Faculty of Medical Sciences, Fukuoka, Japan.* Stem cell maintenance in embryos and adults—coauthored.

Isarakorn, Dr. Don. *Department of Instrumentation and Control Engineering, King Mongkut's Institute of Technology Ladkrabang, Bangkok, Thailand.* Harvesting waste energy—coauthored.

J

Jacobs, Dr. Gerald H. *Department of Psychology, University of California, Santa Barbara.* Primate color vision.

Jayasinghe, Dr. Chaminda. *Department of Chemical and Materials Engineering, University of Cincinnati, Ohio.* Carbon nanotube responsive materials and applications—coauthored.

Johnson, Dr. Sylvia M. *NASA Ames Research Center, Moffett Field, California.* Ultrahigh temperature ceramics.

Jones, Dr. J. T. *Atkins and Pearce, Covington, Kentucky.* Carbon nanotube responsive materials and applications—coauthored.

Jones, Dr. Kate L. *Department of Physics and Astronomy, University of Tennessee, Knoxville.* Doubly magic tin-132.

K

Key, Dr. Joey Shapiro. *Montana Space Grant Consortium, Montana State University, Bozeman.* The shape of the universe.

Khosravi, Dr. Ezat. *Department of Chemistry, Durham University, United Kingdom.* Ring-opening metathesis polymerization of cycloolefins—coauthored.

Kirshner, Prof. Robert P. *Department of Astronomy, Harvard University, Cambridge, Massachusetts.* Hubble constant and dark energy.

Klein, Dr. Raymond M. *Department of Psychology, Dalhousie University, Halifax, Nova Scotia, Canada.* Cell-phone use and driving.

Kluener, Joe. *Mechanical Engineering, University of Cincinnati, Ohio.* Carbon nanotube responsive materials and applications—coauthored.

Koenig, Dr. Robert. *Atkins and Pearce, Covington, Kentucky.* Carbon nanotube responsive materials and applications—coauthored.

Kristensen, Dr. Reinhardt M. *Department of Invertebrate Zoology, Natural History Museum of Denmark, University of Copenhagen, Denmark.* Anoxyphilic Loricifera—coauthored.

Krock, Dr. Richard E. *Alcatel-Lucent, Naperville, Illinois.* Priority emergency communications.

Kuhlmann, Julia. *Department of Chemistry, University of Cincinnati, Ohio.* Carbon nanotube responsive materials and applications—coauthored.

Kulkeaw, Dr. Kasem. *Division of Hematopoietic Stem Cells, Department of Advanced Medical Initiatives, Kyushu University Faculty of Medical Sciences, Fukuoka, Japan.* Stem cell maintenance in embryos and adults—coauthored.

Kumta, Dr. Prashant. *Department of Bioengineering, University of Pittsburgh, Pennsylvania.* CARBON NANOTUBE RESPONSIVE MATERIALS AND APPLICATIONS—coauthored.

Kuntner, Dr. Matjaž. *Scientific Research Centre, Institute of Biology, Slovenian Academy of Arts and Sciences, Ljubljana, Slovenia.* DARWIN'S BARK SPIDER—coauthored.

L

Lane, Dr. Todd. *School of Earth Sciences, University of Melbourne, Australia.* THUNDERSTORM-GENERATED TURBULENCE—coauthored.

Lebow, Dr. Stan T. *Forest Products Laboratory, USDA Forest Service, Madison, Wisconsin.* WOOD PRESERVATIVE TESTING—coauthored.

Lee, Dr. Jun Hee. *Department of Pharmacology, University of California, San Diego, La Jolla.* SESTRINS.

Li, Dr. Weifeng. *Mechanical Engineering, University of Cincinnati, Ohio.* CARBON NANOTUBE RESPONSIVE MATERIALS AND APPLICATIONS—coauthored.

Libault, Dr. Marc. *Division of Plant Sciences, National Center for Soybean Biotechnology, University of Missouri, Columbia.* PLANT ROOT HAIRS AS A MODEL CELL SYSTEM—coauthored.

Lombard, Dr. Marlize. *Department of Anthropology, University of Johannesburg, Gauteng, South Africa.* HOWIESON'S POORT.

Luthar, Dr. Suniya S. *Department of Counseling and Clinical Psychology, Columbia University's Teachers College, New York.* DEVELOPMENT AND RESILIENCE—coauthored.

M

Mast, Dr. David. *Department of Physics, University of Cincinnati, Ohio.* CARBON NANOTUBE RESPONSIVE MATERIALS AND APPLICATIONS—coauthored.

Mata-Toledo, Dr. Ramon A. *Professor of Computer Science, James Madison University, Harrisonburg, Virginia.* WEB 2.0 TECHNOLOGIES—coauthored.

Medina, Dr. Victor F. *Environmental Laboratory, Army Engineer Research & Development Center, Vicksburg, Mississippi.* WASTEWATER TREATMENT FOR DEPLOYED MILITARY FORCES AND DISASTER RESPONSE—coauthored.

Mikhailine, Alexander A. *Department of Chemistry, University of Toronto, Ontario, Canada.* IRON CATALYSIS—coauthored.

Miloslavich, Dr. Patricia. *Departamento de Estudios Ambientales, Universidad Simón Bolívar, Caracas, Venezuela.* CENSUS OF MARINE LIFE—coauthored.

Morris, Prof. Robert H. *Department of Chemistry, University of Toronto, Ontario, Canada.* IRON CATALYSIS—coauthored.

Munné-Bosch, Dr. Sergi. *Departament de Biologia Vegetal, Universitat de Barcelona, Spain.* VITAMINS IN PLANTS—coauthored.

Murrow, Kurt. *Advanced Technology and Preliminary Design, GE Aviation, Cincinnati, Ohio.* UNDUCTED FANS FOR AIRCRAFT PROPULSION—coauthored.

Mykytyn, Dr. Kirk. *Departments of Pharmacology and Internal Medicine, Division of Human Genetics, Ohio State University, Columbus.* PRIMARY CILIA—coauthored.

N

Nasrallah, Dr. June B. *Department of Plant Biology, Cornell University, Ithaca, New York.* PLANT REPRODUCTIVE INCOMPATIBILITY.

Naus, Dr. G. J. L. *Department of Mechanical Engineering, Eindhoven University of Technology, Netherlands.* ROBOTIC SURGERY—coauthored.

Neumann, Dr. Nadja. *Department of Molecular Biology and Functional Genomics, Stockholm University, Sweden.* EVOLUTION OF PHAGOTROPHY—coauthored.

O

Oberacker, Prof. Volker E. *Department of Physics and Astronomy, Vanderbilt University, Nashville, Tennessee.* MICROSCOPIC CALCULATIONS OF HEAVY-ION FUSION REACTIONS—coauthored.

Oganessian, Dr. Yuri Ts. *Joint Institute for Nuclear Research, Dubna, Russian Federation.* DISCOVERY OF ELEMENT 117—coauthored.

Olsen, Dr. Glenn H. *Patuxent Wildlife Research Center, United States Geological Survey, Laurel, Maryland.* WHOOPING CRANE RESTORATION.

Osborn, Dr. Karen J. *Department of Invertebrate Zoology, National Museum of History, Smithsonian Institution, Washington, District of Columbia.* SQUIDWORM.

Ozerskaya, Dr. Svetlana M. *All-Russian Collection of Microorganisms, G. K. Skryabin Institute of Biochemistry and Physiology of Microorganisms RAS, Moscow, Russia.* FUNGAL DIVERSITY IN GENBANK.

P

Paraszczak, Prof. Jacek. *Department of Mining, Metallurgical and Materials Engineering, Universite Laval, Quebec, Canada.* MINING AUTOMATION.

Parcy, Dr. François. *Laboratoire de Physiologie Cellulaire Végétale, Centre National de la Recherche Scientifique, Institut National de la Recherche Agronomique, Université Joseph Fourier, Grenoble, France.* LEAFY: A MASTER REGULATOR OF FLOWER DEVELOPMENT—coauthored.

Pawlikowski, James J. *Associate Director, Skidmore, Owings & Merrill LLP, Chicago, Illinois.* CHARACTERISTICS OF SUPERTALL BUILDING STRUCTURES—coauthored.

Pawlowski, Dr. Wojciech P. *Department of Plant Breeding and Genetics, Cornell University, Ithaca, New York.* ROLE OF TELOMERES AND CENTROMERES IN MEIOTIC CHROMOSOME PAIRING—coauthored.

Peterson, Dr. Jonathan M. *Department of Physiology, The Johns Hopkins University School of Medicine, Baltimore, Maryland.* CTRPs: NOVEL ADIPOKINES—coauthored.

Pierce, Dr. Marcia M. *Department of Biological Sciences, Eastern Kentucky University, Richmond.* CHOLERA IN HAITI; HUMAN SUSCEPTIBILITY TO STAPHYLOCOCCUS AUREUS.

Pirmoradian, Mahdi. *School of Computing and Information Systems, Penrhyn Road Campus, Kingston University London, Kingston upon Thames, United Kingdom.* COGNITIVE RADIO—coauthored.

Pixley, Dr. Sarah. *Department of Cancer and Cell Biology, University of Cincinnati, Ohio.* CARBON NANOTUBE RESPONSIVE MATERIALS AND APPLICATIONS—coauthored.

Platt, Dr. Donald. *Micro Aerospace Solutions, Inc., Melbourne, Florida.* SPACE FLIGHT, 2010.

Politis, Dr. Christos. *School of Computing and Information Systems, Penrhyn Road Campus, Kingston University London, Kingston upon Thames, United Kingdom.* COGNITIVE RADIO—coauthored.

Poole, Dr. Anthony M. *School of Biological Sciences, University of Canterbury, Christchurch, New Zealand.* EVOLUTION OF PHAGOTROPHY—coauthored.

Pusceddu, Prof. Antonio. *Department of Marine Science, Polytechnic University of Marche, Ancona, Italy.* ANOXYPHILIC LORICIFERA—coauthored.

R

Ramage, Dr. Gordon. *School of Medicine, College of Medicine, Veterinary and Life Sciences, University of Glasgow, United Kingdom.* FUNGAL BIOFILMS—coauthored.

Ramanathan, Dr. Madhumati. *Department of Bioengineering, University of Pittsburgh, Pennsylvania.* CARBON NANOTUBE RESPONSIVE MATERIALS AND APPLICATIONS—coauthored.

Ramírez, Dr. Cristina M. *Department of Medicine, New York University School of Medicine.* MICRORNA-33 (MIR-33)—coauthored.

Rau, Dr. Devaki. *Department of Management, College of Business, Northern Illinois University, DeKalb.* STRATEGIC DECISION MAKING—coauthored.

Reinhardt, Dr. Klaus. *Department of Animal and Plant Sciences, University of Sheffield, United Kingdom.* BEDBUG INFESTATIONS.

Riggio, Dr. Ronald E. *Department of Psychology, Claremont McKenna College, Claremont, California.* ETHICAL LEADERSHIP.

Rizzolatti, Dr. Giacomo. *Dipartimento di Neuroscienze, and Istituto Italiano di Tecnologia, Brain Center for Social and Motor Cognition, University of Parma, Italy.* MIRROR NEURONS AND SOCIAL NEUROSCIENCE—coauthored.

Roeder, Dr. Peter. *Taurus Animal Health, Headley Down, Bordon, Hampshire, United Kingdom.* RINDERPEST ERADICATION.

Rogan, Dr Yulia. *Department of Chemistry, Durham University, United Kingdom.* RING-OPENING METATHESIS POLYMERIZATION OF CYCLOOLEFINS—coauthored.

Romeo, Prof. Giulio. *Department of Aerospace Engineering, Politecnico di Torino, Italy.* ELECTRIC AIRPLANE.

Ronsheim, Dr. Margaret L. *Department of Biology, Vassar College, Poughkeepsie, New York.* INVASIVE SPECIES AND THEIR EFFECTS ON NATIVE SPECIES.

Rothstein, Jeremy D. *Research Assistant, Yale Child Study Center, Yale University School of Medicine, New Haven, Connecticut.* DEVELOPMENT AND RESILIENCE—coauthored.

Ruff, Brad. *Mechanical Engineering, University of Cincinnati, Ohio.* CARBON NANOTUBE RESPONSIVE MATERIALS AND APPLICATIONS—coauthored.

S

Schmidt, Prof. David. *Department of Computing and Information Sciences, Kansas State University, Manhattan.* DENOTATIONAL SEMANTICS.

Schulz, Dr. Mark. *Mechanical Engineering, University of Cincinnati, Ohio.* CARBON NANOTUBE RESPONSIVE MATERIALS AND APPLICATIONS—coauthored.

Shaffer, Julie. *Printing Industries of America, Sewickley, Pennsylvania.* WEB TO PRINT.

Shanov, Dr. Vesselin. *Department of Chemical and Materials Engineering, University of Cincinnati, Ohio.* CARBON NANOTUBE RESPONSIVE MATERIALS AND APPLICATIONS—coauthored.

Sharman, Dr. Robert. *Research Applications Laboratory, National Center for Atmospheric Research, Boulder, Colorado.* THUNDERSTORM-GENERATED TURBULENCE—coauthored.

Shereen, Dr. Duke. *Department of Physics, University of Cincinnati, Ohio.* CARBON NANOTUBE RESPONSIVE MATERIALS AND APPLICATIONS—coauthored.

Shors, Dr. Teri. *Department of Biology and Microbiology, University of Wisconsin-Oshkosh.* HIV AND BONE MARROW TRANSPLANTATION.

Shuldiner, Dr. Albert D. *iBiquity Digital Corporation, Columbia, Maryland.* HD RADIO TECHNOLOGY.

Simmons, Dr. Alan H. *Department of Anthropology, University of Nevada, Las Vegas.* ORIGINS OF AGRICULTURE.

Simmons, Dr. Kristin. *Atkins and Pearce, Covington, Kentucky.* CARBON NANOTUBE RESPONSIVE MATERIALS AND APPLICATIONS—coauthored.

Sirigu, Dr. Angela. *Center for Cognitive Neuroscience, Institute for Cognitive Science, Centre National de la Recherche Scientifique (CNRS), Bron, France.* OXYTOCIN AND AUTISM—coauthored.

Smith, Dr. Krister T. *Department of Paleoanthropology and Messel Research, Senckenberg Research Institute and Natural History Museum, Frankfurt am Main, Germany.* FOSSIL RECORD OF VERTEBRATE RESPONSES TO GLOBAL WARMING.

Snyder, Dr. Michael. *Department of Genetics, Stanford University, California.* RNA-SEQ TRANSCRIPTOMICS—coauthored.

Soderblom, Dr. David R. *Space Telescope Science Institute, Baltimore, Maryland.* THE STAR THAT CHANGED THE UNIVERSE.

Song, Yi. *Mechanical Engineering, University of Cincinnati, Ohio.* CARBON NANOTUBE RESPONSIVE MATERIALS AND APPLICATIONS—coauthored.

Sowani, Anshuman. *Mechanical Engineering, University of Cincinnati, Ohio.* CARBON NANOTUBE RESPONSIVE MATERIALS AND APPLICATIONS—coauthored.

Spiropulu, Dr. Maria. *Division of Physics, Mathematics and Astronomy, California Institute of Technology, Pasadena.* LARGE HADRON COLLIDER: FIRST YEAR.

Stacey, Dr. Gary. *Division of Plant Sciences, University of Missouri, Columbia.* PLANT ROOT HAIRS AS A MODEL CELL SYSTEM—coauthored.

Stanhope, Dr. Kimber L. *Department of Molecular Biosciences, University of California, Davis,* DIETARY FRUCTOSE AND THE PHYSIOLOGY OF BODY WEIGHT REGULATION.

Steinbuch, Prof. M. *Department of Mechanical Engineering, Eindhoven University of Technology, Netherlands.* ROBOTIC SURGERY—coauthored.

Stiavelli, Dr. Massimo. *Space Telescope Science Institute, Baltimore, Maryland.* HUBBLE ULTRA DEEP FIELD.

Strickland, Dr. James R. *Forage-Animal Production Research Unit, United States Department of Agriculture-Agricultural Research Service, University of Kentucky Campus, Lexington.* ENDOPHYTE-ASSOCIATED ERGOT ALKALOIDS.

Sugiyama, Dr. Daisuke. *Division of Hematopoietic Stem Cells, Department of Advanced Medical Initiatives, Kyushu University Faculty of Medical Sciences, Fukuoka, Japan.* STEM CELL MAINTENANCE IN EMBRYOS AND ADULTS—coauthored.

T

Tiang, Dr. Choon-Lin. *Department of Plant Breeding and Genetics, Cornell University, Ithaca, New York.* ROLE OF TELOMERES AND CENTROMERES IN MEIOTIC CHROMOSOME PAIRING—coauthored.

Tichtinsky, Dr. Gabrielle. *Laboratoire de Physiologie Cellulaire Végétale, Centre National de la Recherche Scientifique, Institut National de la Recherche Agronomique, Université Joseph Fourier, Grenoble, France.* LEAFY: A MASTER REGULATOR OF FLOWER DEVELOPMENT—coauthored.

Tjaden, Dr. Brett C. *Department of Computer Science, James Madison University, Harrisonburg, Virginia.* MALWARE.

Tsai, Dr. Hsinhan. *Chemistry Division, Los Alamos National Laboratory, New Mexico.* TRANSPARENT LIGHT-HARVESTING MATERIALS—coauthored.

Tsuru, Prof. Toshinori. *Department of Chemical Engineering, Hiroshima University, Higashi-Hiroshima, Japan.* MEMBRANE REACTOR.

U

Umar, Prof. A. Sait. *Department of Physics and Astronomy, Vanderbilt University, Nashville, Tennessee.* MICROSCOPIC CALCULATIONS OF HEAVY-ION FUSION REACTIONS—coauthored.

Utyonkov, Dr. Vladimir K. *Joint Institute for Nuclear Research, Dubna, Russian Federation.* DISCOVERY OF ELEMENT 117—coauthored.

Uusijärvi, Dr. Richard. *SP Swedish National Testing and Research Institute, Building Technology and Mechanics-Trätek, Stockholm, Sweden.* FROM FOREST LOG TO PRODUCTS.

V

Vachon, Dr. Gilles. *Laboratoire de Physiologie Cellulaire Végétale, Centre National de la Recherche Scientifique, Institut National de la Recherche Agronomique, Université Joseph Fourier, Grenoble, France.* LEAFY: A MASTER REGULATOR OF FLOWER DEVELOPMENT—coauthored.

Vallero, Dr. Daniel A. *U.S. Environmental Protection Agency, National Exposure Research Laboratory, Research Triangle Park, North Carolina.* NANOPARTICLE RISK INFORMATICS—coauthored.

van den Bedem, Dr. L. J. M. *Department of Mechanical Engineering, Eindhoven University of Technology, Netherlands.* ROBOTIC SURGERY—coauthored.

Venkatasubramanian, Rajiv. *Mechanical Engineering, University of Cincinnati, Ohio.* CARBON NANOTUBE RESPONSIVE MATERIALS AND APPLICATIONS—coauthored.

Vennemeyer, John. *Department of Biomedical Engineering, University of Cincinnati, Ohio.* CARBON NANOTUBE RESPONSIVE MATERIALS AND APPLICATIONS—coauthored.

W

Waern, Mr. Karl. *Department of Genetics, Stanford University, California.* RNA-SEQ TRANSCRIPTOMICS—coauthored.

Waikel, Dr. Rebekah L. *Department of Biological Sciences, Eastern Kentucky University, Lexington.* ARSENIC-EATING BACTERIA—coauthored.

Waisner, Scott A. *Environmental Laboratory, Army Engineer Research & Development Center, Vicksburg, Mississippi.* WASTEWATER TREATMENT FOR DEPLOYED MILITARY FORCES AND DISASTER RESPONSE—coauthored.

Wang, Dr. Hsing Lin. *Chemistry Division, Los Alamos National Laboratory, New Mexico.* TRANSPARENT LIGHT-HARVESTING MATERIALS—coauthored.

Wang, Dr. Zhong Lin. *School of Material Science and Engineering, Georgia Institute of Technology, Atlanta.* PIEZOTRONICS.

Williams, Dr. Craig. *Department of Microbiology, Royal Hospital for Sick Children, Glasgow, United Kingdom.* FUNGAL BIOFILMS—coauthored.

Wojno, Dr. John. *Advanced Technology and Preliminary Design, GE Aviation, Cincinnati, Ohio.* UNDUCTED FANS FOR AIRCRAFT PROPULSION—coauthored.

Wong, Dr. G. William. *Department of Physiology, The Johns Hopkins University School of Medicine, Baltimore, Maryland.* CTRPS: NOVEL ADIPOKINES—coauthored.

Worzyk, Dr. Thomas. *ABB AB High Voltage Cables, Karlskrona, Sweden.* SUBMARINE POWER CABLES.

Wright, Dr. Richard N. *ASCE Volunteer, Montgomery Village, Maryland.* ASCE PROGRAM FOR SUSTAINABLE INFRASTRUCTURE.

X

Xu, Dr. Zhihua. *Center for Functional Nanomaterials, Brookhaven National Laboratory, Upton, New York.* TRANSPARENT LIGHT-HARVESTING MATERIALS—coauthored.

Index

Index